Complementarity, Equilibrium, Efficiency and Economics

Nonconvex Optimization and Its Applications

Volume 63

The titles published in this series are listed at the end of this volume.

Complementarity, Equilibrium, Efficiency and Economics

By

G. Isac
Department of Mathematics and Computer Science,
Royal Military College of Canada,
Kingston, Ontario, Canada

V.A. Bulavsky

and

V.V. Kalashnikov
Central Economics Institute (CEMI) of Russian Academy of Sciences,
Moscow, Russia

KLUWER ACADEMIC PUBLISHERS
DORDRECHT / BOSTON / LONDON

A C.I.P. Catalogue record for this book is available from the Library of Congress.

ISBN 978-1-4419-5223-3

Published by Kluwer Academic Publishers,
P.O. Box 17, 3300 AA Dordrecht, The Netherlands.

Sold and distributed in North, Central and South America
by Kluwer Academic Publishers,
101 Philip Drive, Norwell, MA 02061, U.S.A.

In all other countries, sold and distributed
by Kluwer Academic Publishers,
P.O. Box 322, 3300 AH Dordrecht, The Netherlands.

Printed on acid-free paper

This volume is dedicated to our families.

"Mon ami, chacun de ces hommes t'a raconté
ce qui fut vraiment; mais peu d'entre nous
sont capables d'ajouter un fait à un autre et
d'en faire une verité."

(Khalil Gibran: L'ERRANT)

Contents

ACKNOWLEDGEMENT

The research activity of the monograph's authors Vladimir Bulavsky and Vyacheslav Kalashnikov was financially supported by the Russian Foundation for Basic Research (RFBR), project No. 00-01-00404.

Sections 7.5 and 7.6 of Chapter 7 were prepared by George Isac and Vyacheslav Kalashnikov with the financial support by the Volkswagen-Stiftung, during the authors' stay in Oberwolfach in frames of the Research-in-Pairs Program (RiP), June, 1999.

PREFACE

Now many books and papers are dedicated to the study of equilibrium for economic systems. It is well-known that the concept of equilibrium is central to the understanding of competitive systems in different domains of Economics. We can cite as examples: general economic equilibrium problems in which all the commodity prices are to be determined, general financial equilibrium of markets in which firms compete to determine their profit-maximizing production outputs etc. Many authors studied equilibrium of economic systems by several mathematical methods and from several points of view. The recent development of Complementarity Theory helps us to understand better some aspects of equilibrium in Economics. Another concept, strongly related to Economics and considered by many authors is the mathematical notion of efficiency in Pareto's sense. Although the list of papers dedicated to this notion is huge, it seems that there are not many books dealing with this concept. Now, in this book we put together the complementarity, the equilibrium and the efficiency in the context of economic systems.

An idea of this book occured to us after several years of the fruitful cooperation of the former author, Prof.Dr. George Isac, Royal Military College, Kingston, Canada, with the latter ones, Prof.Dr. Vladimir Bulavsky and Dr. Vyacheslav Kalashnikov, Central Economics and Mathematics Institute (CEMI), Russian Academy of Sciences, Moscow, Russian Federation. Both parties have been engaged in research of complementarity problems of various kinds, including those involved in mathematical models of Economics. The results of almost eight years of this cooperation are presented in Parts I, II and III of the book. We should also note that one of the most interesting results of the joint research, the notion of an exceptional family of elements and its relation to existence of solutions to complementarity problems, has become very popular and widely used in Optimization Theory by many authors.

Chapters 1, 2 and 3 of Part I are based mainly upon the papers by Vladimir Bulavsky and are dedicated to Optimization Models and General Economic Equilibrium. Chapter 1 deals with variational inequalities defined by set-valued operators, and the equivalence of the latter to fixed-point problems is showed. The notion of an exceptional family of points is introduced in a very general form, and the theorem of alternative of existence either of solutions or an exceptional family of elements, is proved. These results serve as tools for investigations presented in Parts II and III of the book.

Chapter 2 is dedicated to optimization models involving single- and multi-valued functions whose differential mapping is strictly quasi-monotone. Mathematical programs with feedback between the primal and dual variables, as well as multicriterial problems constitute a considerable part of the chapter.

In Chapter 3, models of decentralized economy with competitive equilibrium and Arrow-Debreu models are presented, and theorems of existence of equilibria are proved.

Chapter 4 is dedicated to generalized models of the single-product market. The models are based upon the conjectural equilibria and propose the whole spectre of intermediate models between the models of perfect competition on one side, and the Cournot and Stackelberg models, on the other side. The complementarity theory tools are used to obtain conditions guaranteeing existence and uniqueness of equilibrium in these generalized models.

Chapter 5 scrutinizes in more detail the generalized Stackelberg model, allowing several leaders, instead of a single leader in the classical Stackelberg model. Existence of stationary points in thus extended model is established. Chapters 4 and 5 follow mainly the third author Vyacheslav Kalashnikov's Habilitation Thesis.

Part II that includes Chapters 6 and 7 applies the general notion of an exceptional family of elements to various particular classes of complementarity problems having obtained both well-known and brand new sufficient conditions of solvability of the complementarity problems. The majority of these results have already been published in previous papers and books by all three authors, but this is the first time that they all have been collected together. However, some results on solvability of Implicit Complementarity Problem presented in the last sections of Chapter 7 have not been yet published in papers.

Part III of the book is dedicated to description and properties of some numerical methods solving both linear and nonlinear complementarity problems. Chapter 8 presents various versions of the Lemke's algorithm and its extensions by M.J. Todd and M.M. Kostreva. Chapter 9 deals with Scarf type algorithms solving linear complementarity problems, and follows mainly the interesting work by P.J.-J. Herings, A.A.J. Talman and Z. Yang (1999). A Newton-like method for solving nonlinear complementarity problems and thus finding an equilibrium in the generalized Cournot model is examined in Chapter 10. Chapter 11 contains results of two types. First, some results on the parametrization approach to solving the nonlinear complementarity problem proposed by L. McLinden (1980), is examined, a numerical method is presented and its convergence is demonstrated. Second, quite promising algorithms using the Fischer-Burmeister's merit functions developed in a series of papers by C. Kanzow, F. Facchinei, M.C. Ferris, M.S. Gowda, and others, are also described.

We do not have a chapter on the interior point method because now there is a plenty of books on this subject. Among others, we should cite the following ones:

Yinyu Ye, *Interior Point Algorithms. Theory and Analysis*. John Wiley and Sons, Inc., 1997;

Tamas Terlaky (Ed.) *Interior Point Methods of Mathematical Programming*. Kluwer Academic Publishers, 1996;

Jiming Peng, "New Design and Analysis of Interior Point Methods', PhD. Thesis , University Delft, 2001.

Part IV is dedicated to efficiency problems. In Chapter 12, we present the most important notions of efficiency and we study the relations between these notions. We establish several existence theorems related to efficiency and examine some properties of solution sets. Some Arrow-Barankin-Blackwell type theorem are also presented. The notion of nuclear cone is applied to the study of the efficiency. Finally, we consider several topological properties of efficient point sets and we finish the chapter with a constructive test for Pareto efficiency.

Chapter 13 is dedicated to the study of several notions of approximative efficiency. We investigate the most important properties of each kind of approximative efficiency. We consider also the efficiency in product spaces and some interesting relations with Ekeland's variational principle. Chapters 12 and 13 are prepared by the first author George Isac.

It is a pleasure for us to acknowledge the excellent assistance that the staff of Kluwer Academic Publishers has provided in the publication of this book.

George Isac, Vladimir A. Bulavsky, Vyacheslav V. Kalashnikov
November 2001

PART I.

ECONOMIC MODELS AND COMPLEMENTARITY

Chapter 1

INTRODUCTION

1.1 MULTI-VALUED MAPPINGS AND VARIATIONAL INEQUALITIES

To a certain degree, the concept of complementarity is analogous to the concept of a stationary point in the extremum problems. If the point $\bar{z} = 0$ is a (local) minimum of a real differentiable function f defined over the positive half-axis $R_+ = [0, +\infty)$, then the inequality $f'(0) \geq 0$ is the necessary condition of that. However, if the point in question $\bar{z}$ is strictly positive then the necessary minimum condition reduces to the equality $f'(\bar{z}) = 0$. These two cases can be described uniformly as follows:

$$\bar{z} \geq 0, \qquad f'(\bar{z}) \geq 0, \qquad \bar{z} f'(\bar{z}) = 0.$$

More generally, when a differentiable function f is defined over a vector space R^n, and we have to find a point $\bar{z} = (\bar{z}_1, \bar{z}_2, \ldots, \bar{z}_n)$ providing for a local minimum of the function f over the nonnegative ortant R^n_+, the necessary conditions of that are:

$$\frac{\partial f(\bar{z})}{\partial z_i} \geq 0, \text{ if } \bar{z}_i = 0, \text{ and } \frac{\partial f(\bar{z})}{\partial z_i} = 0, \text{ if } \bar{z}_i > 0,$$

for all $i = 1, \ldots, n$. If we set $F_i(z) = \partial f(z)/\partial z_i$, we can rewrite the latter relationships in the equivalent form:

$$F_i(\bar{z}) \geq 0, \ z_i \geq 0, \ i = 1, \ldots, n, \ \sum_{i=1}^{n} \bar{z}_i F_i(\bar{z}) = 0. \tag{1.1}$$

If we "forget" about the origin of the functions F_i and consider them as arbitrary ones, we call problem (1.1) the nonlinear *complementarity* problem (in the narrow meaning of this term).

By arranging the values $F_i(z)$, $i = 1, \ldots, n$, in the vector form $F(z)$, we obtain the (generally) nonlinear mapping F that associates $z \in R^n$ with $F(z) \in R^n$. Conceptually, it is convenient to consider these spaces R^n as two distinct ones, each serving as the space of linear forms defined on the other. Making use of the standard notation of a canonical bilinear form, we have

$$\langle F(z), z\rangle = \sum_{i=1}^{n} z_i F_i(z).$$

Let $K = R_+^n$ be the convex cone of nonnegative vectors in R^n. Problem (1.1) can be represented as follows:

$$\bar{z} \in K, \ \langle F(\bar{z}), z - \bar{z}\rangle \geq 0 \text{ for all } z \in K. \tag{1.2}$$

Indeed, if the vector $\bar{z}$ satisfies (1.1) then for any $z \in K$ it holds that

$$\langle F(\bar{z}), z\rangle = \sum_{i=1}^{n} z_i F_i(\bar{z}) \geq 0.$$

Hence $\langle F(\bar{z}), z - \bar{z}\rangle = \langle F(\bar{z}), z\rangle \geq 0$, that is, $\bar{z}$ satisfies (1.2) as well. Conversely, if (1.2) is valid, then by taking $z = 0$ and $z = 2\bar{z}$ we obtain $\langle F(\bar{z}), \bar{z}\rangle = 0$. Furthermore, if we had $F_i(\bar{z}) < 0$ for some i, we would be able to choose the value of z_i large enough to violate the inequality in (1.2). Therefore, $F_i(\bar{z}) \geq 0$, $i = 1, \ldots, n$, and the vector $\bar{z}$ solves problem (1.1).

Problem (1.2) is usually referred to as a *variational inequality*, even if the set K is not a convex cone. As a rule, this set is assumed to be nonempty, convex and closed; in existence theorems, in addition, it is supposed to be bounded (hence, compact) one.

We will also use another form of problem (1.2). Denote by $N_K(\bar{z})$ the *normal cone* of the set K at the point $\bar{z}$, i.e. the collection of all linear forms that attain their maximum values over the set K at the point $\bar{z}$. According to (1.2), the linear form $q = -F(\bar{z})$ belongs to the normal cone, that is, problem (1.2) reduces to

$$\bar{z} \in K, \quad -F(\bar{z}) \in N_K(\bar{z}).$$

Actually, we will consider a more general problem. The mapping F will not be assumed single-valued: for each $z \in K$, the image $F(z)$ will be a set of linear forms. This generalization is motivated, for instance, by the fact that in Convex Analysis, instead of the classical derivative of a convex function, one has to consider its subgradients constituting the function's subdifferential. For a multi-valued F, problem (1.2) transforms into the following problem: Find a vector $\bar{z}$ and a linear form $\bar{p}$ such that

$$\bar{z} \in K, \ \bar{p} \in F(\bar{z}), \ \langle \bar{p}, z - \bar{z}\rangle \geq 0 \text{ for all } z \in K. \tag{1.3}$$

Making use of the concept of normal cone, we can rewrite the latter inequality as $(-\bar{p}) \in N_K(\bar{z})$. We will also represent problem (1.3) in the equivalent form that follows:

$$\bar{z} \in K, \quad 0 \in F(\bar{z}) + N_K(\bar{z}). \tag{1.4}$$

Here, as usual, the sum of two sets is understood as a collection of all sums of pairs of elements from these sets, that is,

$$F(\bar{z}) + N_K(\bar{z}) = \{p + q \mid p \in F(\bar{z}), \ q \in N_K(\bar{z})\}.$$

Similarly to single-valued mappings, a multi-valued mapping F can map $K \subset R^n$ to a vector space of other dimension. In what follows, we will not indicate explicitly the range space of multi-valued mappings, except for some special cases.

DEFINITION 1.1. A multi-valued mapping F is called *upper semicontinuous* at a point $\hat{z}$ if and only if, for any neighborhood V of the set $F(\hat{z})$ there is a neighborhood U of the point $\hat{z}$ such that $F(z) \subset V$ for all $z \in U \cap K$. A mapping F is referred to as upper semicontinuous over the set K if and only if it is upper semicontinuous at every point of K.

Note that the neighborhoods U and V can always be considered as open convex subsets. In case of a single-valued mapping, Definition 1.1 turns into the standard definition of continuity. The concept of upper semicontinuity is very close to the concept of closedness of a mapping.

DEFINITION 1.2. A multi-valued mapping F is called *closed* on a (closed) set K, if and only if its graph

$$G_F = \{(z, y) \mid z \in K, y \in F(z)\}$$

is a closed set in the Cartesian product of the domain and range spaces of the mapping F.

The concept of an *image* of a set $U \subset K$ for multi-valued mappings is introduced exactly as that for single-valued ones:

$$F(U) = \bigcup_{z \in U} F(z).$$

LEMMA 1.1. *Let K be a closed subset. If F is an upper semicontinuous mapping and the image $F(z)$ is closed for each $z \in K$, then the mapping F is closed. Conversely, if the mapping F is closed and the image $F(U)$ of every bounded subset $U \subset K$ is bounded, then the mapping F is upper semicontinuous over K.*

PROOF. Let F be an upper semicontinuous mapping and the sets $F(z)$ be closed for all $z \in K$. In order to prove the closedness of the graph G_F of the mapping F we have to show that for each pair $(\bar{z}, \bar{y}) \notin G_F$, there exist neighborhoods U of the point $\bar{z}$ and W of the point $\bar{y}$ such that $(U \times W) \cap G_F = \emptyset$. If $\bar{z} \notin K$ then the closedness of K implies the existence of a neighborhood U of the point $\bar{z}$ which does not intersect with K, hence $(U \times W) \cap G_F = \emptyset$ for any neighborhood W of the point $\bar{y}$. Otherwise, if $\bar{z} \in K$, then necessarily $\bar{y} \notin F(\bar{z})$. As $F(\bar{z})$ is a closed set, there is a neighborhood W of the point $\bar{y}$ such that $W \cap F(\bar{z}) = \emptyset$. Without lack of generality, we can consider the neighborhood W being closed. Then the complement V to the subset W is an open set that contains $F(\bar{z})$. Since F is upper semicontinuous, there exists a neighborhood U of the point $\bar{z}$ such that $F(U \cap K) \subset V$. As V

is the complement to W, then $(U \times W) \cap G_F = \emptyset$ which proves the closedness of the graph of the mapping F.

Now let the graph of the mapping F be closed and the image of every bounded subset of the set K also be bounded. We are going to show that the mapping F is upper semicontinuous over K. Choose an arbitrary $\bar{z} \in K$ and an open subset V such that $F(\bar{z}) \subset V$. Define the subset $M = \{z \in K \mid F(z) \not\subset V\}$, whence $\bar{z} \notin M$. Denote by U the complement to the subset M in the space R^n, hence $F(U \cap K) \subset V$. Now we show that the subset U is a neighborhood of the point $\bar{z}$. On the contrary, suppose that there exists a sequence $\{z^\nu\} \subset M$ converging to $\bar{z}$. The inclusion $z^\nu \in M$ means that there exist points $y^\nu \in F(z^\nu)$ that do not belong to the open subset V. Because $\{y^\nu\} \subset F(\{z^\nu\})$ and the convergent sequence $\{z^\nu\}$ is bounded, the sequence $\{y^\nu\}$ is also bounded, hence it has a limit point $\bar{y}$. As y^ν belongs to the complement to the open subset V, then $\bar{y} \notin V$, too. Nevertheless, due to the closedness of the graph of the mapping F, we must have the inclusion $\bar{y} \in F(\bar{z}) \subset V$. Thus obtained contradiction proves that the subset U is indeed a neighborhood of the point $\bar{z}$, which implies the upper semicontinuity of the mapping F and completes the proof. ∎

For a single-valued continuous mapping, the image of a compact subset is also compact. The lemma below gives us an analogue of the latter fact for upper semicontinuous multi-valued mappings.

LEMMA 1.2. *Let F be an upper semicontinuous mapping, and the image $F(z)$ be compact for every $z \in K$. If the set K itself is compact then its image $F(K)$ also is a compact set.*

PROOF. We need to show that from any system of open subsets $\{W_\sigma \mid \sigma \in S\}$ covering the set $F(K)$, one can extract a finite subsystem that also covers $F(K)$. For each $z \in K$, a given cover of the set $F(K)$ also covers the compact subset $F(z)$. Hence, there exists a neighborhood V_z of the subset $F(z)$ obtained as the union of a finite number of the subsets W_σ. Since the mapping F is upper semicontinuous, then for every $z \in K$ one can find a neighborhood U_z such that $F(K \cap U_z) \subset V_z$. Without lack of generality, we can consider the neighborhoods U_z as open ones. As they constitute an open cover of the compact K, we can extract a finite number of the subsets U_z covering K. Then the finite number of the corresponding subsets V_z constitute a cover for $F(K)$. Since each V_z is the union of the finite number of the subsets W_σ, we obtain the finite cover of the set $F(K)$ composed by the subsets W_σ and complete the proof of the lemma. ∎

COROLLARY 1.1. *If the set K is compact, then the following two assertions are equivalent:*

1) the graph of the mapping F is a compact set;

2) the mapping F is upper semicontinuous, and the set $F(z)$ is compact for any $z \in K$.

PROOF. Let 1) be valid. Then the mapping F is closed, and the image of each $U \subset K$ is bounded being a subset of the projection of the compact graph of the mapping F. Due to Lemma 1.1, the mapping F is hence upper semicontinuous, and for any $z \in K$, the subset $F(z)$ is compact as a section of the compact graph.

Conversely, if we have 2), then Lemma 1.1 implies the closedness of the graph G_F of the mapping F, and the set $F(K)$ is compact according to Lemma 1.2. Since $G_F \subset K \times F(K)$, then G_F is compact as a closed subset of the compact set. The proof is complete. ■

1.2 VARIATIONAL INEQUALITIES ON CONVEX COMPACT SETS

In this section, we consider variational inequality (1.3)–(1.4) when the set K is nonempty, convex and compact. The case of a closed noncompact set K will be studied in the next section. In both cases, we will use the following assumptions.

A1. The graph of the mapping F is closed.

A2. For all $z \in K$, the subsets $F(z)$ are nonempty and convex.

A3. The image $F(U)$ of each bounded subset $U \subset K$ is bounded, too.

In order to establish the existence of a solution to the variational inequality, we construct an auxiliary mapping. For every linear form p, denote by $H(p)$ the collection of vectors of the set K such that the form p attains its minimum value at these vectors.

Choose a nonempty convex compact set P of linear forms such that $F(K) \subset P$. If K is compact and the mapping F satisfies assumption A3 then the latter operation is possible. Since $F(K)$ is a bounded set, we can take its closed convex hull for the set P.

LEMMA 1.3. *The following assertions are valid.*

1) *The mapping H defined over the set P satisfies assumptions* A1–A3 *in which K and F are substituted by P and H, respectively.*

2) *A pair $(\bar{p}, \bar{z}) \in P \times K$ solves variational inequality* (1.3)–(1.4) *if and only if $\bar{p} \in F(\bar{z})$ and $\bar{z} \in H(\bar{p})$.*

PROOF. Properties A2 and A3 are direct consequences of the set K being nonempty, convex and compact. Now we verify the closedness of the graph of the mapping H. Consider a sequence of pairs $\{(p^\nu, z^\nu)\}$ contained in the graph G_H of the mapping H, and suppose that the sequence converges to a pair $(\hat{p}, \hat{z})$. This means that $p^\nu \in P$, $z^\nu \in K$, and $\langle p^\nu, z^\nu \rangle \le \langle p^\nu, z \rangle$ for each $z \in K$. As the sets P and K are closed, and the canonical bilinear form is continuous with respect to the variable pair (p, z), we obtain the limit relationships $\hat{p} \in P$, $\hat{z} \in K$, and $\langle \hat{p}, \hat{z} \rangle \le \langle \hat{p}, z \rangle$ for all $z \in K$. Hence, $\hat{z} \in H(\hat{p})$ which implies

$(\hat{p}, \hat{z}) \in G_H$ and thus completes the proof of the first assertion of the lemma.

Now we prove assertion 2). If a pair $(\bar{p}, \bar{z})$ satisfies relationships (1.3) then $\bar{p} \in P$, and according to the definition of the mapping H this implies $\bar{z} \in H(\bar{p})$. Conversely, if $(\bar{p}, \bar{z}) \in P \times K$, $\bar{p} \in F(\bar{z})$, and $\bar{z} \in H(\bar{p})$, then the linear form $\bar{p}$ attains its minimum value at the point $\bar{z}$, i.e. relationships (1.3) take place. This completes the proof of the lemma. ∎

Suppose for a while that the considered mappings F and H are single-valued and continuous. In this case we can construct their superposition $g(z) = H(F(z))$ and thus obtain a continuous mapping $g : K \to K$ of the nonempty convex compact to itself. If we find a fixed point $\bar{z}$ of this mapping, such that $g(\bar{z}) = \bar{z}$, then we set $\bar{p} = F(\bar{z})$ and thus obtain that $H(\bar{p}) = \bar{z}$. Due to Lemma 1.3, the pair $(\bar{p}, \bar{z})$ solves variational inequality (1.3). The existence of the fixed point $\bar{z}$ is guaranteed here by the following *Brouwer Theorem.*

THEOREM 1.1. *Let K be a nonempty convex compact set, and $g : K \to K$ be a continuous mapping. Then the mapping g has a fixed point.* ∎

In the general case of multi-valued mappings F and H, we will realize the following scheme. For each of the mappings F and H, we will construct single-valued continuous mappings f and h, respectively, that approximate the multi-valued ones in an appropriate sense. Based upon the Brouwer Theorem, we will get a common point of the graphs of the mappings f and g. Then making use of a limit argument, we will obtain a common point of the graphs of the mappings F and H.

DEFINITION 1.3. Let K and P be two subsets of the space R^n, F be a multi-valued mapping for which $F(z) \subset P$ is valid for all $z \in K$. Finally, let U and V be two neighborhoods of zero in R^n. A single-valued mapping $f : K \to P$ is called a (U, V)*-approximative selection* for the multi-valued mapping F if and only if, for each $z \in K$, there exist $\hat{z} \in K$ and $\hat{p} \in F(\hat{z})$ such that $(z - \hat{z}) \in U$ and $(f(z) - \hat{p}) \in V$.

PROPOSITION 1.1. *Let K be a nonempty convex compact set, and let F be a multi-valued mapping defined on K satisfying assumptions* A1–A3. *Then for any pair (U, V) of neighborhoods of zero in R^n, there exists a continuous (U, V)-approximative selection for the mapping F.*

PROOF. Consider two arbitrary neighborhoods of zero U and V. Without loss of generality, we can reduce (if needed) the neighborhood V to obtain an open and convex one. Now we set $V_z = F(z) + V$ for all $z \in K$. The set V_z is convex as the sum of convex sets, and open because V is open. Since $0 \in V$ then $F(z) \subset V_z$. Thus, V_z is an open convex neighborhood of $F(z)$. Due to Lemma 1.1, the mapping F is upper semicontinuous, hence every point $z \in K$ has a neighborhood U_z such that $F(U_z) \subset V_z$. Again without affecting generality, we can consider U_z being so small that $U_z \subset z + U = \{z + u \mid u \in U\}$. Introduce an arbitrary norm on the space R^n, and for $z \in K$

choose a number $r(z) > 0$ so small that all the points $v \in K$, satisfying the inequality $\|v - z\| < 2r(z)$, belong to the neighborhood V_z. It is easy to see that for all these points $v \in K$ the inclusion $F(v) \subset V_z$ is valid. The open balls $B_z = \{v \in R^n \mid \|v - z\| < r(z)\}$ cover the compact K, whence there is a finite system of points $\{z^i,\ i = 1, \dots, m\} \subset K$ such that the system of open balls $\{B_{z^i},\ i = 1, \dots, m\}$ also covers K.

Introduce continuous functions $s_i : K \to R$, $i = 1, \dots, m$, by setting for $z \in K$

$$s_i(z) = \max\left\{0;\ r\left(z^i\right) - \|z - z^i\|\right\}.$$

It is evident that $s_i(z) \geq 0$; moreover, at each point $z \in K$, at least one of the values $s_i(z)$ is strictly positive. Thus, for each $z \in K$,

$$s(z) = \sum_{i=1}^{m} s_i(z) > 0.$$

Therefore the functions $\alpha_i : K \to R$ defined by the rule $\alpha_i(z) = s_i(z)/s(z)$ are nonnegative and continuous over K. Besides, at every $z \in K$, the following equality is clearly valid

$$\sum_{i=1}^{m} \alpha_i(z) = 1.$$

Now we fix arbitrary linear forms $p^i \in F\left(z^i\right)$, $i = 1, \dots, m$, and compose a single-valued mapping $f : K \to P$ by setting for every $z \in K$:

$$f(z) = \sum_{i=1}^{m} \alpha_i(z) p^i.$$

The mapping f is evidently continuous due to the fact that the functions α_i, $i = 1, \dots, m$ are continuous. Now we show that f satisfies Definition 1.3.

Let us fix a point $z \in K$ and determine $J = \{i \mid s_i(z) > 0\}$. The set of indices J is non-empty; moreover, $\|z - z^i\| < r(z^i)$ for $i \in J$. Pick up a number $j \in J$ such that $r\left(z^j\right) \geq r\left(z^i\right)$ for all $i \in J$. Thus, for all $i \in J$ we have

$$\|z^i - z^j\| \leq \|z - z^i\| + \|z - z^j\| < r\left(z^i\right) + r\left(z^j\right) \leq 2r\left(z^j\right),$$

that is, $z^i \in U_{z^j}$. Hence, $p^i \in F\left(z^i\right) \subset V_{z^j}$. As the set V_{z^j} is convex and $f(z)$ is a convex combination of the points p^i, $i \in J$, then $f(z) \in V_{z^j} = F\left(z^j\right) + V$. Therefore, there exists a $\hat{p} \in F\left(z^j\right)$ such that $(f(z) - \hat{p}) \in V$. To finish the proof, we only have to note that $z \in U_{z^j} \subset z^j + U$, whence $\left(z - z^j\right) \in U$ and we can take $\hat{z} = z^j$ for Definition 1.3. The proposition is proved completely. ■

Now we are in a position to establish the existence of solutions to variational inequality (1.3) or, what is the same, to problem (1.4).

THEOREM 1.2. *If K is a nonempty, convex and compact set, and F is a multi-valued mapping satisfying assumptions* A1–A3, *then variational inequality* (1.3)–(1.4) *is solvable.*

PROOF. Choose a convex compact P comprising the set $F(K)$ and construct the mapping H as demonstrated above. According to Lemma 1.3, we only need showing that the graphs of the mappings F and H have a common point. As it was demonstrated by Lemma 1.3, the mapping H also satisfies assumptions A1–A3. Since both graphs in question are closed and contained in the compact subset $P \times K$ then the graphs G_F and G_H are compact, too.

On the contrary, suppose that $G_F \cap G_H = \emptyset$. Then the pair $(0,0)$ cannot belong to the set $(G_F - G_H)$ where

$$G_F - G_H = \{(z-u, p-q) \mid p \in F(z), u \in H(q), z \in K, q \in P\}.$$

Being a difference of two compact sets, $G_F - G_H$ is closed, and hence there exist neighborhoods of zero $\tilde{U}$ and $\tilde{V}$ in R^n such that

$$\left(\tilde{U} \times \tilde{V}\right) \cap (G_F - G_H) = \emptyset. \tag{1.5}$$

We can choose open symmetrical neighborhoods of zero U and V in such a manner that $U + U \subset \tilde{U}$ and $V + V \subset \tilde{V}$. According to Proposition 1.1, there exist a continuous (U, V)-approximative selection f for the mapping F and a continuous (V, U)-approximative selection h for the mapping H. Their superposition $g = h \circ f$ is a continuous mapping of the nonempty convex compact K to itself. Due to Theorem 1.1, there exists a point $z \in K$ with $g(z) = z$. We put $p = f(z)$ and obtain that $p \in P$ and $z = h(p)$. According to Definition 1.3 there are $\hat{z} \in K$ and $\hat{p} \in F(\hat{z})$ for which the inclusions $(\hat{z} - z) \in U$ and $(\hat{p} - p) = (\hat{p} - f(z)) \in V$ take place. Similarly, there exist $\hat{q} \in P$ and $\hat{u} \in H(\hat{q})$ such that $(p - \hat{q}) \in V$ and $(z - \hat{u}) = (h(p) - \hat{u}) \in U$. These relationships imply the inclusions

$$(\hat{z} - \hat{u}) \in U + U \subset \tilde{U}, \qquad (\hat{p} - \hat{q}) \in V + V \subset \tilde{V}.$$

Since $(\hat{z}, \hat{p}) \in G_F$ and $(\hat{u}, \hat{q}) \in G_H$, the previous inclusions contradict equality (1.5) and thus complete the proof of the theorem. ■

By analyzing the above proof we notice that the nature of the multi-valued mapping H is not important. In order to guarantee that the graphs of the mappings F and H intersect it is sufficient to show that the compacts K and P are nonempty and convex, and assumptions A1–A3 are valid for both mappings. This assertion is well-known as *von Neumann Theorem* and is usually formulated as follows.

THEOREM 1.3. *Let K and P be nonempty convex compact sets, and F and H be closed subsets of the Cartesian product $K \times P$. If the sets*

$$F(z) = \{p \mid (z,p) \in F\}, \quad H(p) = \{z \mid (z,p) \in H\}$$

are nonempty and convex for any $z \in K$ and $p \in P$, then the intersection $F \cap H$ is not empty. ■

In order to represent Theorem 1.3 in terms of mappings F and H one only needs to denote their graphs by the same symbols F and H, respectively.

1.3 NONCOMPACT CASE

Now we suppose that the set K is nonempty, closed and convex but, in general, not bounded. In particular, K may be a closed convex cone. In this case, the variational inequality is usually called a *complementarity problem.* The classical variant of complementarity problem has $K = R^n_+$. At last, the set K may coincide with the whole space R^n. In the last case, for any point $\bar{z}$, it turns out that $N_K(\bar{z}) = \{0\}$ and hence, problem (1.4) reduces to the following one:

$$\bar{z} \in R^n, \quad 0 \in F(\bar{z}).$$

The general scheme of analysis of problem (1.4) for an unbounded set K may be as follows. We choose a closed convex subset S such that the intersection $K \cap S$ is nonempty and compact. For example, one can pick up a closed ball in R^n, with its radius large enough, for the set S, and apply Theorem 1.2 to the nonempty convex compact $K \cap S$ instead of the set K. Then in (1.4), the normal cone to the set K will be repaced by the normal cone to the intersection $K \cap S$. In order to obtain conditions that guarantee existence of solutions on the set K, we can impose some additional restrictions on the mapping F which imply that for some S, any solution with respect to the set $K \cap S$ is also a solution for the set K. Universal conditions demand the latter property to be valid for wide enough classes of sets S.

To realize the above-mentioned plan, we need the following theorem of Convex Analysis.

THEOREM 1.4. *Let M^i, $i = 0, \dots, m$, be convex sets and M be their intersection. If there is a point in the intersection M which is an interior point for all the sets $M^1, \dots, M^m$ (that is, for all with the only possible exception M^0), then*

$$N_M(z) = N_{M^0}(z) + N_{M^1}(z) + \cdots + N_{M^m}(z)$$

for any point $z \in M$. ∎

In the other words, the normal cone to an intersection of sets is equal to the sum of the normal cones to these sets.

We recall that a point is referred to as an *interior* point of a set if and only if the set itself is a neighborhood of this point.

PROPOSITION 1.2. *Let K be a nonempty closed subset of R^n, and the mapping F satisfy assumptions* A1–A3. *If a closed convex subset $S \subset R^n$ is such that the intersection $K \cap S$ is bounded and contains an interior point of the subset S, then there exists a $\bar{z} \in K \cap S$ such that*

$$0 \in F(\bar{z}) + N_K(\bar{z}) + N_S(\bar{z}). \tag{1.6}$$

PROOF. It is evident that the restriction of the mapping F to the subset $K \cap S$ keeps properties A2 and A3 valid. In order to obtain the validity of

A1, note that the graph of the restriction coincides with the intersection of the graph of F and the Cartesian product $S \times R^n$ which is closed, due to the closedness of the subset S. According to Theorem 1.2, there exists a $\bar{z} \in K \cap S$ such that

$$0 \in F(\bar{z}) + N_{K \cap S}(\bar{z}).$$

Since the subset $K \cap S$ contains an interior point of the set S, then Theorem 1.4 implies (1.6) and completes the proof. ■

Proposition 1.2 allows one to obtain sufficient conditions for existence of solutions to variational inequality (1.4) in case of an unbounded set K. Indeed, if for some subset S it turns out that the point $\bar{z}$ from (1.6) is an interior point of this subset, then $N_S(\bar{z}) = \{0\}$. Therefore, the last term in (1.6) can be omitted, with $\bar{z}$ satisfying relationship (1.4) as a consequence. However, before obtaining sufficient conditions of solvability of variational inequality (1.4), we consider the problem of *uniqueness of solution.*

Suppose we have two distinct solutions $\bar{z}$ and $\tilde{z}$ with corresponding points $\bar{p} \in F(\bar{z})$ and $\tilde{p} \in F(\tilde{z})$. By substituting $z = \tilde{z}$ in (1.3) and then commutating $\bar{z}$ and $\tilde{z}$ there, we come to

$$\langle \bar{p}, \tilde{z} - \bar{z} \rangle \geq 0, \qquad \langle \tilde{p}, \bar{z} - \tilde{z} \rangle \geq 0. \tag{1.7}$$

By summing up these two inequalities we obtain the relationships

$$\langle \tilde{p} - \bar{p}, \tilde{z} - \bar{z} \rangle \leq 0, \; \tilde{p} \in F(\tilde{z}), \quad \bar{p} \in F(\bar{z}). \tag{1.8}$$

Thus, if the mapping F is subject to an additional constraint that contradicts relationships (1.8), this leads to the uniqueness of solution.

DEFINITION 1.4. A mapping F is referred to as *strictly monotone* if for any two distinct points $\tilde{z} \in K$, $\bar{z} \in K$, and arbitrary $\tilde{p} \in F(\tilde{z})$, $\bar{p} \in F(\bar{z})$, the strict inequality $\langle \tilde{p} - \bar{p}, \tilde{z} - \bar{z} \rangle > 0$ takes place. The mapping F is called *monotone* if only the non-strict inequality is valid, i.e. $\langle \tilde{p} - \bar{p}, \tilde{z} - \bar{z} \rangle \geq 0$.

In the latter terms, we can formulate the following evident uniqueness condition.

PROPOSITION 1.3. *If F is a strictly monotone mapping then variational inequality* (1.3)–(1.4) *cannot have more than one solution.* ■

In spite of its elegance, the condition of Proposition 1.3 has a certain minor point. The concept of solution to variational inequality (1.3) does not depend upon the norms of the linear forms $p \in F(z)$, it is only their "directions" that matter. If we pick up a strictly positive function $r : K \to R$ and replace the mapping F with the mapping $\tilde{f}$ whose values are defined by the formula $\tilde{F}(z) = r(t)F(z)$, then the new problem will have the same solutions set. At the same time, the concepts of monotone and strictly monotone mapping essentially depend upon the relative norms of the linear forms $p \in F(z)$ for different $z \in K$.

DEFINITION 1.5. A mapping F is called *strictly quasi- monotone* if for any two distinct points $\tilde{z} \in K$, $\bar{z} \in K$, and arbitrary $\tilde{p} \in F(\tilde{z})$, $\bar{p} \in F(\bar{z})$, at least one of the following strict inequality takes place:

$$\langle \tilde{p}, \bar{z} - \tilde{z} \rangle < 0, \qquad \langle \bar{p}, \tilde{z} - \bar{z} \rangle < 0.$$

The property of strict quasi-monotonicity is weaker than that of the strict monotonicity and does not depend upon the norms of the linear forms $p \in F(z)$. Notice that in Economics Theory, the assumption of strict quasi-monotonicity of certain mappings is referred to as the *weak axiom of revealed preference.* Since the strict quasi-monotonicity clearly contradicts (1.7), the following proposition obtains.

PROPOSITION 1.4. *If a mapping F is strictly quasi-monotone, then variational inequality* (1.3)–(1.4) *can have at most one solution.* ■

Let Σ be a system of sets in R^n satisfying the following assumptions.

B1. Every set $S \in \Sigma$ is closed and convex.

B2. For each $S \in \Sigma$ the subset $K \cap S$ is a nonempty (convex) compact set.

B3. For each bounded subset $B \subset R^n$ there is a $S \in \Sigma$ such that $B \subset S$.

A simple example of the system Σ is a collection of closed balls (with respect to a certain norm) having nonempty intersection with K. However, other variants are possible, too.

DEFINITION 1.6. Let a system of sets Σ possess properties B1–B3. A system of points $\{z_S,\ S \in \Sigma\}$ is called an *exceptional family of elements* for variational inequality (1.3)–(1.4) if and only if, for any $S \in \Sigma$, the following inclusions are valid:

$$z_S \in K \cap S, \quad 0 \in F(z_S) + N_K(z_S) + N_S(z_S), \tag{1.9}$$

and the point z_S is a boundary point of the subset S.

Recall that a point is referred to as a *boundary* one of a set S if and only if it is not an interior point of either S or its complement.

PROPOSITION 1.5. *Let K be a nonempty closed convex subset of R^n and a mapping F satisfy assumptions* A1–A3. *If a system of sets Σ possesses properties* B1–B3, *then variational inequality* (1.3)–(1.4) *has either a solution or an exceptional family.*

PROOF. According to Proposition 1.2, for each $S \in \Sigma$ there exists a point $z_S \in K \cap S$ satisfying (1.9). If the point z_S turns out to be an interior point for at least one set S, then $N_S(z_S) = \{0\}$ which means that z_S solves variational inequality (1.4). Otherwise, the system of points $\{z_S,\ S \in \Sigma\}$ constitutes an exceptional family of elements which completes the proof. ■

Notice that in general case, the alternative mentioned in Proposition 1.5 is not strict: it is possible that there exist simultaneously both a solution to the variational inequality and an exceptional family. However, if the properties of the mapping F provide for the uniqueness of solution over the subset $K \cap S$ for all $S \in \Sigma$ then the alternative becomes a strict one. Indeed, if there exists a solution $\bar{z}$ to variational inequality (1.4) then, due to assumption B3, there is a set $S \in \Sigma$ for which $\bar{z}$ is an interior point. The latter solves the variational inequality at the subset $K \cap S$ as well. Due to the uniqueness property, the point $\bar{z}$ must coincide with the point z_S and hence, the latter cannot be a boundary point for S. Therefore, no exceptional family exists. In particular, the alternative is strict when the mapping F is strictly quasi-monotone (strictly monotone). In Chapter 7, it will be demonstrated that the alternative is strict for pseudomonotone mappings, too.

Now we consider some variants of conditions that guarantee the existence of solutions to the variational inequality. We will use here a norm in the space R^n without defining the norm precisely, as the latter does not matter.

THEOREM 1.5. *Let K be a nonempty closed convex subset in R^n, and a mapping F satisfy assumptions* A1–A3. *Suppose that there exist $z_0 \in K$ and $r \geq 0$ such that $\langle p, z - z_0 \rangle \geq 0$ for $z \in K$ and $p \in F(z)$ if $\|z - z_0\| > r$. Then variational inequality* (1.4) *is solvable.*

PROOF. On the contrary, suppose that there is no solution to variational inequality (1.4). For the system Σ, we choose a collection of closed balls centered at the point z_0 and having radii greater than r. It is clear that assumptions B1–B3 are valid. Then according to Proposition 1.5, there exists an exceptional family $\{z_S,\ S \in \Sigma\}$. Due to (1.9), one can find elements $p \in F(z_S)$, $q \in N_K(z_S)$, and $h \in N_S(z_S)$ such that $p + q + h = 0$ and hence,

$$\langle p, z_S - z_0 \rangle + \langle q, z_S - z_0 \rangle + \langle h, z_S - z_0 \rangle = 0. \tag{1.10}$$

Since $z_0 \in K$ and $z_0 \in S$, by the definition of a normal cone we have $\langle q, z_S - z_0 \rangle \geq 0$ and $\langle h, z_S - z_0 \rangle \geq 0$. If $h = 0$ then $p + q = 0$ and $0 \in F(z_S) + N_K(z_S)$, that is, z_S solves variational inequality (1.4). Now let $h \neq 0$. As $h \in N_S(z_S)$, the point z_S is a boundary point of the set S and hence, $\|z_S - z_0\| > r$. At the same time, the point z_0 being an interior point of S, the non-trivial linear form h cannot attain its maximum over the set S at the point z_0. Therefore $\langle h, z_S - z_0 \rangle > 0$ since z_S is the maximum point of the linear form h on the set S. Now (1.10) implies that $\langle p, z_S - z_0 \rangle < 0$ although $\|z_S - z_0\| > r$, which contradicts the assumptions of the theorem and thus completes the proof. ■

By imposing somewhat stronger restrictions on the mapping F, we can rewrite the just obtained existence condition in another form.

DEFINITION 1.7. Let assumptions A1–A3 be valid. Set

$$s(z, z_0) = \min\{\langle p, z - z_0\rangle \mid p \in F(z)\},$$

where $z \in K$ and z_0 is a fixed point of the set K. A mapping F is called *coercive* with respect to the point z_0 if and only if

$$\liminf_{z \in K, \|z\| \to \infty} s(z, z_0) > 0.$$

COROLLARY 1.2. *Let K be a nonempty closed convex subset in R^n and F be a mapping satisfying assumptions* A1–A3. *If F is coercive with respect to a point $z_0 \in K$ then variational inequality* (1.4) *is solvable.* ∎

We can deduce an existence condition in terms of monotonicity of the mapping, too. However, we need first replace the strict monotonicity by a stronger requirement.

DEFINITION 1.8. A mapping F is referred to as *uniformly* (or *strongly*) monotone if and only if there is a constant $\gamma > 0$ such that for any $z \in K$, $u \in K$, $p \in F(z)$, $q \in F(u)$ the following inequality is valid:

$$\langle p - q, z - u\rangle \geq \gamma \|z - u\|^2.$$

COROLLARY 1.3. *Let K be a nonempty closed convex set in R^n and F be a mapping satisfying* A1–A3. *If the mapping F is uniformly monotone, then variational inequality* (1.4) *has a solution.*

PROOF. Let us fix a point $z_0 \in K$ and a linear form $p_0 \in F(z_0)$. For every $z \in K$ and $p \in F(z)$ we have

$$\langle p, z - z_0\rangle = \langle p - p_0, z - z_0\rangle + \langle p_0, z - z_0\rangle \geq (\gamma\|z - z_0\| - \|p_0\|)\,\|z - z_0\|.$$

If $\|z - z_0\| > \|p_0\|/\gamma = r$ then $\langle p, z - z_0\rangle > 0$. Due to Theorem 1.5, there exists a solution to variational inequality (1.4). The corollary is proved. ∎

Analyzing the latter argument we see that it is only the unboundedness of the ratio $\langle p - p_0, z - z_0\rangle/\|z - z_0\|$ that matters. Therefore, we can formulate the following existence result as well.

COROLLARY 1.4. *Let K be a nonempty closed convex set, and F be a mapping satisfying assumptions* A1–A3. *Furthermore, let there be a function $\beta : R_+ \to R_+$, a point $z_0 \in K$ and a linear form $p_0 \in F(z_0)$ such that*

$$\lim_{t \to +\infty} \beta(t) = +\infty, \quad \langle p - p_0, z - z_0\rangle \geq \beta(\|z - z_0\|)\,\|z - z_0\|$$

for all $z \in K$ and $p \in F(z)$. Then variational inequality (1.4) *is solvable.* ∎

Notice that all the existence and uniqueness results for variational inequalities obtained in this chapter are valid for single-valued mappings F, too. In order to obtain them, one only need substituting $F(z)$ everywhere in place of $p \in F(z)$.

At the beginning of the chapter, we already compared the complementarity problem solution with the stationary point in the extremum problem. It is well known that the concept of a stationary point is local: it does not depend upon the size of the neighborhood of the stationary point on which the extremum problem is defined (if the domain is convex). The proposition below shows that the solution of a variational inequality has the same feature. By the way, we used this property implicitly while proving the existence results in case of an unbounded set K.

PROPOSITION 1.6. *Let $K \subset R^n$ be a closed convex set and F be a multi-valued mapping which maps every point $z \in K$ to the subset $F(z)$ of linear forms on R^n. Further, let $\bar{z} \in K$, and U be a neighborhood of the point $\bar{z}$. Then the following two assertions are equivalent.*

1) The point $\bar{z}$ is a solution of the variational inequality with the mapping F on the set K.

2) The point $\bar{z}$ solves the variational inequality with the mapping F over the subset $K \cap U$.

PROOF. It is clear that 1) implies 2) trivially: if the inequality in (1.3) is valid for $z \in K$ then it is also valid for $z \in K \cap U$. Conversely, let the inequality in (1.3) take place for all $z \in K \cap U$. Pick up $u \in K$, $\varepsilon \in (0,1)$ and define $z_\varepsilon = \varepsilon u + (1-\varepsilon)\bar{z}$. For ε sufficiently small we get $z_\varepsilon \in U$, and the convexity of the set K also implies $z_\varepsilon \in K$. Hence, $z_\varepsilon \in K \cap U$, that is, $\langle \bar{p}, z_\varepsilon - \bar{z} \rangle \geq 0$. By substituting here the expression of z_ε in terms of u and $\bar{z}$ and by dividing both parts by $\varepsilon > 0$ we obtain that $\langle \bar{p}, u - \bar{z} \rangle \geq 0$, too. Thus we have demonstrated that the inequality in (1.3) is valid for all the vectors $z \in K$ which implies that $\bar{z}$ solves the variational inequality on the whole set K and completes the proof.

■

1.4 COMMENTS

A strong interest in the study of Complementarity Problem was justified by the dramatic development of Games Theory and Mathematical Programming in the 1950s, although some elements of Complementarity Theory can be found in the earlier works by Farkas (1902), Motzkin (1936), von Neumann and Morgenstern (1944). The research activity that created Duality Theory for linear and nonlinear programming problems, as well as Equilibrium Theory in Mathematical Models of Economics, served as a serious stimul to advance in this direction. As an example, we mention the works by Tucker (1956), Lemke and Howson (1964). Further generalizations of the complementarity problem arised from the study of a generalized liner complementarity problem in Cottle and Dantzig (1970) and some more general ones; *cf.*, for instance, Karamardian (1972). The number of papers and books that followed those ones was enormous.

Originally, the problem in form of variational inequality was characteristic for the research on the control of the mathematical physics equations (Lions, 1968,1969). These problems were naturally considered in the special functional spaces related to the examined equations. The cornerstone of the study of finite-dimensional variational inequalities and their relations to the complementarity problem, was laid by the publication of the famous Proceedings volume by Cottle, Giannessi and Lions as editors, in 1980. For the further acquaintance with the problem history and the bibliography, we can recommend the works by Harker and Pang (1990), and by Isac (1992).

As we have demonstrated, the question of existence of solution to variational inequality on a compact set is actually equivalent to the corresponding fixed point theorems. For the single-valued mappings it is Brouwer Theorem, whereas for the multi-valued ones, von Neumann Theorem is essential. Proofs of the above mentioned theorems can be found, for example, in Nikaido (1968). We also mention that von Neumann Theorem was obtained while studying the equilibrium models in economics. In Nikaido (1968), on can also find some other results concerning the multi-valued mappings.

As for the noncompact case, the assumptions of coercivity and monotonicity of the mapping are utilized in various forms by a series of authors (Ortega and Rheinboldt, 1970; Moré, 1974). The scheme with exceptional families is used in Isac, Bulavsky and Kalashnikov (1997).

REFERENCES

Cottle RW, Dantzig GB. A generalization of the linear complementarity problem. J. of Combinatorial Theory 1970; 8: 79-90.

Cottle, R.W., Giannessi, F., Lions, J.-L. (eds). *Variational inequalities and complementarity problems*. New York: Academic Press, 1980.

Farkas J. Über die Theorie der einfachen Ungleichungen. J. Reine Angew. Math. 1902; 124: 1-24.

Harker PT, Pang J-S. Finite-dimensional variational inequalities and nonlinear complementarity problems: a survey of theory, algorithms and applications. Math. Programming 1990; 48: 161-220.

Isac, G. *Complementarity Problems*. Lecture Notes in Mathematics. Berlin-Heidelberg: Springer-Verlag, 1992.

Isac G, Bulavsky VA, Kalashnikov VV. Exceptional families, topological degree and complementarity problems. J. Global Optim. 1997; 10: 207-225.

Karamardian S. The complementarity problem. Math. Programming 1972; 2: 107- 129.

Lemke CE, Howson JJ. Equilibrium points of bimatrix games. SIAM J. 1964; 12: 413-423.

Lions, J.-L.

1. *Contrôle optimal de systèmes gouvernès par des équations aux dérivées partielles*. Paris, 1968.

2. *Quelques méthodes de résolution des problèmes aux limites non linéares*. Paris, 1969.

Moré JJ. Coercivity conditions in nonlinear complementarity problems. Math. Programming 1974; 5: 327-338.

Motzkin, T.S. *Beiträge zur Theorie der linearen Ungleichungen.* (Dissertation, Basel, 1933), Jerusalem, 1936.

von Neumann, J., and Morgenstern, O. *Theory of Games and Economic Behavior.* Princeton: Princeton University Press, 1944.

Nikaido, H. *Convex Structures and Economic Theory.* New York - London: Academic Press, 1968.

Ortega, J.M, Rheinboldt, W.C. *Iterative Solutions of Nonlinear Equations of Several Variables.* New York - London: Academic Press, 1970.

Tucker, A.W. "Dual systems of homogeneous linear relations".- In: *Linear Inequalities and Related Systems*, H.W. Kuhn and A.W. Tucker, eds. Princeton: Princeton University Press, 1956.

Chapter 2

OPTIMIZATION MODELS

2.1 MINIMIZATION PROBLEM: GENERAL FORM

This section deals mainly with the problem of minimization of the function $f : D \to R$ over the closed convex subset $K \subset D$. The domain $D \subset R^n$, in general, may not coincide with the whole space R^n. However, in order to avoid considering the boundary effects, we will always assume that D is an open set. Thus, the closed subset K is contained in the interior of D. It is traditional to consider two classes of these problems: the case of continuously differentiable function f and the case of convex function f.

Let $f : D \to R$ be continuously differentiable, and we are to find a point $\bar{z} \in K$ at which the function f attains its minimal value over the subset K. Thus, the inequality $f(\bar{z}) \leq f(z)$ must hold for all $z \in K$. The subset K being convex, the point $z_\lambda = \bar{z} + \lambda(z - \bar{z})$ belongs to K for $\lambda \in (0, 1)$. Therefore, for the derivative $f'(\bar{z})$ of the function f at the point $\bar{z}$, we have

$$\langle f'(\bar{z}), z - \bar{z} \rangle = \lim_{\lambda \to +0} \frac{f(z_\lambda) - f(\bar{z})}{\lambda} \geq 0.$$

This inequality shows that the point $\bar{z}$ and the linear form $\bar{p} = f'(\bar{z})$ constitute a solution to variational inequality (1.3) in which $F : D \to R^n$ coincides with the (single-valued) differential mapping of the function f, i.e. if we set $F(z) = f'(z)$ for all $z \in D$. The function f being continuously differentiable, assumptions A1–A3 are clearly valid.

Of course, it is not true that each solution of the variational inequality provides the minimum to the function f over the whole subset K. The solutions of variational inequality (1.3) with the differential mapping F satisfy the first order optimality conditions for the problem of minimization of the function f. Therefore, these solutions correspond to the concept of stationary points, the collection of which comprises the needed minimum point (if the latter exists at all).

According to Proposition 1.6, the concept of solution to the variational inequality is local. There exists an analogous concept for the problem of minimization of function f, too. Recall that a point $\bar{z} \in K$ is a *local minimum* point (on the subset K) if and only if there is a neighborhood U of the point $\bar{z}$ such that $f(\bar{z}) \leq f(z)$ for all $z \in K \cap U$. Therefore, the set of solutions to the above variational inequality comprises all the local minimum points of the function f on K.

Sufficient conditions of the local minimum usually exploit the second derivative of the function f and guarantee that the local minimum point is *isolated*. This means that in some neighborhood of the local minimum point $\bar{z}$, there is no other local minimum points. The isolated local minimum point can be characterized in terms of the differential mapping, too.

PROPOSITION 2.1. *Let $K \subset R^n$ be a closed convex set and $\bar{z} \in R^n$ be a solution of variational inequality* (1.3) *with the differential mapping F of the function f. The point $\bar{z}$ is an isolated local minimum point of the function f if there exists a neighborhood U of the point $\bar{z}$ such that the differential mapping F is strictly quasi-monotone on the subset $K \cap U$.*

PROOF. As $\bar{z}$ solves variational inequality (1.3) then we have $\langle F(\bar{z}), z - \bar{z}\rangle \geq 0$ for all $z \in K \cap U$. According to Definition 1.5, the latter means that $\langle F(z), \bar{z} - z\rangle < 0$ for all $z \in K \cap U$. Thus the points $z \in K \cap U$ do not solve the variational inequality on the subset $K \cap U$. Since the neighborhood U can always be considered as open, it is a neighborhood for all the points $z \in K \cap U$ as well. Then by virtue of Proposition 1.6, the points $z \in K \cap U$ do not solve the variational inequality over the whole set K either, which completes the proof of the proposition. ■

Now we are going to interpret the concept of the strict quasi-monotonicity of the differential mapping of the function f. According to Definition 1.5, if $\langle f'(\bar{z}), u\rangle \geq 0$ then $\langle f'(\bar{z} + \varepsilon u), u\rangle > 0$ for $\varepsilon > 0$. In the other words, if the function does not decrease along the direction u at the point $\bar{z}$, then it strictly increases along the same direction on the ray $\{\bar{z} + \varepsilon u \mid \varepsilon > 0\}$.

The local strict quasi-monotonicity is sufficient to guarantee that the solution to variational inequality (1.3) is isolated not only for the differential mapping but also for other variants chosen for the mapping F. Now we consider the following model. We need to make a choice among possible alternatives described by a set K. An estimate of a chosen alternative $\bar{z}$ depends upon the context. Namely, in order to compare a chosen alternative $\bar{z}$ to other points of the set K we use a linear form $\bar{p} \in F(\bar{z})$. The latter is treated as an estimate of some factors modeled by the components of vectors from K. We can think that these estimates represent our losses implied by our choice. In the other words, we are inclined to consider the vector $\bar{z}$ as acceptable if $\langle \bar{p}, z\rangle \geq \langle \bar{p}, \bar{z}\rangle$ for all the alternatives $z \in K$. Thus, the problem is equivalent to variational inequality (1.3).

Suppose that the mapping F is strictly quasi-monotone at the neighborhood U of the point $\bar{z}$ (i.e., the weak axiom of revealed preference is assumed). This means that for any point $z \in K$ close enough to $\bar{z}$, there is no estimates $p \in F(z)$ for which the point z proves to be preferable to $\bar{z}$. In the other words, the local attempt to change the made choice will persuade us to return to the choice $\bar{z}$. Thus, the local strict quasi-monotonicity of the mapping F means the local stability of the decision made. If F has this property on the whole K, then the stability is global. Note that in the latter case, due to Proposition 1.4, the solution $\bar{z}$ is unique.

In Definition 1.4, the notions of monotone and strictly monotone mappings are introduced. Now we interpret these concepts in case when F is the differential mapping of a function f. Let us fix a point $\bar{z}$ and pick up another point $z \in K$. If the differential mapping of the funtion f is monotone, then for $\alpha \in (0,1)$ we get

$$\langle f'(\bar{z}+\alpha(z-\bar{z})), z-\bar{z}\rangle = \langle f'(\bar{z}+\alpha(z-\bar{z}))-f'(\bar{z}), z-\bar{z}\rangle + \langle f'(\bar{z}), z-\bar{z}\rangle \geq$$

$$\geq \langle f'(\bar{z}), z-\bar{z}\rangle.$$

At the same time, for the differentiable function, the following equality holds:

$$f(z) = f(\bar{z}) + \int_0^1 \langle f'(\bar{z}+\alpha(z-\bar{z})), z-\bar{z}\rangle \, d\alpha.$$

Hence, the monotonicity of the differential mapping means that $f(z) \geq f(\bar{z}) + \langle f'(\bar{z}), z-\bar{z}\rangle$. This inequality is well known as the *differential inequality* for the convex (differentiable) function f. From the differential inequality, one can easily deduce the definition of the convex function without use of derivatives:

$$f(\lambda z + (1-\lambda)\bar{z}) \leq \lambda f(z) + (1-\lambda) f(\bar{z})$$

for all $z \in D$, $\bar{z} \in D$, and $\lambda \in (0,1)$. We remark that convex functions are usually considered on convex sets D. It is easy to verify that the strict monotonicity of the differential mapping implies the strict convexity of the function f, with the latter inequality becoming strict.

Now we consider the minimization problem for the convex function $f : D \to R$ where $D \subset R^n$ is an open convex subset. Recall some facts from Convex Analysis. First of all we note that in the finite-dimensional space, any convex function is continuous at every interior point of its domain, hence in our case, the function f is continuous over D.

DEFINITION 2.1. A linear form p is called a *subgradient* of a convex fuction f at a point $\bar{z}$ if and only if $f(z) \geq f(\bar{z}) + \langle p, z-\bar{z}\rangle$ for all $z \in D$. The set of all subgradients of the function f at the point $\bar{z}$ is referred to as its *subdifferential* at this point and is denoted by $\partial f(\bar{z})$.

It is easily verified that the notion of subgradient is local: if the inequality from Definition 2.1 is valid for z from a neighborhood of the point $\bar{z}$ then, due to the convexity of the function f, it keeps being valid for all $z \in D$. Note that if (and only if) the function f is differentiable at the point $\bar{z}$ then there is the unique subgradient coinciding with $f'(\bar{z})$. Then the inequality from Definition 2.1 reduces to the above-mentioned differential inequality. On the other hand, if $\partial f(\bar{z})$ contains more than one linear form, then the ordinary derivative does not exist. In the latter case, there exist only derivatives with respect to each direction ζ:

$$f'(\bar{z};\zeta) = \lim_{\lambda \to +0} \frac{f(\bar{z}+\lambda\zeta) - f(\bar{z})}{\lambda}.$$

The relation between the subdifferential and the *directional derivative* is demonstrated by the formula:

$$f'(\bar{z};\zeta) = \sup\{\langle p,\zeta\rangle \mid p \in \partial f(\bar{z})\}. \tag{2.1}$$

Note that in case of the finite-dimensional space and the open set D, the subdifferential is a nonempty convex compact set. Hence, one can use the symbol "max" instead of "sup" in (2.1).

The subgradient plays the role of the derivative in the problem of minimization of the convex function. One can see immediately from Definition 2.1 that the point $\bar{z}$ is a minimum point of the convex function f on the whole its domain D if and only if $0 \in \partial f(\bar{z})$. Due to the localness of the notion of subgradient, the latter assertion also remains true in case when $\bar{z}$ is an interior minimum point on a subset $K \subset D$.

In order to examine the problem of minimization of a convex function over a closed convex subset $K \subset D$, we need the following theorem of separability of convex sets, known as the Separation Theorem.

THEOREM 2.1. *Let A and B be convex sets, B being an open set. If $A \cap B = \emptyset$ then there exists a non-zero linear form q that separates these two sets, i.e. $\langle q,u\rangle > \langle q,v\rangle$ for all $u \in A$ and $v \in B$.* ■

To establish a relationship between the problem of minimization of a convex function and the variational inequality problem, we will also utilize the following lemma (of *supporting functional*).

LEMMA 2.1. *Let the point $\bar{z} \in D$ be no minimum point of the convex function f on the whole set D, and the linear form h have the following property: $\langle h, z-\bar{z}\rangle < 0$ for all z belonging to the level set $\{z \in D \mid f(z) < f(\bar{z})\}$. Then there is a number $\gamma > 0$ such that the linear form $p = \gamma h$ is a subgradient of the function f at the point $\bar{z}$.*

PROOF. In the space $R^{n+1} = R^n \times R$, consider the subsets

$$A = \{(y,\beta) \mid y \in R^n, \langle h, y-\bar{z}\rangle \geq 0, \beta = f(\bar{z})\}$$

and

$$B = \{(z,\alpha) \mid z \in D, \alpha > f(z)\}.$$

The form h being linear and the function f being convex imply immediately that the sets A and B are convex. The set B also is open being defined by a strict inequality with continuous right-hand and left-hand parts. Finally, were $(z, \alpha) \in A \cap B$, the following inequalities would be valid: $f(z) < f(\bar{z})$ and $\langle h, z - \bar{z} \rangle \geq 0$, which would contradict the lemma's assumptions. Therefore, $A \cap B = \emptyset$.

Now by virtue of Theorem 2.1, there exists a non-trivial linear form $q = (p, r)$ over $R^n \times R$ such that

$$\langle p, z \rangle + r\alpha < \langle p, y \rangle + r\beta \tag{2.2}$$

for all $(z, \alpha) \in B$ and $(y, \beta) \in A$. Since the number α in (2.2) can be chosen arbitrarily large, we necessarily have $r \leq 0$. If $r = 0$ then for $y = \bar{z}$ and each $z \in D$ we obtain $\langle p, z \rangle < \langle p, \bar{z} \rangle$. However the latter implies $p \neq 0$ and hence contradicts the fact that $\bar{z}$ is an interior point of D. Thus obtained contradiction shows that $r < 0$. As the linear form $q = (p, r)$ may be normalized by an arbitrary positive factor without affecting its properties, we can set $r = -1$. Then we rewrite inequality (2.2) as follows:

$$\alpha > f(\bar{z}) + \langle p, z - y \rangle$$

for all $z \in D$ such that $\alpha > f(z)$ and $\langle h, y - \bar{z} \rangle \geq 0$.

Having set $y = \bar{z}$ and tended α to $f(z)$, we obtain $p \in \partial f(\bar{z})$. Moreover, $p \neq 0$, because $\bar{z}$ is no minimum point of the function f over D. Having taken $z = \bar{z}$ and tended α to $f(\bar{z})$ we will find that $\langle p, y - \bar{z} \rangle \geq 0$ for vectors y satisfying the inequality $\langle h, y - \bar{z} \rangle \geq 0$. For non-trivial linear forms p and h, the latter is possible only when the forms can be equalized by a positive multiplier. Therefore, there is a $\gamma > 0$ such that $p = \gamma h$ which completes the proof. ■

DEFINITION 2.2. The multi-valued mapping F such that $F(z) = \partial f(z)$, $z \in D$, is called the *subdifferential mapping* of the convex function f.

PROPOSITION 2.2. *The point $\bar{z} \in K$ is a minimum point of the convex function $f : D \to R$ on the convex subset K if and only if $\bar{z}$ solves variational inequality* (1.3) *with the subdifferential mapping F.*

PROOF. The necessary part of the assertion is very easy to prove. Indeed, if $\bar{z} \in K$ and $\bar{p} \in \partial f(\bar{z}) = F(\bar{z})$ satisfy relationships (1.3) then the definition of subgradient implies

$$f(z) \geq f(\bar{z}) + \langle \bar{p}, z - \bar{z} \rangle \geq f(\bar{z})$$

for all $z \in K$, i.e. $\bar{z}$ is a minimum point.

Now we prove the sufficient part of the assertion. Let $\bar{z} \in K$ be a minimum point of f on K. If it also is a minimum point of the function f over the whole D then $0 \in \partial f(\bar{z})$, and we can take $\bar{p} = 0$ satisfying conditions (1.3). Otherwise, if $0 \notin \partial f(\bar{z})$ then the set $B = \{z \in D \mid f(z) < f(\bar{z})\}$ is nonempty and open, due to the continuity of the function f. Moreover, $B \cap K = \emptyset$.

According to Theorem 2.1, there exists a non-trivial linear form h such that $\langle h, v-u\rangle < 0$ for $v \in B$ and $u \in K$. Having set $v = z$ and $u = \bar{z}$, we are under the assumptions of Lemma 2.1. Therefore, there is a number $\gamma > 0$ such that the linear form $\bar{p} = \gamma h$ is a subgradient of the function f at the point $\bar{z}$. Since $\gamma > 0$, we have $\langle \bar{p}, v - z\rangle < 0$ for $v \in B$ and $z \in K$. Now we fix a vector $v \in B$ and set $v_\varepsilon = \varepsilon v + (1-\varepsilon)\bar{z}$ for $\varepsilon \in (0,1)$. Because $f(v_\varepsilon) \le \varepsilon f(v) + (1-\varepsilon) f(\bar{z}) < f(\bar{z})$ then $v_\varepsilon \in B$. Therefore, $\langle \bar{p}, v_\varepsilon - z\rangle < 0$ for $z \in K$ and $\varepsilon \in (0,1)$. By tending ε to zero we find that $\langle \bar{p}, z - \bar{z}\rangle \ge 0$ for all $z \in K$, which means that $\bar{z}$, together with $\bar{p} \in \partial f(\bar{z})$, solve variational inequality (1.3) which completes the proof. ■

COROLLARY 2.1. *At any point $\bar{z} \in D$, there exists a subgradient of the convex function f, i.e. $\partial f(\bar{z}) \neq \emptyset$.*

PROOF. Indeed, to prove the corollary, it is sufficient to take $K = \{\bar{z}\}$ in Proposition 2.2. ■

Now we will establish that the subdifferential mapping of a convex function satisfies assumptions A1–A3.

PROPOSITION 2.3. *If the convex function f has the open convex set $D \subset R^n$ as its domain, then for any closed (convex) subset $K \subset D$, the subdifferential mapping F over K possesses properties* A1–A3.

PROOF. The closedness of the graph of F follows immediately from the definition of subgradient and the continuity of the function f. Indeed, the graph of the subdifferential mapping F consists of the pairs $(z,p) \in K \times R^n$ satisfying the inequalities $f(u) \ge f(z) + \langle p, u - z\rangle$, $u \in K$. Each inequality defines a closed subset of pairs (z,p) in the closed set $K \times R^n$, and the intersection of all those closed subsets coincides with the graph of the subdifferential mapping on K. The convexity of the subdifferential $\partial f(z)$ follows immediately from Definition 2.1, whereas its nonemptiness is confirmed by Corollary 2.1.

Now we prove property A3 by the exclusion method. On the contrary, let $U \subset K$ be a bounded subset and there exist a sequence of pairs $\{(z^\nu, p^\nu)\} \subset U \times R^n$ such that $p^\nu \in \partial f(z^\nu)$ and $\|p^\nu\| \to \infty$. As the subset U is bounded, without loss of generality we can accept that $z^\nu \to \bar{z}$ and $q^\nu = p^\nu/\|p^\nu\| \to \bar{q}$. For a fixed $z \in D$ we have

$$f(z) \ge f(z^\nu) + \langle p^\nu, z - z^\nu\rangle.$$

Since $f(z^\nu) \to f(\bar{z})$ then by dividing this equality by $\|p^\nu\|$ we obtain the limit relationship $0 \ge \langle \bar{q}, z - z^\nu\rangle$ for all $z \in D$. Because the point $\bar{z}$ is an interior point of the set D, the latter relationship can take place only if $\bar{q} = 0$. However, $\bar{q}$ is the limit of the normed sequence so that $\|q\| = 1$. Thus obtained contradiction shows that $\partial f(U)$ is bounded and completes the proof. ■

As we showed previously for the differentiable case, the monotonicity of the differential mapping implies the convexity of the function f. It is easy to verify that in general case, too, the subdifferential mapping of the convex function is monotone. Indeed, if $z \in D$ and $\bar{z} \in D$, $p \in \partial f(z)$ and $\bar{p} \in \partial f(\bar{z})$, then

$$f(z) \geq f(\bar{z}) + \langle \bar{p}, z - \bar{z} \rangle, \; f(\bar{z}) \geq f(z) + \langle p, \bar{z} - z \rangle.$$

Summing up these inequalities we come to $\langle p - \bar{p}, z - \bar{z} \rangle \geq 0$.

The coincidence of the concepts of local and global minima is another important property of the above considered problem of minimization of the convex function on the convex set. Indeed, if a point $\bar{z}$ is a local minimum of the function f over K then there is a convex neighborhood U of the point $\bar{z}$ such that $\bar{z}$ is a minimum point on $K \cap U$. Having applied Proposition 2.2 with K substituted by $K \cap U$, we will find that $\bar{z}$ solves the variational inequality with the differential mapping over the subset $K \cap U$. By virtue of Proposition 1.6, this feature of the point $\bar{z}$ is valid for the whole set K. Applying again Proposition 2.2, we get that the point $\bar{z}$ is a minimum point of the function f on the set K.

One has often to deal with not a distinct minimum point but the whole set of minimum points (when the uniquness of solution is not the case). The following assertion is appropriate for such situations.

PROPOSITION 2.4. *If K is a closed convex set and f a convex continuous function, then the set of all minimum points of the function f on the set K is closed and convex.*

PROOF. As the empty set is closed and convex, we may consider only the case when the minimum points exist. Then $\min\{f(z) \mid z \in K\} = \varphi$ is a finite value, and the set of minimum points coincides with the set $\{z \in K \mid f(z) \leq \varphi\}$. This set is closed and convex because the set K is convex, the function f is convex and continuous. This completes the proof. ■

We conclude the section with the following remark. One frequently needs considering functions with the property opposite to the convexity:

$$f(\lambda z + (1 - \lambda)\bar{z}) \geq \lambda f(z) + (1 - \lambda) f(\bar{z})$$

for $z \in D$, $\bar{z} \in D$, and $\lambda \in (0, 1)$. Such functions are usually referred to as *concave* ones. By replacing f with $-f$, the concave function becomes convex, and vice versa. Therefore, all the results obtained above for the convex functions can be easily fitted to the case of the concave function f. In particular, the concept of subgradient is replaced by that of *supergradient.*

DEFINITION 2.3. The linear form p is called a *supergradient* of the concave function $f : D \to R$ at the point $\bar{z}$ if and only if $f(z) \leq f(\bar{z}) + \langle p, z - \bar{z} \rangle$ for all $z \in D$. The collection of all the supergradients is referred to as the *superdifferential* at the point $\bar{z}$ and denoted by $\partial f(\bar{z})$.

The identical notation for the subdifferential and superdifferential does not lead to confusion, because it is always contextually clear what is the case. Note that for the concave function f, the inclusion $0 \in \partial f(\bar{z})$ takes place for the maximum points of the function f on D. At last, the inclusion $0 \in \partial f(\bar{z}) + N_K(\bar{z})$ which is analogous to variational inequality (1.4), means that the point $\bar{z}$ is a maximum point of the function f over the (closed) convex subset $K \subset D$.

2.2 CONVEX PROGRAMMING PROBLEM

Suppose that an open set $D \subset R^n$ is the domain of convex functions $f : D \to R$ and $g_i : D \to R^n$, $i = 1, \dots, m$, and $M^0 \subset D$ is a nonempty closed convex subset. The following problem

$$\min \left\{ f(z) \mid z \in M^0, g_i(z) \le 0, \; i = 1, \dots, m \right\} \tag{2.3}$$

is usually referred to as a *convex programming* problem.

Define

$$K = \left\{ z \in M^0 \mid g_i(z) \le 0, \; i = 1, \dots, m \right\}.$$

Due to the properties of the subset M^0 and the functions g_i, $i = 1, \dots, m$, the subset K is closed and convex. As a rule, the subset K is also assumed to be nonempty. We will do that in form of the standard *regularity condition.*

Slater's Condition. There is a point $\hat{z} \in M^0$ such that $g_i(\hat{z}) < 0$, $i = 1, \dots, m$.

The point $\hat{z}$ in the Slater's condition is often called the *Slater's point* or the *interior point,* although it may not be an interior point for the subset M^0.

Problem (2.3) can be formulated in the form presented in Section 2.1. Hence, all the results previously obtained for the convex problem are valid for convex programming problem (2.3) as well. Our aim is to interpret those results in terms of the subset M^0 and the functions g_i, $i = 1, \dots, m$. Note that the points of the subset K are frequently called *feasible* points, whereas the solution of problem (2.3) is usually referred to as an *optimal* point, an *optimum* point, or simply an *optimum.*

According to Proposition 2.2, the feasible point $\bar{z}$ is an optimum if and only if $0 \in \partial f(\bar{z}) + N_K(\bar{z})$. This condition is interpreted by the following assertion that is well-known as the *Karush-Kuhn-Tucker Theorem.*

THEOREM 2.2. *For convex programming problem* (2.3), *the feasible point* $\bar{z}$ *is an optimum if, and under the Slater's condition, only if there exist subgradients* $\bar{p} \in \partial f(\bar{z})$, $p_i \in \partial g_i(\bar{z})$, $i = 1, \dots, m$, *and numbers* $\bar{\lambda}_i$, $i = 1, \dots, m$, *satisfying the following relationships:*

1) $\bar{\lambda}_i g_i(\bar{z}) = 0, \quad i = 1, \dots, m;$

2) *the linear form* $q = \bar{p} + \sum_{i=1}^{m} \bar{\lambda}_i p_i$ *attains its minimum value over the subset* M^0 *at the point* $\bar{z}$.

PROOF. Begin with the sufficiency. Let there exist the above-mentioned numbers and subgradients. Then for any $z \in K$, it holds

$$f(z) \geq f(\bar{z}) + \langle \bar{p}, z - \bar{z} \rangle, \; 0 \geq g_i(z) \geq g_i(\bar{z}) + \langle p_i, z - \bar{z} \rangle, \tag{2.4}$$

$i = 1, \ldots, m$. Due to condition 1) and the assumption $\bar{\lambda}_i \geq 0$, $i = 1, \ldots, m$, we come to the inequality $f(z) \geq f(\bar{z}) + \langle q, z - \bar{z} \rangle$. As $z \in M^0$, condition 2) implies $f(z) \geq f(\bar{z})$ which means that $\bar{z}$ is an optimum point.

Now let the Slater's condition take place and the point $\bar{z}$ be an optimum. According to Proposition 2.2, $0 \in \partial f(\bar{z}) + N_K(\bar{z})$. Hence there exist $\bar{p} \in \partial f(\bar{z})$ and $h \in N_K(\bar{z})$ such that $0 = \bar{p} + h$. Define

$$M^i = \{z \in D \mid g_i(z) \leq 0\}, \quad i = 1, \ldots, m.$$

Then $K = M^0 \cap M^1 \cap \ldots \cap M^m$, and the continuity of functions g_i implies that the Slater's point $\hat{z}$ belongs to K and moreover, it is an interior point for the subsets $M^1, \ldots, M^m$. Due to Theorem 1.4, then

$$N_K(\bar{z}) = N_{M^0}(\bar{z}) + \cdots + N_{M^m}(\bar{z}),$$

hence there exist $h_i \in N_{M^i}(\bar{z})$, $i = 0, 1, \ldots, m$, such that $\bar{p} + h_0 + \cdots + h_m = 0$.

If there happens that $g_i(\bar{z}) < 0$ for some $i \geq 1$ then the point $\bar{z}$ is an interior point for the subset M^i, hence the normal cone to M^i at the point $\bar{z}$ contains only the trivial (zero) linear form. Therefore $h_i = 0$ when $g_i(\bar{z}) < 0$. We notice that the linear form h_i may happen to be trivial when $g_i(\bar{z}) = 0$, too. Such a situation can be treated as a casual coincidence (degeneracy). Anyway, when $h_i = 0$, we set $\bar{\lambda}_i = 0$ and take an arbitrary subgradient p_i in $\partial g_i(\bar{z})$ to satisfy condition 1).

Now we consider any index i such that $g_i(\bar{z}) = 0$ but $h_i \neq 0$. According to the definition of normal cone, then $\langle h_i, z - \bar{z} \rangle \leq 0$ for all $z \in M^i$. Hence, the latter inequality is valid on the subset $\{z \in D \mid g_i(z) < 0\}$. However, since the latter subset is open, the non-trivial linear form h_i cannot attain its maximal value $\langle h_i, \bar{z} \rangle$ on this subset. Therefore, in fact, the strict inequality $\langle h_i, z - \bar{z} \rangle < 0$ must take place over the whole subset $\{z \in D \mid g_i(z) < g(\bar{z}) = 0\}$. We also note that the point $\bar{z}$ is no minimum point of the function g_i on the set D as $g_i(\hat{z}) < 0 = g_i(\bar{z})$. Therefore, we are under assumptions of Lemma 2.1 which assures that there exist a number $\gamma_i > 0$ and a linear form $p_i \in \partial g_i(\bar{z})$ such that $p_i = \gamma_i h_i$. Having set $\bar{\lambda}_i = 1/\gamma_i$ we obtain $h_i = \bar{\lambda}_i p_i$.

Now we can compose the linear form q from thus constructed subgradients $\bar{p}$, p_i, $i = 1, \ldots, m$, and numbers $\bar{\lambda}_i \geq 0$, according to condition 2). Furthermore, $h_0 + q = 0$, and the linear form h_0 from the normal cone of the subset M^0 attains its maximal value on M^0 at the point $\bar{z}$. Consequently, the form $q = -h_0$ attains its minimal value on the subset M^0 at the same point $\bar{z}$ which completes the proof of the theorem. ∎

The numbers $\bar{\lambda}_i$, $i = 1, \ldots, m$, introduced in Theorem 2.2, are referred to as *Lagrange multipliers*. If $\bar{z}$ is an interior point of the subset M^0 then the linear form q must be trivial. This is the case, for example, if $M^0 = D = R^n$.

When the functions f, g_i, $i = 1, \ldots, m$, are differentiable, all the subgradients are replaced by the ordinary derivatives. In this case, the linear form below must be trivial, or, in general, attain its minimum at the point $\bar{z}$:

$$f'(\bar{z}) + \sum_{i=1}^{m} \bar{\lambda}_i g_i'(\bar{z}).$$

If we consider the *Lagrange function* (or the *Lagrangian*)

$$\mathcal{L}(z, \lambda) = f(z) + \sum_{i=1}^{m} \lambda_i g_i(z), \tag{2.5}$$

then for $\lambda = \bar{\lambda} = (\lambda_1, \ldots, \lambda_m)$, the function $\mathcal{L}(z, \bar{\lambda})$ is convex with respect to the variable z (because of nonnegativity of the Lagrange multipliers $\bar{\lambda}_i$). By virtue of the results of the previous section, condition 2) means that the function $\mathcal{L}(z, \bar{\lambda})$ attains its minimum over the subset M^0 at the point $\bar{z}$. In order to obtain the same result for convex functions in general, we need the following assertion well-known in Convex Analysis as the *Moreau-Rockafellar Theorem.* We formulate this theorem for a particular case when the convex functions are defined over an open domain.

THEOREM 2.3. *Let the convex functions f, $g_1, \ldots, g_m$ be defined on an open set $D \subset R^n$, and the numbers $\bar{\lambda}_i$, $i = 1, \ldots, m$, be nonnegative. Then*

$$\partial \left(f + \bar{\lambda}_1 g_1 + \cdots + \bar{\lambda}_m g_m\right)(\bar{z}) = \partial f(\bar{z}) + \bar{\lambda}_1 \partial g_1(\bar{z}) + \cdots + \bar{\lambda}_m \partial g_m(\bar{z}). \quad \blacksquare$$

If we denote by $\partial_z \mathcal{L}(\bar{z}, \bar{\lambda})$ the subdifferential of the convex function $\mathcal{L}(z, \bar{\lambda})$ with respect to z at the point $z = \bar{z}$, then the assertion of Theorem 2.3 can be rewritten as follows:

$$\partial_z \mathcal{L}(\bar{z}, \bar{\lambda}) = \partial f(\bar{z}) + \bar{\lambda}_1 \partial g_1(\bar{z}) + \cdots + \bar{\lambda}_m \partial g_m(\bar{z}).$$

The collection $\lambda = (\lambda_1, \ldots, \lambda_m)$ of Lagrange multipliers can also be treated as a vector in the space R^m. When the vector z is fixed, the Lagrange function is linear (more precisely, affine) with respect to the variable λ, i.e., it is both convex and concave. However, it is convenient to treat the Lagrange function as being concave with respect to λ, with $\partial_z \mathcal{L}(\bar{z}, \bar{\lambda}) = (g_1(\bar{z}), \ldots, g_m(\bar{z}))$. Therefore, the inequalities $g_i(\bar{z}) \le 0$, $i = 1, \ldots, m$, together with condition 1) of Theorem 2.2, mean that the function $\mathcal{L}(\bar{z}, \lambda)$ over the set of vectors $\lambda \in R_+^m$ attains its maximal value at the point $\bar{\lambda} \in R_+^m$. On the other hand, condition 2) of Theorem 2.2, due to Theorem 2.3, means that the function $\mathcal{L}(z, \bar{\lambda})$ attains its minimal value on the subset M^0 at the point $\bar{z} \in M^0$. Now we are in a position to introduce the following notion.

DEFINITION 2.4. For the function of two variables $\mathcal{L}(z, \lambda)$ defined over the direct product $M^0 \times \Lambda$, the point $(\bar{z}, \bar{\lambda}) \in M^0 \times \Lambda$ is called a *saddle point,* if $\mathcal{L}(\bar{z}, \lambda) \le \mathcal{L}(\bar{z}, \bar{\lambda}) \le \mathcal{L}(z, \bar{\lambda})$ for all $z \in M^0$ and $\lambda \in \Lambda$.

Summing up, we can reformulate Theorem 2.2 in form of the assertion below which is also reffered to as the Karush-Kuhn-Tucker Theorem.

THEOREM 2.4. *The point $\bar{z} \in M^0$ solves the convex programming problem* (2.3) *if, and under the Slater's condition, only if there exist Lagrange multipliers $\bar{\lambda} \in R^m_+$ such that the pair $(\bar{z}, \bar{\lambda})$ is a saddle point of the Lagrange function* (2.5) *on the direct product $M^0 \times R^m_+$.* ■

The problems of Theorem 2.4 can be represented again as a variational inequality. In order to do that, we must be able to construct the normal cone to a direct product of sets. Suppose that subsets $M^i \subset R^{n_i}$, $i = 1, \dots, s$, are given (with possibly different dimensions n_i of the Euclidean spaces). Thus, the direct product $M = N^1 \times \dots \times M^s$ is a subset of the direct product $R^{i_1} \times \dots \times R^{i_s}$. If h_i is a linear form over R^{n_i}, $i = 1, \dots, s$, then the collection $h = (h_1, \dots, h_s)$ provides a linear form over $R^{i_1} \times \dots \times R^{i_s}$ defined by the rule below:

$$\langle h, x \rangle = \sum_{i=1}^{s} \langle h_i, x_i \rangle,$$

where $x = (x_1, \dots, x_s)$. We have already utilized this representation while proving Lemma 2.1. One can verify immediately that

$$N_M(\bar{x}) = N_{M^1}(\bar{x}_1) \times \dots \times N_{M^s}(\bar{x}_s). \tag{2.6}$$

In the other words, the normal cone to a direct product is the direct product of the corresponding normal cones.

Actually, the pair $(\bar{z}, \bar{\lambda})$ being a saddle point of the Lagrange function over the direct product $M^0 \times R^m_+$ is equivalent to the following relationships

$$0 \in \partial_z \mathcal{L}(\bar{z}, \bar{\lambda}) + N_{M^0}(\bar{z}), \quad 0 \in -\partial_\lambda \mathcal{L}(\bar{z}, \bar{\lambda}) + N_{R^m_+}(\bar{\lambda}). \tag{2.7}$$

Let us construct the multi-valued mapping $F : M^0 \times R^m_+ \to R^n \times R^m$ having defined

$$F(z, \lambda) = \partial_z \mathcal{L}(z, \lambda) \times (-\partial_\lambda \mathcal{L}(z, \lambda)). \tag{2.8}$$

Due to (2.6), inclusions (2.7) can be replaced by the inclusion

$$0 \in F(\bar{z}, \bar{\lambda}) + N_{M^0 \times R^m_+}(\bar{z}, \bar{\lambda}). \tag{2.9}$$

As this variational inequality is completely analogous to problem (1.4), we have thus obtained the following result.

PROPOSITION 2.5. *The vectors $\bar{z} \in M^0$ and $\bar{\lambda} \in R^m_+$ are, respectively, a solution of the convex programming problem* (2.3) *and a vector of Lagrange multipliers corresponding to this solution, if the pair $(\bar{z}, \bar{\lambda})$ solves variational inequality* (2.9) *with the mapping F specified by formula* (2.8). ■

In the particular case when $M^0 = R^n_+$ and the functions f, g_i, $i = 1, \dots, m$, are convex and differentiable, mapping (2.8) is single-valued, with

$$F(z, \lambda) = \left(f'(z) + \sum_{i=1}^{m} \lambda_i g_i'(z); -g_1(z), \dots, -g_m(z) \right).$$

Then variational inequality (2.9) over the cone of non-negative pairs (z, λ) takes the following form:

$$F(z,\lambda) \geq 0, \ (z,\lambda) \geq 0, \ \langle F(z,\lambda), (z,\lambda) \rangle = 0.$$

Actually, the latter is a standard complementarity problem of type (1.1).

2.3 MARGINAL VALUES AND FEEDBACK IN OPTIMIZATION PROBLEMS

One of the standard interpretations of the convex programming problem (2.3) is as follows. The vector $z \in M^0$ is treated as a plan, the value of the minimized function $f(z)$ as the cost of realization of this plan, whereas the inequalities $g_i(z) \leq 0$, $i = 1, \ldots, m$, are interpreted as constraints with respect to a series of factors. In this section, we will extend the model by indroducing auxiliary variables y_i, $i = 1, \ldots, m$ treated as parameters. After that, problem (2.3) comes to the form

$$\min \left\{ f(z) \mid z \in M^0, g_i(z) + y_i \leq 0, \ i = 1, \ldots, m \right\}. \tag{2.10}$$

The act of adding the terms y_i can be treated as imposing a certain load (perturbation) upon the model. This load may be either positive or negative. Actually, some load could be present in problem (2.3) as well, but there it was reckoned as a constant one, and hence, it was included into the functions g_i. In contrast to that, here the parameters y_i will be subject to variations. Therefore, we are going to answer two questions: first, how the optimal cost value $f(z)$ depends upon the load y, and second, how one can model the load selection process.

The set of feasible vectors in problem (2.10) clearly depends upon the load values y. Define

$$K(y) = \left\{ z \in M^0 \mid g_i(z) + y_i \leq 0, \ i = 1, \ldots, m \right\}.$$

For some values of the parameters y the set $K(y)$ may happen to be empty. Now let Y denote the collection of y such that $K(y) \neq \emptyset$. It is easy to see that the set Y is convex and unbounded, as, together with every $\bar{y}$, the set Y comprises all the vectors y such that $y \leq \bar{y}$. We assume, of course, that $M^0 \neq \emptyset$. In general, the set Y may not be closed. We also note that the interior points of the set Y are exactly those $y \in Y$ for which the Slater's condition for problem (2.3) is valid. Now we make the following assumption.

C. The sets $K(y)$ are bounded for $y \in Y$.

The latter assumption may be provided either by the boundedness of the set M^0 or by the configuration of inequalities in (2.10). From the point of view of a modeler, assumption C does not look too heavy a burden. At the same time, it helps us to avoid discussing the question of existence of an optimum point in problem (2.10). Moreover, as the argument below shows, the boundedness of at least one of the nonempty sets $K(y)$ is enough for assumption C to be valid.

PROPOSITION 2.6. *The nonempty sets $K(y)$ are either all bounded, or all unbounded.*

PROOF. Let the set $K(y)$ be unbounded and $\bar{z} \in K(\bar{y})$. We will prove that $K(\bar{y})$ also is unbounded. Consider an unbounded sequence $\{z^\nu\} \subset K(y)$. Without loss of generality we can accept that the normed sequence $x^\nu = (z^\nu - \bar{z})/\|z^\nu - \bar{z}\|$ converges to a vector x with $\|x\| = 1$. Select an arbitrary $\mu > 0$ and set $\lambda_\nu = \mu/\|z^\nu - \bar{z}\|$, so that $\lambda_\nu \to 0$. For ν large enough, we have $\lambda_\nu \in (0,1)$. Since

$$g_i(\bar{z}) + \bar{y}_i \leq 0, \; g_i(z^\nu) + y_i \leq 0, \; i = 1, \ldots, m,$$

we deduce that

$$g_i(\bar{z} + \mu x^\nu) + (\lambda_\nu y_i + (1 - \lambda_\nu)\bar{y}_i) \leq 0, i = 1, \ldots, m,$$

where $\bar{z} + \mu x^\nu = \lambda_\nu z^\nu + (1 - \lambda_\nu)\bar{z}$. As $(\bar{z} + \mu x^\nu) \in M^0$ then we can obtain the limit inclusion $(\bar{z} + \mu x) \in K(\bar{y})$ for all $\mu > 0$ and thus complete the proof. ∎

Now consider

$$\varphi(y) = \min\{f(z) \mid z \in K(y)\}, \; y \in Y. \tag{2.11}$$

The function φ defined by (2.11) is referred to either as a *perturbation* function, or a *marginal* function. We will use the second term. In our interpretation, the marginal function indicates the minimal possible costs under the perturbation (load) vector y.

It is easily verified that the marginal function is convex. Indeed, let $z_1 \in K(y_1)$, $z_2 \in K(y_2)$, $\varphi(y_1) = f(z_1)$, $\varphi(y_2) = f(z_2)$. Then for $\lambda \in (0,1)$, it turns out that $(\lambda z_1 + (1-\lambda)z_2) \in K(\lambda y_1 + (1-\lambda)y_2)$. Therefore,

$$\varphi(\lambda y_1 + (1-\lambda)y_2) \leq f(\lambda y_1 + (1-\lambda)y_2) \leq \lambda f(z_1) + (1-\lambda)f(z_2) =$$

$$= \lambda\varphi(y_1) + (1-\lambda)\varphi(z_y).$$

Consequently, the marginal function is continuous and has a nonempty subdifferential at every interior point of the set Y. In the literature on Economics, subgradients (or derivatives, in the differentiable case) of the marginal function with respect to the parameters y are sometimes called *marginal values.* The following theorem of marginal values establishes that these values are exactly the Lagrange multipliers corresponding to the optimum point of problem (2.10) with the given values of parameters y.

THEOREM 2.5. *Let $\bar{y}$ be an interior point of the set Y and $\bar{z} \in K(\bar{y})$ be an optimum point of problem* (2.10) *with $y = \bar{y}$. The pair $(\bar{z}, \bar{\lambda})$ is a saddle point of the Lagrange function*

$$\mathcal{L}(z, \lambda; \bar{y}) = f(z) + \sum_{i=1}^{m} \lambda_i (g_i(z) + \bar{y}_i) \tag{2.12}$$

on the set $M^0 \times R^m_+$, if and only if $\lambda \in \partial\varphi(\bar{y})$.

PROOF. Let $\lambda \in \partial\varphi(\bar{y})$. The latter means that

$$\varphi(y) \geq \varphi(\bar{y}) + \langle \lambda, y - \bar{y} \rangle = \varphi(\bar{y}) + \sum_{i=1}^{m} \bar{\lambda}_i (y_i - \bar{y}_i) \tag{2.13}$$

for all $y \in Y$. We need showing that

$$\mathcal{L}(z, \bar{\lambda}; \bar{y}) \geq \mathcal{L}(\bar{z}, \bar{\lambda}; \bar{y}) \geq \mathcal{L}(\bar{z}, \lambda; \bar{y}) \tag{2.14}$$

for all $z \in M^0$ and $\lambda \in R^m_+$. We begin with the right inequality in (2.14).

First, we choose an arbitrary index k and compose the vector $y \in Y$ by setting $y_i = \bar{y}_i$ for $i \neq k$, and $y_k < \bar{y}_k$. Since $K(y) \supset K(\bar{y})$ then $\varphi(y) \leq \varphi(\bar{y})$. Therefore, $\langle \lambda, y - \bar{y} \rangle = \bar{\lambda}_k (y_k - \bar{y}_k) \leq 0$, whence $\bar{\lambda}_k \geq 0$. Consequently, the subgradient $\bar{\lambda}$ is nonnegative, i.e. $\bar{\lambda} \in R^m_+$. If for the optimum point $\bar{z}$ it happens that $(g_k(z) + \bar{y}_k) < 0$, then we select another point $y \in Y$ having set $y_i = \bar{y}_i$ for $i \neq k$, and $\bar{y}_k < y_k < -g_k(\bar{z})$. Then $K(y) \subset K(\bar{y})$, but $\bar{z} \in K(y)$. Therefore, $\varphi(y) = \varphi(\bar{y})$ and again, $\langle \lambda, y - \bar{y} \rangle = \bar{\lambda}_k (y_k - \bar{y}_k) \leq 0$ which implies $\bar{\lambda}_k \leq 0$. As $\bar{\lambda} \in R^m_+$ then $\bar{\lambda}_k = 0$. Thus, we have proved that $\bar{\lambda}_i (g_i(z) + \bar{y}_i) = 0$ for each $i = 1, \ldots, m$. Due to (2.12) we finally have that

$$\mathcal{L}(\bar{z}, \bar{\lambda}; \bar{y}) = f(\bar{z}) \geq f(\bar{z}) + \sum_{i=1}^{m} \lambda_i (g_i(z) + \bar{y}_i) = \mathcal{L}(\bar{z}, \lambda; \bar{y})$$

for all $\lambda \in R^m_+$ which proves the right inequality in (2.14).

Now we establish the left inequality in (2.14). Let $z \in M^0$, and take $y_i = -g_i(z)$ for $i = 1, \ldots, m$. Then $z \in K(y)$ and hence,

$$f(z) \geq \varphi(y) \geq \varphi(\bar{y}) + \langle \lambda, y - \bar{y} \rangle = f(\bar{z}) - \sum_{i=1}^{m} \lambda_i (g_i(z) + \bar{y}_i).$$

Thus we obtain

$$\mathcal{L}(z, \bar{\lambda}; \bar{y}) \geq f(\bar{z}) = \mathcal{L}(\bar{z}, \bar{\lambda}; \bar{y}),$$

which completes the proof of the sufficiency part of the assertion.

In the other words, we have shown that any subgradient λ of the marginal function is composed of the Lagrange multipliers. Conversely, let it be known that the pair $(\bar{z}, \bar{\lambda})$ is a saddle point of the Lagrange function. Select an arbitrary $y \in Y$ and denote by z the optimum point of problem (2.10). Since for the saddle point one has $\mathcal{L}(\bar{z}, \bar{\lambda}; \bar{y}) = f(\bar{z})$, $\bar{\lambda} \geq 0$, $(g_i(z) + y_i) \leq 0$, $i = 1, \ldots, m$, and $z \in M^0$, then

$$f(z) + \sum_{i=1}^{m} \bar{\lambda}_i (g_i(z) + \bar{y}_i) = \mathcal{L}(\bar{z}, \bar{\lambda}; \bar{y}) \geq f(\bar{z}) \geq f(\bar{z}) + \sum_{i=1}^{m} \bar{\lambda}_i (g_i(z) + y_i).$$

Thus we come to the inequality $f(z) \geq f(\bar{z}) + \langle \lambda, y - \bar{y} \rangle$. The latter, together with the equalities $f(z) = \varphi(y)$ and $f(\bar{z}) = \varphi(\bar{y})$, implies (2.13) and completes the proof of the theorem. ■

Therefore, we have established that if $\bar{y}$ is an interior point of the set Y (the region of solvability of problem (2.10)), then all the vectors of Lagrange multipliers for its solution $\bar{z}$ are described by the condition $\bar{\lambda} \in \partial\varphi(\bar{y})$. Proposition 2.3 implies that the set $\partial\varphi(\bar{y})$ is a nonempty convex compact set. We note that even though an optimum point $\bar{z}$ of problem (2.10) with $y = \bar{y}$ is not unique, nevertheless the set $\partial\varphi(\bar{y})$ of Lagrange multipliers is the same for all the optimum points. This fact leads to the idea to try revealing a problem in which the Lagrange multipliers would be determined immediately by $\bar{y}$, without selecting an optimum point $\bar{z}$.

We notice that the inclusion $\bar{\lambda} \in \partial\varphi(\bar{y})$ is tantamount to the fact that the point $\bar{y}$ is a maximum point of the function $\langle \bar{\lambda}, y \rangle - \varphi(y)$, which is concave with respect to the variable y. Let us introduce the function

$$\varphi^*(\lambda) = \sup_{y \in Y} \{\langle \lambda, y \rangle - \varphi(y)\}. \tag{2.15}$$

The function φ^* defined by formula (2.15) is referred to as the *conjugate* one to the function φ. It can be verified immediately that the function φ^* is always convex, independently of the properties of the original function φ. The conjugate function may not be defined over the whole R^m, as the supremum in (2.15) may be equal to $+\infty$. The domain of the function φ^* will be denoted by Λ. In particular, we have that $\partial\varphi(y) \subset \Lambda$ for all the interior points y of the set Y.

PROPOSITION 2.7. *If $\partial\varphi(\bar{y}) \neq \emptyset$, then $\lambda \in \partial\varphi(\bar{y})$ if and only if $\bar{y} \in \partial\varphi^*(\bar{\lambda})$.*

PROOF. Let $\lambda \in \partial\varphi(\bar{y})$. Then $\varphi^*(\bar{\lambda}) = \langle \bar{\lambda}, \bar{y} \rangle - \varphi(\bar{y})$. However, for every $\lambda \in \Lambda$, the following inequality holds: $\varphi^*(\lambda) \geq \langle \lambda, \bar{y} \rangle - \varphi(\bar{y})$. Consequently,

$$\varphi^*(\lambda) \geq \varphi^*(\bar{\lambda}) + \langle \lambda - \bar{\lambda}, \bar{y} \rangle, \quad \lambda \in \Lambda. \tag{2.16}$$

But the latter simply means that $\bar{y} \in \partial\varphi^*(\bar{\lambda})$.

Conversely, let $\bar{y} \in \partial\varphi^*(\bar{\lambda})$, i.e. (2.16) is valid. Choose a $\lambda \in \partial\varphi(\bar{y})$. It means that

$$\langle \lambda, y \rangle - \varphi(y) \leq \langle \lambda, \bar{y} \rangle - \varphi(\bar{y}), \quad y \in Y.$$

Having taken the supremum in both parts with respect to $y \in Y$, we get $\varphi^*(\lambda) = \langle \lambda, \bar{y} \rangle - \varphi(\bar{y})$. At the same time, $\varphi^*(\bar{\lambda}) \geq \langle \bar{\lambda}, y \rangle - \varphi(y)$ for all $y \in Y$.

Combined with (2.16), the latter two inequalities lead to

$$\varphi(y) \geq \varphi(\bar{y}) + \langle \bar{\lambda}, y - \bar{y} \rangle, \quad y \in Y,$$

which confirms that $\lambda \in \partial\varphi(\bar{y})$ and thus completes the proof. ■

Proposition 2.7 shows that if $\bar{y}$ is an interior point of the solvability region Y of problem (2.10) then the Lagrange multipliers can be found (at least, in theory) by solving the problem

$$\sup\{\langle \lambda, \bar{y} \rangle - \varphi^*(\lambda) \mid \lambda \in \Lambda\}. \tag{2.17}$$

Problem (2.17) is called the *dual* one to problem (2.10) for $y = \bar{y}$. In the classical case where $\bar{y} = 0$, problem (2.17) reduces to

$$\sup\{-\varphi^*(\lambda) \mid \lambda \in \Lambda\}$$

which is the dual problem to the convex programming problem (2.3).

A strictly fixed load y (as in (2.3) with $y = 0$) supposes certain targets to be completely definite. However the choice of y may depend upon the cost of each extra unit of every component y_i, $i = 1, \ldots, m$. By virtue of Theorem 2.5 these marginal costs are determined by the Lagrange multipliers which, it their turn, depend upon the load y. Now we consider a model in which we introduce, in addition, a feedback between the Lagrange multipliers and the load y.

Assume that we are given a nonempty convex compact set $M \subset Y$ that can model a really possible choice of loads. Not to violate the solvability boundary of problem (2.10), we will suppose that all the points of the subset M are interior in Y. Due to Proposition 2.3, the subset

$$\partial\varphi(M) = \bigcup_{y \in M} \partial\varphi(y)$$

is bounded and nonempty. Let $\partial\varphi(M) \subset N$ where $N \subset R^m$ is a convex compact.

In the simplest situation when there exist exterior prices $c = (c_1, \ldots, c_m)$ for each component of the load, and moreover, when a decision maker is interested only in the pure income in monetary terms, the problem is solved by maximizing the function $\langle c, y \rangle - \varphi(y)$ over the set M. In the latter case, we should immediately include the parameters y into the collection of variables, together with the variables z, and replace the objective function $f(z)$ by the new one $f(z) - \langle c, y \rangle$.

In more complex cases, the decision maker may also be interested in the factors themselves that form the load y. These may include, for instance, the level of emergency security, the ability of defence, the environment conditions, the social factors etc. In such a situation, one can try describing relationships which the pairs (y, λ) must satisfy to be feasible from the point of view of the decision maker. In framework of the examined mathematical techniques, the latter can be realized by the following way.

Let us define a (single-valued or multi-valued) mapping H that associates each linear form $\lambda \in N$ with a subset $H(\lambda) \subset M$ of acceptable values of the load. Certainly, we suppose that the mapping H satisfies assumptions A1–A3 of Section 1.2. If we select the subdifferential mapping of the marginal function for F, i.e. define $F(y) = \partial\varphi(y)$, $y \in M$, then all the assumptions of Theorem 1.3 (von Neumann Theorem) are valid. Therefore, there exist $\bar{y} \in M$ and $\bar{\lambda} \in \partial\varphi(\bar{y})$ such that $\bar{y} \in H(\bar{\lambda})$.

From the point of view of the person who makes a decision concerning the load y, the choice of $\bar{y}$ is acceptable if $\bar{\lambda}$ coincides with the marginal costs related to the load $\bar{y}$. However, the set $\partial\varphi(\bar{y})$ may contain other subgradients as well. If we try to change the load by moving from $\bar{y}$ along a direction η, i.e., if we take $y = \bar{y} + \varepsilon\eta$, then we obtain the following formula for the derivative of the marginal function φ with respect to the direction η which is analogous to (2.1):

$$\varphi'(\bar{y};\eta) = \sup\{\langle\lambda,\eta\rangle \mid \lambda \in \partial\varphi(\bar{y})\} \geq \langle\bar{\lambda},\eta\rangle.$$

Thus we have shown that, from the point of view of the decision maker, there exists no direction η along which it would make sense to move from the point $\bar{y}$.

Note that the inclusion $\bar{\lambda} \in \partial\varphi(\bar{y})$ means that

$$\varphi(y) + \langle\bar{\lambda},-y\rangle \geq \varphi(\bar{y}) + \langle\bar{\lambda},-\bar{y}\rangle, \quad y \in M \subset Y.$$

Consequently, we see that the point $\bar{y}$ is a maximum point of the function $\varphi(y) + \langle\bar{\lambda},-y\rangle$ over the subset $M \subset Y$. This scalar criterion can be considered as a weighted vector criterion $(\varphi(y),-y)$. Moreover, the weight coefficients $(1,\bar{\lambda})$ are related to the minimum point $\bar{y}$ by the relationship $\bar{y} \in H(\bar{\lambda})$ that is treated as a *feedback*. In the next section, we will consider a more general version of the multi-objective optimization.

2.4 MULTI-OBJECTIVE PROBLEM

In optimization models of Sections 2.1 and 2.2, we considered problems in which we needed to find such a point of a feasible set K that would be the best choice according to a certain quality criterion. This criterion could be defined either as a value of some function $f : K \to R$ (the objective function), or as a mapping F replacing the subdifferential mapping of the convex function f. Section 2.3 dealt with a model in which the decision maker treated a much less definite problem. There we had the collection of criteria $(\varphi(y),-y)$ for the criterion of quality, where $\varphi(y)$ were the minimal costs necessary to make the model free from the utility load y. Recall that for $y = \bar{y}$, the linear form $\bar{\lambda} \in \partial\varphi(\bar{y})$ (Lagrange multipliers) played the role of the weight coefficients for reducing the several criteria to a single one. If the linear form $\bar{\lambda}$ is strictly positive then an increase in any component of the load $\bar{y}$ leads to the costs $\varphi(\bar{y})$ strictly increasing, too. Conversely, the costs $\varphi(\bar{y})$ cannot be decreased without decreasing some component of the load. Such a situation is usually referred to as the *Pareto optimality*. This concept is related to the multi-objective optimization problem studied below.

Let an open convex subset $D \subset R^n$ be the domain of the convex functions $f_i : D \to R$, $i = 1, \ldots, m$. These functions are considered as the quality criteria to estimate a decision, which is made in order to diminish the value of each of the functions. Furthermore, the choice is restricted to a set of feasible alternatives $K \subset D$. It is customary to suppose that the subset K is nonempty, convex and compact.

In reality, our desires often happen to be contradictory. Therefore, two problems usually arise. First, we need to describe such a subset of alternatives from K, selection of which would not be a priori unreasonable. These alternatives conform the so called *Pareto boundary*. Second, it is necessary to construct a model of choice of a particular alternative from the Pareto boundary; or, in the other words, a scalarization model.

DEFINITION 2.5. A point $\bar{z} \in K$ is called *efficient* if and only if, there is no point $z \in K$ such that $f_i(z) \le f_i(\bar{z})$, $i = 1, \ldots, m$, and $f_j(z) < f_j(\bar{z})$ for at least one index j. The collection of all the efficient points is reffered to as the *efficient set* or the *Pareto boundary.*

DEFINITION 2.6. A point $\bar{z} \in K$ is called *weakly efficient* if and only if, there is no point $z \in K$ such that $f_i(z) < f_i(\bar{z})$, $i = 1, \ldots, m$. The collection of all the weakly efficient points is called the *weakly efficient set* or the *weak Pareto boundary.*

In general, the weakly efficient set is wider than the efficient one. It is a rare case that the efficient set (and even rarer, the weakly efficient one) consists of a single point. A standard way to choose a particular alternative (more generally, a narrower subset of the alternatives) consists in the use of the weights $\lambda_i \ge 0$, $i = 1, \ldots, m$, which cannot be all equal to zero. After having selected the weights, we solve the following problem

$$\min\left\{f(z) \mid z \in K\right\}, \quad f(z) = \sum_{i=1}^{m} \bar{\lambda}_i f_i(z). \tag{2.18}$$

PROPOSITION 2.8. *The point $\bar{z} \in K$ is a weakly efficient one if and only if there exist nonnegative multipliers $\bar{\lambda}_i$, $i = 1, \ldots, m$, not all of which are equal to zero, and such that the point $\bar{z}$ solves problem* (2.18). *If $\bar{\lambda}_i > 0$, $i = 1, \ldots, m$, then each solution of problem* (2.18) *is an efficient point.*

PROOF. Let $\bar{z} \in K$ be a weakly efficient point. Define the set

$$B = \left\{v = (v_1, \ldots, v_m) \mid \exists z \in K,\ f_i(z) < v_i, i = 1, \ldots, m\right\}.$$

The set B is clearly nonempty and open. Moreover, the set B is convex due to the convexity of the set K and the functions f_i, $i = 1, \ldots, m$. If $\bar{v} = (f_1(\bar{z}), \ldots, f_m(\bar{z}))$, then $\bar{v} \notin B$ because of the weak efficiency of the point $\bar{z}$. By having chosen $A = \{\bar{v}\}$ and applied Theorem 2.1 (the Separation Theorem) we find a linear form $q \neq 0$ such that $\langle 1, \bar{v} \rangle > \langle q, v \rangle$ for all $v \in B$.

After setting $\bar{\lambda} = -q$ we obtain

$$f(\bar{z}) = \sum_{i=1}^{m} \bar{\lambda}_i f_i(\bar{z}) < \sum_{i=1}^{m} \bar{\lambda}_i v_i, \quad v \in B. \tag{2.19}$$

Since B is a nonempty set and every component v_i of the vector v can be increased infinitely without v quitting the set B, then clearly $\bar{\lambda}_i \geq 0$, $i = 1, \ldots, m$.

Now let $z \in K$. For an arbitrary $\varepsilon > 0$, we consider $v_i = f_i(z) + \varepsilon$, $i = 1, \ldots, m$. Then evidently $v \in B$. By tending $\varepsilon \to 0$ in (2.19), we obtain the limit relationship $f(\bar{z}) \leq f(z)$, that is, the point $\bar{z}$ solves problem (2.18).

Conversely, if a point $\bar{z}$ solves problem (2.18) for certain $\bar{\lambda}_i \geq 0$, $i = 1, \ldots, m$, not all of which are equal to zero, then the weak efficiency of $\bar{z}$ follows immediately from Definition 2.6. Similarly, Definition 2.5 implies the efficiency of the point $\bar{z}$ when $\bar{\lambda}_i > 0$, $i = 1, \ldots, m$, which completes the proof of the proposition. ■

Due to Proposition 2.2, a point $\bar{z}$ solves problem (2.18) if and only if $0 \in \partial f(\bar{z}) + N_K(\bar{z})$. On the other hand, by virtue of Theorem 2.3, $\partial f(\bar{z}) = \bar{\lambda}_1 \partial f_1(\bar{z}) + \cdots + \bar{\lambda}_m \partial f_m(\bar{z})$. However, we note that the solution of problem (2.18) does not depend upon the norm of the row $\bar{\lambda}$ of the weight coefficients. Therefore, we can restrict ourselves to the rows $\lambda = (\lambda_1, \ldots, \lambda_m)$ satisfying the conditions

$$\lambda_i \geq 0, \ i = 1, \ldots, m; \quad \sum_{i=1}^{m} \lambda_i = 1.$$

Now we compose the convex hull of the subgradients $\partial f_i(z)$, $i = 1, \ldots, m$, as follows

$$F(z) = \left\{ \sum_{i=1}^{m} \lambda_i p_i \;\middle|\; p_i \in \partial f_i(z), \ \lambda_i \geq 0, \ \sum_{i=1}^{m} \lambda_i = 1 \right\}. \tag{2.20}$$

Then we can reformulate the part of Proposition 2.8 concerning the weak efficiency, in the following form: the point $\bar{z} \in K$ is a weakly efficient one if and only if $0 \in F(\bar{z}) + N_K(\bar{z})$. Thus the weakly efficient set consists of the solutions of variational inequality (1.4) with the mapping F defined by formula (2.20).

We can also proceed in an alternative manner. Let us set

$$\Phi(z) = \partial f_1(z) \times \cdots \times \partial f_m(z). \tag{2.21}$$

If we treat a linear form over R^n as a row, then each element of product (2.21) con be considered as an $m \times n$-matrix. Hence, each linear form $\bar{p} \in \partial f(\bar{z})$ can be represented as $\bar{p} = \lambda P$ where $P \in \Phi(\bar{z})$.

Now we again are going to "forget" about the origin of the mapping Φ and consider it as a given (in general) multi-valued mapping of a convex compact K to the space of $m \times n$-matrices. As an analogue of variational inequality

(1.3), we examine the following *multi-objective variational inequality*: Find a vector $\bar{z}$ such that

$$\bar{z} \in K, \ \bar{P} \in \Phi(\bar{z}), \bar{P}(z - \bar{z}) \not< 0 \text{ for all } z \in K. \tag{2.22}$$

Here, for convenience purpose, the vector $(z - \bar{z})$ is treated as a column, whereas the sign $\not<$ means that the components of the column $\bar{P}(z - \bar{z})$ cannot be all negative.

PROPOSITION 2.9. *The point $\bar{z} \in K$ is a weakly efficient one if and only if $\bar{z}$ solves the multi-objective variational inequality* (2.22) *with the mapping Φ defined by* (2.21).

PROOF. Let $\bar{z}$ be a weakly efficient point. According to Proposition 2.8, there exists $\bar{\lambda} = (\bar{\lambda}_1, \dots, \bar{\lambda}_m) \geq 0$, $\bar{\lambda} \neq 0$, for which $\bar{z}$ solves problem (2.18), and hence, also is a solution to variational inequality (1.4) with the subdifferential mapping $F = \partial f$. Consequently, there is a linear form $\bar{p} \in \partial f(\bar{z})$ such that $\langle \bar{p}, z - \bar{z} \rangle \geq 0$ for all $z \in K$. Due to (2.21) and Theorem 2.3, we can find a matrix $\bar{P} \in \Phi(\bar{z})$ such that $\bar{p} = \bar{\lambda}\bar{P}$. If $\bar{P}(z - \bar{z}) < 0$ for some $z \in K$, we obtain $\langle \bar{p}, z - \bar{z} \rangle = \bar{\lambda}\bar{P}(z - \bar{z}) < 0$ that contradicts (1.4). Thus the point $\bar{z}$ and the matrix $\bar{P} \in \Phi(\bar{z})$ satisfy conditions (2.22).

Conversely, let $\bar{z} \in K$ solve problem (2.22). For the subgradients $p_i \in \partial f_i(\bar{z})$, $i = 1, \dots, m$, that compose the matrix $\bar{P}$, the following inequalities hold:

$$f_i(z) \geq f_i(\bar{z}) + \langle \bar{p}_i, z - \bar{z} \rangle.$$

If for some $z \in K$, the inequalities $f_i(z) < f_i(\bar{z})$, $i = 1, \dots, m$ are valid, then for all i we get $\langle \bar{p}_i, z - \bar{z} \rangle < 0$, i.e., $\bar{P}(z - \bar{z}) < 0$ that contradicts (2.22) and thus completes the proof. ∎

We can scalarize the multi-objective variational inequality by the standard way. Namely, we can select a nonnegative and nonzero row $\bar{\lambda}$ and construct a mapping F as follows:

$$F(z) = \{p = \bar{\lambda}P \mid P \in \Phi(z)\}, \quad z \in K.$$

Then every solution to variational inequality (1.3) also will solve problem (2.22). Conversely, for each solution of problem (2.22), there exists an appropriate row $\bar{\lambda}$. The proof of this fact is completely analogous to that of Proposition 2.8.

Now we consider a more general way of scalarization. Suppose that the mapping Φ is obtained by the following formula:

$$\Phi(z) = \Phi_1(z) \times \cdots \times \Phi_m(z),$$

where Φ_i, $i = 1, \dots, m$, are certain mappings that associate each $z \in k$ with a set $\Phi_i(z)$ of linear forms over R^n. In contrast to (2.21), the mappings Φ_i are not assumed to be subdifferential ones. For thus constructed mapping Φ, we examine the multi-objective variational inequality (2.22). Besides, we select a scalarizing mapping H which associates each $z \in K$ with a subset $H(z)$ of nonnegative rows of length m.

PROPOSITION 2.10. *Let each of the mappings H, Φ_i, $i = 1,\dots,m$, be upper semicontinuous on a nonempty convex compact K. Furthermore, let the images $H(z)$ and $\Phi(z)$, $i = 1,\dots,m$, be nonempty convex compact sets. Then there exist $\bar{z} \in K$ and $\bar{P} \in \Phi(\bar{z})$ satisfying relationships* (2.22), *and a row $\bar{\lambda} \in H(\bar{z})$ such that $\bar{\lambda}\bar{P}(z - \bar{z}) \geq 0$ for all $z \in K$.*

PROOF. Let us construct the convolution F of the mappings H and Φ as follows

$$F(z) = \{p = \lambda P \mid \lambda \in H(z),\ P \in \Phi(z)\}, \quad z \in K.$$

The nonemptyness and compactness of the sets $F(z)$ follow immediately from the analogous properties of the sets $H(z)$ and $\Phi(z)$. Now we verify that $F(z)$ also is convex. If $p^1 \in F(z)$, $p^2 \in F(z)$ and $\alpha \in (0,1)$ then there exist rows $\lambda^1 = (\lambda_1^1, \dots, \lambda_m^1) \in H(z)$, $\lambda^2 = (\lambda_1^2, \dots, \lambda_m^2) \in H(z)$, and matrices $P^1 \in \Phi(z)$, $P^2 \in \Phi(z)$ such that $p^1 = \lambda^1 P^1$ and $p^2 = \lambda^2 P^2$. Let us set $\lambda_i = \alpha\lambda_i^1 + (1-\alpha)\lambda_i^2$, $i = 1,\dots,m$. As the rows λ^1 and λ^2 are nonnegative, then $\lambda_i \geq 0$ for all i. We set $\alpha_i = 0$ if $\lambda_i = 0$ and $\alpha_i = \alpha\lambda_i^1/\lambda_i$, if $\lambda_i > 0$. It is evident that $\alpha_i \in (0,1)$. Moreover, $\alpha\lambda_i^1 = \alpha_i\lambda_i$ and $(1-\alpha)\lambda_i^2 = (1-\alpha_i)\lambda_i$. Since the set $H(z)$ is convex, then $\lambda \in H(z)$. Denote by $p_i^1 \in \Phi_i(z)$ and $p_i^2 \in \Phi_i(z)$ the rows of the matrices P^1 and P^2. Then the convexity of the sets $\Phi_i(z)$ implies the inclusions $\alpha_i p_i^1 + (1-\alpha_i)p_i^2 = p_i \in \Phi_i(z)$, $i = 1,\dots,m$. Consequently, the matrix P composed of the rows p_i belongs to the set $\Phi(z)$. Thus we obtain

$$\alpha p^1 + (1-\alpha)p^2 = \alpha\sum_{i=1}^{m}\lambda_i^1 p_i^1 + (1-\alpha)\sum_{i=1}^{m}\lambda_i^2 p_i^2 =$$

$$= \sum_{i=1}^{m}\lambda_i\left(\alpha_i p_i^1 + (1-\alpha_i)p_i^2\right) = \sum_{i=1}^{m}\lambda_i p_i = \lambda P \in F(z).$$

Hence, for each $z \in K$, the set $F(z)$ is a nonempty convex compact.

As the graphs of the mappings H and Φ_i, $i = 1,\dots,m$, are compact due to Corollary 1.1, then the graph of the convolution F is also compact. Therefore assumptions A1–A3 of Section 1.2 are valid. According to Theorem 1.2, variational inequality (1.3) is then solvable. Hence, there exist $\bar{z} \in K$, $\bar{\lambda} \in H(\bar{z})$, and $\bar{P} \in \Phi(\bar{z})$ such that $\bar{\lambda}\bar{P}(z - \bar{z}) \geq 0$ for all $z \in k$. The proposition is proved completely. ■

Finally, we note that, just as in the proof of Proposition 2.8, the found here $\bar{z} \in K$ and $\bar{P} \in \Phi(\bar{z})$ are solutions to the multi-objective variational inequality (2.22).

2.5 COMMENTS

Optimization problems on convex sets belong to the area of mathematics known as Convex Analysis. Being a part of Nonlinear Analysis, Convex Analysis has deep roots in the part of Functional Analysis related to the Separation Theorems. Convex Analysis has been separated and transformed into a special, dramatically developing research area after the arising of Games Theory and Mathematical Programming. We indicate the work by Fenchel (1951) as one of the first special monographs dedicated to Convex Analysis. Convex Analysis in its mature form can be also found, for example, in the books by Rockafellar (1970), Ekeland and Temam (1976), Clarke (1983).

The seminal paper by Kuhn and Tucker (1951) is one of the cornerstones of both Nonlinear and Convex Programming. The great number of works devoted to various aspects of the problem appeared later. Among them, we mention the book by Arrow e.a.(1958), in which the ideas of duality were further developed, and Mangasarian (1969) where the Theory of Nonlinear Programming was presented as a system.

The fact that the Lagrange multipliers coincide with the marginal values of production costs (in economic terms) was noticed in the monograph by Samuelson (1948). The modern view of this aspect can be found, for instance, in Rockafellar (1970), and Ekeland, Temam (1976). The role of the Lagrange multipliers as the variables of the dual problem was established in Kuhn and Tucker (1951). The feedback between the Lagrange multipliers and the variables of the optimization model was introduced in Bulavsky (1988, 1993).

Problems with several optimality criteria were first examined for the linear case by Zeleny (1974), and then their more general forms were considered by Keeny and Raiffa (1976), Cohon (1978). Their generalization to the case of multi-objective variational inequalities follows mainly the works by Bulavsky (1981,1988,1993).

REFERENCES

Arrow, K.J., Hurwicz, L., Uzawa, H. *Studies in Linear and Nonlinear Programming.* Stanford: Stanford University Press, 1958.

Bulavsky, VA.

1. 'Feedback in Optimization Problems'. In: *Proceedings of Institute of Mathematics of the USSR Academy of Sciences*, Novosibirsk, 1988; 10: 55-63 (*in Russian*).

2. 'Local Estimates and the Selection Problem'. In: *Modern Mathematical Methods of Optimization*, K.-H. Elster, ed., Berlin: Academic-Verlag, 1993.

3. Quasilinear programming and vector optimization. Soviet Math. Dokl. 1981; 23(2): 328-332.

Clarke, F.H. *Optimization and Non-Smooth Analysis.* New York: Wiley Interscience, 1983.

Cohon, J.L. *Multiobjective Programming and Planning.* New York - London: Academic Press, 1978.

Ekeland, I., and Temam, R. *Convex Analysis and Variational Problems*. New York: Academic Press, 1976.

Fenchel, W. *Convex Cones, Sets, and Functions*. Notes of Lectures at Princeton University, 1951.

Keeney, R.L., Raiffa, H. *Decisions with Multiple Objectives: Preferences and Value Tradeoffs*. New York: Wiley, 1976.

Kuhn, HW, Tucker, AW. 'Nonlinear Programming'. In: *Proceedings of the Second Berkeley Symposium on Mathematics, Statistics and Probability*. Berkeley: University of California Press, 1951.

Mangasarian, O.L. *Nonlinear Programming*. New York - London: McGraw-Hill, 1969.

Rockafellar, R.T. *Convex Analysis*. Princeton: Princeton University University, 1970.

Samuelson, P.A. *Foundations of Economic Analysis*. Harvard: Harvard University Press, 1948.

Zeleny, M. *Linear Multiobjective Programming*. New York-Berlin: Springer-Verlag, 1974.

Chapter 3

GENERAL ECONOMIC EQUILIBRIUM

3.1 ELEMENTS OF GAMES THEORY

The multi-objective models analyzed in Section 2.4 and, partially, in Section 2.3, dealt mainly with the case of a single decision-maker. Even when we supposed that each criterion described the aim of a distinct subject, these aims were uniform enough. Otherwise, the selection of efficient points and the scalarization methods could be strongly criticized. In this and in the next sections of Chapter 3, we will consider a thoroughly different situation.

Suppose that there are m *subjects* (*game participants*, *gamblers*) who have absolutely independent interests. The state of a system (say, an economic one) in which they operate, just as before, will be described by a point in the space R^n. However, now the n variables are splitted into m groups: each of the game participants may choose the values of variables of his group. Technically, the latter means that the space R^n is decomposed into the direct product $R^n = R^{n_1} \times \cdots \times R^{n_m}$, and each vector $x \in R^n$ is represented as a collection $x = (x_1, \ldots, x_m)$ where $x_i \in R^{n_i}$, $i = 1, \ldots, m$. For convenience purpose, we will denote by x_{-j} the collection of all x_i for $i \neq j$. Thus, we have $x = (x_{-j}, x_j)$ for any j. Finally, the notation $(x_{-j}, \bar{x}_j)$ will be used for the vector obtained from x by replacing the part of variables x_j by some fixed values $\bar{x}_j$.

Gambler i can choose values of his part of variables x_i (his *strategy*) within a subset $K^i \subset R^{n_i}$ called the *strategy set* of gambler i. The strategy sets will be always supposed to be nonempty, convex, and compact. We define

$$K = \prod_{i=1}^{m} K^i, \quad K^{-j} = \prod_{i \neq j} K^i. \tag{3.1}$$

Due to the assumptions made about the subsets K^i, sets (3.1) also are nonempty, convex, and compact. Of course, we always suppose that $m \geq 2$.

A quite general way to describe the interests of each game participant is to define multi-valued mappings C_i over the subsets K^{-i}, $i = 1, \ldots, m$. The image sets $C_i(x_{-i}) \subset R^{n_i}$ of the latter mappings are interpreted as collections of acceptable responses of player i to the choice x_{-i} of the other participants of the game. The mappings C_i can be collected to a united mapping by setting

$$C(x) = \prod_{i=1}^{m} C_i(x_{-i}). \tag{3.2}$$

DEFINITION 3.1. A collection of strategies $x \in K$ is called a *non-cooperative equilibrium* (or a *Nash equilibrium*) if and only if $x_i \in C_i(x_{-i})$, $i = 1, \dots, m$, or, what is the same, $x \in C(x)$.

If the collection of strategies $x = (x_1, \dots, x_m)$ is a non-cooperative equilibrium then each of the participants of game reckons his choice as acceptable. Thus, none of the gamblers has an incentive to change the choice made by him. It is important to underline that we do not discuss the ways that have led the players to the non-cooperative equilibrium.

Before examining the problem of existence of the non-cooperative equilibrium, we present a version of Theorem 1.3 (von Neumann Theorem) well-known as the *Kakutani Theorem.*

THEOREM 3.1. *Let a multi-valued mapping C map a nonempty convex compact K into itself, i.e. $C(x) \subset K$ for each $x \in K$. If the graph of the mapping C is closed, and for every $x \in K$, the image set $C(x)$ is nonempty and convex, then there exists a fixed point of the mapping C, that is, a point $x \in K$ satisfying the condition $x \in C(x)$.* ∎

One can deduce this theorem from Theorem 1.3 by taking $P = K$, replacing the subset $F \subset K \times K$ by the graph of the mapping C, and the subset H – by the diagonal of the set $K \times K$, i.e. $H = \{(x, x) \mid x \in K\}$.

COROLLARY 3.1. *If every subset K^i, $i = 1, \dots, m$, is nonempty, convex and compact, the mappings C_i are closed, and all the image sets $C_i(x_{-i})$, $x \in K$, $i = 1, \dots, m$, are nonempty and convex, then there exists a non-cooperative equilibrium.* ∎

In order to prove this corollary it is sufficient to note that the mapping C defined by equality (3.2), clearly satisfies the assumptions of Theorem 3.1.

Now suppose that for each gambler, a function $f_i : D \to R$ is given, $i = 1, \dots, m$, which is called a *payment function* (or a *loss function*). Just as before, we suppose that the subset $D \subset R^n$ is open and convex, with $K \subset D$. In relation with representation (3.1), it will be convenient for us to suppose that $D = D_1 \times \cdots \times D_m$, where D_i are open sets in the spaces R^{n_i}, and that $K^i \subset D$, $i = 1, \dots, m$. Similarly to the previous agreements, we will denote by D_{-j} the direct (Cartesian) product of all D_i, $i \neq j$.

The payment functions model the game in which each agent i, after having chosen his strategy x_i (without informing the others), pays (say, to a common bank) the sum $f_i(x)$, where $x = (x_1, \dots, x_m)$. The payment also may be negative, that is, the gambler actually may win. In general, we do not suppose that the total sum of payments is equal to zero. Let us define

$$C_i(x_{-i}) = \{x_i \in K^i \mid f_i(x) \leq f_i(x_{-i}, u) \text{ for all } u \in K^i\}. \tag{3.3}$$

Thus, an agent considers his choice acceptable if his payment is the minimal one of all the possible payments, for a given choice of all the other agents. When the mappings C_i are defined by (3.3), the game is called a *normal* (or *strategic*) game.

PROPOSITION 3.1. *For each $i = 1, \ldots, m$, let the set K^i be nonempty, convex and compact, the function f_i be continuous with respect to the whole collection of variables x, and convex with respect to the variables x_i, for any fixed x_{-i}. Then the strategic game has a non-cooperative equilibrium.*

PROOF. It is sufficient to verify the assumptions of Corollary 3.1. The convexity and closedness of subsets (3.3) is implied by Proposition 2.3, whereas their being nonempty and bounded are guaranteed by the fact that the subsets K^i, $i = 1, \ldots, m$, are nonempty and compact. Furthermore, the closedness of the set K and continuity of the function f_i with respect to all the variables imply the closedness of graph of each mapping C_i. Indeed, from the inequalities in (3.3), we can obtain the needed limit inequalities by taking limits with respect to both x and x_{-i}, for each fixed value of u. The proposition is proved completely. ■

Now we are going to represent the non-cooperative equilibrium problem for a strategic game in the form of variational inequality. Denote by $\partial_i f_i(x)$ the subdifferential of the function f_i with respect to the variables x_i. Hence, the inclusion $p_i \in \partial_i f_i(x)$ means that

$$f_i(x_{-i}, u) \geq f_i(x) + \langle p_i, u - x_i \rangle, \ u \in D_i.$$

Due to Proposition 2.2, the inclusion $x_i \in C_i(x_{-i})$ is tantamount to the relationship

$$0 \in \partial_i f_i(x) + N_{K^i}(x_i).$$

From $K = K^1 \times \cdots \times K^m$ it follows that, for $x \in K$,

$$N_K(x) = N_{K^1}(x_1) \times \cdots \times N_{K^m}(x_m).$$

Now we set

$$F(x) = \partial_1 f_1(x) \times \cdots \times \partial_m f_m(x)$$

and conclude that the inclusion $x \in C(x)$ is equivalent to the condition $0 \in F(x) + N_K(x)$. This variational inequality, analogous to (1.4), is equivalent to the definition of non-cooperative equilibrium if we add to it the condition $x \in K$.

In conclusion of the section, we consider a particular case of the two persons game with the zero sum. In the other words, we have $m = 2$ and $f_1(x_1, x_2) + f_2(x_1, x_2) = 0$ for all $x_1 \in K^1$ and $x_2 \in K^2$. If we change the notation by setting $\varphi(x_1, x_2) = f_1(x_1, x_2)$ then we come to $f_2(x_1, x_2) = -\varphi(x_1, x_2)$. Since each function f_i must be convex with respect to the corresponding variable x_i, the function φ has to be convex by x_1 and concave by x_2. According to Definition 2.4, the pair $\bar{x} = (\bar{x}_1, \bar{x}_2)$ is a non-cooperative equilibrium if and only if it is a saddle point of the function $\varphi(x_1, x_2)$ on the set $K^1 \times K^2$. For this particular case, Proposition 3.1 reduces to the following theorem.

THEOREM 3.2. *Let K^1 and K^2 be nonempty convex compact sets, and $\varphi : K^1 \times K^2 \to R$ be a function continuous by the pair of variables, convex by x_1 and concave with respect to x_2. Then the function φ has a saddle point.* ■

3.2 MODEL OF DECENTRALIZED ECONOMY, AND PERFECT COMPETITION EQUILIBRIUM

The previous section dealt with the situation where every participant (gambler) estimated the quality of his choice by observing immediately the choice of the other participants. Such a direct observation may not be possible in the economic systems. More frequently, the participants have access to a system of prices of the goods that are involved in the model. Each participant undertakes some acts as the response to the observed prices. In the models below, these acts will be formed as the participant's supply at the goods markets. For some goods, the supply may be negative that reveals the participant's intention to acquire these goods. The supply vector is usually subject to constraints of two types.

Technical constraints reflect the limited capacities of production or consumption, the limited resources at the participants' disposal (for example, the working hours, the natural resources). Financial constraints presuppose the existence of the circulating money, and model the absence of the budget deficit for every subject of the model. Of course, the budget must take into account both the subject's debts and his possible revenues.

Now we describe the model precisely. Let there be n types of *goods* and m economic *subjects* (*agents*) in the model. Agent i can choose his *supply vector* z_i in its production set $K^i \subset R^n$ that models the technical constraints of the participant. The set K^i may be described by different ways which will not be fixed here. As the participant of the money circulation, agent i must pay a sum $r_i(p)$ depending upon the prices established in the system: $p = (p_1, \ldots, p_n) \in R^n_+$. Some prices may be zeros but the whole vector cannot be zero: $p \neq 0$. The payment $r_i(p)$ may be negative which means that actually the given agent gets a transfer. It is important to underline that the sum $r_i(p)$ does not include the revenue gained by selling the supply z_i at the goods markets. This sum is supposed to take into account other relationships existing in the system: property taxes, stocks dividends, pension payments, salary of state employees, etc.

The absence of the deficit of the individual budgets is represented by the inequalities

$$\langle p, z_i \rangle \geq r_i(p), \quad i = 1, \ldots, m. \tag{3.4}$$

These inequalities are referred to as the *budget constraints*. Hence, agent i makes his choice within the subset K^i subject to the corresponding budget constraint (3.4). To examine the choice in more detail, we have to use the agents' preferences that can also be described in different manners. We will not fix the ways to define the agents' preferences in this section. We only suppose that, as a result of taking the preferences into account, we obtain some mappings $p \to S_i(p)$ describing the agents' choice depending upon the established prices. The mappings S_i may be single-valued as well, but we will consider the more general case of multi- valued mappings. Note that always $S_i(p) \subset K^i$, and for each $z_i \in S_i(p)$ the budget constraint (3.4) holds.

In such models, it is usually assumed that the general price scale does not affect the economic situation. Therefore, the payments $r_i(p)$ are positive homogeneous functions (of order 1) of the prices, that is, $r_i(\alpha p) = \alpha r_i(p)$ for $\alpha > 0$. This assumption reflects the static nature of the model: the prices p are those of the long-term equilibrium. The choice of the economic subject also is considered as independent of the price scale, as the half-space (3.4) does not depend upon the latter. Thus the mappings S_i prove to be positive homogeneous of order 0, i.e. $S_i(\alpha p) = S_i(p)$ for $\alpha > 0$. By summing up the mappings S_i we obtain the summary mapping S:

$$S(p) = \sum_{i=1}^{m} S_i(p) = \left\{ \sum_{i=1}^{m} z_i \mid z_i \in S_i(p), \ i = 1, \ldots, m \right\}. \tag{3.5}$$

The mapping S is called the *excess supply mapping.*

The economic system is supposed very often to be closed in the financial aspect which is expressed by the assumption

$$\sum_{i=1}^{m} r_i(p) = 0 \quad \text{for all} \quad p. \tag{3.6}$$

However, in many cases, it is sufficient to demand the absence of outside financial interventions, i.e.

$$\sum_{i=1}^{m} r_i(p) \geq 0 \quad \text{for all} \quad p. \tag{3.7}$$

Together with the budget constraints (3.4), condition (3.7) provides for the validity of the *Extended Walras Law*:

$$\langle p, z \rangle \geq 0, \quad z \in S(p). \tag{3.8}$$

If we postulate the financial closedness by (3.6), and the properties of the subsets K^i and the preferences of the economic agents lead to the equalities in the budget constraints for the vectors $z_i \in S_i(p)$, then the *Walras Law* is valid:

$$\langle p, z\rangle = 0, \quad z \in S(p). \tag{3.9}$$

DEFINITION 3.2. The prices $\bar{p} \in R^n_+ \setminus \{0\}$ are referred to as *equilibrium* ones, if these prices allow a non-negative total supply, i.e. if there exists a $\bar{z} \in S(\bar{p})$, $\bar{z} \geq 0$. The collection $(\bar{p}; \bar{z}_1, \dots, \bar{z}_m)$ is called a *competitive equilibrium* if and only if $\bar{z}_i \in S_i(\bar{p})$ and $\bar{z}_1 + \cdots + \bar{z}_m = \bar{z} \geq 0$.

The attribute "competitive" reflects the fact that the economic subjects make their decisions assuming that the prices are constant. Any possibility of individual influence upon the prices is rejected.

As the scale of prices does not matter, then together with the equilibrium prices $\bar{p}$, the prices $\alpha\bar{p}$ will also be equilibrium ones for $\alpha > 0$. Therefore, the region of variation of prices can be restricted by some condition imposed on their norm. Let us fix a vector of goods $a = (a_1, \dots, a_n)$, $a_j > 0$, $j = 1, \dots, n$, and set

$$P = \left\{p \in R^n_+ \mid \langle p, a\rangle = 1\right\}. \tag{3.10}$$

Now we can look for the equilibrium prices within the convex compact P, which allows one to apply the results of the two previous chapters. Denote by $K = K^1 + \cdots + K^m$.

LEMMA 3.1. *Let the subsets K^i, $i = 1, \dots, m$, be nonempty compact convex sets, the mappings S_i, $i = 1, \dots, m$, defined on the nonempty compact convex set P, be closed and every image set $S_i(p) \subset K^i$ be nonempty and convex. Then the mapping S constructed by formula* (3.5) *also is closed over P, and its image sets $S(p) \subset K$ are nonempty and closed subsets of the convex compact K.*

PROOF. The compactness and convexity of the set K follow from the analogous properties of the subsets K^i. Similarly, the nonemptyness and convexity of $S(p)$ are implied by those of $S_i(p)$. Let us verify the closedness of the mapping S. If a sequence $\{(p^\nu, z^\nu)\} \subset P \times K$ converges to a $(p, z) \in P \times K$ and $z^\nu \in S(p^\nu)$, then there exist $z^\nu_i \in S_i(p^\nu)$ such that $z^\nu = z^\nu_1 + \cdots + z^\nu_m$. Having chosen, if necessary, a subsequence, we can accept that $z^\nu_i \to z_i$. Due to the closedness of the mappings S_i we thus obtain $z_i \in S_i(p)$, that is, $z \in S$, which completes the proof of the lemma. ∎

The following theorem of existence of equilibrium prices is well-known as the *Gale-Nikaido-Debreu Theorem*.

THEOREM 3.3. *Let S be a mapping closed over the nonempty compact convex set P, and for each $p \in P$, the image set $S(p)$ be a nonempty convex subset of the compact convex subset K. If the extended Walras law* (3.8) *holds, then there exist equilibrium prices, i.e. there are $\bar{p} \in P$ and $\bar{z} \in S(\bar{p})$, $\bar{z} \geq 0$.*

PROOF. In addition to the mapping S, let us construct another mapping T as follows:

$$T(z) = \{p \in P \mid \langle p, z\rangle \le \langle \tilde{p}, z\rangle,\ \tilde{p} \in P\}, \ z \in K. \tag{3.11}$$

Thus constructed $T(z)$ consists of the minimum points (with respect to p) of the linear form $\langle p, z\rangle$ on the set P. One can immediately verify that the mapping T is closed, and the sets $T(z)$ are nonempty and convex. By virtue of Theorem 1.3 (the von Neumann Theorem), the graphs of the mappings S and T have a common point $(\bar{z}, \bar{p})$. Hence, $\bar{p} \in T(\bar{z})$ and $\bar{z} \in S(\bar{p})$. If we replace $\tilde{p}$ in (3.11) by the row with the components $\tilde{p}_i = 0$, $i \ne j$, and $\tilde{p}_j = 1/a_j$, then for $z = \bar{z}$ and $p = \bar{p}$ we obtain $\langle \bar{p}, \bar{z}\rangle \le \bar{z}_j/a_j$. Since $\bar{z} \in S(\bar{p})$ then the extended Walras law implies $\langle \bar{p}, \bar{z}\rangle \ge 0$. As $a_j > 0$ we get $\bar{z}_j \ge 0$ and complete the proof.

■

By analyzing the latter proof we conclude that the just found equilibrium prices $\bar{p}$ and the corresponding excess supply $\bar{z}$ taken together solve the following variational inequality problem over the compact P: Find $\bar{p}$ and $\bar{z}$ such that

$$\bar{p} \in P, \bar{z} \in S(\bar{p}),\ \langle p - \bar{p}, \bar{z}\rangle \ge 0 \text{ for all } p \in P. \tag{3.12}$$

In comparison to (1.3), the roles of the variables z and p are mutually exchanged here. Variational inequality (3.12) can be rewritten in the equivalent form:

$$\bar{p} \in P, \quad 0 \in S(\bar{p}) + N_P(\bar{p}). \tag{3.13}$$

Since the compact P is the intersection of the cone R^n_+ and the hyperplane

$$\{p \in R^n \mid \langle p, a\rangle = 1\},$$

then according to Theorem 1.4, every vector $u \in N_P(\bar{p})$ can be represented in the form $u = -v - \gamma a$, where $v \ge 0$ and $\langle \bar{p}, v\rangle = 0$. The sign "minus" before the coefficient γ is chosen in order to deal with $\gamma \ge 0$. Thus, (3.13) can be reformulated as follows:

$$\bar{p} \in P, \quad \bar{z} \in S(\bar{p}), \quad \bar{z} = \bar{v} + \bar{\gamma} a, \quad \bar{v} \ge 0, \quad \bar{\gamma} \ge 0,\ \langle \bar{p}, \bar{v}\rangle = 0. \tag{3.14}$$

Nonnegativity of $\bar{\gamma}$ is implied by the fact that $\bar{\gamma} = \langle \bar{p}, \bar{z}\rangle \ge 0$ due to the extended Walras law.

Notice that not all the competitive equilibria are solutions of variational inequality (3.13) and consequently, not all of them may be represented in form (3.14). In general, there may exist other equilibria as well.

Relationship (3.14) can be interpreted as follows. If $\bar{\gamma} > 0$ then the part $\bar{\gamma} a$ of the excess supply $\bar{z}$ is an extraneous complete load (perturbation) in the model. When a component $\bar{z}_s$ of the excess supply exceeds the complete load $\bar{\gamma} a_s$, then $\bar{p}_s = 0$. Of course, the latter is true even for $\bar{\gamma} = 0$. Then the zero prices are assigned to the goods with the strictly positive componenets of the excess supply. Note that the load $\bar{\gamma} a$ can be positive only when the extended Walras law is valid. If we demand that relationships (3.9) hold, then $\bar{\gamma} = \langle \bar{p}, \bar{z}\rangle = 0$.

Having in mind the possibility of a strict positive load to exist under the extended Walras law, it occurs to one's head that the model has too much liberty in choosing the equilibrium. Indeed, if we change the way to norm the prices (3.10) and pick up another collection of goods with other proportions among them, then in order to keep the completeness of the load with this other vector a, we will have to change the equilibrium prices not proportionally to the old ones, either.

Now suppose that besides the normalization (3.10), some further constraints are imposed upon the feasible prices. For example, the government may decide that the cost of the socially important collection of goods $\hat{a} \geq 0$ must not exceed ε when normed by (3.10). Of course, we also suppose that $\varepsilon > 0$ and that the new feasible set of prices

$$\hat{P} = \left\{p \in R_+^m \mid \langle p, a\rangle = 1, \ \langle p, \hat{a}\rangle \leq \varepsilon\right\}$$

contains a collection of prices $p > 0$ such that $\langle p, \hat{a}\rangle < \varepsilon$. For the compact $\hat{P}$, we can repeat the first part of the proof of Theorem 3.3 having replaced the set P by $\hat{P}$. The pair $(\bar{z}, \bar{p})$ for which $\bar{z} \in S(\bar{p}) \subset K$ and $\bar{p} \in T(\bar{z}) \subset \hat{P}$ then solves the variational inequality

$$\bar{p} \in \hat{P}, \quad 0 \in S(\bar{p}) + N_{\hat{P}}(\bar{p}). \tag{3.15}$$

If the additional constraint on the prices turns out to be active, than a new term will appear in the representation of $u \in N_{\hat{P}}(\bar{p})$. Namely, $u = -v - \gamma a + \beta\hat{a}$ where $v \geq 0$, $\langle \bar{p}, v\rangle = 0$ and $\beta \geq 0$. Therefore, (3.15) can be rewritten as follows:

$$\bar{p} \in \tilde{P}, \quad \bar{z} \in S(\bar{p}), \quad \bar{z} = \bar{v} + \bar{\gamma}a - \bar{\beta}\hat{a}, \quad \bar{v} \geq 0,$$

$$\langle \bar{p}, \quad \bar{v}\rangle = 0, \quad \bar{\gamma} - \varepsilon\bar{\beta} = \langle \bar{p}, \quad \bar{z}\rangle \geq 0.$$

Instead of the nonnegativity of the excess supply $\bar{z}$, we can now only guarantee the inequality

$$\bar{z} \geq \beta(\varepsilon a - \hat{a}).$$

As the vector $(\varepsilon a - \hat{a})$ must have negative components (otherwise, the additional constraint on the prices does not make sense), then at the equilibrium, one has to expect the deficit of some of the goods, namely those that are present in the socially important collection $\hat{a}$ in the positive quantities. It does not look very natural to consider such a state as an equilibrium.

We can amend the situation by introducing an *extraneous exchange (trade)* with the balance of goods $\bar{\beta}(\varepsilon a - \hat{a})$. If the *extraneous prices* of goods are denoted by q, then the nonnegative account of the extraneous trade implies the inequality $\langle q, \varepsilon a - \hat{a}\rangle \geq 0$. Therefore, a socially oriented policy should not violate the inequality

$$\varepsilon \geq \frac{\langle q, \hat{a}\rangle}{\langle q, a\rangle}.$$

3.3 ARROW-DEBREU MODEL

In this section, a version of the model of decentralized economy will be considered in more detail. All the model subjects are divided into two groups: the subjects with the indices $i \in I$ are treated as firms-*producers*, whereas the subjects with the indices $j \in J$ are considered as *consumers* (*households*). Hence, $I \cup J = \{1, \dots, m\}$. All the subsets K^i, $i = 1, \dots, m$, are assumed to be nonempty, closed and convex.

For each firm $i \in I$, its production set K^i contains the zero vector, i.e. the firm may not produce. Under given prices p, the single aim of the firm is to maximize its *income* $\langle p, z_i \rangle$ with respect to $z_i \in K^i$. Since the subsets K^i, in general, are not assumed to be bounded, the maximal income may not be attained for arbitrary prices p. Therefore, we postpone for a while the description of the way of generating the payments $r_i(p)$ for the producers and consumers.

Each consumer $j \in J$ is usually characterized by its *consumption set* X^j and some *inicial property* (*endowment*) $a_j \in R^n$. The vector a_j is customarily supposed to be nonnegative, although the latter is not always necessary. The subsets X^j are assumed to be nonempty, closed, and convex. Consumer j consumes $x_j \in X^j$ and proposes the vector $z_j = a_j - x_j$ as its supply to the goods market. Thus, for consumer $j \in J$, we have

$$K^j = \{a_j\} - X^j = \left\{a_j - x_j \mid x_j \in X^j\right\}.$$

In the analyzed economy, the financial relations are based upon the buying and selling goods freely, and the division of the production capacities among the consumers. The capacities division techniques consist in defining the values $\alpha_{ij} \geq 0$ which indicate the consumer $j \in J$ *share* of the property of firm i. We suppose that

$$\sum_{j \in J} \alpha_{ij} = 1, \quad i \in I, \tag{3.16}$$

i.e., the total property is divided among the consumers.

According to the above-mentioned assumptions, given a production plan $u = \left\{z_i \in K^i, \ i \in I\right\}$ and prices p, the income $\langle p, z_i \rangle$ of firm $i \in I$ is distributed as follows:

$$R_j(p, u) = \sum_{i \in I} \alpha_{ij} \langle p, z_i \rangle, \ j \in J. \tag{3.17}$$

Since recently, the consumer's preferencies are usually described axiomatically. However, we will restrict ourselves to the classic way (which actually is equivalent to the axiomatic one) that uses the *utility functions*. Suppose that for each $j \in J$, we have an open closed subset $D_j \subset R^n$ and a concave function $\psi_j : D_j \to R$. We also assume that $X^j \subset D_j$ for all $j \in J$. Consumer j chooses a vector $x_j \in X^j$, and therewith, its supply $z_j = a_j - x_j$ to the goods market, by solving the problem

$$\max \left\{\psi_j(x_j) \mid x_j \in X^j, \langle p, x_j \rangle \leq \langle p, a_j \rangle + R_j(p, u)\right\}.$$

Nevertheless, it is more convenient for us to operate with the variables z_j rather than x_j. Therefore, on the subsets $K^j = \{a_j\} - X^j$, we define the functions φ_j by the rule $\varphi_j(z_j) = \psi_j(a_j - z_j)$, $z_j \in K^j$. By virtue of this rule, the concave function φ_j is also defined on the open set $\{a_j\} - D_j$ that comprises K^j. Hence, the problem to determine the j–th consumer's supply z_j, $j \in J$, reduces to

$$\max\left\{\varphi_j(z_j) \mid z_j \in K^j,\ \langle p, z_j\rangle + R_j(p,u) \geq 0\right\}.$$

Within the examined model, Definition 3.2 can be replaced by the following one.

DEFINITION 3.3. A collection $\{z_1, \ldots, z_m\}$ is called a *feasible distribution* if and only if

$$\sum_{i=1}^{m} z_i \geq 0, \quad z_i \in K^i,\ i = 1, \ldots, m.$$

Furthermore, the prices $\bar{p} \in R^n_+ \setminus \{0\}$ and the feasible distribution $\{\bar{z}_1, \ldots, \bar{z}_m\}$ constitute a *competitive equilibrium* if for $\bar{u} = \{\bar{z}_i, i \in I\}$ we have

$$\langle \bar{p}, \bar{z}_i\rangle = \max\left\{\langle \bar{p}, z_i\rangle \mid z_i \in K^i\right\}, \quad i \in I, \tag{3.18}$$

$$\varphi_j(\bar{z}_j) = \max\left\{\varphi_j(z_j) \mid z_j \in K^j,\ \langle \bar{p}, z_j\rangle + R_j(\bar{p}, \bar{u}) \geq 0\right\}, j \in J. \tag{3.19}$$

Here the values of $R_j(\bar{p}, \bar{u})$ are calculated according to (3.17) with $p = \bar{p}$ and $u = \bar{u}$.

We will require the following regularity assumption which can be interpreted as the *individual welfare assumption.*

D1. For each $j \in J$ there exist $\hat{z}_j \in K^j$ and $z_i^j \in K^i$, $i \in I$, such that

$$\hat{z}_j + \sum_{i \in I} \alpha_{ij} z_i^j > 0.$$

Assumption D1 means that if the firms' property were divided among the consumers proportionally to their shares in it, then each consumer would have obtained strictly more than enough to be self-maintained.

On the first step, we will examine the problem of existence of the competitive equilibrium in the case of bounded subsets K^i, $i = 1, \ldots, m$. As before, we select first a collection of goods $a > 0$, define the convex compact P according to (3.10), and for each $i \in I$ and $p \in P$ we determine

$$r_i(p) = \max\left\{\langle p, z_i\rangle \mid z_i \in K^i\right\},$$

$$S_i(p) = \left\{z_i \in K^i \mid \langle p, z_i\rangle = r_i(p)\right\}.$$

For every consumer $j \in J$, we calculate

$$r_j(p) = \sum_{i \in I} \alpha_{ij} r_i(p) \tag{3.20}$$

and define the mapping S_j as follows:

$$M_j(p) = \left\{z_j \in K^i \mid \langle p, z_j \rangle + r_j(p) \geq 0\right\},$$

$$\tilde{\varphi}_j(p) = \max\left\{\varphi_j(z_j) \mid z_j \in M_j(p)\right\},$$

$$S_j(p) = \left\{z_j \in M_j(p) \mid \varphi_j(z_j) = \hat{\varphi}_j(p)\right\}.$$

LEMMA 3.2. *If the subsets K^i, $i = 1, \ldots, m$, are nonempty, compact, and convex, and assumption* D1 *holds, then the multi-valued mappings S_i, $i = 1, \ldots, m$, are closed and have nonempty and convex image sets $S_i(p)$.*

PROOF. Let $i \in I$. The function r_i is well-defined for all $p \in R^n$, because K^i is nonempty, compact, and convex. Moreover, the function r_i is convex. Indeed, if $p^1 \in R^n$, $p^2 \in R^n$ and $\lambda \in [0,1]$, then there exists a $z^\lambda \in K^i$ such that

$$r_i\left(\lambda p^1 + (1-\lambda)p^2\right) = \langle \lambda p^1 + (1-\lambda)p^2, z^\lambda \rangle =$$

$$= \lambda \langle p^1, z^\lambda \rangle + (1-\lambda)\langle p^2, z^\lambda \rangle \leq \lambda r_i(p^1) + (1-\lambda) r_i(p^2).$$

Therefore, the function r_i is continuous as well. Now we show that $S_i(p) = \partial r_i(p)$. If $z \in S_i(p)$ then $z_i \in K^i$ and $r_i(p) = \langle p, z \rangle$. Hence, for an arbitrary q, we get the relationship $r_i(q) \geq \langle q, z \rangle = r_i(p) + \langle q - p, z \rangle$, i.e. $z \in= \partial r_i(p)$.

Conversely, let $z \in \partial r_i(p)$. The definition of subgradient for $q = 2p$ and $q = p/2$, together with the positive homogeneity of the function r_i, imply the inequalities

$$2r_i(p) \geq r_i(p) + \langle p, z \rangle \quad \text{and } r_i(p)/2 \geq r_i(p) - \langle p, z \rangle / 2,$$

that is, $r_i(p) = \langle p, z \rangle$. If $z \notin K^i$ then there is an open neighborhood B of the point z having the empty intersection with K^i. Due to Theorem 2.1 (Separation Theorem), there exists a linear form q such that $\langle q, z \rangle > \langle q, z_i \rangle$ for all $z_i \in K_i$. Now we can pick up the vector z_i for which $r_i(q) = \langle q, z_i \rangle$ and obtain the inequality

$$r_i(q) = \langle q, z_i \rangle < \langle q, z \rangle = r_i(p) + \langle q - p, z \rangle$$

which contradicts the definition of subgradient. Hence, $z \in S_i(p)$. According to Proposition 2.3, the mapping S_i is closed and has nonempty convex image sets.

Now let $j \in J$. For every $p \in P$, for a point $\hat{z}_j$ we have from assumption D1 that

$$0 < \langle p, \hat{z}_j \rangle + \sum_{i \in I} \alpha_{ij} \langle p, \hat{z}_i \rangle \leq \langle p, \hat{z}_j \rangle + r_j(p). \tag{3.21}$$

The latter implies that the convex compact set $M_j(p)$ is nonempty. Due to Proposition 2.4, the sets $S_j(p)$ are nonempty and convex. Now we will prove that the mapping S_j is closed.

Let a sequence $\{(p^\nu, z^\nu)\}$ converge to (p, z), with $z^\nu \in S_j(p^\nu)$ for all ν. In particular, the latter means that $z^\nu \in K^j$ and $\langle p^\nu, z^\nu \rangle + r_j(p^\nu) \geq 0$. As the functions r_j, $j \in J$ are continuous due to the continuity of r_i, $i \in I$, and (3.20), then we can obtain the limit relationship $z \in M_j(p)$. Now let us consider an arbitrary point $\tilde{z} \in M_j(p)$. According to (3.21), for any $\varepsilon \in (0,1)$, the point $z_\varepsilon = \varepsilon \tilde{z}_j + (1-\varepsilon)\tilde{z}$ belongs to K^j and satisfies the strict inequality $\langle p, z_\varepsilon \rangle + r_j(p) > 0$. As $p^\nu \to p$, the same inequality is valid for the ν large enough: $\langle p^\nu, z_\varepsilon \rangle + r_j(p^\nu) > 0$. Therefore, we have the inequality $\varphi_j(z_\varepsilon) \leq \varphi_j(z^\nu)$, and when $\nu \to \infty$ we get $\varphi_j(z_\varepsilon) \leq \varphi_j(z)$. By tending ε to zero, we come to the inequality $\varphi_j(\tilde{z}) \leq \varphi_j(z)$, and $\tilde{z}$ being an arbitrary point of the set $M_j(p)$, we obtain $z \in S_j(p)$ which completes the proof of the lemma. ∎

THEOREM 3.4. *Under assumptions of Lemma* 3.2, *there exists a competitive equilibrium.*

PROOF. Let us construct the mapping S according to (3.5). Due to Lemma 3.1, this mapping is closed and has nonempty convex image sets $S(p)$, $p \in P$. Moreover, the inclusion $S(p) \subset K = K^1 + \cdots + K^m$ holds. Now we are going to verify that the extended Walras law is valid. For $z \in S(p)$ there are $z_i \in S_i(p)$, $i = 1, \ldots, m$, such that $z = z_1 + \cdots + z_m$. Since $r_i(p) = \langle p, z_i \rangle$, $i \in I$, and $\langle p, z_j \rangle \geq -r_j(p)$, $j \in J$, then (3.20) implies that $\langle p, z \rangle \geq 0$. According to Theorem 3.3, there exist $\bar{p} \in P$ and $\bar{z} \in S(\bar{p})$, $\bar{z} \geq 0$. Therefore, the vectors $\bar{z}_i \in S_i(\bar{p})$, $i = 1, \ldots, m$, for which $\bar{z} = \bar{z}_1 + \cdots + \bar{z}_m$, and the prices $\bar{p}$ constitute a competitive equilibrium which proves the theorem. ∎

Now let us consider the case when some (or even all) of the subsets K^i, $i = 1, \ldots, m$, are unbounded. Denote by Y the summary production set of the firms-producers, that is,

$$Y = \sum_{i \in I} K^i.$$

Because $0 \in K^i$ for all $i \in I$, we have $0 \in Y$. Let us introduce the following additional assumption.

D2. Assume that

$$Y \cap R^n_+ = \{0\}, \qquad Y \cap (-Y) = \{0\}. \tag{3.22}$$

The first assumption in (3.22) is interpreted as the non-existence of the *"Land of Cocaigne"*, i.e. the impossibility to produce something without expenses. The second assumption means the *irreversibility* of production.

The next additional assumption provides the upper bounds for the possible consumers' supplies at the goods markets.

D3. For each $j \in J$ there exists a vector c_j such that $z_j \leq c_j$ for all $z_j \in K^j$.

In terms of the consumption sets X^j, the latter means that the possible levels of consumption $x_j \in X^j$ are lower bounded by the vectors $a_j - c_j$.

LEMMA 3.3. *Let the subsets K^i, $i = 1, \dots, m$, be nonempty, closed, and convex. If assumptions* D1 *and* D2 *are valid, then the set of feasible distributions is bounded.*

PROOF. First, we will prove that the components z_i, $i \in I$, of the feasible distributions are bounded. Since $z_1 + \cdots + z_m \geq 0$ then according to assumption D3,

$$\sum_{i \in I} z_i + \sum_{j \in J} c_j \geq 0. \tag{3.23}$$

Suppose that there is a sequence of feasible distributions $\{z_1^\nu, \dots, z_m^\nu\}$ such that the numerical sequence

$$v_\nu = \max \{ \|z_i^\nu\|, \ i \in I \} \tag{3.24}$$

is unbounded. Having chosen, if necessary, a subsequence, we can accept that there exist limits $\bar{z}_i = \lim (z_i^\nu / v_\nu)$ for each $i \in I$, and furthermore, the maximum in (3.24) is attained at the same number $i = s$ for all $\nu = 1, 2, \dots$. Hence, $\|\bar{z}_s\| = 1$. Since $0 \in K^i$, $i \in I$, then by setting $\alpha_\nu = 1/v_\nu$, the convexity of K^i implies

$$z_i^\nu / v_\nu = \alpha_\nu z_i^\nu + (1 - \alpha_\nu) \cdot 0 \in K^i$$

for all ν large enough. As the subsets K^i are closed, then the limit inclusions $\bar{z}_i \in K^i$, $i \in I$, are valid.

Substituting z_i^ν in (3.23) instead of z_i and dividing by v_ν, we come to the limit inequality

$$\bar{z} = \sum_{i \in I} \bar{z}_i \geq 0. \tag{3.25}$$

Because $\bar{z} \in Y$, assumption D2 implies $\bar{z} = 0$, but then (3.25) implies $-\bar{z}_s \in Y$. However, as also $\bar{z}_s \in Y$, we use (3.22) to obtain $\bar{z}_s = 0$, whereas $\|z_s\| = 1$. Thus obtained contradiction establishes the boundedness of the components z_i, $i \in I$, of the feasible distributions. Therefore, there exists such a positive real number $\mathcal{L}$ such that, for any feasible distribution $\{z_1, \dots, z_m\}$ the following estimates hold: $-\mathcal{L}e \leq z_i \leq \mathcal{L}e$, $i \in I$; here e is the vector with all its components equal to 1. Now the boundedness of the components z_j, $j \in J$, of the feasible distribution follows from the estimates

$$c_j \geq z_j \geq -\left(\sum_{i \in I} z_i + \sum_{k \in J; k \neq j} z_k \right) \geq -\left(|I|\mathcal{L}e + \sum_{k \in J; k \neq j} c_k \right),$$

and hence, completes the proof of the lemma. ■

THEOREM 3.5. *Let the subsets K^i, $i = 1, \dots, m$, be nonempty, closed, and convex. If assumptions* D1–D3 *are valid, then there exists a competitive equilibrium.*

PROOF. According to Lemma 3.3, we can select a convex compact subset B such that all the vectors z_i, $i = 1, \dots, m$, belonging to at least one feasible distribution, are interior points of B. If necessary, we can extend the compact set B so that it also contains all the vectors $\hat{z}_j$ and z_i^j, $i \in I$, $j \in J$, from assumption D1. If we restrict our model to the subsets $K^i \cap B$ instead of K^i, then condition D1 keeps holding. By virtue of Theorem 3.4, thus modified model has a competitive equilibrium. Consequently, there exist $\bar{p} \in P$ and $\bar{z}_i \in K^i \cap B$, $i = 1, \ldots, m$, satisfying the inequality $\bar{z}_1 + \cdots + \bar{z}_m \geq 0$, and the points $\bar{z}_i$ solve the maximization problems

$$\max \left\{ \langle \bar{p}, z_i \rangle \mid z_i \in K^i \cap B \right\},$$

$$\max \left\{ \varphi_j(z_j) \mid z_j \in K^j \cap B,\ \langle \bar{p}, z_j \rangle + r_j(\bar{p}) \geq 0 \right\},\ j \in J.$$

However, the set B is a neighborhood of the points $\bar{z}_i$, $i = 1, \ldots, m$. It is evident that, for the problem of maximization of the concave function over the convex set, the concepts of local and global solutions coincide (just as it is the case for the problem of minimization of the convex function). Therefore, $\bar{z}_i$, $i = 1, \ldots, m$, are the solutions of the maximization problems over the sets K^i, $i \in I$, and K^j, $j \in J$ as well. Hence, the prices $\bar{p}$ and the feasible distribution $(\bar{z}_1, \ldots, \bar{z}_m)$ also constitute a competitive equilibrium in the initial model which completes the proof of the theorem. ■

In conclusion of this section, we consider a particular case of the examined model that can be treated as a *division* model. Suppose that there is no production. Formally, we can assume that the summary production set Y contains only the zero vector. Therefore, the market supply consists of only the goods constituting the consumers' initial endowments. For the simplicity purpose, we will assume that $a_j \geq 0$, whereas their sum $a = a_1 + \cdots + a_m$ is a strictly positive vector. At last, we suppose that the consumption sets X^j are all equal to R^n_+.

If, for given prices p, consumer j had the sum $\langle p, a_j \rangle$ at his disposal for consumption, that would be the so-called *model of pure exchange* (or pure exchange model). However, we will suppose that agent j possesses the share γ_j of the total property. Hence, normalized by (3.10), its budget does not depend upon p and is equal to γ_j, which allows one to call the latter as the *fixed budget* model. Therefore, the choice of agent j, under the prices p given, is described by the problem:

$$\max \left\{ \psi_j(x_j) \mid x_j \geq 0,\ \langle p, x_j \rangle \leq \gamma_j \right\}.$$

We assume, of course, that $\gamma_1 + \cdots + \gamma_m = 1$, $\gamma_j \geq 0$, $j = 1, \ldots, m$.

Let us accept an additional assumption concerning the utility functions.

D4. For every $j = 1, \ldots, m$, the concave function ψ_j is defined upon an open convex set D containing the set $R^n_+ \setminus \{0\}$. Furthermore, ψ_j also is positively homogeneous of order 1 and strictly positive over the set $R^n_+ \setminus \{0\}$.

Consider the following convex programming problem:

$$\min \left\{ -\sum_{j=1}^{m} \gamma_j \ln \psi_j(x_j) \mid x_j \geq 0, j = 1, \ldots, m,\ \sum_{j=1}^{m} x_j - a \leq 0 \right\}. \qquad (3.26)$$

For this problem, the Slater's condition is clearly satisfied (although the latter is not needed in case of linear constraints). By virtue of Theorem 2.4 and Theorem 2.5, the feasible point $\bar{x}_j$, $j = 1, \ldots, m$, is an optimum point if and only if there exist a linear form $\bar{p}$ (the Lagrange multipliers for the problem constraints) and subgradients $h_j \in \partial(-\ln \psi_j(\bar{x}_j))$ such that:

1) $\sum_{j=1}^{m} \langle \bar{p}, \bar{x}_j \rangle - \langle \bar{p}, a \rangle = 0$;

2) the linear form $q = (\gamma_1 h_1 + \bar{p}, \ldots, \gamma_m h_m + \bar{p})$ attains its minimum over the set $\left(R_+^n\right)^m = R_+^n \times \cdots \times R_+^n$ at the point $(\bar{x}_1, \ldots, \bar{x}_m)$.

Condition 2) means that each linear form $(\gamma_j h_j + \bar{p})$ attains its minimum on R_+^n at the point $\bar{x}_j$. The latter implies that

$$\gamma_j h_j + \bar{p} \geq 0, \ \langle \gamma_j h_j + \bar{p}, \bar{x}_j \rangle = 0, j = 1, \ldots, m. \tag{3.27}$$

Let us calculate the value $\langle h_j, \bar{x}_j \rangle$. According to the definition of subgradient, for any $\varepsilon \in (-1, +1)$, we have the inequality

$$-\ln \psi_j\left((1+\varepsilon)\bar{x}_j\right) \geq -\ln \psi_j(\bar{x}_j) + \varepsilon \langle h_j, \bar{x}_j \rangle.$$

Since assumption D4 implies $\psi_j\left((1+\varepsilon)\bar{x}_j\right) = (1+\varepsilon)\psi_j(\bar{x}_j)$, then we deduce from the latter inequality that

$$-\ln(1+\varepsilon) \geq \varepsilon \langle h_j, \bar{x}_j \rangle.$$

Dividing thus obtained inequality by $\varepsilon > 0$ and by $\varepsilon < 0$ and passing to limit as $\varepsilon \to \pm 0$, we come to the inequalities $-1 \geq \langle h_j, \bar{x}_j \rangle$ and $-1 \leq \langle h_j, \bar{x}_j \rangle$, whence $\langle h_j, \bar{x}_j \rangle = -1$. Now we deduce from (3.27) that

$$\langle \bar{p}, \bar{x}_j \rangle = \gamma_j, \quad j = 1, \ldots, m. \tag{3.28}$$

Therefore, the point $\bar{x}_j$ is feasible for the problem

$$\min\left\{-\gamma_j \ln \psi_j(x_j) \mid x_j \geq 0, \ \langle \bar{p}, x_j \rangle \leq \gamma_j \right\}. \tag{3.29}$$

Since, according to Theorem 2.4, relationships (3.27) are the sufficient optimality conditions, then the point $\bar{x}_j$ solves problem (3.29). The logarithm function being monotone, $\bar{x}_j$ simultaneously maximizes the utility function ψ_j on the same set of variables x_j. Taking into account (3.28) and condition 1), we get the equality $\langle \bar{p}, a \rangle = \gamma_1 + \cdots + \gamma_m = 1$. Hence, the solutions of problem (3.26) and the Lagrange multipliers $\bar{p}$ taken together constitute an equilibrium in the examined division model. ■

3.4 COMMENTS

Although some elements of Games Theory can be detected in a series of earlier works, the value of this theory for Economics was revealed in the book by von Neumann and Morgenstern (1944). Since then, Games Theory has been developing intensely both in its practical part and in its theoretical plane. In

McKinsey (1952), one can find the elements of Games Theory presented as a system, whereas the book by Luce and Raiffa (1957) contains the critical analysis of its postulates and methods. In the monograph by Aubin (1984), Games Theory is elucidated from the point of view of Convex Analysis.

The origins of the modern Economic Equilibrium Theory are traced back to Walras (1874) who applied Mathematics to Economics without having at his disposal appropriate mathematical tools for strict proofs. As soon as the Fixed Point Theorems for Nonlinear Mappings were discovered and Games Theory arised, the existence theorems for Economic Equilibrium Models in the systems of decentralized economics were obtained.

The review of problems of Welfare Economy with their economic interpretations can be found in Samuelson (1955). The modern aspects of the theory are presented in Karlin (1959), Nikaido (1968), as well as in the book by Aubin (1984) mentioned above. The paper by Bulavsky (1998) is dedicated to reduction of the economic equilibrium problems to the convex programming and complementarity problems.

REFERENCES

Aubin, J.P. *L'analyse Non-Linéaire es sus Motivations Économiques.* Paris: Masson, 1984.

Bulavsky, VA. On the solution of a type of equilibrium problems. Siberian Mathematical Journal, 1997; 35(5): 990-999. *(in Russian)*

Karlin, S. *Mathematical Methods and Theory in Games, Programming and Economics.* London–Paris: Pergamon Press, 1959.

Luce, R.D., and Raiffa, H. *Games and Decisions.* New York–London: Academic Press, 1957.

McKinsey, J.C.C. *Introduction to the Theory of Games.* New York–Toronto–London: Academic Press, 1952.

von Neumann, J., and Morgenstern, O. *Theory of Games and Economic Behavior.* Princeton: Princeton University Press, 1944.

Nikaido, H. *Convex Structures and Economic Theory.* New York: Academic Press, 1968.

Samuelson, P.A. *Foundations of Economic Analysis.* Cambridge (Mass): Harvard University Press, 1955.

Walras, L. *Éléments d'Économie Politique Pure ou Théorie de la Richesse Sociale.* Lausanne, 1874.

Chapter 4

MODELS OF OLIGOPOLY

As an example of an important complementarity problem, a model of an oligopolistic market with a homogeneous product is examined in this chapter. In Section 4.2, a generalized Cournot model is introduced. Each subject of the model uses a conjecture about the market response to variations of its production volume. The conjecture value depends upon both the current total volume of production at the market and the subject's contribution into it. Under general enough assumptions, the equilibrium existence and uniqueness theorems are proven. When analyzing the network equilibrium model, the topological degree theory is used to prove the existence theorem.

The Chapter is organized as follows. After a short introduction (Section 4.1), Section 4.2 presents the extended problem specification, describes and partially discusses assumptions concerning the functions relevant in the model, and justifies the equilibrium notion (in Subsection 4.2.1). Subsection 4.2.2 is dedicated to the equilibrium existence theorem, whereas Subsection 4.2.3 deals with theorems stating the uniqueness of the equilibrium total volume and the proper equilibrium. In Subsection 4.2.4, a particular case is examined, which comprises a version of the model in which each agent supposes the elasticity of the total market volume, with respect to his output, to remain constant. Subsection 4.2.5 deals with the particular case of influence quotients being functions of the total bargain volume only.

The chapter is closed by Section 4.3 that examines the network equilibrium problems in oligopolistic models. These problems are also generalized by introducing influence coefficients of producers at different markets. Existence and uniqueness theorems are obtained, and a particular case of human migration problem is examined.

4.1 INTRODUCTION

In this chapter, some well-known models of the oligopolistic market with a homogeneous product are extended to the case when the producers, instead of the standard Cournot model assumptions, use the following more general ones:

$$G_i(\eta) = G + (\eta - q_i)w_i(G, q_i). \tag{4.1}$$

Here G is the current total volume of the product sold at the market, q_i and η are the current and expected supplies, respectively, of the i-th producer, whereas $G_i(\eta)$ is the total market volume conjectured by the i-th agent as a response to changing his/her own supply from q_i to η. The *conjecture function* w_i will be referred to as the i-th agent's *influence quotient*. The standard Cournot model assumes $w_i \equiv 1$ for all i.

There exists a plethora of works on imperfect competition (*cf.* Bresnahan (1981), Friedman (1977), Klemperer and Meyer (1985), Kreps and Wilson (1982), Kreps (1990), Laitner (1980), Riordan (1985), Spiller and Favaro (1984), Ulph (1983)) in which the questions of the rationality of the assumptions are discussed. But the purpose of our chapter is somewhat different. We aim at finding some very general forms of possible conjectures that allow the equilibrium to exist. The main difference of our framework from those of the above cited papers consists in that we permit the conjectures being functions of two variables (unlike in Bresnahan (1981), Spiller and Favaro (1984) where they are functions of one variable). Moreover, we do not demand that the influence quotients be differentiable or even continuous; the agents need not being identical (unlike Novshek (1985)).

Another essential motivation for developing these results was a desire to formally embed the Stackelberg model into the generalized Cournot framework, at least in terms of the first order necessary conditions. This embedding is realized in Section 4.3 through the explicit construction of the (one or multiple) leaders' influence quotients. Since each leading agent of the (extended) Stackelberg model takes into account the other agents' optimal responses, the leader's conjecture inevitably loses its smoothness if one of the followers stops producing. That leads not only to rejecting the previously assumed continuity of the factors w_i in (4.1), but also to the splitting of conjecture (4.1) into two parts. More precisely, we allow the factors w_i to take different values for $\eta > q_i$ and $\eta < q_i$. These extentions lead to modifying some other assumptions, too.

4.2 EXTENDED COURNOT MODELS

4.2.1 PROBLEM SPECIFICATION

Consider n firms producing a homogeneous product and denote by q_i the i-th firm current output and by $f_i(q_i)$ its cost function, $i = 1, \ldots, n$. Let G be the total market output volume, and $p(G)$ be the inverse demand function value, i.e. the price of the product unit established at the market with the total bargain volume G. We also allow a constant volume Q of an extraneous supply of the product. Therefore, the balanced market implies the equality $G = Q + \sum_{i=1}^{n} q_i$.

As mentioned above, instead of (4.1), we use the *conjectures*

$$G_i(\eta) = \begin{cases} G + (\eta - q_i)w_i^+(G, q_i), & \eta > q_i, \\ G + (\eta - q_i)w_i^-(G, q_i), & \eta < q_i. \end{cases} \tag{4.2}$$

From now on, for brevity purposes, we omit the explicit indication of the fact that $G_i(\eta)$ is also a function of G and q_i. Behavior of each subject is determined by its current state (G, q_i) and conjecture (4.2). Namely, if the i-th agent produces η, then its *expected profit* is equal to

$$\mu_i(\eta) = \eta p\left(G_i(\eta)\right) - f_i(\eta), \tag{4.3}$$

and he chooses his optimal output volume that maximizes function (4.3).

Since this chapter deals only with static problems, we are interested in the situations where each of the *expected profit functions* (4.3) attains its (local) maximum at $\eta = q_i$. If the market is balanced, such a situation is naturally treated as a *conjectured equilibrium*. In order not to avoid considering the interesting particular case of subsection 4.2.4, we assume the existence of conjectures (4.2) only for situations with $G > 0$, $q_i > 0$. The expected profit function (4.3) is also defined in this area. Nevertheless, under certain conditions, some agents of the market can have zero outputs. Because of that, the equilibrium will be defined in terms of the first order necessary conditions for the expected profit function (4.3) to attain its maximum. The latter conditions are then extended to the agents with $q_i = 0$ in a natural manner.

Before the formal problem specification, we state the basic assumptions concerning the cost functions f_i, the conjecture derivatives $w_i^{\pm}$ and the inverse demand function p.

E1. Each of the functions $f_i(q_i)$ defined over $q_i \geq 0$, are continuously differentiable, non-decreasing and convex.

E2. The inverse demand function $p(G)$ defined for positive G is continuously differentiable, $p(G) > 0$ and $p'(G) < 0$ for all G.

E3. For each i there exists $H_i > 0$ such that $f_i'(H_i) = p(H_i)$.

E4. For each i, the functions $w_i^{\pm}(G, q_i)$ are defined for $0 < q_i \leq G$, and the product $q_i w_i^{\pm}$ does not decrease over $q_i > 0$. Furthermore, at the domain $(G, q_i) > 0$, the function w_i^+ is upper semicontinuous (u.s.c) with respect to G and right-continuous with respect to q_i; w_i^- is lower semicontinuous (l.s.c.) by G and left-continuous by q_i, and the following relationships take place:

$$0 \leq w_i^-(G, q_i) \leq w_i^+(G, q_i), \tag{4.4}$$

and

$$w_i^+(G, G) = 1. \tag{4.5}$$

E5. Principle of Potential Participation. For some k there exist $G_0 > 0$ and $q_{0k} > 0$ such that the relationship $G < G_0$ implies the inequality

$$p(G) + p'(G)G - f_k'(q_{0k}) > 0. \tag{4.6}$$

E6. The function $p(G)G$ is concave over $G \geq 0$.

Assumption E4 allows the following interpretation. The equality $G = q_i$ means that the i-th agent produces the whole output in the monopolistic manner, and condition (4.5) reflects that. In the classical Cournot model, $w_i^-(G, q_i) = w_i^+(G, q_i) = 1$ for all (G, q_i), which indicates the absence of the other agents' reaction to the i-th subject's output variations. If $w_i^{\pm}(G, q_i) > 1$, then the i-th agent is expecting the others' total output to increase or decrease simultaneously with its own volume produced. Moreover, the rate of the expected reaction grows together with the value $w_i^{\pm}$. Therefore, in this case, the right inequality in (4.4) signifies that the rate of the reaction expected by the i-th agent to an increase in his output is not lower than that to a decrease in it. When $w_i^{\pm}(G, q_i) < 1$, the situation is reverse. The others' common output variation is opposite to that of i-th subject, and the rate of the former is growing when the value of the function $w_i^{\pm}$ decreases. However, the nonnegativity of the latter values implies the shift of the total production volume in the same direction as that of the i-th agent. At last, the right inequality in (4.4) indicates that the market abandoned by the i-th subject is conquered by his rivals with the rate equal or greater than that of giving it up (by the rivals) when conquered by the same agent i.

Monotonicity of the products $q_i w_i^{\pm}(G, q_i)$ with respect to q_i can also be interpreted in economic terms. Consider the following identities

$$\Delta(G, q_i, \eta) = \frac{\partial \ln G_i(\eta)}{\partial \ln \eta} = \frac{\eta}{G_i(\eta)} \cdot \frac{\partial G_i(\eta)}{\partial \eta}.$$

The value $\Delta(G, q_i, \eta)$ is called an *expected elasticity* of the total volume with respect to the subject i output. For $\eta = q_i \pm 0$ one obtains $\Delta(G, q_i, q_i \pm 0) = q_i w_i^{\pm}(G, q_i)/G$. Therefore, the assumed monotonicity of $q_i w_i^{\pm}$ by q_i for any fixed value G means that the expected (by the i-th subject) elasticity of the total output does not decrease along with the subject's contribution increasing. In addition, this monotonicity, together with condition (4.5), imply the inequality $w_i^+(G, q_i) \leq G/q_i$ for $0 < q_i < G$.

Now consider the functions $\sigma_i^+(G, q_i) = q_i w_i^+(G, q_i)/G$ and $\sigma_i^-(G, q_i) = q_i w_i^-(G, q_i)/G$. Since they are monotone by q_i and non-negative, we can define their limit values as follows:

$$\sigma_i^+(G, 0) = \lim_{q_i \to 0} \sigma_i^+(G, q_i), \quad \text{and} \quad \sigma_i^-(G, 0) = \lim_{q_i \to 0} \sigma_i^-(G, q_i).$$

The functions $\sigma_i^{\pm}(G, q_i)$ have a clear economic sense. Namely, under conjecture (4.2), $\sigma_i^+(G, q_i)$ presents the expected (by the i-th agent) elasticity of the total market output with respect to his/her own output q_i when the latter increases. The function $\sigma_i^-(G, q_i)$ can be treated as the elasticity of the same type when q_i decreases. In the elasticity terms, assumption E4 allows the following reformulation.

E4′. For each i, the functions $\sigma_i^{\pm}(G, q_i)$ are defined for $G > 0$, $0 \leq q_i \leq G$, and do not decrease with respect to q_i. Furthermore, for $q_i > 0$, the function $\sigma_i^+(G, q_i)$ is upper semicontinuous by G and right-continuous with respect to q_i, the function $\sigma_i^-(G, q_i)$ is lower semicontinuous by G and left-continuous with respect to q_i, and the following relationships are valid

$$0 \leq \sigma_i^-(G, q_i) \leq \sigma_i^+(G, q_i) \leq \sigma_i^+(G, G) = 1. \tag{4.7}$$

The monotonicity of $\sigma_i^+(G, q_i)$ by q_i means that at a fixed total market volume, each subject expects his influence to grow (more precisely, not to decrease) as his contribution increases. The last equality in (4.7), similarly to (4.5), indicates that the subject does not expect the other suppliers to arise when he is increasing his output after having conquered the market completely.

Principle E5 means that for G sufficiently small (and hence, $p = p(G)$ sufficiently high) there exists at least one agent striving to increase his output volume over q_{0k}.

Differentiating the expected profit function (4.3) by η, one obtains the relationships

$$\mu_i'(\eta) = \begin{cases} p\left(G_i(\eta)\right) + \eta w_i^+(G, q_i)p'\left(G_i(\eta)\right) - f_i'(\eta), & \eta > q_i, \\ p\left(G_i(\eta)\right) + \eta w_i^-(G, q_i)p'\left(G_i(\eta)\right) - f_i'(\eta), & \eta < q_i. \end{cases} \tag{4.8}$$

At the point $\eta = q_i$, the function $\mu_i(\eta)$ has the leftside derivative $\mu_i^-(G, q_i)$ and the rightside derivative $\mu_i^+(G, q_i)$ calculated by the formulae

$$\mu_i^-(G, q_i) = p(G) + \sigma_i^-(G, q_i)p'(G)G - f_i'(q_i), \tag{4.9}$$

$$\mu_i^+(G, q_i) = p(G) + \sigma_i^+(G, q_i)p'(G)G - f_i'(q_i). \tag{4.10}$$

Using the first order necessary local maximum conditions for each subject's expected profit function, we specify our problem of *generalized equilibrium*, or conjectured equilibrium, as follows. Given $Q \geq 0$, find a vector $Z = (G, q_1, \ldots, q_n) \in R_+^{n+1}$ such that the market is balanced:

$$\sum_{i=1}^{n} q_i + Q = G, \tag{4.11}$$

and for each i, the following inequalities are valid:

$$\mu_i^+(G, q_i) = p(G) + \sigma_i^+(G, q_i)p'(G)G - f_i'(q_i) \leq 0, \tag{4.12}$$

and if $q_i > 0$, then

$$\mu_i^-(G, q_i) = p(G) + \sigma_i^-(G, q_i)p'(G)G - f_i'(q_i) \geq 0. \tag{4.13}$$

If $\sigma_i^+(G,q_i) = \sigma_i^-(G,q_i) = \sigma_i(G,q_i)$ and $q_i > 0$, inequalities (4.12) and (4.13) reduce to the single equality

$$p(G) + \sigma_i(G,q_i)p'(G)G - f_i'(q_i) = 0,$$

that describes the classical Cournot equilibrium when $\sigma_i(G,q_i) = q_i/G$, $i = 1,\dots,n$.

In order to justify the interpretation of a solution to problem (4.12)–(4.13) as an equilibrium for the agents with $q_i > 0$, we need showing the concavity of each subject's expected profit function as it is done by the following theorem.

THEOREM 4.1. *Under assumptions* E1–E6, *for each agent with* $q_i > 0$, *his expected profit function* $\mu_i(\eta)$ *is concave over* $\eta \geq 0$.

PROOF. In order to prove the theorem, it suffices to verify that both terms at the righthand side in (4.8) do not increase by η and that $\mu_i^+(G,q_i) \leq \mu_i^-(G,q_i)$. The latter inequality follows from (4.7) and E2 immediately. As for the terms at the righthand side in (4.8), we show the upper one to be non-increasing. The same property of the lower one can be shown in a similar manner.

Consider $q_i \leq \eta_1 < \eta_2$. Since the value $w_i^+(G,q_i)$ is nonnegative, it follows from (4.2) that $G \leq G_i(\eta_1) \leq G_i(\eta_2)$. Hence, making use of E2, we get $p\left(G_i(\eta_1)\right) \geq p\left(G_i(\eta_2)\right)$. Moreover, E1 implies $f_i'(\eta_1) \leq f_i'(\eta_2)$. In addition, if $p'\left(G_i(\eta_1)\right) \geq p'\left(G_i(\eta_2)\right)$, then $\eta_1 p'\left(G_i(\eta_1)\right) \geq \eta_2 p'\left(G_i(\eta_2)\right)$ due to the negativity of p'. All the above inequalities taken together imply $\mu_i'(\eta_2) \leq \mu_i'(\eta_1)$.

Otherwise, if $p'\left(G_i(\eta_1)\right) < p'\left(G_i(\eta_2)\right)$, we use (4.2) and rearrange the difference $\mu_i'(\eta_1) - \mu_i'(\eta_2)$ in the form

$$\begin{aligned}\mu_i'(\eta_1) - \mu_i'(\eta_2) = \left[p\left(G_i(\eta_1)\right) - p\left(G_i(\eta_2)\right)\right] + \left[G_i(\eta_1)p'\left(G_i(\eta_1)\right) - \right. \\ \left. - G_i(\eta_2)p'\left(G_i(\eta_2)\right)\right] + \left\{\left(1 - \sigma_i^+(G,q_i)\right) G \left[p'\left(G_i(\eta_2)\right) - p'\left(G_i(\eta_1)\right)\right]\right\} + \\ + \left[f_i'(\eta_2) - f_i'(\eta_1)\right].\end{aligned}$$

It is obvious from E6 that the sum of expressions in two former square brackets takes nonnegative values, and the last term is nonnegative due to the convexity of f_i. The sign of the term in the curly brackets coincides with that of the factor $\left(1 - \sigma_i^+(G,q_i)\right)$ which is nonnegative according to (4.7). Therefore, we get the inequality $\mu_i'(\eta_1) \geq \mu_i'(\eta_2)$ and thus complete the proof of the theorem. ∎

In what follows we justify the definition of equilibrium from the point of view of the agent with $q_i = 0$. The necessary and sufficient maximum conditions (4.12)–(4.13) for the expected profit function can be rewritten in the form of the pair of inequalities

$$-\sigma_i^-(G,q_i)p'(G)G \leq p(G) - f_i'(q_i) \leq -\sigma_i^+(G,q_i)p'(G)G. \tag{4.14}$$

The concluding (righthand side) term in (4.14) can be treated as the threshold that the difference between the price and the cost of production of an extra product unit (i.e. the *marginal cost*) of the i-th agent must overcome to make him increase his output. The first (lefthand side) term is interpreted in a similar manner. As the concluding righthand side term in (4.14) tends to the limit value $-\sigma_i^+(G,0)p'(G)G$ as q_i vanishes, it is natural enough to postulate the following behavior of the agent with $q_i = 0$. Namely, he starts producing whenever the difference between the price $p(g)$ and $f_i'(0)$ exceeds that threshold, i.e. the inequality $p(G) - f_i'(0) > -\sigma_i^+(G,0)p'(G)G$ holds. However, condition (4.12) means just the opposite at $q_i = 0$. Therefore, agent i has no reason to shift from zero when (4.12) takes place.

Notice that if conjecture (4.2) is valid for $q_i = 0$, too, i.e. if the value $w_i^+(G,0)$ is well-defined, then the right continuity of w_i^+ implies $\sigma_i^+(G,0) = 0$. Then it follows from (4.12) that $p(G) \leq f_i'(0)$, and the i-th subject has no incentive to start producing, either.

4.2.2 EXISTENCE THEOREM

Assumptions E4 and E4′ postulate the only one-sided continuity of the functions $w_i^\pm$ and $\sigma_i^\pm$ with respect to q_i for fixed G, as well as their semi-continuity with respect to G for a fixed $q_i > 0$. However, when proving the existence theorem, we need passing to limit with respect to the pair of variables (G, q_i), including the case $q_i \to 0$. The following lemmas not only help to avoid complicated arguments in the main theorem but also are interesting by themselves.

First, we extend the domains of the functions $w_i^\pm$ having set $w_i^\pm(G, q_i) = w_i^\pm(G, G) = 1$ for $q_i > G$ which obviously does not affect the continuity and monotonicity assumptions from E4. As a consequence, the domains of the functions $\sigma_i^\pm(G, q_i)$ and $\mu_i^\pm(G, q_i)$ are then extended without breaking assumption E4′.

LEMMA 4.1. *Functions $\mu_i^+(G, q_i)$ are right-continuous with respect to q_i and lower semicontinuous by G for $q_i \geq 0$, whereas functions $\mu_i^-(G, q_i)$ are left-continuous with respect to q_i and upper semicontinuous by G for $q_i > 0$.*

PROOF. The above-mentioned partial continuity properties of functions $\mu_i^\pm$ with respect to q_i and G when $q_i > 0$ follow immediately from E1, E2 and E4 (E4′). As for the lower semicontinuity of $\mu_i^+(G, q_i)$ at $q_i = 0$, it is equivalent to the upper semicontinuity of the function $\sigma_i^+(G,0)$ by G. However, the latter feature does take place because of the obvious identity

$$\sigma_i^+(G,0) = \inf_{q_i>0} \sigma_i^+(G, q_i)$$

and the function $\sigma_i^+(G, q_i)$ upper semicontinuity by G for any $q_i > 0$. This completes the proof. ∎

LEMMA 4.2. *Under assumptions* E4 (E4′) *functions* μ_i^+, μ_i^-, $i = 1, \ldots, n$, *have the following properties:*

$$\liminf_{(G,q_i)\to(\tilde{G},\tilde{q}_i)} \mu_i^+(G, q_i) = \mu_i^+(\tilde{G}, \tilde{q}_i), \quad \forall \tilde{G} > 0,\ \tilde{q}_i \geq 0,$$

$$\limsup_{(G,q_i)\to(\tilde{G},\tilde{q}_i)} \mu_i^-(G,q_i) = \mu_i^-(\tilde{G},\tilde{q}_i), \quad \forall \tilde{G} > 0, \ \tilde{q}_i > 0.$$

PROOF. Let us fix a positive $\tilde{G}$ and a non-negative $\tilde{q}_i$ and consider an arbitrary sequence of points $(G^k, q_i^k) > 0$ convergent to $(\tilde{G}, \tilde{q}_i)$. For an arbitrary $\varepsilon > 0$, there exists a number K such that $q_i^k \leq \tilde{q}_i + \varepsilon$ for each $k > K$. The function $\mu_i^+(G, q_i)$ being monotone non-increasing by q_i (as it follows from E1 and E4′), we obtain $\mu_i^+(G^k, q_i^k) \geq \mu_i^+(G^k, \tilde{q}_i + \varepsilon), \quad \forall k > K$. Since $\mu_i^+(G, q_i)$ is lower semicontinuous by G for positive q_i, the latter inequality implies the following relationship

$$\liminf \mu_i^+(G^k, q_i^k) \geq \liminf \mu_i^+(G^k, \tilde{q}_i + \varepsilon) \geq \mu_i^+(\tilde{G}, \tilde{q}_i + \varepsilon). \tag{4.15}$$

From the right continuity of $\mu_i^+(G, q_i)$ by q_i, we have the relationship $\lim_{\varepsilon\to+0} \mu_i^+(\tilde{G}, \tilde{q}_i + \varepsilon) = \mu_i^+(\tilde{G}, \tilde{q}_i)$. Hence, inequalities (4.15) imply (as $\varepsilon \to +0$)

$$\liminf \mu_i^+(G^k, q_i^k) \geq \mu_i^+(\tilde{G}, \tilde{q}_i). \tag{4.16}$$

Since the sequence (G^k, q_i^k) was selected arbitrarily, (4.16) leads to

$$\mu_i^+(\tilde{G}, \tilde{q}_i) = \lim_{q_i\to\tilde{q}_i+0} \mu_i^+(\tilde{G}, q_i) \geq \liminf_{(G,q_i)\to(\tilde{G},\tilde{q}_i)} \mu_i^+(G, q_i) \geq \mu_i^+(\tilde{G}, \tilde{q}_i). \tag{4.17}$$

The first assertion of the Lemma follows from inequalities (4.17). The second equality is obtained in a similar way, thus completing the proof. ■

THEOREM 4.2. *Let* $Q > 0$ *and assumptions* E1–E4 *hold. Then problem* (4.11)–(4.13) *has a solution. In addition, if assumption* E5 *is valid, then the problem is solvable in case* $Q = 0$*, too.*

PROOF. For each i, we construct a multi-valued mapping $G \mapsto [q_i^-, q_i^+]$ that associates each $G > 0$ with the segment, the endpoints of which are defined as follows:

$$q_i^- = \inf \left\{ q_i \mid 0 \leq q_i \leq H_i, \mu_i^+(G, q_i) \leq 0 \right\}, \tag{4.18}$$

$$q_i^+ = \sup \left\{ q_i \mid 0 \leq q_i \leq H_i, \mu_i^-(G, q_i) \geq 0 \right\}; \tag{4.19}$$

here $H_i > 0$ is the scalar from condition E3. Moreover, we accept $q_i^- = H_i$ or $q_i^+ = 0$, if the set in (4.18) or (4.19), respectively, is empty.

Now we prove that $q_i^- \leq q_i^+$. On the contrary, suppose that $q_i^- > q_i^+$. Then $q_i^- > 0$ and, by virtue of (4.19), $\mu_i^-(G, q_i^-) < 0$. In view of Lemma 4.1, the functions $\mu_i^-(G, q_i)$ are left-continuous by q_i, hence $\mu_i^-(G, q_i^- - \delta) < 0$. As $\mu_i^-(G, q_i) \geq \mu_i^+(G, q_i)$ for each pair (G, q_i), then $\mu_i^+(G, q_i^- - \delta) < 0$, too. The latter inequality contradicts the definition of q_i^- by (4.18). Therefore, $q_i^- \leq q_i^+$.

If both feasible sets in definitions (4.18) and (4.19) are nonempty, then Lemma 4.1 and the monotonicity of functions $\mu_i^\pm(G, q_i)$ imply that all the points of the closed segment $[q_i^-, q_i^+]$ satisfy inequalities (4.12) and (4.13). If the feasible set in (4.19) is empty, then $\mu_i^+(G, q_i) \leq \mu_i^-(G, q_i) < 0$ for all $0 \leq q_i \leq H_i$, i.e. $q_i^- = q_i^+ = 0$, and $q_i = 0$ satisfies (4.12). On the other hand, if the feasible set in (4.18) is empty, then $\mu_i^-(G, q_i) \geq \mu_i^+(G, q_i) > 0$ for all $0 \leq q_i \leq H_i$, i.e. $q_i^- = q_i^+ = H_i$. Thus we have shown that in each case, except for $q_i^- = q_i^+ = H_i$, all the points of the segment $[q_i^-, q_i^+]$ solve problem (4.12)–(4.13).

Suppose that $Q > 0$. Consider the point-to-set mapping $G \mapsto [q_i^-, q_i^+]$ on the halfline $[Q, +\infty)$ and show it to be closed. Since the graph of the mapping is the intersection of the epigraph of the function $q_i^-(G)$ and the undergraph of the function $q_i^+(G)$ (i.e. the set of points $\{(G, \lambda) \mid G > 0, \lambda \leq q_i^+(G)\}$), it suffices to establish that the two sets are closed. In view of their symmetric construction, we do that only for the epigraph of the function $q_i^-(G)$. Let a sequence $\{(G^k, q_i^k)\}$ with $G^k \geq Q$, $q_i^k \geq q_i^-(G)$ be convergent to a point $(\tilde{G}, \tilde{q}_i)$. Two outcomes are possible. If there exists an infinite subsequence with $q_i^k = H_i$, then $\tilde{q}_i = H_i$ and the point $(\tilde{G}, \tilde{q}_i)$ obviously belongs to the epigraph. Otherwise, if $q_i^k < H_i$ from some k onward, then $q_i^-(G^k) < H_i$ which implies $\mu_i^+(G^k, q_i^-(G^k)) \leq 0$. Since $\mu_i^+(G, q_i)$ is monotone with respect to q_i, then $\mu_i^+(G^k, q_i^k) \leq 0$, too. In view of Lemma 4.2, we obtain the limit relationship $\mu_i^+(\tilde{G}, \tilde{q}_i) \leq 0$, which means that $(\tilde{G}, \tilde{q}_i)$ belongs to the epigraph of the function $q_i^-(G)$ and thus proves the closedness of the epigraph.

The closedness of the undergraph of the function $q_i^+(G)$ is established in a similar manner. Thus the closedness of the mapping $G \mapsto [q_i^-, q_i^+]$ is verified.

Now consider the sum of the above constructed mappings and the value Q, i.e. the multi-valued mapping that associates each $G > 0$ with the segment $A(G) = \left[Q + \sum q_i^-, Q + \sum q_i^+\right]$. We examine this mapping on the segment $[Q, Q + \sum H_i]$. It is easily verified that all the conditions of the Kakutani theorem are satisfied in this case, so there exists a fixed point $G \in A(G)$. In the other words, there is a $(n+1)$-tuple $Z = (G, q_1, \ldots, q_n)$ satisfying the balance equality (4.11) and the restrictions $q_i^-(G) \leq q_i \leq q_i^+(G)$, $i = 1, \ldots, n$. Now we show that $q_i^-(G) < H_i$ holds. Indeed, if $q_i^-(G) = H_i$ for some i, then $q_i = H_i$, hence (4.11) and $Q > 0$ imply $G > H_i$. However, the latter, together with A3, yields $p(G) - f_i'(H_i) < 0$ which means that $\mu_i^-(G, H_i) < 0$. Since $\mu_i^-(G, q_i)$ is left-continuous by q_i, we come to $q_i^+(G) < H_i = q_i^-(G)$ which contradicts the above obtained estimate $q_i^- \leq q_i^+$. Thus, the inequality $q_i^-(G) < H_i$ really holds, whence q_i solves problem (4.12)– (4.13), and the whole $(n+1)$-tuple $Z = (G, q_1, \ldots, q_n)$ solves problem (4.11)–(4.13). This completes the proof in case $Q > 0$.

Now let assumption E5 be valid. For each $Q > 0$, fix a $(n+1)$-tuple $Z(Q) = \left[G(Q), q_1(Q), \ldots, q_n(Q)\right]$ solving problem (4.11)–(4.13), and tend Q to zero. As the estimates $Q \leq G(Q) \leq Q + \sum H_i$, $0 \leq q_i < H_i$ hold true, there exists an accumulation point $\tilde{Z} = (\tilde{G}, \tilde{q}_1, \ldots, \tilde{q}_n)$. Suppose that $\tilde{G} = 0$. Then there exists a positive Q such that $G(Q) < G^0$ and $q_k(Q) < q_{0k}$ which is prohibited by E5. Hence $\tilde{G} > 0$, and we can use Lemma 4.2 implying that the $(n+1)$–tuple $\tilde{Z}$ solves problem (4.11)–(4.13) in case of $Q = 0$. The theorem is proved completely.

■

Remark. In the particular case when $w_i^- = w_i^+ \equiv 1$ (the classical Cournot model), assumptions E1–E3 are enough to guarantee the existence of an equilibrium.

4.2.3 UNIQUENESS OF SOLUTION

The question of the equilibrium uniqueness can be divided into two parts. First, from the general modeling point of view, it is interesting to examine uniqueness of the equilibrium total bargain volume G. This aspect is thoroughly investigated in the section. However in general, at the (possibly multiple) equlilibria, this total bargain volume can be distributed among the active producers in various proportions. This question is connected with the uniqueness of solutions to the local problems (4.12)–(4.13) and discussed briefly in the end of the section.

To examine conditions ensuring the uniqueness of the solution to problem (4.11)–(4.13), we keep assumptions E1–E6 valid and add the following one.

E7. For each i, the relationships $0 < G_1 < G_2$ and $0 < \delta < 1$ imply the inequality

$$w_i^+(G_1, \delta G_1) \le w_i^-(G_2, \delta G_2). \tag{4.20}$$

Moreover,

$$w_i^-(G, q_i) < \frac{G}{q_i}, \quad \text{if} \quad 0 < q_i < G. \tag{4.21}$$

Note that condition (4.20) means that the functions w_i^+ and w_i^- are conformed in their non-decreasing along the straight rays starting at the origin and contained in the positive orthant. The condition can be rewritten also as $\sigma_i^+(G_1, \delta G_1) \le \sigma_i^-(G_2, \delta G_2)$ which implies the limit relationship $\sigma_i^+(G_1, 0) \le \sigma_i^-(G_2, 0)$ when δ vanishes. Thus, under assumption E7, the inequality

$$\sigma_i^+(G_1, \delta G_1) \le \sigma_i^-(G_2, \delta G_2) \tag{4.22}$$

is valid for $\delta \ge 0$.

DEFINITION 4.1. Let $Z(Q) = \left[G(Q), q_1(Q), \dots, q_n(Q)\right]$ be an equilibrium, i.e. a solution to problem (4.11)–(4.13) for a given value of the extraneous supply Q. We call it *non-monopolistic*, if $q_i < G$, $i = 1, \dots, n$.

THEOREM 4.3. *Under assumptions* E1–E7, *for each* $Q \ge 0$, *the total bargain value* $G(Q)$ *is the same for all the non-monopolistic equilibria.*

PROOF. For a given $Q \ge 0$, suppose on the contrary that there exist two non-monopolistic equilibria

$$Z_1(Q) = \left[G_1(Q), q_1^1(Q), \dots, q_n^1(Q)\right],$$

$$Z_2(Q) = \left[G_2(Q), q_1^2(Q), \dots, q_n^2(Q)\right]$$

with $G_1 = G_1(Q) < G_2(Q) = G_2$. Define

$$I(G_1) = \left\{1 \le i \le n \mid q_i^1(Q) > 0\right\},$$

$$I(G_2) = \left\{1 \le i \le n \mid q_i^2(Q) > 0\right\}.$$

Observe that $G_2 > Q$ and consequently, $I(G_2) \neq \emptyset$. Now we prove that for each $i \in I(G_2)$ the following inequality holds

$$\frac{q_i^2(Q)}{G_2} < \frac{q_i^1(Q)}{G_1}. \tag{4.23}$$

Indeed, denote by $q_1 = q_i^1(Q)$ and $q_2 = q_i^2(Q)$ and suppose, on the contrary, that $q_2/G_2 \geq q_1/G_1$. Then obviously $q_2 > q_1$ and there exists $\tilde{q}$ such that $q_1 \leq \tilde{q} \leq q_2$ and $\tilde{q}/G_2 = q_1/G_1 = \delta$. Using (4.22) and E4′, we obtain

$$0 \leq z_1 = \sigma_i^+(G_1, q_1) \leq \sigma_i^-(G_2, \tilde{q}) \leq \sigma_i^-(G_2, q_2) = z_2.$$

Since $Z_2(Q)$ is a non-monopolistic equilibrium, then $q_2 < G_2$, and condition (4.21) implies $z_2 < 1$. Moreover, relationships (4.12),(4.13) for this i yield

$$p(G_1) + G_1 p'(G_1) z_1 - f_i'(q_1) \leq p(G_2) + G_2 p'(G_2) z_2 - f_i'(q_2).$$

The convexity of f_i allows one to omit the last terms in both parts without violating the inequality:

$$p(G_1) + G_1 p'(G_1) z_1 \leq p(G_2) + G_2 p'(G_2) z_2. \tag{4.24}$$

If $z_1 = 0$, then (4.24) is impossible as the function $p(G)$ is strictly decreasing. Let $z_1 > 0$. Subtract from (4.24) the inequality $(1 - z_2)p(G_1) > (1 - z_2)p(G_2)$, which is valid due to E2 and $z_2 < 1$, and obtain the relationship $z_2 p(G_1) + G_1 p'(G_1) z_1 < z_2 p(G_2) + G_2 p'(G_2) z_2$. Taking into account that $z_2 \geq z_1 > 0$ and $p'(G_1) < 0$, we deduce from the latter the following inequality

$$p(G_1) + G_1 p'(G_1) < p(G_2) + G_2 p'(G_2).$$

Since $G_1 < G_2$, the last inequality contradicts E6. Thus, we have proved that (4.23) holds for $i \in I(G_2)$.

Inequality (4.23) shows that $I(G_2) \subseteq I(G_1)$. Consequently, the balance equalities for Z_1 and Z_2 imply the following series of relationships

$$1 = \sum_{i \in I(G_2)} \frac{q_i^2(Q)}{G_2} + \frac{Q}{G_2} < \sum_{i \in I(G_2)} \frac{q_i^1(Q)}{G_1} + \frac{Q}{G_1} \leq$$

$$\leq \sum_{i \in I(G_1)} \frac{q_i^1(Q)}{G_1} + \frac{Q}{G_1} = 1,$$

which leads to the impossible inequality $1 < 1$. It means that the assumption about the existence of two non-monopolistic equilibria with different total bargain volumes was wrong, which completes the proof of the Theorem. ∎

If $Q > 0$, then each equilibrium is non-monopolistic. In this case, according to Theorem 4.3, the value of equilibrium total bargain volume $G(Q)$ is unique. However if $Q = 0$, both monopolistic and non-monopolistic equilibria with different total bargain values may occur in the framework of a single problem.

For instance, that situation is possible when functions f_i and $\zeta(G) = p(G)G$ are piecewise linear. Nevertheless, the following result obtains.

COROLLARY 4.1. *In addition to Theorem 4.3 assumptions, suppose that for each i either f_i is strictly convex, or for all $G > 0$ the relationships $0 \le q_i^1 < q_i^2$ imply the inequality*

$$\sigma_i^+(G, q_i^1) < \sigma_i^-(G, q_i^2). \tag{4.25}$$

Then the non-monopolostic equilibrium vector $Z(Q)$ is unique.

PROOF. On the contrary, suppose that for the equilibrium total bargain volume G and for some $i \in I(G)$, there exist two distinct values $0 \le q_i^1 < q_i^2$ satisfying inequalities (4.12) and (4.13). Then making use of either the strict convexity of f_i or inequality (4.25), we can write down the series of relationships

$$0 \le \mu_i^-(G, q_i^2) = p(G) + \sigma_i^-(G, q_i^2)p'(G)G - f_i'(q_i^2) <$$

$$< p(G) + \sigma_i^+(G, q_i^1)p'(G)G - f_i'(q_i^1) = \mu_i^+(G, q_i^1) \le 0,$$

which implies the impossible inequality $0 < 0$. This contradiction completes the proof. ∎

Remark. Under assumptions of Corollary 1 and with $Q > 0$, the (non-monopolistic) equilibrium exisits uniquely. In order to get a similar result in case of $Q = 0$, we need specifying the form of the influence coefficients $w_i^\pm$, as it is done in Subsection 4.2.4.

Remark. Monopolistic equilibria (when $Q = 0$) can be found by maximizing the functions $\mu_k(G, G) = p(G)G - f_k(G)$ along the halfline $(0, +\infty)$ under the conditions $\mu_i^+(G, 0) \le 0$ for $i \ne k$.

Remark. In the particular case when $w_i^+ = w_i^- \equiv 1$ (Cournot model), assumptions E1–E3 with the functions f_i' being monotone increasing suffice to prove the existence of a unique equilibrium.

4.2.4 CASE OF CONSTANT ELASTICITY

In this subsection, we discuss in more detail the particular case when the functions f_i and p are twice continuously differentiable, the functions w_i^+ and w_i^- coincide over the positive orthant $(G, q_i) > 0$ except for the diagonal ray $q_i = G,\ q_i > 0$, and they take the following form:

$$w_i^-(G, q_i) = w_i^+(G, q_i) = \alpha_i + \sigma_i \frac{G}{q_i}, \quad \text{if } 0 < q_i < G, \tag{4.26}$$

$$w_i^-(G, G) = \alpha_i + \sigma_i, \qquad w_i^+(G, G) = 1. \tag{4.27}$$

Here $0 \le \sigma_i < 1,\ \ 0 \le \alpha_i \le 1 - \sigma_i$. It is easily verified that these w_i^- and w_i^+ satisfy conditions E4 and E7. Therefore, Theorems 4.1 and 4.2 are valid for problem (4.11)–(4.13) with these functions. As p is twice differentiable, condition E6 reduces to the inequality

$$2p'(G) + p''(G)G \leq 0, \qquad \forall G > 0. \tag{4.28}$$

Moreover, we replace condition (4.25) from Corollary 4.1 by the following (a bit stronger) assumption.

E8. The functions f_i and p are twice continuously differentiable, and for each i either $f_i''(q_i) > 0$ for all $q_i > 0$ and $\lim\limits_{q_i \to +\infty} f_i'(q_i) = +\infty$, or $\alpha_i > 0$.

The case $\alpha_i = 0$ corresponds to the constant elasticity (equal to σ_i) of the variation of the expected total bargain volume G with respect to the variation of q_i. Lemma 4.3 proven below is an instrument for investigating the generalized Stackelberg model in Section 4.3.

Under assumption E8, we will examine the behavior of solutions $q_i = q_i(G)$ to problem (4.12)–(4.13) which reduces to the following form: Find a $q_i \geq 0$ satisfying the conditions below:

$$q_i = 0, \qquad \text{if} \quad f_i'(0) - \sigma_i p'(G)G - p(G) \geq 0, \tag{4.29}$$

and

$$f_i'(q_i) - \alpha_i q_i p'(G) - \sigma_i p'(G)G - p(G) = 0, \quad \text{if } q_i > 0. \tag{4.30}$$

First, we extend the functions f_i smoothly to the negative half-axis $(-\infty, 0)$ having set $f_i(q_i) = f_i(0) + f_i'(0)q_i + f_i''(0)q_i^2/2$ at the points $q_i < 0$. Under assumptions E1-E3, E5-E6, E8, the solution $\tilde{q}_i = \tilde{q}_i(G)$ of equation (4.30) is determined uniquely and it is continuously differentiable for all $G > 0$. If $\tilde{q}_i(G) > 0$ at $G > 0$, then the function $q_i(G)$ coincides with $\tilde{q}_i(G)$. Otherwise, if $\tilde{q}_i(\tilde{G}_i) = 0$ at some $\tilde{G}_i$, i.e. $f_i'(0) \geq \sigma_i p'(\tilde{G}_i)\tilde{G}_i + p(\tilde{G}_i)$, then due to the inequalities $0 \leq \sigma_i < 1$ and assumptions E2, E6, we get the inequality $f_i'(0) > \sigma_i p'(G)G + p(G)$ for all $G > \tilde{G}_i$. Therefore, $\tilde{q}_i(G) < 0$ for $G > \tilde{G}_i$ whereas $q_i(G) = 0$ at $G > \tilde{G}_i$. Thus, we have proved that $q_i(G) = \max\{\tilde{q}_i(G), 0\}$, and the functions $q_i(G)$ are continuously differentiable everywhere, except for the single point $\tilde{G}_i$. At the point $\tilde{G}_i$, the function $q_i(G)$ has the one-sided derivatives $q_i'(\tilde{G}_i+0) = 0$ and $q_i'(\tilde{G}_i-0) \leq 0$. The derivatives of the functions $q_i(G)$ at the points $G < \tilde{G}_i$ can take either positive or negative values. However, below we show that at the points G satisfying the inequality

$$Q + \sum_{i=1}^{n} q_i(G) \leq G, \tag{4.31}$$

(for instance, at the equilibrium), there is at most one subject with the positive derivative value. Indeed, let us define the index subsets

$$I(G) = \{i \mid q_i(G) > 0\}, \qquad I^+(G) = \{i \in I(G) \mid q_i'(G) > 0\}.$$

LEMMA 4.3. *Let assumptions* E1-E3, E5-E6, E8 *be valid and conjectures* $w_i^{\pm}$ *be defined by* (4.26)–(4.27). *Then for any* $G > 0$ *and* $Q \geq 0$ *satisfying inequality* (4.31), *the following estimates hold:*

$$\sum_{i=1}^{n} q_i'(G-0) \leq \sum_{i=1}^{n} q_i'(G+0) \leq \begin{cases} 1 - \frac{2Q}{G}, & \text{if } I^+(G) \neq \emptyset, \\ 0, & \text{if } I^+(G) = \emptyset. \end{cases} \tag{4.32}$$

Moreover, under condition (4.31), *(for example, at an equilibrium), there can be at most one producer with the positive derivative value* $p_i'(G) > 0$.

PROOF. As equality (4.30) is valid for $i \in I(G)$, we can use the Implicit Function Theorem and express the implicit function's derivative as follows:

$$q_i'(G) = \frac{p'(1+\sigma_i) + \sigma_i p'' G + \alpha_i q_i p''}{f_i''(q_i) - \alpha_i p'}, \qquad i \in I(G). \tag{4.33}$$

Here the variable G is omitted at the righthand side. From E2 and E8, it follows that the denominator of (4.33) always takes positive values. Therefore, the sign of the derivative $q_i'(G)$ coincides with that of the numerator of (4.33).

It is readily verified that if $\alpha_i = 0$, then E2, (4.28) and $0 \leq \sigma_i < 1$ imply

$$q_i'(G) = \frac{1}{f_i''(q_i)} \left[(1-\sigma_i)p' + \sigma_i(2p' + p''G) \right] < 0.$$

However, if $\alpha_i > 0$, the situation is more complicated since $q_i'(G)$ can take both positive and negative values . Suppose that $q_i'(G) > 0$. In this case, the numerator in (4.33) is obviously positive. Since $f_i'' \geq 0$, then by omitting it we obtain from (4.33) and (4.28)

$$0 < q_i'(G) \leq -\frac{1-\sigma_i}{\alpha_i} + \frac{2}{G} q_i(G) \leq -1 + \frac{2}{G} q_i(G). \tag{4.34}$$

From (4.34) it follows that $q_i'(G) > 0$ can only hold if $q_i(G) > G/2$. Thus we have shown that under condition (4.31) (e.g., at the equilibrium) there can be at most one subject with a positive derivative value $q_i'(G) > 0$.

Now we can evaluate $\sum\limits_{i \in I(G)} q_i'(G)$ in case when $\sum q_i(G) \leq G - Q$. Indeed, if $I^+(G) = \emptyset$, then

$$\sum_{i \in I(G)} q_i'(G) \leq 0. \tag{4.35}$$

Otherwise, if $I^+(G) = \{k\}$, $1 \leq k \leq n$, then $q_k(G) \leq G - Q$, and (4.34) implies

$$\sum_{i \in I(G)} q_i'(G) \leq q_k'(G) \leq -1 + \frac{2q_k(G)}{G} \leq 1 - \frac{2Q}{G}. \tag{4.36}$$

Since $q_i'(G+0) = 0$ and $q_i'(G-0) \leq 0$ for $i \notin I(G)$, i.e. when $G \geq \tilde{G}_i$, the Lemma's assertion follows from inequalities (4.35) and (4.36). ■

THEOREM 4.4. *Let assumptions* E1-E3, E5-E6, E8 *be valid and conjectures* $w_i^{\pm}$ *be defined by* (4.26)–(4.27). *In addition, when* $Q = 0$, *let* $f_i''(q_i) > 0$ *for all* i *and* $q_i \geq 0$. *Then the equilibrium* $Z(Q)$ *exists uniquely.*

PROOF. The existence of equilibrium follows directly from Theorem 4.2. If $Q > 0$, then Lemma 4.3 implies $\sum_{i \in I(G)} q_i'(G) < 1$. Otherwise, if $Q = 0$, the positivity of f_i'' allows one to rewrite (4.34) with the strict inequality:

$$0 < q_i'(G) < -\frac{1-\sigma_i}{\alpha_i} + \frac{2}{G} q_i(G) \leq -1 + \frac{2}{G} q_i(G),$$

that again leads to the estimate $\sum_{i \in I(G)} q_i'(G) < 1$. Therefore, in both cases, one can conclude that if balance (4.11) occurs for some value $G^* > 0$, then the difference $G - Q - \sum q_i(G)$ strictly increases with G increasing beyond G^*. Therefore, the balance equality can never occur for a value G greater than G^*, which means that problem (4.11)–(4.13) has a unique solution. This completes the proof. ■

4.2.5 INFLUENCE QUOTIENT AS FUNCTION OF TOTAL BARGAIN VOLUME

In this subsection, we consider another particular case of the generalized Cournot model in which the agents' influence quotients are continuous functions of the total bargain volume only: $w_i = w_i(G)$, $i = 1, \ldots, n$. In that case, the equilibrium problem reduces to the form: given a $Q \geq 0$, find an n-tuple $[q_1(Q), \ldots, q_n(Q)]$ such that, for each agent $i = 1, \ldots, n$, under the assumption that $q_j(Q)$, $j \neq i$, are fixed, the value $q_i = q_i(Q) \geq 0$ solves the following (univariate) complementarity problem

$$\varphi_i(q_i) = f_i'(q_i) - q_i w_i(G) p'(G) - p(G) \geq 0, \tag{4.37}$$

$$\varphi_i(q_i) = 0, \qquad \text{if} \qquad q_i > 0; \tag{4.38}$$

here G is defined as the sum

$$G = Q + \sum_{i=1}^{n} q_i. \tag{4.39}$$

Complementarity problem (4.37)–(4.39) shows that the agent i need not varying its output q_i provided that he uses the conjecture $w_i(G)$.

We begin with conditions that guarantee the existence of solution to problem (4.37)–(4.39). Assume that the functions f_i, w_i and p satisfy the following assumptions.

F1. Each of the functions $f_i(q_i)$, $i = 1, \ldots, n$, defined at $q_i \geq 0$ are convex, twice continuously differentiable and non-decreasing.

F2. The inverse demand function $p(G)$ defined for positive values of G is twice continuously differentiable, takes positive values, and $p'(G) < 0$ for each $G > 0$.

F3. For each $i = 1, \dots, n$, there exists a scalar $H_i > 0$ such that

$$f_i'(H_i) = p(H_i). \tag{4.40}$$

F4. Each of the functions $w_i(G)$ defined at $G > 0$ are continuous and take non-negative values.

F5. For every $i = 1, \dots, n$, and $G > 0$ small enough, the following inequality holds:

$$f_i'(G) - \frac{1}{n} w_i(G) p'(G) G - p(G) < 0.$$

Note that assumption F5 is surely valid if the right-side derivative $p'(+0)$ is finite and assumptions F1–F4 hold. Also notice that condition (4.40) is natural enough. Indeed, if the latter fails for some agent i, then one of the following relationships holds: either

a) $f_i'(0) \geq \lim_{G \to +0} p(G)$; in this case, the agent i produces nothing and can be excluded from the model; or

b) $f_i'(G) < p(G)$ for every $G > 0$; moreover, if $\lim_{G \to +\infty} [p(G) - f_i'(G)] > 0$, then agent i can produce an arbitrary large amount of the good with a positive profit which means that problem (4.37)–(4.39) is unsolvable. Only when

$$\lim_{G \to +\infty} [p(G) - f_i'(G)] = 0 \quad \text{and} \quad \limsup_{G \to +\infty} [-p'(G)/p''(G)] < +\infty$$

the problem has a solution, but we do not consider that extremal case here.

Since the problem considered here is a particular case of the generalized Cournot model examined in the previous sections, the proofs of the theorems below are omitted.

THEOREM 4.5 *Let $Q > 0$ and assumptions* F1–F4 *hold. Then problem* (4.37)–(4.39) *is solvable. In addition, with assumption* F5 , *the problem has a solution for $Q = 0$, too.* ∎

In order to establish the solution uniqueness, we keep assumptions F1–F5 valid and add the following ones.

F6. For each $i = 1, \dots, n$, function w_i is differentiable and non-decreasing. Moreover, it takes only positive values not exceeding 1.

F7. The function $p(G)G$ is concave.

THEOREM 4.6. *If assumptions* F1–F7 *are valid and $f_i''(G) > 0$ for every i and $G \geq 0$, then problem* (4.37) – (4.39) *is solvable uniquely.* ∎

Remark. In the particular case when $w_i \equiv 0$ for every i (the perfect competition), conditions F1–F3 are sufficient for the existence of solution to problem (4.37)-(4.39). To guarantee the solution's uniqueness, we demand the derivative f_i', $i = 1, \dots, n$, to be strictly increasing.

4.3 NETWORK MODELS OF OLIGOPOLY

In this section, the generalized network models of oligopolistic markets of homogeneous product and of human migration are developed. In framework of the first model, agents sell their product at several independent markets taking into account the prices of the product unit at different markets, production expenditures, and transportation costs. The unit price at a market depends upon the total supply, whereas the production expenditures may grow along with the total volume of output by all producers. The latter ones choose production volumes and distribution of their outputs to be sold at the different markets. In order to do that, each agent uses conjectures about the total market supply variations depending upon those of his own supply. Under general enough assumptions concerning the market inverse demand functions and the producers' cost functions, the equilibrium existence and uniqueness theorems are formulated and proven. The techniques of the proof is based upon the very recent progress in existence theory for variational inequalities and complementarity problems. The similar techniques can be applied to the human migration model.

4.3.1 INTRODUCTION

First, this subsection offers an attempt to modify the oligopoly model using both the network approach exploited when examining human migration models; *cf.* Dafermos and Nagurney (1986), Nagurney e.a. (1992), Nagurney (1990). Second, the idea of production cost depending upon the demand for the raw materials is realized. Such modifications should increase the model's flexibility and its adaptation facilities.

In our previous works, we did not exploit any market mechanisms affecting the production costs. More precisely, each agent's costs depended only upon his own output which means that each subject has a separate production factors market, if any. Of course, such individual markets of the production factors do exists and must be taken into account when modeling. They comprise the markets of local resources with limited transportability such as local raw materials, working hands etc. However, there are also some well-transportable production factors, and their price must depend upon not only a single agent's demand for them (and hence, upon his output), but upon the aggregate demand by all the subjects. The latter quality is proposed to model by the cost functions depending upon two variables: an individual output, and the total production output by all the oligopoly agents.

As for the market sale process, it is modeled with the aid of the classical inverse demand functions. However, instead of a single market, a network of markets with different demand functions is considered. Each producer can distribute his output among the sale markets taking into account his/her conjectures of his/her influence degree at each market.

The crucial difference of this section from the previous work by Bulavsky and Kalashnikov (1998) lies in the new techniques of proof of the main existence results. These techniques are based upon the recent papers by the authors

Bulavsky, Isac and Kalashnikov (1998), Isac, Bulavsky and Kalashnikov (1997), in which very general existence theorems are presented. The latter allows one to weaken the conditions that guarantee the equilibrium existence in the previous paper by Bulavsky and Kalashnikov (1998).

This section is organized as follows. Subsection 4.3.2 describes the model with the sale markets network and individual markets of production factors, and the equilibrium problem for such a model is specified. Subsection 4.3.3 is dedicated to establishing the equilibrium existence and uniqueness theorems. Subsection 4.3.4 examines the model with the market of production factors in which the producers' expenditures depend upon not only their own outputs but also on the aggregate output of all the subjects of the model. For this model, the equilibrium existence and uniqueness theorems also are proved. Notice that the model described in Subsection 4.3.2 is a particular case of that of Subsection 4.3.4. However, restrictions on the relevant functions in the former model are weaker than those in the latter one. Therefore the theorems of Subsection 4.3.3 cannot be formally deduced from those of Subsection 4.3.4. At last, Subsection 4.3.5 deals with the model of human migration and reduces it to the oligopolistic model with conjectures.

4.3.2 MODEL WITH INDIVIDUAL MARKETS OF PRODUCTION FACTORS

Consider M firms producing a single homogeneous product with their current outputs denoted by q_i, and their cost functions by $f_i(q_i)$, $i = 1, \dots, M$. The product is supposed to sell at N different markets. Therefore, if s_{ij} is the quantity of the product supplied by i-th producer for sale at market j, then the equality $\sum\limits_j s_{ij} = q_i$ holds. Besides, producer i may suffer supplementary expenses (for instance, transportation costs) at amount of $c_{ij} \geq 0$ per unit of the product. Denote by G_j the total bargain volume at market j, and by $p_j(G_j)$ the inverse demand function, i.e. the price of a unit of product at the market with the total bargain volume G_j. For each place j, we also allow some local sources of the product to exist and supply $Q_j \geq 0$ units at the market. Thus, the following equalities must hold $G_j = Q_j + \sum\limits_i s_{ij}$, $j = 1, \dots, N$.

Furthermore, with each pair "producer–consumer" (i, j), we associate a (conjectured by producer i) influence quotient $w_{ij} = w_{ij}(s_{ij}, G_j) \geq 0$ at market j as follows. Producer i assumes that his ***supply variation*** from s_{ij} to η_{ij} implies the total sale (bargain) volume at market j to become equal to

$$G_{ij}(\eta_{ij}) = G_j + w_{ij}(\eta_{ij} - s_{ij}). \tag{4.41}$$

Arrange thus introduced values of sales and local supplies into vectors the $G, Q \in R^N$, respectively, and form the product flows from producer i as the vector $S_i = (s_{ij}, j = 1, \dots, N) \in R^N$. The expected profit of each agent depends upon the current situation at the market, his output $\eta_i = (\eta_{ij}, j = 1, \dots, N)$, and conjecture (4.41):

$$\mu_i(\eta_i) = \sum_{j=1}^{N} \eta_{ij} \left[p_j\left(G_{ij}(\eta_{ij})\right) - c_{ij}\right] - f_i\left(\sum_{j=1}^{N} \eta_{ij}\right). \tag{4.42}$$

The agent considers his choice to be correct if his expected profit (4.42) is maximal at $\eta_{ij} = s_{ij}$.

At the balanced market, such a situation is naturally treated as an equilibrium. The rigorous definition of the equilibrium will be given by making use of the first order optimality condition for the expected profit function (4.42). Before doing that formally, we list the assumptions concerning the functions f_i, p_j.

G1. Each of the functions $f_i(q_i)$ are defined over $q_i \geq 0$, continuously differentiable, convex and non-decreasing.

G2. Every inverse demand function $p_j(G_j)$ assumes positive values at nonnegative G_j, and it is continuously differentiable with $p_j'(G_j) \leq 0$ for $G_j \geq 0$.

G3. For each i there exists an $H_i > 0$ such that

$$f_i'(H_i) \geq \max_j \left[p_j(H_i) - c_{ij}\right]. \tag{4.43}$$

G4. For each pair i, j, the quotients $w_{ij} = w_{ij}(s_{ij}, G_j)$ are continuous functions satisfying the constraints

$$0 \leq w_{ij} \leq 1. \tag{4.44}$$

G5. For every $j = 1, \ldots, N$, the function $p_j(G_j)G_j$ is concave.

Now differentiate the expected profit function (4.42) by η_{ij} and obtain

$$\frac{\partial \mu_i}{\partial \eta_{ij}}(\eta_i) = \mu_{ij}'(\eta_i) = \left[p_j\left(G_{ij}(\eta_{ij})\right) + \eta_{ij} w_{ij} p_j'\left(G_{ij}(\eta_{ij})\right) - c_{ij}\right] -$$

$$-f_i'\left(\sum_{j=1}^{N} \eta_{ij}\right). \tag{4.45}$$

If $\eta_{ij} = s_{ij}$, the derivative of the function μ_i can be written in the form

$$\mu_{ij}'(S_i) = \left[p_j\left(G_j\right) + s_{ij} w_{ij} p_j'\left(G_j\right) - c_{ij}\right] - f_i'\left(q_i\right). \tag{4.46}$$

Remark. In the network models with fixed transportation routes it is frequently supposed that the quotients c_{ij} are not constants but (non-decreasing) functions of the transported volumes s_{ij}; *cf.* Dafermos and Nagurney (1986), Nagurney e.a. (1992), Nagurney (1990). However, we do not use this assumption here not to exclude the possibility for different suppliers to share the same transportation routes.

Making use of the first order extremum conditions for the expected profit functions of each agent, we specify the *extended network equilibrium problem* as follows. Given $Q \geq 0$, find a set of $M+1$ nonnegative vectors $(G, S_1, \dots, S_M)$ such that

$$\sum_{i=1}^{M} s_{ij} + Q_j = G_j, \quad j = 1, \dots, N, \tag{4.47}$$

$$\sum_{j=1}^{N} s_{ij} = q_i, \quad i = 1, \dots, M, \tag{4.48}$$

and for each pair (i, j) the relationships hold:

$$\mu'_{ij}(S_i) = \left[p_j(G_j) + s_{ij} w_{ij} p'_j(G_j) - c_{ij}\right] - f'_i(q_i) \leq 0, \quad s_{ij}\mu'_{ij} = 0. \tag{4.49}$$

In order to treat the solution to complementarity problem (4.47)-(4.49) as an equilibrium for the agents with $q_i > 0$, it suffices to check the concavity of the expected profit function (4.42) with respect to η_i. In the other words, we show that the derivative mapping $\mu'_i : R^N \to R^N$ with the components $\mu'_{ij}(\eta_i)$ is antitone over R^N_+. Indeed, consider $\eta^1_i, \eta^2_i \in R^N_+$ and examine the scalar product

$$\left[\mu'_i(\eta^1_i) - \mu'_i(\eta^2_i)\right]^T (\eta^1_i - \eta^2_i) =$$

$$= \sum_{j=1}^{N} \left[p_j\left(G_{ij}\left(\eta^1_{ij}\right)\right) - p_j\left(G_{ij}\left(\eta^2_{ij}\right)\right)\right] \cdot \left(\eta^1_{ij} - \eta^2_{ij}\right) +$$

$$+ \sum_{j=1}^{N} w_{ij}\left[\eta^1_{ij} p'_j\left(G_{ij}\left(\eta^1_{ij}\right)\right) - \eta^2_{ij} p'_j\left(G_{ij}\left(\eta^2_{ij}\right)\right)\right] \cdot \left(\eta^1_{ij} - \eta^2_{ij}\right) -$$

$$- \left[f'_i\left(\sum_{j=1}^{N} \eta^1_{ij}\right) - f'_i\left(\sum_{j=1}^{N} \eta^2_{ij}\right)\right] \cdot \left(\sum_{j=1}^{N} \eta^1_{ij} - \sum_{j=1}^{N} \eta^2_{ij}\right). \tag{4.50}$$

The latter term in (4.50) is nonpositive due to the convexity of f_i. As for the former two sums, we investigate each term in them separately. Fix an arbitrary j and assume for certainty that $\eta^1_{ij} \leq \eta^2_{ij}$. Then $G_{ij}\left(\eta^1_{ij}\right) \leq G_{ij}\left(\eta^2_{ij}\right)$, and condition G2 (about p_j being monotone non-increasing functions) implies that

$$\left[p_j\left(G_{ij}\left(\eta^1_{ij}\right)\right) - p_j\left(G_{ij}\left(\eta^2_{ij}\right)\right)\right] \cdot \left(\eta^1_{ij} - \eta^2_{ij}\right) \leq 0. \tag{4.51}$$

Besides, if $p'_j\left(G_{ij}\left(\eta^1_{ij}\right)\right) \geq p'_j\left(G_{ij}\left(\eta^2_{ij}\right)\right)$, then $\eta^1_{ij} p'_j\left(G_{ij}\left(\eta^1_{ij}\right)\right) \geq$ $\geq \eta^2_{ij} p'_j\left(G_{ij}\left(\eta^2_{ij}\right)\right)$ since the derivative p'_j is nonpositive; hence

$$w_{ij}\left[\eta^1_{ij} p'_j\left(G_{ij}\left(\eta^1_{ij}\right)\right) - \eta^2_{ij} p'_j\left(G_{ij}\left(\eta^2_{ij}\right)\right)\right] \cdot \left(\eta^1_{ij} - \eta^2_{ij}\right) \leq 0. \tag{4.52}$$

Otherwise, if $p'_j\left(G_{ij}\left(\eta^1_{ij}\right)\right) < p'_j\left(G_{ij}\left(\eta^2_{ij}\right)\right)$, we make use of (4.41) and transform the sum of the first factors in the left-hand sides of (4.51) and (4.52) as follows:

$$\begin{aligned}&\left[p_j\left(G_{ij}\left(\eta_{ij}^1\right)\right)-p_j\left(G_{ij}\left(\eta_{ij}^2\right)\right)\right]+\\&+\left[G_{ij}\left(\eta_{ij}^1\right)p_j'\left(G_{ij}\left(\eta_{ij}^1\right)\right)-G_{ij}\left(\eta_{ij}^2\right)p_j'\left(G_{ij}\left(\eta_{ij}^2\right)\right)\right]+\\&+\left(G_j-w_{ij}s_{ij}\right)\left[p_j'\left(G_{ij}\left(\eta_{ij}^2\right)\right)-p_j'\left(G_{ij}\left(\eta_{ij}^1\right)\right)\right].\end{aligned} \tag{4.53}$$

The sum of the first two terms in the square brackets in (4.53) is nonnegative due to G5, and the sign of the last term coincides (as (4.43) and (4.47) imply that) with that of the difference $p_j'\left(G_{ij}\left(\eta_{ij}^2\right)\right)-p_j'\left(G_{ij}\left(\eta_{ij}^1\right)\right)$, which is positive according to our assumption. Thus we have shown that the sum of the left-hand sides of (4.51) and (4.52) is nonpositive in all cases which implies that term (4.50) is nonpositive, too. But the latter means that the derivative mapping μ_i' is antitone over R_+^N, and consequently, the expected profit function μ_i is concave with respect to $\eta_{i1},\dots,\eta_{iN}$.

In order to justify our definition of equilibrium from the point of view of agents with $q_i=0$, we remark that the necessary and sufficient maximum condition (4.49) for the expected profit function can be rearranged in the following form

$$p_j\left(G_j\right)-c_{ij}-f_i'\left(q_i\right)\le -s_{ij}w_{ij}p_j'\left(G_j\right). \tag{4.54}$$

The right-hand side in (4.54) can be treated as a threshold that the "price–cost" difference for an extra product unit should overcome, in order to make agent i increase his output.

Since the right-hand side in (4.54) vanishes as q_i tends to zero, it is natural enough to postulate the following behavior of the passive agent (with $q_i=0$). Namely, he has an incentive to start producing only if, at least for one j, the difference between the price $p_j(G_j)$ and his expenditures $f_i'(0)+c_{ij}$ is positive, i.e. $p_j(G_j)-c_{ij}-f_i'(0)>0$. But condition (4.54) at $q_i=0$ is exactly opposite to that, hence agent i will not produce.

4.3.3 EXISTENCE AND UNIQUENESS THEOREMS

In order to establish the existence of an equilibrium defined in the previous subsection, we are going to use the theorem of alternative proven in Chapter 1 of this book. We first recall the needed results of that chapter applied to the considered case. Namely, let us select the system of closed balls $\{B_r\}$ centered at the origin of R^n and having radius $r>0$ as the system Σ. It is easy to see that thus selected system Σ satisfies assumptions B1–B3. It is also evident that the following notion of an exceptional family of elements is a particular case of the concept introduced by Definition 1.6.

DEFINITION 4.2. For a mapping $F: R_+^n\to R^n$, we refer to a family of elements $\{t^r\}_{r>0}\subset R_+^n$ as *exceptional*, if:

$$(i)\qquad \|t^r\|\to\infty \quad\text{as}\quad r\to+\infty;$$

$$(ii)\qquad \forall r>0\ \exists\mu_r>0\ \text{ such that }\ \begin{cases}F_i(t^r)=-\mu_r t_i^r, & \text{if } t_i^r>0;\\ F_i(t^r)\ge 0, & \text{if } t_i^r=0.\end{cases}$$

Now consider the standard complementarity problem for the mapping F: find a vector $x^* \in R^n$ such that

$$x^* \geq 0, \quad F(x^*) \geq 0, \quad \text{and} \quad (x^*)^T F(x^*) = 0. \tag{4.55}$$

Here the inequality $x \geq 0$ for the vector $x \in R^n$ means that $x_i \geq 0$ for all $i = 1, \ldots, n$.

The theorem below is a corollary of Proposition 1.5, as the single-valued mapping F clearly satisfies assumptions A1–A3.

THEOREM OF ALTERNATIVE. *For each continuous mapping $F : R^n_+ \to R^n$, either complementarity problem* (4.55) *is solvable, or there exists an exceptional family of elements (EFE).* ■

Making use of this theorem one easily obtains

THEOREM 4.7. *If a continuous mapping $F : R^n_+ \to R^n$ has no exceptional family of elements, then problem* (4.55) *with F is solvable.* ■

COROLLARY 4.2. *Consider a continuous mapping $F : R^n_+ \to R^n$ and a bounded subset $C \subset R^n_+$ such that $F(y) \geq 0 \ \forall y \in R^n_+ \setminus C$. Then problem* (4.55) *is solvable. Moreover, if $0 \in C$ and the strict inequality $F(y) > 0$ holds for every $y \in R^n_+ \setminus C$, then all the solutions of problem* (4.55) *belong to C.*

PROOF. It is not difficult to show that the conditions of the corollary exclude the possibility of existence of an exceptional family of elements for the mapping F, which implies the solvability of problem (4.55) due to Theorem 4.7 and completes the proof. ■

Now we are able to establish the existence of equilibrium (4.47)–(4.49).

LEMMA 4.4. *Consider a family of elements $s^r = \{s^r_{ij}\}_{i=1}^{M}{}_{j=1}^{N} \subset R^{MN}_+$ such that*

$$\|s^r\| = \left[\sum_{i=1}^{M}\sum_{j=1}^{N}(s^r_{ij})^2\right]^{1/2} \to \infty \quad as \quad r \to +\infty. \tag{4.56}$$

Then there exist numbers $i' \in \{1, \ldots, M\}$ and $j' \in \{1, \ldots, N\}$ such that

$$\text{(i)} \quad s^r_{i'j'} \to +\infty \ as \ r \to +\infty; \tag{4.57}$$

$$\text{(ii)} \quad \sum_{j=1}^{N} s^r_{i'j} \to +\infty \ as \ r \to +\infty;$$

$$\text{(iii)} \quad \sum_{i=1}^{M} s^r_{ij'} \to +\infty \ as \ r \to +\infty.$$

PROOF. We prove only assertion (i) as (ii) and (iii) follow from (i) and from the nonnegativity of s^r_{ij}. On the contrary, assume that there exists $B > 0$ such that for each $i = 1, \ldots, M$ and $j = 1, \ldots, N$, the inequality $s_{ij}^r \leq B$ holds for all $r > 0$. The latter implies

$$\sum_{i=1}^{M}\sum_{j=1}^{N}(s_{ij}^r)^2 \leq MNB^2 \quad \forall r > 0,$$

which contradicts (4.56).The proof is complete. ■

THEOREM 4.8. *Let assumptions* G1–G4 *be valid. Then problem* (4.47)–(4.49) *is solvable.*

PROOF. First of all, we note that the equilibrium problem (4.47)–(4.49) is equivalent to complementarity problem (4.55) with $n = MN$, $x = \{s_{ij}\}$, and $F = \{F_{ij}\}$, where

$$F_{ij} = f_i'(q_i) + c_{ij} - p_j(G_j) - w_{ij}s_{ij}p_j'(G_j)$$

is continuous due to assumptions G1–G4. In the other words, we have to show the solvability of the problem: find $\{s_{ij}\} \subset R_+^{MN}$ such that

$$F_{ij} \geq 0, \quad s_{ij}F_{ij} = 0, \quad i = 1, \ldots, M; \; j = 1, \ldots, N, \tag{4.58}$$

where q_i and G_j are defined by (4.47) and (4.48), respectively. In order to do that we can apply Theorem 4.7 if we establish the absence of exceptional families of elements.

On the contrary, suppose that there exists an exceptional family of elements $s^r = \{s_{ij}^r\} \subset R_+^{MN}$ satisfying the conditions of Definition 4.2. As $\|s^r\| \to \infty$ when $r \to +\infty$, Lemma 4.4 implies the existence of numbers i' and j' such that relationships (i)–(iii) in (4.57) hold. Without affecting generality assume that $i' = 1$, $j' = 1$. Consider the entry

$$F_{11}^r = F_{11}(s^r) = f_1'(q_1^r) + c_{11} - p_1(G_1^r) - s_{ij}^r w_{ij} p_1'(G_1^r)$$

where $q_1^r = \sum_{j=1}^{N} s_{1j}^r$ and $G_1^r = Q_1 + \sum_{i=1}^{M} s_{i1}^r$. On one hand, assumptions G1–G4 together with (4.47)–(4.49) imply existence of $r_0 > 0$ such that $F_{11}^r \geq 0$ for $r > r_0$. On the other hand, from (4.57) and Definition 4.2 it follows that $F_{11}^r = -\mu_r s_{11}^r < 0$ for r greater than a certain $r_1 > 0$. This contradiction denies the existence of any exceptional family of elements and completes the proof. ■

In order to establish the uniqueness of equilibrium (4.47)–(4.49), we have to sharpen our assumptions. Namely, we assume the inverse demand functions p_j to be strictly decreasing, and we restrict ourselves to the case when the influence coefficients w_{ij} are positive constants. More precisely, we assume the following.

G6. For each $j = 1, \ldots, N$, the inverse demand function p_j has negative derivative $p_j'(G_j) < 0$ for every nonnegative G_j. Moreover, the influence coefficients w_{ij} are constants with the property $0 < w_{ij} \leq 1$ for each pair (i, j).

Now the question of uniqueness of the equilibrium can be reduced to that of the equilibrium total bargain volumes G_j. Indeed, fix an N-tuple $G = (G_1, \ldots, G_N) > 0$ and consider problem (4.49): find a (unique) solution $S(G) = (S_1, \ldots, S_M) \in R_+^{MN}$ of a complementarity problem:

$$s_{ij} \geq 0, \ \varphi_{ij} = f_i'\left(\sum_{k=1}^{N} s_{ik}\right) + c_{ij} - w_{ij} s_{ij} p_j'(G_j) - p_j(G_j) \geq 0,$$

$$\text{and} \quad s_{ij}\varphi_{ij} = 0, \ i = 1, \ldots, M; \ j = 1, \ldots, N. \tag{4.59}$$

Underline that in this problem, G_j and s_{ij} are not bound by constraint (4.47).

Under assumptions G1, G2, G4, G6, problem (4.59) has a unique solution, since the mapping $\varphi = (\varphi_{ij})_{i=1,j=1}^{M \ \ N}$ is strongly monotone over R_+^{MN}. Indeed,

$$\left[\varphi(S^1) - \varphi(S^2)\right]^T (S^1 - S^2) = \sum_{i=1}^{M}\sum_{j=1}^{N} \left[\varphi_{ij}(S^1) - \varphi_{ij}(S^2)\right] \cdot \left(s_{ij}^1 - s_{ij}^2\right) =$$

$$= \sum_{i=1}^{M} \left[f_i'\left(\sum_{k=1}^{N} s_{ik}^1\right) - f_i'\left(\sum_{k=1}^{N} s_{ik}^2\right)\right]\left(\sum_{j=1}^{N} s_{ij}^1 - \sum_{j=1}^{N} s_{ij}^2\right) -$$

$$- \sum_{i=1}^{M}\sum_{j=1}^{N} w_{ij} p_j'(G_j) \cdot \left(s_{ij}^1 - s_{ij}^2\right)^2 \geq \alpha \|S^1 - S^2\|^2$$

due to assumptions G1, G2 and G4; here $\alpha = \min_{ij}\left(-w_{ij} p_j'(G_j)\right) > 0$ due to G6. Thus we have established the following assertion.

LEMMA 4.5. *Given G_j, $j = 1, \ldots, n$, relationships (4.59) determine the values of variables s_{ij} uniquely.* ■

In order to examine the uniqueness of the total sales at the markets, we keep on assumptions G1–G6, and add the following one.

G7. Functions f_i, p_j are twice differentiable, and for all $G_j > 0$ the following inequality holds

$$p_j''(G_j)G_j - 2p_j'(G_j) \geq 0. \tag{4.60}$$

DEFINITION 4.3. An equilibrium $Z(Q) = \left[G(Q), S(Q)\right]$ is referred to as *non-monopolistic* for market j, if $s_{ij} < G_j$, $i = 1, \ldots, M$.

THEOREM 4.9. *Let assumptions* G1–G7 *be valid. Then there can be at most one equilibrium $Z(Q) = \left[G(Q), S(Q)\right]$ that is non-monopolistic for each market j.*

PROOF. Consider the operator $\psi : R^{MN} \to R_+^{MN}$ with the components

$$\psi_{ij} = f_i'\left(\sum_{r=1}^{N} s_{ir}\right) + c_{ij} - w_{ij} s_{ij} p_j'\left(Q_j + \sum_{t=1}^{M} s_{tj}\right) - p_j\left(Q_j + \sum_{t=1}^{M} s_{tj}\right),$$

$i = 1, \ldots, M$, $j = 1, \ldots, N$, and calculate the entries of its Jacobi matrix $D = \left(d_{ij}^{k\ell}\right)$, omitting, for brevity purpose, the arguments of the functions f_i'', p_j', and p_j'':

$$d_{ij}^{ij} = \frac{\partial \psi_{ij}}{\partial s_{ij}} = f_i'' - (1 + w_{ij})p_j' - w_{ij} s_{ij} p_j''; \tag{4.61}$$

$$d_{ij}^{i\ell} = \frac{\partial \psi_{ij}}{\partial s_{i\ell}} = f_i'', \qquad \text{if} \qquad \ell \neq j; \tag{4.62}$$

$$d_{ij}^{kj} = \frac{\partial \psi_{ij}}{\partial s_{kj}} = -p_j' - w_{ij} s_{ij} p_j'', \qquad \text{if} \qquad k \neq i; \tag{4.63}$$

$$d_{ij}^{kl} = \frac{\partial \varphi_{ij}}{\partial s_{k\ell}} = 0, \qquad \text{if} \qquad k \neq i, \ \ell \neq j. \tag{4.64}$$

Now we prove that the matrix D is positive definite. For an arbitrary collection $\alpha = (\alpha_{ij})$, write down the product

$$\alpha^T D \alpha = \sum_{i,k=1}^{M} \sum_{j,\ell=1}^{N} \alpha_{k\ell} d_{ij}^{k\ell} \alpha_{ij}. \tag{4.65}$$

Making use of (4.61)–(4.64) and introducing

$$\beta_j = \sum_{i=1}^{M} \alpha_{ij}, \tag{4.66}$$

rearrange (4.65) as follows

$$\alpha^T D \alpha = \sum_{i=1}^{M} \left[f_i'' \cdot \left(\sum_{j=1}^{N} \alpha_{ij} \right)^2 \right] - \sum_{j=1}^{N} B_j, \tag{4.67}$$

where each B_j, $j = 1, \ldots, N$, can be represented in the form

$$B_j = p_j' \sum_{i=1}^{M} w_{ij} \alpha_{ij}^2 + \beta_j p_j'' \sum_{i=1}^{M} w_{ij} s_{ij} \alpha_{ij} + p_j' \beta_j^2. \tag{4.68}$$

In order to show each B_j to be nonpositive, solve the extremal problem

$$B_j = B_j(\alpha_{1j}, \ldots, \alpha_{Mj}) \longrightarrow \max, \tag{4.69}$$

subject to condition (4.66), by the standard Lagrange multipliers techniques. Consider the Lagrange function of problem (4.66), (4.69)

$$\mathcal{L}(\alpha_{1j}, \ldots, \alpha_{Mj}) = B_j(\alpha_{1j}, \ldots, \alpha_{Mj}) + \lambda \cdot \left(\sum_{i=1}^{M} \alpha_{ij} - \beta_j \right)$$

and set its partial derivatives equal to zero:

$$\frac{\partial \mathcal{L}}{\partial \alpha_{ij}} = 2p_j' w_{ij} \alpha_{ij} + \beta_j p_j'' w_{ij} s_{ij} + \lambda = 0, \ i = 1, \ldots, M. \tag{4.70}$$

Solving the linear equations with respect to α_{ij}, find

$$\alpha_{ij} = -\frac{\lambda + \beta_j p_j'' s_{ij} w_{ij}}{2p_j' w_{ij}}, \ i = 1, \ldots, M. \tag{4.71}$$

Note that B_j are concave with respect to α_{ij} due to the inequality $p_j' < 0$. Therefore, by substituting the terms from (4.71) into (4.68), we find the maximal value $\hat{B}_j = \max B_j$ of the objective function of problem (4.69) subject to (4.66):

$$\hat{B}_j = \sum_{i=1}^{M} \left[\frac{\lambda^2 + 2\lambda\beta_j p_j'' w_{ij} s_{ij} + \beta_j^2 \left(p_j''\right)^2 w_{ij}^2 s_{ij}^2}{4w_{ij}p_j'} - \right.$$

$$\left. -\beta_j \frac{\lambda s_{ij} p_j''}{2p_j'} - \beta_j^2 \frac{w_{ij} s_{ij}^2 \left(p_j''\right)^2}{2p_j'} \right] + p_j' \beta_j^2 =$$

$$= \frac{\lambda^2}{4p_j'} \sum_{i=1}^{M} \frac{1}{w_{ij}} - \beta_j^2 \frac{\left(p_j''\right)^2}{4p_j'} \sum_{i=1}^{M} w_{ij} s_{ij}^2 + p_j' \beta_j^2. \tag{4.72}$$

Again make use of the negativity of p_j', the upper bound 1 of the quotients w_{ij}, and assumption that market j is non-monopolistic. Combined with the nonnegativity of s_{ij}, the latter assumption implies the strict inequality

$$\sum_{i=1}^{M} s_{ij}^2 < \left(\sum_{i=1}^{M} s_{ij} + Q_j \right)^2 = G_j^2. \tag{4.73}$$

Omitting the (nonpositive and even strictly negative when $\lambda \neq 0$) first term in the righthand side of (4.72) and making use of assumptions G5 and G7, finally come to the estimates

$$\hat{B}_j \le -\beta_j^2 \frac{\left(p_j''\right)^2}{4p_j'} \sum_{i=1}^{M} s_{ij}^2 + p_j' \beta_j^2 \le \frac{\beta_j^2}{4p_j'} \left[\left(2p_j'\right)^2 - G_j^2 \left(p_j''\right)^2 \right] =$$

$$= \frac{\beta_j^2}{4p_j'} \left(2p_j' - G_j p_j''\right) \cdot \left(2p_j' + G_j p_j''\right) \le 0; \tag{4.74}$$

moreover, (4.73) implies the strict inequality $\hat{B}_j < 0$, if $\beta_j \neq 0$ or $\lambda \neq 0$. Remark that if $\beta_j = 0$ and $\lambda = 0$, then from (4.71) it follows that $\alpha_{ij} = 0$, $i = 1, \ldots, M$.

Coming back to equality (4.67), we deduce that

$$\alpha^T D \alpha = \sum_{i=1}^{M} \left[f_i'' \cdot \left(\sum_{j=1}^{N} \alpha_{ij} \right)^2 \right] - \sum_{j=1}^{N} B_j \ge 0 \tag{4.75}$$

due to convexity of the functions f_i and to inequalities (4.74). Besides, the inequality in (4.75) is strict unless all of the α_{ij}, $i = 1, \ldots, M$, $j = 1, \ldots, N$ are equal to zero.

Therefore, the matrix D is positive definite, hence the mapping of the complementarity problem (4.47)–(4.49) is strongly monotone. The latter, together with Lemma 4.5, implies that there can exist at most one equilibrium with non-monopolistic markets. The proof is complete. ■

Remark. If $Q_j > 0$, $j = 1, \ldots, N$, then every equilibrium is non-monopolistic. Consequently, in this case, the equilibrium is determined uniquely, according to Theorem 4.9. In other cases, in general, it is possible that a single network has both monopolistic and non-monopolistic equilibria with different sales at the markets (for instance, it can be the case if the functions f_i and $p_j(G)G$ are piecewise linear).

Remark. Inequality (4.60) in Assumption G7 is valid, for example, if function p_j satisfies Assumption G2 and is convex.

4.3.4 MODEL WITH A COMMON MARKET OF PRODUCTION FACTORS

In contrast to Subsection 4.3.2, here we suppose that the cost function f_i of the i-th producer depends upon not only his output $q_i = \sum\limits_{j=1}^{N} s_{ij}$ but also the aggregate output T by all the producers:

$$T = \sum_{i=1}^{M} \sum_{j=1}^{N} s_{ij}. \tag{4.76}$$

For example, this dependence may be of the form

$$f_i(q_i, T) = q_i h_i(T), \tag{4.77}$$

where $h_i(T)$ is the producer i expenditure for production of a unit of a good under the assumption that the *aggregate output* by all the producers is equal to T. In what follows, however, we do not suppose that function f_i has to be represented by equality (4.77).

We keep valid the hypotheses of Subsection 4.3.2 and add the assumption that each producer is aware of his cost as a function of the aggregate output by all the producers. Hence making use of his influence quotients w_{ij} he can predict variations of his expenditures as a result of varation of his output. Namely, agent i predicts the variation of the aggregate output T according to the rule

$$T_i(\eta_i) = T + \sum_{j=1}^{N} w_{ij} \left(\eta_{ij} - s_{ij}\right); \tag{4.78}$$

here, as above, $\eta_i = \left(\eta_{ij}\right)_{j=1}^{N}$. Then the i-th agent expects his profit to constitute

$$\mu_i(\eta_i) = \sum_{j=1}^{N} \eta_{ij} \left[p_j\left(G_j(\eta_{ij})\right) - c_{ij}\right] - f_i\left(\sum_{j=1}^{N} \eta_{ij}, T_i(\eta_i)\right), \tag{4.79}$$

and he makes his choice in order to maximize the value of function (4.79).

Like in Subsection 4.3.2, we define an equilibrium through the first order optimality condition for the expected profit function (4.79). First of all, list the assumptions concerning functions f_i and p_j.

H1. Each of the functions $f_i(q_i, T)$ are defined over R^2_+, continuously differentiable, convex by the pair of variables, and non-decreasing by T.

H3. There exists such scalar $a > 0$ that

$$\frac{\partial f_i}{\partial q_i}(0, T) \geq a \quad \text{at every} \quad T \geq 0. \tag{4.80}$$

H5. For each $j = 1, \ldots, N$, function $p_j(G_j)G_j$ is bounded over $G_j \in [0, +\infty)$.

H2 and **H4** coincide with assumptions **G2** and **G4**, respectively.

Differentiating the expected profit function (4.79) with respect to η_{ij}, we get

$$\frac{\partial \mu_i}{\partial \eta_{ij}}(\eta_i) = \mu'_{ij}(\eta_i) = \left[p_j\left(G_j(\eta_{ij})\right) + \eta_{ij} w_{ij} p'_j\left(G_j(\eta_{ij})\right) - c_{ij}\right] -$$

$$-\frac{\partial f_i}{\partial q_i}\left(\sum_{j=1}^{N} \eta_{ij}, T_i(\eta_i)\right) - w_{ij}\frac{\partial f_i}{\partial T}\left(\sum_{j=1}^{N} \eta_{ij}, T_i(\eta_i)\right). \tag{4.81}$$

At the point $S = (S_i)_{i=1}^{M}$, the derivative of function μ_i can be represented in the form

$$\mu'_{ij}(S) = \left[p_j\left(G_j\right) + s_{ij} w_{ij} p'_j\left(G_j\right) - c_{ij}\right] -$$

$$-\frac{\partial f_i}{\partial q_i}\left(q_i, T\right) - w_{ij}\frac{\partial f_i}{\partial T}\left(q_i, T\right). \tag{4.82}$$

Now making use of the first order optimality conditions for the expected profit functions of each agent, we specify the equilibrium problem as follows. Given $Q_j \geq 0$, $j = 1, \ldots, N$, find a set of nonnegative vectors $(G, S_1, \ldots, S_M)$ such that

$$\sum_{i=1}^{M} s_{ij} + Q_j = G_j, \quad j = 1, \ldots, N, \tag{4.83}$$

$$\sum_{j=1}^{N} s_{ij} = q_i, \quad i = 1, \ldots, M, \tag{4.84}$$

$$\sum_{i=1}^{M}\sum_{j=1}^{N} s_{ij} = T, \tag{4.85}$$

and for each pair (i, j) the following conditions hold

$$\mu'_{ij}(S) = \left[p_j(G_j) + s_{ij}w_{ij}p'_j(G_j) - c_{ij}\right] - \frac{\partial f_i}{\partial q_i}(q_i, T) -$$
$$-w_{ij}\frac{\partial f_i}{\partial T}(q_i, T) \le 0, \quad s_{ij}\mu'_{ij}(S) = 0. \tag{4.86}$$

In order to treat a solution to complementarity problem (4.85) – (4.86) as an equilibrium for agents with $q_i > 0$, it again suffices to verify the concavity of the expected profit function (4.79) with respect to η_i. But the latter is established immediately since the first sum in (4.79) coincides exactly with that of Subsection 4.3.2, whereas the second term is a concave function being the superposition of a concave and a linear ones.

THEOREM 4.10. *Let assumptions* H1–H5 *be valid. Then problem* (4.83)–(4.86) *is solvable.*

PROOF. According to H5, there exists a constant B such that

$$\sum_{j=1}^{N} p_j(G_j)G_j \le B$$

for all $G_j \ge 0$. Estimate the sum

$$V = \sum_{i=1}^{M}\sum_{j=1}^{N} \mu'_{ij}(S)s_{ij}. \tag{4.87}$$

Due to (4.86) and assumptions H1–H5,

$$V = \sum_{i=1}^{M}\sum_{j=1}^{N}\left\{ \left[p_j(G_j) + s_{ij}w_{ij}p'_j(G_j) - c_{ij}\right] - \right.$$
$$\left. -\frac{\partial f_i}{\partial q_i}(q_i, T) - w_{ij}\frac{\partial f_i}{\partial T}(q_i, T)\right\} s_{ij} \le$$
$$\le \sum_{j=1}^{N} p_j(G_j)G_j - \sum_{i=1}^{M} q_i \frac{\partial f_i}{\partial q_i}(q_i, T) \le B - a\sum_{i=1}^{M}\sum_{j=1}^{N} s_{ij}. \tag{4.88}$$

Thus, $V < 0$ if the vector $\{s_{ij}\}_{i,j=1}^{M,N}$ belongs to the subset

$$\left\{ s_{ij} \ge 0 \mid \sum_{i=1}^{M}\sum_{j=1}^{N} s_{ij} > \frac{B}{a} \right\}. \tag{4.89}$$

According to Corollary 4.2 of the previous Subsection, there exists a solution to the complementarity problem

$$\varphi_{ij}(S) = -\mu'_{ij}(S) \ge 0, \ s_{ij} \ge 0, \ -s_{ij}\mu'_{ij}(S) = 0. \tag{4.90}$$

The proof is complete. ■

4.3.5 HUMAN MIGRATION MODEL

In this subsection, the human migration model examined in Nagurney (1992) is generalized by introducing the influence factor, and then reduced to the network oligopoly model discussed in previous subsections.

As in Nagurney (1992), consider a closed economy in which there are n locations, typically denoted by i, and J classes, typically denoted by k. Further, assume that the attractiveness of any location i as perceived by class k is represented by a utility u_i^k. Let $\bar{Q}_i^k$ denote the initial fixed population of class k in location i, and let Q_i^k denote the population of class k in location i. Group the utilities into a row vector $u \in R^{J\times n}$ and the populations into a column vector $Q \in R^{J\times n}$. As in Nagurney (1990), assume the situation in which there are no births and no deaths in the economy.

Associate with each class k and each pair of locations i, j a nonnegative cost of migration c_{ij}^k and let the migration flow of class k from origin i to destination j be denoted by s_{ij}^k. The migration costs are grouped into a row vector $c \in R^{J\times n\times(n-1)}$ and the flows into a column vector $s \in R^{J\times n\times(n-1)}$. Assume that the migration cost reflects not only the cost of physical movement but also the personal and psychic cost as perceived by a class in moving between locations.

The conservation of flow equations, given for each class k and each location i, assuming no repeat or chain migration, are

$$Q_i^k = \bar{Q}_i^k + \sum_{\ell\neq i} s_{\ell i}^k - \sum_{\ell\neq i} s_{i\ell}^k \tag{4.91}$$

and

$$\sum_{\ell\neq i} s_{i\ell}^k \leq \bar{Q}_i^k, \tag{4.92}$$

with $s_{i\ell}^k \geq 0$, $\forall k = 1,\ldots,J$, $\ell \neq i$. Denote

$$K = \{(Q,s) \mid s \geq 0,\ (Q,s) \text{ satisfy (4.91),(4.92)}\}.$$

Equation (4.91) states that the population at location i of class k is given by the initial population of class k at location i plus the migration flow into i of that class minus the migration flow out of i for that class. Equation (4.92) affirms that the flow out of i by class k cannot exceed the initial population of class k at i, since no chain migration is allowed.

We are now ready to state the equilibrium conditions. As in Nagurney (1992), assume that migrants are rational and that migration continue until no individual has any incentive to move since a unilateral decision will no longer yield a positive net gain (the gain in expected utility minus the migration cost). In contrast to Nagurney (1992,1999), here we assume that the utility $u_i^k = u_i^k(Q_i^k)$ is a differentiable and monotone decreasing function of the total population Q_i^k of class k at location i, for all $i = 1,\ldots,n$ and $k = 1,\ldots,J$. Moreover, we assume that each individual of class k, when considering his possibility of moving from location i to location j, takes into account not only the current utility value $u_j^k = u_j^k(Q_j^k)$ for location j but also the expected (negative) increment $s_{ij}^k w_{ij}^k (u_j^k)'$ of the utility caused by the migration of s_{ij}^k representatives of class k from location i to location j. Here $w_{ij}^k \geq 0$ is an influence quotient taken in account by the individual of class k moving from i to j; this quotient is defined by his assumption that after moving of s_{ij}^k individuals of class k from i to j the total population of class k at j will become equal to $\bar{Q}_j^k + w_{ij}^k s_{ij}^k$.

Therefore, the multiclass equilibrium conditions are stated as follows. A multiclass population and flow pattern $(Q^*, s^*) \in K$ is in *equilibrium*, if for each class $k = 1, \ldots, J$, and for each pair of locations $i, j = 1, \ldots, n; j \neq i$:

$$u_i^k + c_{ij}^k \begin{cases} = u_j^k + {s_{ij}^k}^* w_{ij}^k (u_j^k)' - \lambda_i^k, & \text{if } {s_{ij}^k}^* > 0; \\ \geq u_j^k + {s_{ij}^k}^* w_{ij}^k (u_j^k)' - \lambda_i^k, & \text{if } {s_{ij}^k}^* = 0; \end{cases} \tag{4.93}$$

and

$$\lambda_i^k \begin{cases} \geq 0, & \text{if } \sum_{\ell \neq i} {s_{i\ell}^k}^* = \bar{Q}_i^k; \\ = 0, & \text{if } \sum_{\ell \neq i} {s_{i\ell}^k}^* < \bar{Q}_i^k. \end{cases} \tag{4.94}$$

Assume, as in Nagurney (1992, 1999), that the utility associated with a particular location and class can depend upon the population associated with every class and every location, that is,

$$u = u(Q). \tag{4.95}$$

Assume also that the cost associated with migrating between two locations as perceived by a particular class can depend, in general, upon the flow of every class between every pair of location, that is

$$c = c(s). \tag{4.96}$$

Finally, group the terms ${s_{ij}^k}^* w_{ij}^k (u_j^k)'(Q^*)$ into a row vector $d \in R^{J \times n \times (n-1)}$. The variational inequality formulation of the migration equilibrium condition is then given by:

THEOREM 4.11. *A population and migration flow pattern* $(Q^*, s^*) \in K$ *satisfies equilibrium conditions* (4.93) *and* (4.94) *if and only if it satisfies the variational inequality problem*

$$\langle -u(Q^*), Q - Q^* \rangle + \langle c(s^*) - d(s^*), s - s^* \rangle \geq 0, \quad \forall (Q, s) \in K. \tag{4.97}$$

PROOF. We must prove that if a pattern $(Q^*, s^*) \in K$ satisfies equilibrium conditions (4.93) and (4.94) (subject to constraints (4.91) and (4.92)), then it also satisfies the variational inequality in (4.97). Suppose that (4.93) and (4.94) hold. Then

$${s_{ij}^k}^* \geq 0 \quad \text{and} \quad \sum_{\ell \neq i} {s_{i\ell}^k}^* \leq \bar{Q}_i^k, \ \forall i, j, k.$$

For a fixed class k we define $\Gamma_1^k = \Gamma_1^k(i) = \{\ell \mid s_{i\ell}^{k^*} > 0\}$ and $\Gamma_2^k = \Gamma_2^k(i) = \{\ell \mid s_{i\ell}^{k^*} = 0\}$. Then

$$\sum_{\ell \neq i} \left[u_i^k(Q^*) + c_{i\ell}^k(s^*) - u_\ell^k(Q^*) - s_{i\ell}^{k^*} w_{i\ell}^k (u_\ell^k)'(Q^*)\right] \times \left[s_{i\ell}^k - s_{i\ell}^{k^*}\right] =$$

$$\sum_{\ell \in \Gamma_1^k} \left[u_i^k(Q^*) + c_{i\ell}^k(s^*) - u_\ell^k(Q^*) - s_{i\ell}^{k^*} w_{i\ell}^k (u_\ell^k)'(Q^*)\right] \times \left[s_{i\ell}^k - s_{i\ell}^{k^*}\right] +$$

$$\sum_{\ell \in \Gamma_2^k} \left[u_i^k(Q^*) + c_{i\ell}^k(s^*) - u_\ell^k(Q^*) - s_{i\ell}^{k^*} w_{i\ell}^k (u_\ell^k)'(Q^*)\right] \times \left[s_{i\ell}^k - s_{i\ell}^{k^*}\right] \geq$$

$$-\lambda_i^k \sum_{\ell \in \Gamma_1^k} \left(s_{i\ell}^k - s_{i\ell}^{k^*}\right) + \sum_{\ell \in \Gamma_2^k} (-\lambda_i^k) s_{i\ell}^k =$$

$$-\lambda_i^k \left(\sum_{\ell \neq i} s_{i\ell}^k - \sum_{\ell \neq i} s_{i\ell}^{k^*}\right) \begin{cases} \geq 0, & \text{if } \sum_{\ell \neq i} s_{i\ell}^{k^*} = \bar{Q}_i^k; \\ = 0, & \text{if } \sum_{\ell \neq i} s_{i\ell}^{k^*} < \bar{Q}_i^k; \end{cases}$$

holds for all locations i.

Therefore, for this class k and all locations i, $s_{i\ell}^{k^*} \geq 0$, $\sum_{\ell \neq i} s_{i\ell}^{k^*} \leq \bar{Q}_i^k$, and

$$\sum_{i=1}^{n} \sum_{\ell \neq i} \left[u_i^k(Q^*) + c_{i\ell}^k(s^*) - u_\ell^k(Q^*) - s_{i\ell}^{k^*} w_{i\ell}^k (u_\ell^k)'(Q^*)\right] \times \left[s_{i\ell}^k - s_{i\ell}^{k^*}\right] \geq 0. \tag{4.98}$$

But inequality (4.98) holds for each k; hence,

$$\sum_{k=1}^{J} \sum_{i=1}^{n} \sum_{\ell \neq i} \left[u_i^k(Q^*) + c_{i\ell}^k(s^*) - u_\ell^k(Q^*) - s_{i\ell}^{k^*} w_{i\ell}^k (u_\ell^k)'(Q^*)\right] \times \left[s_{i\ell}^k - s_{i\ell}^{k^*}\right] \geq 0. \tag{4.99}$$

Observe now that inequality (4.99) can be rewritten as:

$$\sum_{k=1}^{J} \sum_{\ell=1}^{n} u_\ell^k(Q^*) \times \left[\left(\sum_{j \neq \ell} s_{\ell j}^k - \sum_{j \neq \ell} s_{j\ell}^k\right) - \left(\sum_{j \neq \ell} s_{\ell j}^{k^*} - \sum_{j \neq \ell} s_{j\ell}^{k^*}\right)\right] +$$

$$\sum_{k=1}^{J} \sum_{i=1}^{n} \sum_{\ell \neq i} \left[c_{i\ell}^k(s^*) - s_{i\ell}^{k^*} w_{i\ell}^k (u_\ell^k)'(Q^*)\right] \times \left[s_{i\ell}^k - s_{i\ell}^{k^*}\right] \geq 0. \tag{4.100}$$

Now making use of constraint (4.91) and substituting it into (4.100), one concludes that

$$-\sum_{k=1}^{J} \sum_{\ell=1}^{n} u_\ell^k(Q^*) \times \left(Q_\ell^k - Q_\ell^{k^*}\right) +$$

$$\sum_{k=1}^{J} \sum_{i=1}^{n} \sum_{\ell \neq i} \left[c_{i\ell}^k(s^*) - s_{i\ell}^{k^*} w_{i\ell}^k (u_\ell^k)'(Q^*)\right] \times \left[s_{i\ell}^k - s_{i\ell}^{k^*}\right] \geq 0, \tag{4.101}$$

or, equivalently, in vector notation,

$$\langle -u(Q^*), Q - Q^* \rangle + \langle c(s^*) - d(s^*), s - s^* \rangle \geq 0, \ \forall (Q, s) \in K. \tag{4.102}$$

Now we show that if a pattern $(Q^*, s^*) \in K$ satisfies variational inequality (4.97), then it also satisfies equilibrium conditions (4.93) and (4.94). Suppose that (Q^*, s^*) solves variational inequality (4.97). Then

$$\langle -u(Q^*), Q \rangle + \langle c(s^*) - d(s^*), s \rangle \geq$$

$$\langle -u(Q^*), Q^* \rangle + \langle c(s^*) - d(s^*), s^* \rangle, \ \forall (Q, s) \in K.$$

Hence, (Q^*, s^*) solves the minimization problem

$$\min_{(Q,s)\in K} \left[\langle -u(Q^*), Q \rangle + \langle c(s^*) - d(s^*), s \rangle \right], \tag{4.103}$$

or, equivalently, (4.103) may be expressed solely in terms of s, that is,

$$\min_{s\in K_1} \left[\langle -\hat{u}(As^*), As \rangle + \langle c(s^*) - d(s^*), s \rangle \right], \tag{4.104}$$

where $K_1 = \{s \mid s \geq 0, \text{ satisfying (4.92) } \}$, A is the arc-node matrix in (4.92), and $\hat{u}(As) \equiv u(Q)$, $\forall s \in K_1$.

Since the constraints in K_1 are linear, one can apply the Karush-Kuhn-Tucker theorem: There exist

$$\lambda = (\lambda_i^k) \geq 0, \tag{4.105}$$

such that

$$\lambda_i^k \left(\sum_{\ell \neq i} s_{i\ell}^{k\,*} - \bar{Q}_i^k \right) = 0, \ i = 1, \ldots, n, \tag{4.106}$$

and

$$u_i^k - u_j^k + c_{ij}^k - s_{ij}^{k\,*} w_{ij}^k (u_j^k)' + \lambda_i^k \geq 0, \tag{4.107}$$

$$\left[u_i^k - u_j^k + c_{ij}^k - s_{ij}^{k\,*} w_{ij}^k (u_j^k)' + \lambda_i^k \right] s_{ij}^{k\,*} = 0. \tag{4.108}$$

Clearly, equilibrium conditions (4.93) and (4.94) follow from (4.105)–(4.106). The proof is complete. ∎

The existence of at least one solution to variational inequality (4.97) follows from the general theory of variational inequalities, under the sole assumption of continuous differentiability of the utility functions u and continuity of migration cost functions c, since the feasible convex set K is compact. Uniqueness of the equilibrium population and migration flow pattern (Q^*, s^*) follows under the assumption that the compound operator

$$(-u(Q), c(s) - d(s)) : R^{J\times n} \times R^{J\times n\times(n-1)} \rightarrow R^{J\times n} \times R^{J\times n\times(n-1)},$$

involving the utility and migration cost functions, is strictly monotone over the set K, that is

$$-\langle u(Q^1) - u(Q^2), Q^1 - Q^2\rangle + \langle c(s^1) - c(s^2), s^1 - s^2\rangle -$$
$$-\langle d(s^1) - d(s^2), s^1 - s^2\rangle > 0, \tag{4.109}$$

for all $(Q^1, s^1), (Q^2, s^2) \in K$ such that $(Q^1, s^1) \neq (Q^2, s^2)$.

We now interpret monotonicity condition (4.109) in terms of applications. Essentially, it is assumed that the system is subject to congestion; hence, the utility are decreasing with larger populations, and the movement costs are increasing with larger migration flows. Therefore, the strict monotonicity condition will hold, if the operators $-\nabla u$ and $(\nabla c - \nabla d)$ are (strictly) positive definite. For example, it is the case, if each utility function $u_i^k(Q)$ depends only on the population Q_i^k, is decreasing, and the product $Q_i^k u_i^k(Q_i^k)$ is a convex function, whereas each movement cost $c_{ij}^k(s)$ depends mainly on the flow s_{ij}^k. Then it can be shown (in the same way as it was done in Subsection 4.3.2), that under assumption $0 \leq w_{ij}^k < 1$ the operators $-\nabla u(Q)$ and $(\nabla c(s) - \nabla d(s))$ are positive definite.

4.3.6 CONCLUDING REMARKS

In conclusion we note that the human migration model can be further extended and examined in the case when constraint (4.92) is replaced by condition

$$Q_i^k \geq 0, \tag{4.110}$$

which allows in principle the repeat or chain migration. In this case the set K stops being compact (however remaining convex), which makes insufficient the usage of the general theory of variational inequalities in order to demonstrate existence of solution to (4.97). Then the subtler results obtained by the authors Isac, Bulavsky and Kalashnikov (1997), Bulavsky, Isac and Kalashnikov (1998), can be used to that effect. Indeed, the existence will be guaranteed for various classes of utility functions and migration costs that are free of the exceptional families of elements (EFE).

Results of this chapter allow one to consider not only two separate models: Walras (perfect competition) and Cournot ones, but also the whole series of intermediate models determined by different values of the conjectural influence quotients $w_i \in [0, 1]$. Moreover, some brand new models of oligopoly appear when one allows the factors w_i to assume values greater than 1.

Lastly, the influence quotients help one in conducting efficient comparative statics analysis for various models of oligopoly.

REFERENCES

Bulavsky, V.A., Isac, G., and Kalashnikov, V.V. 'Application of topological degree theory to complementarity problems'.- In: *Multilevel Optimization: Algorithms and Applications*, A. Migdalas, P.M. Pardalos and P. Värbrand, eds. Dordrecht-Boston-London: Kluwer Academic Publishers, 1998.

Bulavsky VA, Kalashnikov VV. One-parametric driving method to study equilibrium. Ekonomika i Matematicheskie Metody (*in Russian*) 1994; 30:129-138.

Bulavsky VA, Kalashnikov VV. Equilibria in generalized Cournot and Stackelberg models. Ekonomika i Matematicheskie Metody (*in Russian*), 1995; 31:164-176.

Bulavsky, V.A. and Kalashnikov, V.V.'An alternative model of spatial competition'.- In: *Operations Research and Decision Aid Methodologies in Traffic and Transportation management*, M. Labbé, G. Laporte, K. Tanczos, P. Toint, eds. NATO ASI Series, Series F: Computer and Systems Science, Vol. 166, Springer-Verlag Berlin Hedelberg, 1998.

Bresnahan TF. Duopoly models with consistent conjectures. American Economic Review 1981; 71:934-945.

Dafermos S, Nagurney A. A network formulation of market equilibrium problems and variational inequalities. Operations Research Letters 1986; 3:247-250.

Friedman, J.W. *Oligopoly and the Theory of Games.* Amsterdam: North-Holland, 1977.

Harker PT, Pang JS. Finite-dimensional variational inequalities and nonlinear complementarity problems: a survey of theory, algorithms and applications. Math. Programming 1990; 48: 161-220.

Isac G, Bulavsky VA, and Kalashnikov VV. Exceptional families, topological degree and complementarity problems. Journal of Global Optimization 1997; 10: 207-225.

Kalashnikov, V.V. "Fixed Point Existence Theorems Based upon Topological Degree Theory", Working Paper, Moscow, Central Economics and Mathematics Institute (*in Russian*),1995.

Klemperer P, Meyer M. Price competition vs. quantity competition: the role of uncertainty. The Rand Journal of Economics 1985; 17: 618-638.

Kreps DM, Wilson R. Sequential equilibria. Econometrica 1982; 50: 863-894.

Kreps, D. *A Course in Microeconomic Theory.* Harvester Wheatsheaf, 1990.

Laitner J. "Rational" duopoly equilibria. The Quarterly Journal of Economics 1980; 95: 641-662.

Nagurney A. A network model of migration equilibrium with movement costs. Mathematical and Computer Modelling 1990; 13:79-88.

Nagurney, A. *Network Economics: a Variational Inequality Approach* (revised second edition), Boston: Kluwer Academic Publishers, 1999.

Nagurney A, Jie Pan, Lan Zhao. Human migration network. European Journal of Operational Research 1992; 59: 262-274.

Novshek W. On the existence of Cournot equilibrium. Review of Economic Studies 1985; 52(1): 85-98.

Outrata JV. On necessary optimality conditions for Stackelberg problems. J. Optim. Theory Appl. 1993; 76:305-320.

Riordan MH. Imperfect information and dynamic conjectural variations. The Rand Journal of Economics 1985; 16: 41–50.

Sherali HD, Soyster AL and Murphy FH. Stackelberg-Nash-Cournot equilibria: characterizations and computations. Operations Research 1983; 31: 171-186.

Spiller PT, Favaro E. The effects of entry regulation on oligopolistic interaction: the Uruguayan banking sector. The Rand Journal of Economics 1984;15: 244-254.

Stackelberg, H. *Marktform und Gleichgewicht.* Vienna: Julius Springer, 1934.

Ulph D. Rational conjectures in the theory of oligopoly. International Journal of Industrial Organization 1983;1:131-154.

Chapter 5

OLIGOPOLY WITH LEADERS

This chapter is dedicated to the Stackelberg model which is a particular case of bilevel problems with equilibrium constraints. In Section 5.1, the Stackelberg model is extended to the case of several leaders, and the theorem of existence of a stationary point is obtained. In Section 5.2, we compare the equilibria in the Stackelberg and Cournot models. Section 5.3 presents simple examples of comparison of equilibria in different models: Cournot model, high expectations model, Stackelberg model, and the perfect competition one. These examples illustrate results of Section 5.2. At last, problems of efficient computation of the equilibrium are considered in Section 5.4.

5.1 MARKET WITH SEVERAL LEADERS

In this section, we generalize the Stackelberg model (Stackelberg, 1934) and embed it into the frameworks developed in the previous chapter. In the classical Stackelberg model, one of the firms takes into account the optimal responses of the other market agents to its output variations, and maximizes its profit. In the paper by Sherali e.a. (1983), such a firm is called a *leader*. Now we examine the case of several leaders, who act toward each other as the classical Cournot agents, whereas the rest of the agents *followers* behave according to the model of Section 4.2 of the previous chapter. As it was mentioned before, we are going to embed the generalized Stackelberg model into the general framework making use of the first order optimality conditions. It means that we are looking for such a situation where the leaders'profit functions have *stationary points*. So, from now on, we will talk not about equilibria of the generalized Stackelberg model but of its stationary points.

Consider again the model described in Section 4.2 of the previous chapter and recall assumptions E1-E3, E5-E6, E8.

Let the former s firms ($1 \le s < n$) behave as leaders, i.e. they take into acccount the latter $(n-s)$ agents' (followers's) reaction to variations of the leaders' total output $Q = \sum_{i=1}^{s} q_i$.

Towards each other they act as the classical Cournot subjects, i.e. they suppose the other leaders not to change their outputs. At last, the latter $(n-s)$ firms look for the Nash-Cournot equilibrium among themselves with the extraneous supply Q and functions w_j^-, w_j^+ defined by (4.26)–(4.27), that is

$$w_i^-(G,q_i) = w_i^+(G,q_i) = \alpha_i + \sigma_i \frac{G}{q_i}, \quad \text{if } 0 < q_i < G, \tag{5.1}$$

$$w_i^-(G,G) = \alpha_i + \sigma_i, \qquad w_i^+(G,G) = 1. \tag{5.2}$$

The classical Stackelberg model requires $s = 1$, $\alpha_i = 1$, $\sigma_i = 0$, $i = 2,\ldots,n$. In view of E1-E3, E5-E6, E8, for each $Q \ge 0$, the problem: Find a vector $Z = (G, q_1, \ldots, q_n)$ such that

$$\sum_{i=s+1}^{n} q_j + Q = G, \tag{5.3}$$

and for each i, the following inequalities are valid:

$$\mu_i^+(G,q_i) = p(G) + \sigma_i^+(G,q_i)p'(G)G - f_i'(q_i) \le 0, \tag{5.4}$$

and if $q_i > 0$, then

$$\mu_i^-(G,q_i) = p(G) + \sigma_i^-(G,q_i)p'(G)G - f_i'(q_i) \ge 0, \tag{5.5}$$

has a unique solution. Moreover, all the solution's components, as functions of Q, are differentiable everywhere except for a finite number of points at which only left- and rightside derivatives exist. Differentiating the balance equality

$$G = Q + \sum_{j=s+1}^{n} q_j(G(Q)) \tag{5.6}$$

by Q and using estimates from Lemma 4.3, one can easily show that the function $G(Q)$ is strictly increasing, $G(0) > 0$, and $G(Q) \ge Q$ for all $Q \ge 0$.

THEOREM 5.1. *Let assumptions* E1-E3, E5-E6, E8 *be valid and the functions* $w_j^{\pm}(G,q_j)$, $j = s+1,\ldots,n$, *be defined by equalities* (5.1)–(5.2). *Then the generalized Stackelberg model has a stationary point.*

PROOF. In order to embed our problem into the general model examined in the previous chapter, we need defining functions w_j^+ and w_j^- for $j = 1,\ldots,s$, or conjecture functions for the leaders. According to the supposed behavior of the leaders, the j-th leader's conjecture indicates that its output variation $(\eta - q_j)$ leads to the extraneous (for the followers) supply variation by the same value $(\eta - q_j)$. Hence, the j–th leader assumes that the balance equality (5.3) remains valid after replacing Q by $(Q+\eta-q_j)$. Differentiating this equality by

η from the left and right at the point $\eta = q_j$ and finding the one-side derivatives $dG_j/d\eta(q_j \pm 0)$, we construct the needed functions w_j^+ and w_j^- for $j = 1, \ldots, s$. Namely, if $G > G(0)$, then the value $Q(G) > 0$ (i.e. such that $G(Q(G)) = G$) is determined uniquely, and for $0 < q_j \le Q(G)$ we obtain

$$w_j^-(G, q_j) = \frac{1}{1 - \sum\limits_{m=s+1}^{n} q_m'(G-0)} = u^-(G), \tag{5.7}$$

$$w_j^+(G, q_j) = \frac{1}{1 - \sum\limits_{m=s+1}^{n} q_m'(G+0)} = u^+(G). \tag{5.8}$$

In view of Lemma 4.3,

$$0 < u^-(G) \le u^+(G) \le \max\left\{1, \frac{G}{2Q(G)}\right\}. \tag{5.9}$$

Furthermore, these functions u^- and u^+ are continuous and coincide everywhere except for the points $\tilde{G}_i$, $i = s+1, \ldots, n$. As for the latter break points, the function $u^-(G)$ is left-continuous and $u^+(G)$ is right-continuous at these points. This is in line with assumption E4. In order to apply Theorem 4.1, we need defining functions $w_j^{\pm}$, $j = 1, \ldots, s$, at all the points with $G > 0$, $0 < q_j \le G$, without violating condition E4.

Let us begin with the points (G, q_j) for $G > G(0)$ and $Q(G) < q_j \le G$. To preserve the monotonicity of products $q_i w_j^{\pm}(G, q_j)$ and ensure condition (4.5) to hold at the considered points (G, q_j), we set

$$w_j^{\pm}(G, q_j) = \alpha^{\pm}(G) + \frac{\sigma^{\pm}(G)}{q_j}, \tag{5.10}$$

having determined the quotients from the equations

$$\alpha^{\pm}(G) + \sigma^{\pm}(G)/Q(G) = u^{\pm}(G), \;\; \alpha^{\pm}(G) + \sigma^{\pm}(G)/G = 1. \tag{5.11}$$

Due to $Q(G) < q_i \le G$, the values of $w_j^{\pm}(G, q_j)$, $j = 1, \ldots, s$, defined by (5.10) lie between $u^{\pm}(G) > 0$ and 1, i.e. they are positive. Solving system (5.11) and substituting the values $\alpha^{\pm}$ and $\sigma^{\pm}$ into (5.10), we obtain

$$w_j^{\pm}(G, q_j) = \frac{G\,[1 - Q(G)/q_j] + [G/q_j - 1]\,Q(G)u^{\pm}(G)}{G - Q(G)}. \tag{5.12}$$

Since the coefficient before $u^{\pm}(G)$ in (5.12) is nonnegative for $q_j \le G$, thus defined functions $w_j^{\pm}$, $j = 1, \ldots, s$, keep having the respective properties of semicontinuity. Besides, in view of (5.9), the following upper limit value is finite and nonnegative:

$$V^+ = \limsup_{G \to G(0)} \frac{Q(G)u^+(G)}{G - Q(G)} \le \frac{1}{2}. \tag{5.13}$$

Passing to the upper limit in (5.12) when $G \to G(0)$, we define

$$w_j^+ \left(G\left(0\right), q_j\right) = \left[1 - V^+\right] + \left[G\left(0\right) V^+\right] / q_j. \tag{5.14}$$

The nonnegativity of (5.12) and (5.14) allow one to define $w_j^+(G, q_j) = 0$ for $G < G(0)$ and $w_j^-(G, q_j) = 0$ for $G \le G(0)$. It is readily verified that thus defined functions w_j^- and w_j^+, $j = 1, \ldots, s$, satisfy all the conditions of Theorem 4.1, and therefore, there exists a stationary point of the generalized Stackelberg model with several leaders. The proof is complete. ■

Remark. Under slightly stricter assumptions, it can be shown that $V^+ = 0$, hence $w_j^+ \left(G\left(0\right), q_j\right) = 1$, $j = 1, \ldots, s$. For instance, to obtain that, it suffices to assume that either f_i'' are strictly positive for all $i = s + 1, \ldots, n$, or $p(G)G$ is strictly concave, or at least two followers have positive outputs q_i at the equilibrium with $Q = 0$ (that is, in the absence of leaders).

5.2 COMPARISON OF COURNOT AND STACKELBERG MODELS

In this section, we consider the classical Stackelberg model, i.e. we take $s = 1$. In that case we have $q_1 = Q$, and the leader maximizes his expected profit:

$$\max\{\mu_1(Q) = p\left(G(Q)\right) Q - f_1(Q) \mid Q \ge 0\}, \tag{5.15}$$

treating the price $p\left(G(Q)\right)$ as the function of his output Q. Here

$$G(Q) = Q + \sum_{j=2}^{n} q_j(Q),$$

where $\sum_{j=2}^{n} q_j(Q)$ is the equilibrium total volume of outputs produced by the agents $j = 2, \ldots, n$, who solve problem (5.3)–(5.5) with the extraneous supply Q, while their influence quotients $w_j^{\pm}$ are defined by (5.1)–(5.2).

In what follows we establish the relationships between the solution Q^* of problem (5.15) and the equilibrium output volume $\hat{Q}$ of the first agent if he/she were not behaving as a leader but would be using, like others, the influence quotients $w_1^{\pm}$ defined by (5.1)–(5.2). Besides, it is interesting to compare the values Q^* and $\hat{Q}$ to the leader's optimal output volume $\bar{Q}$ when he ignores the variation of the price, i.e., if he solves the following complementarity problem: Find a $\bar{Q} \ge 0$ such that

$$\beta_1(\bar{Q}) = f_1'(\bar{Q}) - p\left(G(\bar{Q})\right) \ge 0, \qquad \bar{Q} \cdot \beta_1(\bar{Q}) = 0. \tag{5.16}$$

THEOREM 5.2. *Let assumptions* E1-E3, E5-E6, E8 *hold. Then*

$$\max\left\{\hat{Q}; Q^*\right\} \le \bar{Q} \le H_1; \tag{5.17}$$

here $H_1 > 0$ *is the scalar from assumption* E3, *i.e. such that* $f_1'(H_1) = p(H_1)$.

PROOF. In order to prove the inequality $Q^* \le \bar{Q}$, it suffices to show that $\mu_1'(Q+0) < 0 \; \forall Q > \bar{Q}$. Assumptions E1, E2 and the inequality $G'(Q+0) > 0$ established above yield the estimate

$$\mu_1'(Q+0) = p'\left(G(Q)\right) G'(Q+0)Q - \beta_1(Q) < -\beta_1(Q) \qquad \forall Q > 0.$$

Definition of $\bar{Q}$ implies the inequality $\beta_1(\bar{Q}) \ge 0$. Therefore, it is sufficient to verify that $\beta_1'(Q+0) > 0$ for every $Q > 0$. But we have

$$\beta_1'(Q+0) = f_1''(Q) - p'\left(G(Q)\right) G'(Q+0),$$

and the needed inequality follows from assumptions E1-E2 and the property $G'(Q+0) > 0$. Hence, the estimate $Q^* \le \bar{Q}$ is established.

Now we verify the inequality $\hat{Q} \le \bar{Q}$. On the contrary, suppose that $\hat{Q} > \bar{Q}$. Since the function $G(Q)$ strictly increases, we have $G(\hat{Q}) > G(\bar{Q})$. According to the definition of $\hat{Q}$, we can write down

$$\chi(\hat{Q}) = p\left(G(\hat{Q})\right) + \alpha_1 \hat{Q} p'\left(G(\hat{Q})\right) Q + \sigma_1 G(\hat{Q}) p'\left(G(\hat{Q})\right) - f_1'(\hat{Q}) = 0.$$

Making use of conditions E1, E2, E8 and definition (5.16) of the value $\bar{Q}$, we obtain the series of inequalities

$$0 = \chi(\hat{Q}) < p\left(G(\bar{Q})\right) - f_1'(\bar{Q}) = -\beta_1(\bar{Q}) \le 0,$$

that leads to the impossible inequality $0 < 0$. Therefore, $\hat{Q} \le \bar{Q}$, and the first inequality in series (5.17) is proved.

In order to establish the second inequality in (5.17), we note that $f_1'(\bar{Q}) \le p\left(G(\bar{Q})\right) \le p(\bar{Q})$ due to (5.16) and the function p being strictly decreasing. Now it follows from E1–E3 that $f_1'(Q) > p(Q)$ if $Q > H_1$. Therefore, $\bar{Q} \le H_1$, and proof of the theorem is completed. ∎

Notice that the operation of taking maximum of the two values $\hat{Q}$ and Q^* in estimate (5.17) cannot be avoided. The point is that, in contrast to the paper by Sherali e.a. (1983) in which the estimate $\hat{Q} \le Q^*$ was established, the weaker conditions on the inverse demand curve $p(G)$ used in this chapter do not allow one to deduce the same estimate in general case. Moreover, there exist such examples where the leadership of a weak (in a certain sense) agent leads to a decrease of his own (and therefore, the total) output level, i.e. $Q^* < \hat{Q}$ (and hence, $G(Q^*) < G(\hat{Q})$). Now we will consider a particular case of our problem where more specific assumptions about the functions p and f_i permit us to prove the strict concavity of the leader's profit function $\mu_1(Q)$. After that, we will be able to establish the relationships between $\hat{Q}$ and Q^* making use of the local information only.

Namely, let the function p be three times differentiable and satisfy the condition

$$p'' > 0 \qquad \text{and} \qquad \frac{d}{dG}\left(\frac{p''}{p'}\right) \le -\frac{1}{G}\cdot\frac{p''}{p'} \tag{5.18}$$

for each $G > 0$.

LEMMA 5.1. *Let the inverse demand function $p = p(G)$ satisfy conditions (5.18), all the cost functions be linear, i.e., $f_i(q_i) = c_i q_i$, $c_i > 0$, $i = 1, \dots, n$, and the influence quotients be defined by (5.1)–(5.2), with $\alpha_i > 0$, $i = 1, \dots, n$. Then the leader's profit function $\mu_1(Q)$ is strictly convex over $Q \ge 0$.*

PROOF. Under the lemma conditions, it is easily verified that

$$q_i(G) = \frac{c_i - p(G) - \sigma_i G p'(G)}{\alpha_i p'(G)}$$

for $i \in I = I(G) = \{1 \le i \le n \mid q_i(G) > 0\}$. This implies

$$q_i'(G) = -\frac{1+\sigma_i}{\alpha_i} - \frac{\sigma_i G p''(G) + \alpha_i q_i(G) p''(G)}{\alpha_i p'(G)}, \qquad i \in I.$$

The latter yields the equality

$$\sum_{i\in I} q_i'(G) = -\sum_{i\in I}\frac{1+\sigma_i}{\alpha_i} - \frac{p''(G)}{p'(G)}\sum_{i\in I} q_i(G) - \frac{p''(G)G}{p'(G)}\sum_{i\in I}\frac{\sigma_i}{\alpha_i}.$$

Hence, at the point Q where the derivative $G'(Q)$ exists, it can be represented as follows

$$G'(Q) = \frac{1}{1 - \sum\limits_{i\in I} q_i'(G(Q))} =$$

$$= \frac{1}{1 + \sum\limits_{i\in I}\frac{1+\sigma_i}{\alpha_i} + \frac{p''}{p'}\left[(G-Q) + G\sum\limits_{i\in I}\frac{\sigma_i}{\alpha_i}\right]}. \tag{5.19}$$

Note that (5.19) and concavity of the function $p(G)G$ imply the estimate

$$G'(Q) \le \frac{1}{\sum\limits_{i\in I}\frac{1-\sigma_i}{\alpha_i} - 1 - Q\frac{p''}{p'}} \le \frac{1}{-Q\frac{p''}{p'}}. \tag{5.20}$$

Therefore, $\mu_1'(Q) = p(G(Q)) + p'(G(Q))G'(Q)Q - c_1$, and

$$\mu_1''(Q) = 2p'(G(Q))G'(Q) + p''(G(Q))\left[G'(Q)\right]^2 Q + \\ + p'(G(Q))G''(Q)Q. \tag{5.21}$$

By differentiating the terms in (5.19) we obtain (at the points Q where $G''(Q)$ exists) the following expression

$$G''(Q) = -(G')^2\left[\left(\frac{p''}{p'}\right)'(G-Q)G' + \frac{p''}{p'}(G'-1)\right].$$

Substituting the latter to (5.21) we have

$$\mu_1''(Q) = p'G'\left\{2 + 2Q\frac{p''}{p'}G' - Q\left[\frac{p''}{p'} + \left(\frac{p''}{p'}\right)'(G-Q)\right](G')^2\right\}.$$

Note that the following inequality is implied by (5.18)

$$\frac{p''}{p'} + \left(\frac{p''}{p'}\right)'(G-Q) < 0.$$

Consequently, the second derivative $\mu_1''(Q)$ takes negative values if

$$G' \leq \frac{1}{Q(-p''/p')}.$$

But the last inequality is indeed valid due to (5.18) for each $Q > 0$ such that the derivative $G'(Q)$ exists. On the other hand, if only one-sided derivatives of the function G (and hence, of the function μ_1) exist at the point Q, we will show that $\mu_1'(Q-0) > \mu_1'(Q+0)$. Indeed, if $i \in I(G-0)$ and $i \notin I(G+0)$, then it is readily verified that $q_i'(G-0) < 0$ and $q_i'(G+0) = 0$. Thus, we deduce from (5.19) and the continuity of $G(Q)$ that $G'(Q-0) < G'(Q+0)$. Now making use of the formulae

$$\mu_1'(Q-0) = p(G(Q)) + p'(G(Q))G'(Q-0)Q - f_1'(Q),$$

$$\mu_1'(Q+0) = p(G(Q)) + p'(G(Q))G'(Q+0)Q - f_1'(Q)$$

and the function $p = p(G)$ being strictly decreasing, we obtain

$$\mu_1'(Q-0) > \mu_1'(Q+0)$$

for each $Q > 0$ at which μ_1 is not differentiable. Therefore, the function μ_1 is strictly concave, and the proof is complete. ■

Now making use of the latter result, we obtain the complete classification of the cases based upon the following local rules.

(i) If $G'(\hat{Q}+0) \geq 1$ and $G'(\hat{Q}-0) \leq 1$, then $Q^* = \hat{Q}$, $G(Q^*) = G(\hat{Q})$;

(ii) if $G'(\hat{Q}+0) < 1$, then $Q^* > \hat{Q}$, $G(Q^*) > G(\hat{Q})$;

(iii) if $G'(\hat{Q}-0) > 1$, then $Q^* < \hat{Q}$, $G(Q^*) < G(\hat{Q})$.

For example, if $n = 2$ and $q_2'(\hat{G}) < 0$, then

$$G'\left(\hat{Q}+0\right) = \frac{1}{1 - q_2'(\hat{G})} < 1, \tag{5.22}$$

and according to (ii), $Q^* > \hat{Q}$, consequently $G(Q^*) > G\left(\hat{Q}\right)$. In economic terms, it can be interpreted as follows. Agent i of the Cournot model can be called *weak* if his optimal reply $q_i = q_i(G)$ has nonpositive derivative at the

Nash equilibrium point $\hat{G}$. Indeed, the condition $q_i'(\hat{G}) \le 0$ means that agent i is not ready to compete and increase his output if the price $p(G)$ is going to decrease. In contrast, one can reckon agent i as being *strong* if $q_i'(\hat{G}) > 0$. The latter terms can be justified by Lemma 4.3 which showed that at the Nash equilibrium, at most one agent can have a positive derivative value, as $q_i'(\hat{G}) > 0$ implies $q_i(\hat{G}) > \hat{Q}/2$. In case (ii) where the follower, agent 2, is weak as $q_2'(\hat{G}) < 0$ follows from (5.22) and $G'(\hat{Q}+0) < 1$, the leadership of the first agent leads to the increase of both the leader's and the total output at the Stackelberg equilibrium, in comparison to the Cournot equilibrium values.

On the other hand, if $q_2'(\hat{G}) > 0$ then

$$G'\left(\hat{Q}+0\right) = \frac{1}{1-q_2'(\hat{G})} > 1,$$

and because of (iii), $Q^* < \hat{Q}$, whence $G(Q^*) < G\left(\hat{Q}\right)$. The economic interpretation of the latter inequalities is as follows: if the second agent produces more than a half of the total output at the Cournot equilibrium (which is guaranteed by the prositiveness of his output derivative $q_2'(\hat{G}) > 0$), then the leadership of the weaker agent 1 leads to decreasing of both his own and the total level of production at the Stackelberg equilibrium. Therefore, the leadership of strong producers is preferable for consumers as that reduces the prices of the goods.

5.3 EXAMPLES OF MODELS

In this section, we consider oligopolistic models of different kinds examined in previous sections of this chapter, and compare their equilibrium values.

5.3.1 COURNOT MODEL VERSUS HIGH EXPECTATIONS MODEL

In this subsection, we consider the Cournot model with three firms with linear costs at the market with a hyperbolic inverse demand function. In Example 5.1, we compare it with the generalized model in which all three agents assume $w_i \equiv 2$, $i = 1, 2, 3$. It means that they are more precautious than agents of the standard Cournot model and conjecture almost the same behaviour of their rivals as their own.

EXAMPLE 5.1 ($n \ge 2$).

We assume that

$$f_i(q_i) = c_i q_i, \quad c_i > 0, \quad i = 1, \ldots, n; \quad p(G) = AG^{-1}, A > 0; \quad Q \ge 0.$$

5.1a. COURNOT MODEL

In the Cournot model one has that $w_i(G, q_i) \equiv 1, \quad i = 1, \ldots, n$. The equilibrium problem is as follows: Find a $(G, q_1, \ldots, q_n) \geq 0$ such that

$$Q + \sum_{i=1}^{n} q_i = G; \tag{5.23}$$

$$\varphi_i \equiv c_i + \frac{A}{G^2} \cdot q_i - \frac{A}{G} \geq 0, \quad q_i \varphi_i = 0, \ i = 1, \ldots, n.$$

If $Q > 0$, then

$$G = \frac{(n-1)A}{2\sum_{k=1}^{n} c_k} \left(1 + \sqrt{1 + \frac{4Q\sum_{k=1}^{n} c_k}{A(n-1)^2}} \right).$$

If $Q = 0$ then

$$G = \frac{(n-1)A}{\sum_{k=1}^{n} c_k}, \qquad q_i = \frac{(n-1)A}{\left(\sum_{k=1}^{n} c_k\right)^2} \left[\sum_{j \neq i} c_j - (n-2)c_i \right], \quad i = 1, \ldots, n. \tag{5.24}$$

For instance, if $n = 3$ then (5.24) reduces to

$$G = \frac{2A}{\sum_{k=1}^{3} c_k}, \quad \text{and} \quad q_i = \frac{2A}{\left(\sum_{k=1}^{3} c_k\right)^2} \left(\sum_{j \neq i} c_j - c_i \right), \quad i = 1, 2, 3.$$

Furthermore, the net profit of agent $i = 1, 2, 3$ is expressed by

$$\mu_i = \frac{A}{\left(\sum_{k=1}^{3} c_k\right)^2} \sum_{j \neq i} c_j \left(\sum_{j \neq i} c_j - c_i \right).$$

5.1b. HIGH EXPECTATIONS MODEL

Now we compare the values obtained in the Cornot model with the corresponding ones in case when $w_i(G, q_i) \equiv 2, \ i = 1, 2, 3$. Then it is easy to see that

$$G^*|_{w_i \equiv 2} = \frac{A}{\sum_{k=1}^{3} c_k} < G|_{Cournot},$$

that is, the total bargain volume of the high expectations model is strictly less than that of the Cournot model. The relationships between $q_i^*|_{w_i \equiv 2}$ and $q_i|_{Cournot}$ depend upon the proportions among the marginal cost values c_i, $i = 1, 2, 3$. More precisely, if all the three agents really produce, then

$$q_i^*|_{w_i \equiv 2} \leq q_i|_{Cournot} \quad \text{if and only if} \quad c_i \leq \frac{3}{4} \sum_{j \neq i} c_j. \tag{5.25}$$

For example, if $c_1 < c_2 < c_3$ then (5.25) implies that the strongest firm $i = 1$ decreases its output, whereas the two weaker ones may increase their outputs, in comparison to the classic Cournot equilibrium.

As for the profits, one has

$$\mu_i^*|_{w_i \equiv 2} = \frac{A}{2\left(\sum_{k=1}^{3} c_k\right)^2} \left(\sum_{j \neq i} c_j\right)^2,$$

which implies that $\mu_i^*|_{w_i \equiv 2} \leq \mu_i|_{Cournot}$ if and only if $c_i \leq \frac{1}{2}\sum_{j\neq i} c_j$. For instance, if $c_1 < c_2 < c_3$ then we have $\mu_1^*|_{w_i \equiv 2} < \mu_1|_{Cournot}$ for the strongest firm $i = 1$ but $\mu_3^*|_{w_i \equiv 2} > \mu_3|_{Cournot}$ for the weakest firm $i = 3$. ■

5.3.2 COURNOT MODEL VERSUS MIXED CONJECTURES MODEL

Unlike the previous example, in Example 5.2 below, the Cournot model is compared with the generalized one where each firm uses the mixed conjecture introduced in Section 4.2 with $\alpha_i = \sigma_i = 1/2$.

EXAMPLE 5.2 (n=2).

5.2a. COURNOT MODEL WITH TWO AGENTS

Looking for an equilibrium, one comes to the following results:

$$G_{Cournot} = \frac{A}{c_1 + c_2}; \quad q_1|_{Cournot} = \frac{Ac_2}{(c_1 + c_2)^2}, \quad q_2|_{Cournot} = \frac{Ac_1}{(c_1 + c_2)^2};$$

$$\mu_1|_{Cournot} = \frac{Ac_2^2}{(c_1 + c_2)^2}; \qquad \mu_2|_{Cournot} = \frac{Ac_1^2}{(c_1 + c_2)^2}.$$

Now we compare these values to the corresponding ones in case

$$w_i(G, q_i) = \frac{1}{2} + \frac{1}{2} \cdot \frac{G}{q_i}, \quad i = 1, 2.$$

We get

$$G^* = \frac{A}{2(c_1 + c_2)} = \frac{1}{2} G|_{Cournot}; \quad q_i^* = \frac{1}{2} q_i|_{Cournot}, \ i = 1, 2;$$

but

$$\mu_i^* = \mu_i|_{Cournot} + \frac{Ac_1c_2}{2(c_1 + c_2)^2} > \mu_i|_{Cournot}, \quad i = 1, 2.$$

In the other words, the mixed conjectures lead to lower outputs but higher profits for both agents. ■

5.3.3 COMPARISON OF STACKELBERG AND COURNOT MODELS

Here we illustrate the results of Section 5.2. Consider two firms with linear costs at the market and the hyperbolic inverse demand function:

$$n = 2;\ f_i(q_i) = c_i q_i;\ 0 < c_1 < c_2;\ Q = 0;\ p(G) = AG^{-1}, A > 0.$$

At the Nash-Cournot equilibrium, we have

$$q_i'(\hat{G}) = \frac{c_j - c_i}{c_i + c_j},\quad j \neq i;\ i = 1, 2.$$

Therefore, $q_1'(\hat{G}) > 0$ (strong firm); $q_2'(\hat{G}) < 0$ (weak firm). The total Nash-Cournot equilibrium volume is $\hat{G} = A/(c_1 + c_2)$. Now we consider the following two cases.

5.3a. STACKELBERG MODEL: STRONG LEADER, WEAK FOLLOWER

Making the necessary calculations we obtain

$$G^* = \begin{cases} \frac{A}{2c_1}, & \text{if} \quad c_2 \leq 2c_1; \\ \frac{A}{c_2}, & \text{otherwise;} \end{cases}$$

in both cases, $G^* > \hat{G}$;

$$q_1^* = \begin{cases} \frac{Ac_2}{4c_1^2}, & \text{if} \quad c_2 \leq 2c_1; \\ \frac{A}{c_2}, & \text{otherwise;} \end{cases}$$

in both cases, $q_1^* > \hat{q}_1$;

$$q_2^* = \begin{cases} \frac{A(2c_1 - c_2)}{4c_1^2}, & \text{if} \quad c_2 \leq 2c_1; \\ 0, & \text{otherwise;} \end{cases}$$

in both cases, $q_2^* < \hat{q}_2$;

$$\mu_1^* = \begin{cases} \frac{Ac_2}{4c_1}, & \text{if} \quad c_2 \leq 2c_1; \\ \frac{A(c_2 - c_1)}{c_2}, & \text{otherwise;} \end{cases}$$

in both cases, $\mu_1^* > \hat{\mu}_1$;

$$\mu_2^* = \begin{cases} \frac{A(2c_1 - c_2)^2}{4c_1^2}, & \text{if} \quad c_2 \leq 2c_1; \\ 0, & \text{otherwise;} \end{cases}$$

in both cases, $\mu_2^* < \hat{\mu}_2$.

Therefore, in case of the leadership of the strong firm, both the total output and the individual outputs grow up, hence the price goes down, which is good for consumers. Moreover, the leader increases his profit in comparison to the Cournot case, whereas the profit of the follower decreases.

5.3b. STACKELBERG MODEL: WEAK LEADER, STRONG FOLLOWER

Again, after calculations one easily gets

$$G^* = \frac{A}{2c_2} < \frac{A}{c_1 + c_2} = \hat{G};$$

$$q_1^* = \frac{A}{4c_2^2}(2c_2 - c_1) < \frac{Ac_2}{(c_1 + c_2)^2} = \hat{q}_1;$$

$$q_2^* = \frac{Ac_1}{4c_2^2} < \frac{Ac_1}{(c_1 + c_2)^2} = \hat{q}_2;$$

$$\mu_1^* = \frac{A}{4c_2^2}(2c_2 - c_1)^2 > \frac{Ac_2^2}{(c_1 + c_2)^2} = \hat{\mu}_1;$$

$$\mu_2^* = \frac{Ac_1}{4c_2} > \frac{Ac_1^2}{(c_1 + c_2)^2} = \hat{\mu}_2.$$

Here we see that under the leadership of the weak firm, while the follower is a a strong firm, both the total output and the individual outputs fall down (which is bad for consumers as the price goes up!). As for the profits of the firms, they both gain more in comparison to the standard Cournot case. ■

5.3.4 COURNOT OLIGOPOLY VERSUS PERFECT COMPETITION

In this subsection, we consider the homogeneous good market with two firms that have quadratic cost functions. We compare the standard Cournot oligopoly with the perfect competition case, i.e. when $w_i \equiv 0$, $i = 1, 2$. Hence, we have

$$n = 2; \quad f_i(q_i) = c_i q_i^2, \ c_i > 0, i = 1, 2; \quad p(G) = AG^{-1}, A > 0; \ Q = 0.$$

5.4a. COURNOT MODEL.

Find a $(G, q_1, q_2) \geq 0$ such that

$$q_1 + q_2 = G;$$

$$\varphi_i \equiv 2c_i q_i + \frac{A}{G^2} q_i - \frac{A}{G} \geq 0, \quad q_i \varphi_i = 0, \ i = 1, 2.$$

Then

$$G_C^* = \frac{\sqrt[4]{A}}{\sqrt[4]{4c_1 c_2}};$$

$$q_1^* = \frac{\sqrt{A}}{\sqrt{2}(\sqrt{c_1}+\sqrt{c_2})} \cdot \sqrt[4]{c_2/c_1}; \quad q_2^* = \frac{\sqrt{A}}{\sqrt{2}(\sqrt{c_1}+\sqrt{c_2})} \cdot \sqrt[4]{c_1/c_2};$$

$$\mu_1^* = \frac{A}{2(\sqrt{c_1}+\sqrt{c_2})^2}\left(2c_2+\sqrt{c_1c_2}\right);$$

$$\mu_2^* = \frac{A}{2(\sqrt{c_1}+\sqrt{c_2})^2}\left(2c_1+\sqrt{c_1c_2}\right).$$

5.4b. PERFECT COMPETITION ($w_i \equiv 0$).

Find a $(G, q_1, q_2) \geq 0$ such that

$$q_1 + q_2 = G;$$

$$\varphi_i \equiv 2c_i q_i - \frac{A}{G} \geq 0, \quad q_i\varphi_i = 0, \; i = 1,2.$$

Then

$$G_P^* = \sqrt{\frac{A(c_1+c_2)}{2c_1c_2}} > G_C^*;$$

As for the relationships between $q_{i,P}^*$ and q_i^*, it depends upon the relative value $\gamma = c_1/c_2$. The stronger agent always raises up its output, i.e. if $\gamma \leq 1$ then

$$q_{1,P}^* = \sqrt{\frac{Ac_2}{2c_1(c_1+c_2)}} > q_1^*|_{Cournot},$$

whereas the output of the weaker agent can be less than that of the classic Cournot equilibrum, i.e.

$$q_{2,P}^* = \sqrt{\frac{Ac_1}{2c_2(c_1+c_2)}} < q_2^*|_{Cournot},$$

if $0 < \gamma < \gamma^*$, where γ^* is the unique positive root of the cubic polynomial $y = x^3 - x^2 - x - 1$.

However, for the profits we have

$$\mu_{1,P}^* = \frac{Ac_2}{2(c_1+c_2)} < \mu_1^*,$$

$$\mu_{2,P}^* = \frac{Ac_1}{2(c_1+c_2)} < \mu_2^*.$$

As we see, the perfect competition leads to a higher total output but lower profits for both agents, in comparison to the standard Cournot model. ∎

5.4 COMPUTING THE STACKELBERG EQUILIBRIUM

In order to solve problem (5.15) introduced in Section 5.2 and describing the Stackelberg model with one leader, we need determining numerically the optimal responses of the followers $i = 2, \ldots, n$ to the leader's output $Q \geq 0$. In order to do that, we fix a positive extraneous supply $Q > 0$ and consider the dynamic system defined by the differential equation

$$\dot{G} = -G + Q + \sum_{i \in I(G)} q_i(G) \equiv T(G), \quad G(0) = Q; \tag{5.26}$$

here $q_i(G)$ is the (unique) solution of problem (5.4)–(5.5) with the influence quotients defined by formulas (5.1)–(5.2), i.e. given a $G > 0$, find $q_i = q_i(G)$, $i = 2, \ldots, n$, such that

$$q_i \geq 0, \quad \varphi_i(G, q_i) \equiv f_i'(q_i) - \alpha_i q_i p'(G) - \sigma_i G p'(G) - p(G) \geq 0, \tag{5.27}$$

$$\text{and} \qquad q_i \varphi_i(G, q_i) = 0; \tag{5.28}$$

finally, recall that like in the previous chapter,

$$I(G) = \{2 \leq i \leq n \mid q_i(G) > 0\}.$$

It is easily verified that each solution $G = G(Q)$ of equation (5.6) is a stationary point of dynamic system (5.26); its stability depends upon the sign of the derivative $\frac{dT}{dG}$ of the righthand side in (5.26) at the considered point. By differentiating $T(G)$ we obtain

$$\frac{dT}{dG} = -1 + \sum_{i \in I(G)} q_i'(G). \tag{5.29}$$

As it was established by Lemma 4.3, the inequalities

$$Q > 0 \qquad \text{and} \qquad \sum_{i \in I(G)} q_i(G) \leq G - Q$$

imply the estimate $\sum_{i \in I(G)} q_i'(G) < 1$. The latter, together with (5.29), guarantees that given a fixed extraneous supply $Q > 0$, the solution of problem (5.15) under constraints (5.27)–(5.28) is a *stable* stationary point of dynamic system (5.26).

Dynamic system (5.26) can be integrated numerically, for instance, by the following iteration process:

$$\begin{cases} G^{k+1} = G^k - \delta_k \left[G^k - Q - \sum_{i \in I(G^k)} q_i(G^k) \right], & k = 0, 1, 2, \ldots, \\ G^0 = Q, & \delta_k > 0. \end{cases} \tag{5.30}$$

To prove the convergence of the iteration process (5.30) to the stationary point $\bar{G} = G(Q)$, we assume that the inverse demand function $p = p(G)$ is convex over its domain $G \geq 0$. Then it follows from assumptions E1, E6 and the explicit formula (4.33) of the derivative $q_i'(G)$ that

$$\sum_{i \in I(G)} q_i'(G) \geq - \sum_{i \in I(G)} \frac{1-\sigma_i}{\alpha_i} \geq - \sum_{i=2}^{n} \frac{1-\sigma_i}{\alpha_i}. \tag{5.31}$$

Furthermore, we use the fact that $\bar{G}$ is a fixed point of the mapping involved in the righthand side of (5.30). Hence,

$$G^{k+1} - \bar{G} = G^k - \bar{G} - \delta_k \left[\left(G^k - \bar{G}\right) - \sum_{i \in I(G^k)} q_i(G^k) + \sum_{i \in I(\bar{G})} q_i(\bar{G}) \right] =$$

$$= (1-\delta_k)\left(G^k - \bar{G}\right) + \delta_k \left[\sum_{i \in I(G^k)} q_i(G^k) - \sum_{i \in I(\bar{G})} q_i(\bar{G}) \right]. \tag{5.32}$$

If the value $\bar{G}$ is nonsingular in the sense that $I(\bar{G}-0) = I(\bar{G}+0)$ and G^k is close enough to $\bar{G}$ so that $I(G^k) = I(\bar{G})$, then (5.32) implies

$$G^{k+1} - \bar{G} = (1-\delta_k)\left(G^k - \bar{G}\right) + \delta_k \sum_{i \in I(G_m)} q_i'(G_m)(G^k - \bar{G}) =$$

$$= \left\{ 1 - \delta_k \left[1 - \sum_{i \in I(G_m)} q_i'(G_m) \right] \right\} \left(G^k - \bar{G}\right), \tag{5.33}$$

where G_m is an intermediate point between G^k and $\bar{G}$. Therefore, if we choose

$$0 < \bar{\delta} \leq \delta_k \leq \frac{1}{1 + \sum_{i=2}^{n} \frac{1-\sigma_i}{\alpha_i}}, \qquad k = 0, 1, \ldots,$$

we can expect the locally linear convergence of G^k to $\bar{G} = G(Q)$ from above.

At last, the optimal output Q^* of the leader (extraneous supplier) can be determined by the methods of unvariate maximization of his profit function $\mu_1(Q)$.

5.5 COMMENTS

In case of the leadership in the choice of the quantity of product, one agent makes his/her decision before the others. Now this type of model is called the Stackelberg one, in honor of the German economist Heinrich von Stackelberg who was the first to study the relationships between the leader and the follower. His results were published in 1934 the book the market stucture (Stackelberg, 1934).

Until recently, the Stackelberg model traditionally consisted of only two agents: the leader and the follower. In papers by Sherali e.a. (1983) and

Outrata (1993), the extended model with a single leader but more than one follower was examined.

To our knowledge, the paper by Bulavsky and Kalashnikov (1995) was the first to consider the generalized case of multiple leaders. In this paper, both classic and extended (with several leaders) Stackelberg models were embedded in the generalized oligopoly model by the explicit construction of appropriate influence functions $w_i(G, q_i)$ for the leaders.

REFERENCES

Bulavsky VA, Kalashnikov VV. Equilibria in generalized Cournot and Stackelberg models. Ekonomika i Matematicheskie Metody (*in Russian*) 1995; 31: 164-176.

Harker PT, Pang JS. Finite-dimensional variational inequalities and nonlinear complementarity problems: a survey of theory, algorithms and applications. Math. Programming 1990; 48: 161-220.

Outrata JV. On necessary optimality conditions for Stackelberg problems. J. Optim. Theory Appl. 1993; 76: 305-320.

Sherali HD, Soyster AL and Murphy FH. Stackelberg-Nash-Cournot equilibria: characterizations and computations. Operations Research 1983; 31: 171-186.

Stackelberg, H. *Marktform und Gleichgewicht.* Vienna: Julius Springer, 1934.

PART II.

GENERAL COMPLEMENTARITY PROBLEMS

Chapter 6

COMPLEMENTARITY PROBLEMS WITH RESPECT TO GENERAL CONES

The topological degree theory is a powerful tool to study the problem of existence of solutions to nonlinear equations. Since recently, this theory has been widely used in applications to a more general problem, namely, the complementarity problem (CP). In this chapter, we consider both the complementarity problem with respect to an arbitrary cone of the Euclidean space, and its variations: the standard CP (with the nonnegative orthant R^n_+ as the cone), the implicit CP, the general order CP, and the semidefinite CP defined for the real matrices. Having specified the results of Chapter 1 of this book (Proposition 1.5, Proposition 1.6) and definitions of exceptional family of elements for the particular types of complementarity problems, we obtain existence theorems.

6.1 INTRODUCTION AND PRELIMINARIES

The standard complementarity problem in R^n_+ is usually specified as follows. Given a continuous mapping $f : R^n_+ \to R^n$, find a vector $x \in R^n_+$ such that

$$f(x) \geq 0, \qquad x^T f(x) = 0. \tag{6.1}$$

In many results concerning the existence of solutions to (6.1) it is demanded that the mapping f be more or less monotone. For instance, the existence of a solution is guaranteed if either

(i) the mapping f is strongly monotone over R^n_+, i.e. there exists a scalar $\beta > 0$ such that

$$(x-y)^T\left[f(x)-f(y)\right] \geq \beta\|x-y\|^2, \quad \forall x,y \in R^n_+, \tag{6.2}$$

or (a somewhat weaker condition)

(*ii*) the mapping f is locally strongly monotone, i.e. there exists a (non-increasing) function $\beta : R_+ \to R_+$ such that

$$(x-y)^T[f(x)-f(y)] \geq \beta(t)\|x-y\|^2 \tag{6.3}$$

for every $x, y \in R^n_+ \cap \{z \in R^n : \|z\| \leq t\}$, and

$$\lim_{t\to+\infty} \beta(t)t = +\infty, \tag{6.4}$$

or

(*iii*) mapping f is monotone, i.e.

$$(x-y)^T[f(x)-f(y)] \geq 0 \qquad \forall x, y \in R^n_+, \tag{6.5}$$

and there is a vector $x^0 > 0$ with $f(x^0) > 0$ (cf. McLinden (1980)).

Here we present another approach to problem (6.1) and obtain assertions in the form of a (nonstrict) *existence alternative*: for the continuous mapping f there exists either a solution to (6.1) or a family of elements of a special type. After having got those assertions, one can construct various sufficient existence conditions based upon elimination of the second outcome. Making use of this approach, we deduce easily both the above-mentioned and some new existence results.

The following notion was introduced in Smith (1984).

DEFINITION 6.1. A sequence of elements $\{x^k\}_{k=1}^{\infty} \subset R^n_+$ is called *exceptional*, if for each $k = 1, 2, \ldots,$ the following relationships hold:

1) $\|x^k\| = k$,

2) $f_i(x^k) \begin{cases} = -\lambda_k x_i^k, & \text{if } x_i^k > 0; \\ \geq 0, & \text{if } x_i^k = 0; \end{cases}$

here $\lambda_k > 0$ is a scalar.

In Smith (1984), the above-mentioned alternative is shown to take place for each continuous mapping $f : R^n_+ \to R^n$. Notice that condition 2) in Definition 2.1 can be replaced with a tantamount one as follows: one need not require that the scalar $\lambda_k > 0$ be the same for each i with $x_i^k > 0$; it suffices to assume simply that $f_i(x^k) < 0$ if $x_i^k > 0$.

In order to develop our approach, we pass on from the sequence to a continuum family of elements that looks more natural. Namely, in this chapter, we introduce notions of exceptional and regular exceptional families of elements and reformulate the alternative by Smith (1984) in a more general form: For each continuous mapping f there exists either a solution to problem (6.1) or a (regular) family of elements. Based upon that alternative, we prove both the well-known and new sufficient conditions of existence of solutions to the complementarity problems. Like in Smith (1984), we do not demand that the mapping f be monotone or coercive but in contrast to Isac (1992), Smith (1984), we make use of the topological degree theory.

Recall briefly the notion and some properties of the topological degree of a continuous mapping.

Given a continuous mapping $F : D \subset R^n \rightarrow R^n$ and a vector $y \in R^n$, one often need knowing the number of solutions to the equation $F(x) = y$ belonging to a particular set $C \subset R^n$. A straightforward difficulty arises: the number of solutions in general does not depend continuously upon either F or y. However, one can avoid this difficulty if one counts first the number of solutions $x \in C$ for which the mapping F keeps its "orientation" at some neighborhood of the point x, and then subtracts from it the number of those solutions in C at which the "orientation" switches to the opposite. Thus obtained natural number is called the *(topological) degree* of the mapping F at the point y with respect to the set C, and it is often denoted by the symbol $\deg(F, C, y)$.

The above definition of degree is by no means rigorous, as we have neither defined what the keeping or changing of the "orientation" is, nor described precisely the class of sets C to be considered. Nevertheless, even this nonrigorous definition shows that whenever $\deg(F, C, y) \neq 0$, at least one solution to the system $F(x) = y$ exists and belongs to the set C.

The notion of degree was originally introduced in the combinatorial topology, but its purely analytical definitions also exist (*cf.* Ortega and Rheinboldt (1970), or Lloyd (1978)).

Now we cite some key properties of the topological degree that will be used below. Both the theorems by Poincaré-Bohl and Kronecker can be found in Ortega and Rheinboldt (1970), or in Lloyd (1978).

POINCARÉ–BOHL THEOREM. *Let $C \subset R^n$ be an open bounded subset and $F, G : \mathrm{cl}\ C \rightarrow R^n$ be two continuous mappings. If $y \in R^n$ is an arbitrary vector satisfying the condition*

$$y \neq H(z,t), \qquad z \in \partial C, \quad t \in [0,1], \tag{6.6}$$

where $H : \mathrm{cl}\ C \times [0,1] \rightarrow R^n$ is the homotopy

$$H(z,t) = tG(z) + (1-t)F(z),$$

then $\deg(G, C, y) = \deg(F, C, y)$. ■

KRONECKER THEOREM. *Let $C \subset R^n$ be an open bounded subset and $F : \mathrm{cl}\ C \subset R^n \rightarrow R^n$ be a continuous mapping. If $y \notin F(\partial C)$ and* $\deg(F, C, y) \neq 0$*, then equation $F(x) = y$ has a solution in C.* ■

The chapter is organized as follows. In Section 6.2, definitions of the exceptional and normed exceptional families of elements are given and theorems on the alternative for the general complementarity problem (with respect to a cone in R^n) are proven. Sufficient conditions guaranteeing the absence of exceptional families of elements and thus providing for the existence of solutions to the complementarity problem, are presented in Section 6.3. Sections 6.4, 6.5, 6.6 and 6.7 are dedicated to the analogous investigations of the standard, implicit, and general order complementarity problems, respectively. All the results of this chapter were published in Bulavsky, Isac and Kalashnikov (1998,1999,2001).

6.2 COMPLEMENTARITY PROBLEM WITH RESPECT TO A GENERAL CONE

Let $f : R^n \to R^n$ be a continuous mapping and $K \subset R^n$ be a closed convex cone with the origin as its vertex. Denote by K^* the dual cone to K, i.e.

$$K^* = \{y \in R^n \mid y^T x \geq 0 \quad \forall x \in K\}$$

which also is a closed convex cone. Consider the general complementarity problem: find a vector $x \in K$ such that

$$f(x) \in K^*, \qquad x^T f(x) = 0. \tag{6.7}$$

The cone K being a closed convex set, every vector $z \in R^n$ has the uniquely determined projection onto K:

$$z^+ = \mathrm{Pr}_K z = \arg\min\{\|z - x\|, \ x \in K\}. \tag{6.8}$$

We introduce $z^- = z^+ - z$ and remark that z^+ and z^- depend continuously upon z.

LEMMA 6.1. *For each closed convex cone K with the vertex at the origin and for every vector $z \in R^n$, the following relationships hold: $z^- \in K^*$ and $(z^-)^T z^+ = 0$.*

PROOF. Making use of the first order optimality condition for problem (6.8), we write down the inequality

$$(z - z^+)^T (z^+ - w) \geq 0 \qquad \forall w \in K. \tag{6.9}$$

Substituting $w = 0$ and $w = 2z^+$ into it, we obtain $(z^-)^T z^+ = 0$. Now opening the brackets in (6.9) and using the latter equality we deduce the relationship

$$(z^-)^T w \geq 0 \qquad \forall w \in K,$$

which means that $z^- \in K^*$ and thus completes the proof. ∎

The assertion of Lemma 6.1 means that z^- is the *normal vector to the supporting hyperplane* of the cone K at the point z^+.

DEFINITION 6.2. The set of elements $\{z_r\}_{r>0} \subset K$ is called *an exceptional family of elements (EFE)* with respect to the cone K for the mapping f, if $\|z_r\| \to +\infty$ as $r \to +\infty$, and for each $r > 0$ there exists a scalar $\mu_r > 0$ such that the vector $s_r = f(z_r) + \mu_r z_r$ is the normal one to the supporting hyperplane of the cone K at the point z_r, i.e. $s_r \in K^*$, and $s_r^T z^r = 0$.

Remark. Definition 6.2 reveals a particular case of Definition 1.6 of the exceptional family when the cone is selected for the set K, and the closed balls centered at the origin are selected for the system Σ. Because of that, the theorem below is a direct consequence of Proposition 1.5. However it is methodically interesting to present its previous proof using the Topological Degree Theory.

THEOREM 6.1. *For each continuous mapping $f : K \to R^n$, there exists either a solution to problem* (6.7) *or an exceptional family of elements.*

PROOF. Consider the equation

$$F(z) = f(z^+) - z^- = 0, \tag{6.10}$$

where z^+ is projection (6.8), and $z^- = z^+ - z$. It is easy to see that problems (6.7) and (6.10) are tantamount in the following sense. If x solves (6.7), then $z = x - f(x)$ satisfies equation (6.10). Conversely, if z is a solution to (6.10), then $x = z^+$ solves problem (6.7).

In order to investigate problem (6.10) in detail, consider the family of spheres S_r and open balls B_r:

$$S_r = \{z \in R^n \mid \|z\| = r\},$$

$$B_r = \{z \in R^n \mid \|z\| < r\}.$$

Now examine the standard homotopy $H(z,t)$ of the identity mapping $G(z) = z$ and the continuous mapping $F(z)$:

$$H(z,t) = tz + (1-t)F(z), \quad 0 \le t \le 1.$$

In order to apply the Poincaré–Bohl Theorem with $y = 0$, $C = B_r$ we specify the standard homotopy $H(z,t)$ for an arbitrary $z \in \partial B_r = S_r$ and $t \in [0,1]$:

$$H(z,t) = tz^+ + (1-t)f(z^+) - z^-. \tag{6.11}$$

Now two outcomes are possible.

A) There is a value $r > 0$ such that

$$H(z,t) \neq 0, \quad \forall\, z \in S_r, \quad \forall\, t \in [0,1].$$

Then the Poincaré–Bohl theorem allows one to conclude that

$$\deg(F, B_r, 0) = \deg(G, B_r, 0).$$

The degree of the identity mapping is equal to the unity: $\deg(G, B_r, 0) = 1$, hence $\deg(F, B_r, 0) = 1$. Now as the Kronecker theorem implies the existence of a solution to (1.10) within the ball cl B_r, hence problem (6.7) is solvable, too.

B) For every $r > 0$, there exist a point $z_r \in S_r$ and a scalar $t_r \in [0, 1)$ such that

$$H(z_r, t_r) = 0. \tag{6.12}$$

If $t_r = 0$, then z_r solves (6.10), which again implies the solvability of problem (6.7). Otherwise, if $t_r > 0$, then it follows from (6.11) and (6.12) that

$$t_r z_r^+ + (1 - t_r) f(z_r^+) = z_r^- . \tag{6.13}$$

Dividing both parts of the latter equality by $(1 - t_r)$, we obtain

$$f(z_r^+) + \frac{t_r}{1 - t_r} z_r^+ = \frac{1}{1 - t_r} z_r^- . \tag{6.14}$$

Now we set $\mu_r = t_r/(1 - t_r)$ and verify that $\{z_r^+\}$ is an exceptional family of elements for K and f. Indeed, from (6.14) and Lemma 6.1, it follows that the vector $s_r = f(z_r^+) + \mu_r z_r^+$ is the normal one of the supporing hyperplane to the cone K at the point z_r^+. In order to show that the family $\{z_r^+\}$ is an exceptional family of elements, one need verifying that $\|z_r^+\| \to +\infty$ as $r \to +\infty$. On the contrary, suppose that the family $\{z_r^+\}$ is bounded, hence it has a finite accumulation point. On one hand, the equality $\|z_r^-\| = \sqrt{r^2 - \|z_r^+\|^2}$ implies that the righthand side of (6.13) comprises an unbounded family of points. On the other hand, the respective vectors in the lefthand side of (6.13) compose a bounded family due to the continuity of the mapping f. Thus obtained contradiction completes the proof. ■

DEFINITION 6.3. The exceptional family of elements $\{z_r\}$ is referred to as a *regular* one (REFE), if

$$\|z_r\| = r \qquad \forall r > 0.$$

Modifying the proof of Theorem 6.1, we obtain the following result.

THEOREM 6.2. *For each continuous mapping $f : K \to R^n$ there exists either a solution to problem* (6.7), *or a regular exceptional family of elements (REFE).*

PROOF. Examine again equation (6.10). For $r > 0$ we denote

$$W_r = \{z \in R^n \mid \|z^+\| < r\}, \quad M_r = \sup_{z \in W_r} \|f(z^+)\|,$$

$$P = 1 + \sqrt{(\max\{r, M_r\})^2 + r^2}.$$

As before, z^+ is the projection of z onto cone K, defined by (6.8). It is easy to verify that

$$M_r \le \max_{z \in cl B_r \cap K} \|f(z)\| < +\infty,$$

hence, the scalar P is finite. Now define

$$D_r = W_r \cap B_P,$$

where B_P is the open ball of radius P, and apply the Poincaré–Bohl Theorem to the set $C = D_r$ and homotopy (6.11). Again, it suffices to consider two cases.

A) There exists a $r > 0$ such that

$$H(z,t) \neq 0 \qquad \forall\ z \in \partial D_r, \quad \forall\ t \in [0,1].$$

Now repeating exactly the proof of Theorem 6.1, we show the existence of a solution to equation (6.10), and therefore, the solvability of problem (6.7).

B) For every $r > 0$, there exist an element $z_r \in \partial D_r$ and a scalar $t_r \in [0,1)$ such that

$$H(z_r, t_r) = 0.$$

If $t_r = 0$, then the point z_r satisfies equation (6.10), consequently, the vector $x^r = z_r^+$ solves (6.7). Otherwise, if $t_r > 0$, then from (6.13)–(6.16) and Lemma 6.1, we deduce that $s_r = f(z_r^+) + \mu_r z_r^+$ is the normal vector of the supporting hyperplane to the cone K at the point z_r^+; here $\mu_r = t_r/(1-t_r)$. In order to show that $\|z_r^+\| = r$, examine the boundary ∂D_r in detail. It is straightforward that $\partial D_r = V_r \cup U_P$, where

$$V_r = \left\{ z \in R^n \mid \|z^+\| = r,\ \|z\| \leq P \right\} = \partial W_r \cap \text{cl}\ B_P,$$

and

$$U_P = \text{cl}\ W_r \cap S_P.$$

Now verify that $z_r \notin U_P$. Indeed, equality (6.13), the triangle inequality, and the definitions of W_r and M_r imply the relationships

$$\|z_r^-\| \leq \max\left\{ \|z_r^+\|, \|f(z_r^+)\| \right\} \leq \max\left\{ r, M_r \right\}.$$

Making use of the definition of P, we obtain the estimate

$$\|z_r\|^2 = \|z_r^+\|^2 + \|z_r^-\|^2 \leq r^2 + \left(\max\{r, M_r\}\right)^2 = P^2 - 1,$$

which means that $\|z_r\| < P$. Thus we have demonstrated that $z_r \in V_r$, hence $\|z_r^+\| = r$. The exceptional family of elements $\left\{z_r^+\right\}_{r>0}$ is shown to be regular which completes the proof. ∎

6.3 SUFFICIENT EXISTENCE CONDITIONS

By making use of Theorems 6.1 and 6.2, one can both obtain new proofs of some previously known results and also prove some new sufficient conditions that guarantee the existence of solutions to the complementarity problem. We begin with the following well-known result.

PROPOSITION 6.1 [ISAC]. *Let $f : K \to R^n$ be a continuous mapping and there exist a function $\varphi : [0, +\infty) \to [0, +\infty)$ with the properties*

$$\lim_{t \to +\infty} \varphi(t) = +\infty, \quad \varphi(t) > 0 \textit{ for } t > 0, \textit{ and}$$

$$(x-y)^T \left[f(x) - f(y)\right] \geq \|x-y\| \varphi\left(\|x-y\|\right), \ \forall x, y \in K. \tag{6.15}$$

Then problem (6.7) *with the mapping f is solvable.*

PROOF. Suppose that there exists an exceptional family of elements $\{z_r\}_{r>0} \subset K$ for the mapping f. According to the definition of exceptional family of elements, vectors $s_r = f(z_r) + \mu_r z_r$ and $s_1 = f(z_1) + \mu_1 z_1$ are the normal ones to the supporing hyperplanes for the cone K at the points z_r and z_1, respectively. Due to that fact, one can write down the following sequence of relationships

$$(z_r - z_1)^T [f(z_r) - f(z_1)] = (z_r - z_1)^T (s_r - s_1 - \mu_r z_r + \mu_1 z_1) =$$
$$= (z_r - z_1)^T s_r - (z_r - z_1)^T s_1 - (z_r - z_1)^T (\mu_r z_r - \mu_1 z_1) \leq$$
$$\leq -(z_r - z_1)^T [\mu_r (z_r - z_1) - (\mu_1 - \mu_r) z_1] =$$
$$= -\mu_r \|z_r - z_1\|^2 + (\mu_1 - \mu_r)(z_r - z_1)^T z_1. \tag{6.16}$$

By making use of conditions (6.15), (6.16), and the Cauchy-Buniakowski inequality, we finally obtain

$$\varphi(\|z_r - z_1\|) \leq -\mu_r \|z_r - z_1\| + |\mu_1 - \mu_r| \|z_1\|,$$

which contradicts the condition $\varphi(t) \to +\infty$ as $t \to +\infty$; here $\|z_r - z_1\|$ serves as t when $r \to +\infty$. Thus the cone K and mapping f are without exceptional families of elements. Therefore, problem (6.7) is solvable, and the proposition is proved. ∎

PROPOSITION 6.2 [ISAC]. *Consider a continuous mapping $f : K \to R^n$ and a bounded subset $C \subset K$ such that for every $x \in K \setminus C$ there exists a $y \in C$ with*

$$(x-y)^T f(x) \geq 0. \tag{6.17}$$

Then problem (6.7) *has a solution.*

PROOF. Suppose that there is an exceptional family of elements $\{z_r\} \subset K$. Starting from a certain number $r_0 > 0$ onward, the points z_r belong to $K \setminus C$. Then for the point y_r mentioned in the proposition assumptions and associated with z_r, the following relationships hold:

$$0 \leq (z_r - y_r)^T f(z_r) = (z_r - y_r)^T (s_r - \mu_r z_r) = (z_r - y_r)^T s_r -$$
$$-\mu_r \|z_r\|^2 + \mu_r y_r^T z_r \leq -\mu_r \|z_r\| \left(\|z_r\| - \|y_r\|\right) < 0,$$

and lead to the impossible inequality $0 < 0$. Therefore, no exceptional family of elements exists, and problem (6.7) is solvable by virtue of Theorem 6.1. ∎

EXAMPLE 6.1. The mapping $f : R^2 \to R^2$ with the components

$$f_1(x_1, x_2) = x_1 + 3\sin x_2 - 1, \quad f_2(x_1, x_2) = x_2 + 3\sin x_1 - 1$$

is not monotone, but is continuous and satisfies (6.17) over the nonnegative orthant $K = R^2_+$ (the unit ball B_1 can be picked up as the set C). According to Proposition 1.2, the complementarity problem for this mapping has a solution in the cone K. In particular, the solution set comprises the points $(1; 0)$, $(0; 1)$ and $(\xi; \xi)$, where $0 < \xi < 1$ solves the equation $\xi + 3\sin\xi = 1$. ■

COROLLARY 6.1. *Let $f : K \to R^n$ be a continuous mapping and $C \subset K$ a bounded subset such that $y^T f(y) \geq 0$ $\forall y \in K \setminus C$. Then problem (6.7) is solvable. If the strict inequality $y^T f(y) > 0$ takes place for each $y \in K \setminus C$, then all the solutions of problem (6.7) belong to the subset C.* ■

COROLLARY 6.2. *Let $f : K \to R^n$ be a continuous mapping and $C \subset K$ a bounded subset such that $f(y) \in K^*$ $\forall y \in K \setminus C$. Then problem (6.7) is solvable. Moreover, if $0 \in C$, and the stronger condition holds $f(y) \in \text{int } K^*$ for every $y \in K \setminus C$, then all the solutions of problem (6.7) belong to the subset C.* ■

Now we consider a compact set $C \subset K$ which is *star-shaped* with respect to the origin, i.e. it contains the whole segment $[0, x]$ whenever the vector x belongs to C. Then the function $\eta(x)$ is well-defined that maps $x \in K, x \neq 0$ to the most distant (from the origin) point of the set C in the ray running through 0 and the point x. It is clear that the subset $\Gamma = \{y = \eta(x) \mid x \in K, x \neq 0\}$ constitutes the relative boundary of the subset C with respect to the cone K.

THEOREM 6.3. *Consider a continuous mapping $f : K \to R^n$ and a nonempty subset $C \subset K$ that is compact and star-shaped with respect to the origin. If the function $\eta(x)$ is continuous, $\eta(x) \neq 0$, and $y^T f(y) \geq 0$ for every $y \in \Gamma$, then problem (6.7) has a solution in the subset C.*

PROOF. Define a mapping $g : K \to R^n$ as follows:

$$g(x) = \begin{cases} f(x), & \text{if } x \in C; \\ f(\eta(x)) + \|x - \eta(x)\| x, & \text{if } x \in K \setminus C. \end{cases} \tag{6.18}$$

The mapping g is clearly continuous over K and satisfies the following inequality

$$x^T g(x) > 0 \qquad x \in K \setminus C.$$

The latter provides for the existence of a solution to problem (6.7) with the mapping g on the subset C. However, due to (6.18), mappings g and f coincide on C. Therefore, problem (6.7) with the mapping f is solvable over the subset C, too. This completes the proof. ■

EXAMPLE 6.2. We consider the mapping $f : R^2_+ \to R^2$ with the components

$$f_1(x_1, x_2) = 3x_1 - x_1 x_2 - 1, \qquad f_2(x_1, x_2) = 3x_2 - x_1 x_2 + 1.$$

and take the simplex $\{(x_1, x_2) \in K = R^2_+ \mid x_1 + x_2 \leq 1\}$ for the subset C. It is easy to see that the latter mapping and the subset satisfy the assumptions of Theorem 6.3. Indeed, for the relative boundary of the simplex we obtain

$$(\xi, 1-\xi) \cdot f(\xi, 1-\xi) = (\xi, 1-\xi) \begin{pmatrix} \xi^2 + 2\xi - 1 \\ (1-\xi)^2 + 2(1-\xi) + 1 \end{pmatrix} =$$

$$= \xi^3 + (1-\xi)^3 + 2\left[\xi^2 + (1-\xi)^2\right] + 1 - 2\xi.$$

By transforming the linear and quadratic terms, we come to the following inequality:

$$2\left[\xi^2 + (1-\xi)^2\right] + 1 - 2\xi = 4\xi^2 - 6\xi + 3 \geq 0 \quad \forall 0 \leq \xi \leq 1.$$

Finally, we get

$$(\xi, 1-\xi) \cdot f(\xi, 1-\xi) \geq \xi^3 + (1-\xi)^3 \geq 0 \qquad \forall 0 \leq \xi \leq 1.$$

Therefore, problem (6.7) has a solution in the simplex C, although the mapping f is clearly non-monotone. For instance, the point $\left(\frac{1}{3}; 0\right)$ solves the problem. ∎

Now we extend Theorem 6.3 by giving up the continuity of the function η. In order to do that, we first reduce the domain of the function η to the subset $\Omega = S_1 \cap K$ which is the intersection of the unit sphere (centered at the origin) and the cone K. We assume again that $\eta(x) \neq 0 \ \forall x \in \Omega$ but do not demand the function η to be continuous over Ω. By doing that we allow the subset C to have the boundary which contains segments of straight rays going from the origin.

THEOREM 6.4. *Let $f : K \to R^n$ be a continuous mapping and $C \subset K$ be a nonempty compact subset which also is star-shaped with respect to the origin, and $0 \notin \Gamma$. If $y^T f(y) \geq 0$ for each point y of the relative boundary Γ, then problem (6.7) has a solution in the subset C.*

PROOF. Consider the function $\zeta : \Omega \to R$ defined by $\eta(x) = \zeta(x)x$ for $x \in \Omega$. Since C is compact, the function ζ is upper semicontinuous over Ω. It implies the inequality

$$\limsup_{k \to \infty} \zeta(x^k) \leq \zeta(x)$$

holding for each $x \in \Omega$ and every sequence of points $\{x^k\} \subset \Omega$ convergent to x. Hence, the Lebesgue subsets $\{x \in \Omega \mid \zeta(x) \geq t\}$ of the function ζ are closed for every $t \in R$. Therefore, the function ζ is Lebesgue summable over the subset Ω. Now fix a scalar $\varepsilon > 0$ and consider a family of standard nonnegative finite functions $\{\varphi\}_{x \in S_1}$, each of them being infinitely smooth and taking positive values over the neighborhood $U_\varepsilon(x) = \{y \in S_1 \mid \|x - y\| < \varepsilon\}$ of the point x and equal to zero outside it, and

$$\int_{U_\varepsilon(x)} \varphi(y) dy = 1. \tag{6.19}$$

Furthermore, extend the function ζ to the points y of the subset $S_1 \setminus \Omega$ which belong to the ε-neighborhood of the set Ω, by the rule

$$\zeta(y) = \zeta\left(\mu \mathrm{Pr}_K(y)\right), \quad y \in (S_1 \setminus \Omega) \bigcap \left(\cup_{x \in \Omega} U_\varepsilon(x)\right),$$

where $\mu > 0$ is such a scalar that $\mu \mathrm{Pr}_K(y) \in \Omega$. Now define $\zeta_\varepsilon : \Omega \to R$ by averaging ζ with respect to the standard finite functions φ as follows:

$$\zeta_\varepsilon(x) = \int\limits_{U_\varepsilon(x)} \zeta(y)\varphi(y)dy, \quad \forall x \in \Omega. \tag{6.20}$$

The function ζ_ε is apparently continuous and bounded over Ω and takes (like ζ) only positive values. Now consider the compact star-shaped subset $C_\varepsilon \subset K$ defined by the function $\eta_\varepsilon(x) = \zeta_\varepsilon(x)x$ whose value is the most distant (from the origin) point belonging to the intersection of the subset C_ε and the straight ray going from the origin and running through the point $x \in \Omega$. In the other words,

$$C_\varepsilon = \bigcup_{x \in \Omega} \left[0, \eta_\varepsilon(x)\right].$$

Now we show the distance between the boundaries of the subsets C_ε and C to be small together with small ε. In order to do that, we first prove that for each $x \in \Omega$, there exists a point $y \in \Omega$ in the neighborhood $U_\varepsilon(x)$ such that the point $v = \zeta_\varepsilon(x)y$ belongs to the (relative) boundary Γ of the subset C. On the contrary, suppose the latter not to be true, then the closedness of C implies the following alternative: either all the points v lie outside the subset C for every $y \in U_\varepsilon(x)$, or vice versa, they all belong to the interior of C. In the former case, the inequality $\zeta_\varepsilon(x) > \zeta(y)\ \forall y \in U_\varepsilon(x)$ holds; otherwise, the opposite inequality $\zeta_\varepsilon(x) < \zeta(y)\ \forall y \in U_\varepsilon(x)$ is valid. In both cases, we average these inequalities by the finite function φ over the neighborhood $U_\varepsilon(x)$ and make use of (6.19) and (6.20), which leads to the impossible relationships $\zeta_\varepsilon(x) > \zeta_\varepsilon(x)$ or $\zeta_\varepsilon(x) < \zeta_\varepsilon(x)$, respectively. Thus, our assumption is false, and the needed point $y \in \Omega$ in the neighborhood $U_\varepsilon(x)$ does exist. Also note that

$$\|\eta_\varepsilon(x) - v\| = \zeta_\varepsilon(x)\|x - y\| \le \zeta_\varepsilon(x)\varepsilon. \tag{6.21}$$

Now fix a ball B so large that the subset $D = B \cap K$ comprises C and also all the subsets C_ε for $0 < \varepsilon < \bar{\varepsilon}$. The mapping f being continuous, the scalar product $x^T f(x)$ is a uniformly continuous function over D. Therefore, the latter has a continuity module $\omega : D \to R$ with the following properties: the function ω is continuous and non-decreasing,

$$\lim_{\delta \to +0} \omega(\delta) = 0, \tag{6.22}$$

and the following inequality holds

$$|x^T f(x) - y^T f(y)| \le \omega(\delta) \tag{6.23}$$

for the points $x, y \in D$ such that $\|x - y\| < \delta$. Now define the continuous mapping f_ε over the cone K as follows: $f_\varepsilon(0) = f(0)$, and

$$f_\varepsilon(x) = f(x) + \omega\left(\zeta_\varepsilon\left(\frac{x}{\|x\|}\right)\varepsilon\right)\frac{x}{\rho_\varepsilon^2}, \qquad \forall x \in K \setminus \{0\}; \tag{6.24}$$

here $\rho_\varepsilon = \min\{\|x\| \mid x \in \Gamma_\varepsilon\}$, and Γ_ε is the relative boundary of C_ε with respect to the cone K. It is easy to verify that

$$y^T f_\varepsilon(y) = y^T f(y) + \omega\left(\zeta_\varepsilon\left(\frac{y}{\|y\|}\right)\varepsilon\right)\frac{\|y\|^2}{\rho_\varepsilon^2} \geq$$

$$\geq y^T f(y) + \omega\left(\zeta_\varepsilon\left(\frac{y}{\|y\|}\right)\varepsilon\right) \qquad \forall y = \eta_\varepsilon(x), x \in K \setminus \{0\}. \tag{6.25}$$

According to the theorem assumptions, $v^T f(v) \geq 0$ at every point v of the relative boundary Γ. Hence, for arbitrary $x \in \Omega$ and $\varepsilon < \bar{\varepsilon}$, relationships (6.21), (6.22) and (6.25) imply

$$[\eta_\varepsilon(x)]^T f_\varepsilon(\eta_\varepsilon(x)) \geq 0.$$

The latter allows us to make use of Theorem 6.3 which provides for the existence of a solution $z_\varepsilon \in C_\varepsilon$ to problem (6.7) with the mapping f_ε. Now as $\varepsilon \to +0$ then the distance between the subsets C_ε and C also vanishes. The subset C being closed, every accumulation point of the solution family z_ε belongs to C. Taking that into account, as well as the continuity of the mapping f and the function ζ_ε, property (6.23) of the continuity module ω, and definition (6.24) of function f_ε, we pass to the limit as $\varepsilon \to +0$ in the relationships

$$z_\varepsilon \in K, \quad f_\varepsilon(z_\varepsilon) \in K^*, \quad z_\varepsilon^T f_\varepsilon(z_\varepsilon) = 0.$$

Since the cones K and K^* are closed, we thus obtain

$$z \in K, \quad f(z) \in K^*, \quad z^T f(z) = 0$$

for an arbitrary accumulation point z of the family of solutions $\{z_\varepsilon\}$ to the auxiliary complemetarity problem. The latter means that the point z solves (6.7) which completes the proof. ■

In order to deduce new sufficient conditions of the existence of solutions to the complementarity problem with respect to a general cone, we examine some properties of the exceptional family of elements.

LEMMA 6.2. *For each element of an exceptional family of elements $z_r \in K$, the following relationships are true:*

(i) $u^T f(z_r) \leq 0$ for every non-zero vector u that belongs to the minimal dimension facet of the cone K containing the element z_r, and satisfies the inequality $u^T z_r \geq 0$;

(ii) $v^T f(z_r) \geq 0$ for every (nonzero) feasible direction v from the point z_r (i.e. $z_r + \varepsilon v \in K$ for $\varepsilon > 0$ small enough) and such that $v^T z_r = 0$.

PROOF. We recall that for each element $z_r \in K$ of the exceptional family of elements, there exists a scalar $\lambda_r > 0$ such that the vector $s_r = f(z_r) + \lambda_r z_r$ is the (interior) normal vector of the supporting hyperplane to the cone K at the point z_r. Now we consider a direction u that lies in the minimal dimension facet of the cone containing z_r. The properties of the supporting hyperplane imply the equality $u^T s_r = 0$, that is tantamount to $u^T f(z_r) = -\lambda_r u^T z_r$. Moreover, if $u^T z_r \geq 0$, then assertion (i) follows from the latter equality.

We use again the properties of the normal vector of the supporting hyperplane (*cf.* Lemma 6.1) and obtain the inequality $v^T s_r \geq 0$ for each feasible direction v at the point z_r. In additon, if $v^T z_r = 0$, the required inequality (ii) follows immediately from the structure of vector s_r. The proof is complete. ■

THEOREM 6.5. *Let $f : K \to R^n$ be a continuous mapping and $C \subset K$ a nonempty bounded subset such that for every $x \in K \setminus C$ (at least) one of the following assumptions is valid:*

(i) $u^T f(x) > 0$ for a non-zero vector u that belongs to the minimal dimension facet of the cone K, containing the point x, and satisfies $u^T x \geq 0$;

(ii) $v^T f(x) < 0$ for a non-zero feasible direction v at the point x with $v^T x = 0$.

Then problem (6.7) is solvable. Moreover, if assumption (i) holds for $u = x$, all the solutions of (6.7) belong to C.

PROOF. Suppose that there exists an exceptional family of elements $\{z_r\} \subset K$. Starting from some $\bar{r} > 0$ onward, we have $z_r \in K \setminus C$. According to the theorem assumptions, at least one of the conditions (i) or (ii) is valid for the elements z_r, $r > \bar{r}$ which contradicts the assertion of Lemma 6.2. The latter excludes the existence of exceptional families of elements and therefore establishes the solvability of problem (6.7) (see Theorem 6.1). At last, if either the stronger version of assumption (i), or assumption (ii) takes place, we deduce for every $x \in K \setminus C$ that either $x^T f(x) > 0$, or $f(x) \notin K^*$, respectively, is valid. Consequently, all the solutions of (6.7) belong to C, and the proof is complete. ■

By making use of the techniques from the proofs of Theorem 6.3 and Theorem 6.4, we obtain the following corollary.

COROLLARY 6.3. *Let $f : K \to R^n$ be a continuous mapping, and a nonempty subset $C \subset K$ be compact and star-shaped with respect to the origin with $0 \notin \Gamma$. If for every $x \in K \setminus \{0\}$ (at least) one of the following assumptions concerning $y = \eta(x)$ is valid:*

(i) $u^T f(y) > 0$ for a non-zero vector u that lies at the minimal dimension facet of the cone K, containing the point y, and satisfies $u^T y \geq 0$;

(ii) $v^T f(y) < 0$ for a non-zero feasible direction v at the point y such that $v^T y = 0$,

then problem (6.7) is solvable. Moreover if (i) holds for $u = y$, then there exists a solution of (6.7) belonging to C. ■

6.4 STANDARD COMPLEMENTARITY PROBLEM

This section is dedicated to the standard complementarity problem which is a particular case of problem (6.7) with the nonnegative orthant taken for the cone K, i.e., $K = R^n_+$. Therefore, the standard complementarity problem for a continuous mapping $f : R^n_+ \to R^n$ consists in finding a vector $x \in R^n_+$ such that

$$f(x) \geq 0, \qquad x^T f(x) = 0. \tag{6.26}$$

In this case, the dual cone K^* coincides with R^n_+, hence the condition $f(y) \in K^*$ simply means $f(y) \geq 0$. Moreover, the normality to the supporting hyperplane is easily interpreted by the following version of definition of an exceptional family of elements.

DEFINITION 6.4. A family of elements $\{x^r\}_{r>0} \subset R^n_+$ is referred to as *exceptional* if $\|x^r\| \to +\infty$ when $r \to +\infty$, and for every $r > 0$ the following relationships hold:

$$f_i(x^r) \begin{cases} = -\lambda_r x_i^r, & \text{if} \quad x_i^r > 0; \\ \geq 0, & \text{if} \quad x_i^r = 0; \end{cases}$$

here $\lambda_r > 0$ is a scalar. Like above, the exceptional family is called *regular*, if $\|x^r\| = r$, $r > 0$.

The following theorems are the consequences of Theorem 6.1 and Theorem 6.2, respectively.

THEOREM 6.6. *For each continuous mapping $f : R^n_+ \to R^n$, there exists either a solution to problem* (6.26), *or an exceptional family of elements.* ∎

THEOREM 6.7. *For each continuous mapping $f : R^n_+ \to R^n$, there exists either a solution to problem* (6.26), *or a regular exceptional family of elements.* ∎

Like in the previous section, we obtain both well-known and new sufficient existence conditions for problem (6.26).

PROPOSITION 6.3 [ISAC]. *Consider a continuous mapping $f : R^n_+ \to R^n$ and a function $\varphi : [0, +\infty) \to [0, +\infty)$ with $\lim_{t\to+\infty} \varphi(t) = +\infty$, $\varphi(t) > 0$ for $t > 0$, and*

$$(x-y)^T [f(x) - f(y)] \geq \|x - y\| \varphi(\|x - y\|), \ \forall x, y \in R^n_+.$$

Then problem (6.26) *is solvable.* ∎

PROPOSITION 6.4 [ISAC]. *Consider a continuous mapping $f : R^n_+ \to R^n$ and a bounded subset $C \subset R^n_+$ such that for each $x \in R^n_+ \setminus C$ there exists a $y \in C$ with $(x-y)^T f(x) \geq 0$. Then problem* (6.26) *has a solution.* ∎

COROLLARY 6.4 [ISAC]. *Consider a continuous mapping $f : R^n_+ \to R^n$ and a bounded subset $C \subset R^n_+$ such that $y^T f(y) \geq 0 \ \forall y \in R^n_+ \setminus C$. Then problem (6.26) is solvable. Moreover, if for every $y \in R^n_+ \setminus C$ the strict inequality $y^T f(y) > 0$ holds, then all the solutions of (6.26) belong to C.* ■

Corollary 6.1 implies the following assertion.

COROLLARY 6.5. *Consider a continuous mapping $f : R^n_+ \to R^n$ and a bounded subset $C \subset K$ such that $f(y) \geq 0 \ \forall y \in K \setminus C$. Then problem (6.26) is solvable. Moreover, if $0 \in C$ and the strict inequality $f(y) > 0$ holds for every $y \in K \setminus C$, then all the solutions of problem (6.26) belong to C.* ■

EXAMPLE 6.3. Consider a linear complementarity problem with $f(x) = Mx + q$; here M is a real $n \times n$-matrix. It is clear that if all the entries of the matrix M are positive ($M > 0$), then Corollary 6.5 implies that problem (6.26) possesses a solution for an arbitrary vector $q \in R^n$. However, if the matrix M is nonnegative ($M \geq 0$), then problem (6.26) is solvable if (and only if) the inequality below has a nonempty solution set:

$$Mx + q \geq 0, \qquad x \geq 0.$$

Indeed, consider $x^0 \geq 0$ such that $b = Mx^0 + q \geq 0$. Denote by $y = x - x^0$ and examine the complementarity problem: find $y \geq 0$ such that

$$g(y) = My + b \geq 0, \qquad \text{and} \qquad y^T g(y) = 0.$$

Remark that the cone $y \geq 0$ is a subset of R^n_+; hence, each solution of the latter problem generates a solution to the former. At last, the property $b = Mx^0 + q \geq 0$ allows us to apply Corollary 6.5 and confirm the existence of solution. ■

It follows from Definition 6.4 that the inequality $x_i f_i(x) > 0$ is impossible if the vector x belongs to an exceptional family of elements. Therefore, Proposition 6.5 can be refined as follows.

THEOREM 6.8. *Consider a continuous mapping $f : R^n_+ \to R^n$ and a nonempty bounded subset $C \subset R^n_+$ such that for every $x \in R^n_+ \setminus C$, the inequality $x_i f_i(x) > 0$ is valid for at least one of $i = 1, \ldots, n$. Then problem (6.26) is solvable, and all its solutions belong to C.*

PROOF. Suppose that there exists an exceptional family of elements $\{x^r\}_{r>0} \subset R^n_+$. Starting from some $r > 0$ onward, we have $x^r \in R^n_+ \setminus C$. According to the theorem assumption about these elements, $x_i^r f_i(x^r) > 0$ for at least one of $i = 1, \ldots, n$. On the other hand, by Definition 6.4,

$$x_i^r f_i(x^r) = \begin{cases} -\lambda_r \left(x_i^r\right)^2 < 0, & \text{if} \quad x_i^r > 0, \\ 0, & \text{if} \quad x_i^r = 0. \end{cases}$$

Thus obtained contradiction proves the absence of exceptional families of elements for the mapping f, which guarantees (by Theorem 6.6) the existence of solutions to problem (6.26). Furthermore, the inequality $x_i f_i(x) > 0$ for $x \in R^n_+ \setminus C$ denies the existence of solutions outside the subset C, which completes the proof. ■

EXAMPLE 6.4. Consider $f : R^2_+ \to R^2$ with

$$f_1(x_1, x_2) = x_1 + x_2 - 1, \qquad f_2(x_1, x_2) = 1 - x_1 x_2.$$

Choosing a unit square $\{(x_1, x_2) \mid 0 \le x_i \le 1, \quad i = 1, 2\}$ as a subset C we apply Theorem 6.8 and obtain the existence of a solution to problem (6.26) in C. Indeed, consider a point (x_1, x_2) outside the subset C. If $x_1 > 1$, then $x_1 f_1(x_1, x_2) > 0 \; \forall x_2 \ge 0$. Otherwise, if $0 < x_1 \le 1$, then $x_2 > 1$ and again $x_1 f_1(x, y) > 0$. At last, if $x_1 = 0$, then $x_2 f_2(x_1, x_2) > 0$ since $x_2 > 1$. Indeed, for instance, the point $x^* = (1; 0)$ solves the complementarity problem in question.

However, Proposition 6.4 cannot guarantee the existence of solution in that case. Indeed, it is easily verified that the scalar product $(y - x)^T f(x)$ assumes positive values at the points x with their norm large enough, for any vector y selected from an arbitrary bounded $C \subset R^n_+$. ■

Now consider a compact star-shaped (with respect to the origin) subset $C \subset R^n_+$. Like above, denote by Γ its relative boundary with respect to the orthant R^n_+. Theorem 6.4 implies the following assertion.

COROLLARY 6.6. *Let $f : R^n_+ \to R^n$ be a continuous mapping and $C \subset R^n_+$ be a compact star-shaped (with respect to the origin) subset with $0 \notin \Gamma$. If $y^T f(y) \ge 0$ for every $y = \eta(x)$, $x \in R^n_+ \setminus \{0\}$, then problem* (6.26) *has a solution in subset C.* ■

Making use of Theorem 6.8 we establish the following result.

COROLLARY 6.7. *Consider a continuous mapping $f : R^n_+ \to R^n$ and a nonempty compact subset $C \subset R^n_+$ that is star-shaped with respect to the origin. If the function $\eta(x)$ is continuous, $\eta(x) \ne 0$ and $y_i f_i(y) > 0$ for at least one of $i = 1, \ldots, n$ for each $y = \eta(x)$, $x \in R^n_+ \setminus \{0\}$, then problem* (6.26) *has a solution in the subset C.*

PROOF. Define the function $g : R^n_+ \to R^n$ as follows:

$$g(x) = \begin{cases} f(x), & \text{if } x \in C; \\ f(\eta(x)) + \|x - \eta(x)\| x, & \text{if } x \in R^n_+ \setminus C. \end{cases}$$

The function g is apparently continuous over R^n_+ and has the following feature: $x_i g_i(x) > 0$ for at least one of $i = 1, \ldots, n$ if $x \in R^n_+ \setminus C$. By virtue of Theorem 6.8, the latter property provides for the existence of a solution to problem (6.26) with respect to the mapping g, and all the solutions belong to the subset C. However, functions g and f coincide over C, which also implies the solvability of problem (6.26) with the mapping f, which completes the proof. ■

The techniques of averaging used when proving Theorem 6.4 allows us to extend the previous result to the case of a non-continuous function η.

COROLLARY 6.8. *Let $f : R^n_+ \to R^n$ be a continuous mapping and $C \subset R^n_+$ be a nonempty compact subset, star-shaped with respect to the origin, and $0 \notin \Gamma$. If $y_i f_i(y) > 0$ for at least one of $i = 1, \ldots, n$ and every $y = \eta(x)$, $x \in R^n_+ \setminus \{0\}$, then problem* (6.26) *has a solution belonging to the subset C.* ∎

6.5 IMPLICIT COMPLEMENTARITY PROBLEM

Consider a convex closed cone K in R^n with the origin as its vertex, and continuous mappings $f, g : R^n \to R^n$. The *implicit complementarity problem with respect to the cone* K consists in finding a vector $x \in R^n$ such that

$$g(x) \in K, \quad f(x) \in K^*, \quad f(x)^T g(x) = 0. \tag{6.27}$$

If the function f is defined over the nonnegative orthant R^n_+ and the function g coincides with the identity mapping $g(x) = x \; \forall x \in R^n_+$, then problem (6.27) is the standard complementarity problem.

The majority of the well-known existence theorems for problem (6.27) (*cf.* Isac (1992)) demand that the mappings f and g satisfy some strong restrictions. For example, f must be strongly monotone and Lipschitz continuous with respect to g (in the other words, f must be g-strongly monotone and g-Lipschitz continuous). In this chapter, we make use of the topological degree of those mappings and obtain theorems of the alternative type without exploiting explicitly the above-mentioned properties.

DEFINITION 6.5. We refer to the family of elements $\{x_r\}_{r>0} \subset R^n$ as an *exceptional* one for the pair f and g with respect to the cone K, if the following relationships are valid. First, $\|x_r\| \to +\infty$ as $r \to +\infty$, and second, for every $r > 0$ one has $g(x_r) \in K$, and there exists such a scalar $\mu_r > 0$ that the vector $s_r = f(x_r) + \mu_r g(x_r)$ is the normal one for a supporting hyperplane to the cone K at the point $g(x_r)$.

THEOREM 6.9. *Let the mappings $f, g : R^n \to R^n$ be continuous and the element $b \in R^n$ be a unique solution to the equation $g(x) = 0$. Morever, let g map some neighborhood of the element b homeomorhically onto a neighborhood of the origine. Then there exists either a solution to problem* (6.27), *or an exceptional family of elements for the pair f and g.*

PROOF. Consider the following equation with respect to the pair of variables $(z, x) \in R^n \times R^n$:

$$F(z, x) = \begin{pmatrix} f(x) - z^- \\ g(x) - z^+ \end{pmatrix} = 0; \tag{6.28}$$

here $z^+ = \Pr_K(z)$; $z^- = z - z^+$. It is easy to verify that problems (6.27) and (6.28) are tantamount in the same sense, as problems (6.7) and (6.10) are. Indeed, if (z, x) solves (6.28), then x solves (6.27). Conversely, given a solution x of problem (6.27), one can construct the solution to (6.28) as (z, x) with $z = g(x) - f(x)$. The mapping $F(z, x)$ is clearly continuous over $R^n \times R^n$. Let S_r be a $(2n-1)$-dimensional sphere of the radius r centered at the point $(0; b)$:

$$S_r = \{(z, x) \in R^n \times R^n \mid \|(z, x - b)\| = r\},$$

and B_r be the open ball with the same radius and center, i.e.

$$B_r = \{(z, x) \in R^n \times R^n \mid \|(z, x - b)\| < r\}.$$

Now construct the standard homotopy of the mappings $F(z, x)$ and $G(z, x)$ where

$$G(z, x) = \begin{pmatrix} z \\ g(x) \end{pmatrix},$$

as follows:

$$\begin{aligned} H(z, x, t) &= tG(z, x) + (1-t)F(z, x) = \\ &= \begin{pmatrix} tz + (1-t)f(x) - (1-t)z^- \\ tg(x) + (1-t)g(x) - (1-t)z^+ \end{pmatrix} = \\ &= \begin{pmatrix} tz^+ + (1-t)f(x) - z^- \\ g(x) - (1-t)z^+ \end{pmatrix}. \end{aligned} \tag{6.29}$$

Examine the two possible cases:

A). There exists an $r > 0$ such that

$$H(z, x, t) \neq 0, \quad \forall (z, x) \in S_r, \quad t \in [0, 1].$$

The Poincaré–Bohl Theorem then implies

$$\deg(F, B_r, 0) = \deg(G, B_r, 0). \tag{6.30}$$

Due to the theorem assumptions, it is easily verified that $\deg(G, B_r, 0) = \pm 1$. By taking (6.30) into account one also obtains $\deg(F, B_r, 0) = \pm 1$. Now making use of the Kronecker theorem, one can conclude that the closed ball $\mathrm{cl}B_r$ contains at least one solution of (6.28). Therefore, problem (6.27) is solvable, too.

B). For every $r > 0$, there exist a pair $(z_r, x_r) \in S_r$ and a scalar $t_r \in [0, 1)$ such that

$$H(z_r, x_r, t_r) = 0. \tag{6.31}$$

Note that

$$\|(z_r, x_r - b)\|^2_{R^{2n}} = \|z_r^+\|^2_{R^n} + \|z_r^-\|^2_{R^n} + \|x_r - b\|^2_{R^n} = r^2. \tag{6.32}$$

If $t_r = 0$, then (z_r, x_r) solves equation (6.28) and consequently, x_r is a solution to problem (6.27). Otherwise, if $t_r > 0$, then (6.29) and (6.31) imply the following equalities:

$$t_r z_r^+ + (1 - t_r) f(x_r) = z_r^-, \tag{6.33}$$

$$z_r^+ = \frac{g(x_r)}{(1-t_r)}. \tag{6.34}$$

Substituting (6.34) for z^+ in (6.33), one gets

$$\frac{t_r}{1-t_r} g(x_r) + (1-t_r) f(x_r) = z_r^-.$$

Now we divide both sides of the latter equality by $(1 - t_r)$, denote by $\mu_r = t_r/(1-t_r)^2 > 0$, and come to the equality

$$f(x_r) + \mu_r g(x_r) = \frac{1}{1-t_r} z_r^-.$$

The latter guarantees (due to Lemma 6.1) that the family of elements $\{x_r\}$ is exceptional if $\|x_r\|_{R^n} \to +\infty$ as $r \to +\infty$. In order to prove that, suppose, on the contrary, that the family $\{x_r\}$ is bounded, hence it has a finite accumulation point $\bar{x}$. Note that the respective limit scalar $\bar{t}$ cannot be equal to 1 (otherwise, (6.32) contradicts (6.33)). But if $\bar{t} < 1$, then the continuity of the mappings f and g combined with (6.33) and (6.34) imply the boundedness of the family of elements $\{z_r\}$, which again contradicts (6.32), as $r \to +\infty$. Thus we have shown that $\|x_r\| \to +\infty$ which completes the proof. ■

In what follows, we restrict ourselves to considering a particular case of problem (6.27) in which the nonnegative orthant R^n_+ plays the role of the cone K. Then the implicit complementarity problem assumes the following form: given the mappings $f, g : R^n \to R^n$, find a vector $x \in R^n$ such that

$$g(x) \geq 0, \quad f(x) \geq 0, \quad f(x)^T g(x) = 0. \tag{6.35}$$

In this case, the notion of exceptional family of elements can be transformed as follows.

DEFINITION 6.6. We refer to a family of elements $\{x_r\}_{r>0} \subset R^n$ as to *an exceptional one* for the pair f and g if, first, $\|x_r\| \to +\infty$ as $r \to +\infty$, and second, for each $r > 0$, the inequality $g(x_r) \geq 0$ holds, and there exists a scalar $\mu_r > 0$ such that for $i = 1, \ldots, n$,

$$f_i(x^r) \begin{cases} = -\mu_r g_i(x_r), & \text{if } \ g_i(x_r) > 0; \\ \geq 0, & \text{if } \ g_i(x_r) = 0. \end{cases}$$

Theorem 6.9 allows one to deduce some new existence theorems for problem (6.35). Examine the case when the mapping f is monotone with respect to g.

THEOREM 6.10. *Let the mappings $f, g : R^n \to R^n$ be continuous and satisfy the assumptions of Theorem 6.9. Moreover, let $\|g(x)\| \to +\infty$ as $\|x\| \to +\infty$, and there exist a function $\varphi : [0, +\infty) \to [0, +\infty)$ with $\lim_{t \to +\infty} \varphi(t) = +\infty$, $\varphi(t) > 0 \ \ \forall t > 0$, and*

$$[g(x) - g(y)]^T [f(x) - f(y)] \geq \|g(x) - g(y)\| \varphi\left(\|g(x) - g(y)\|\right), \tag{6.36}$$

for all $x, y \in R^n$. Then problem (6.35) with mappings f and g is solvable.

PROOF. On the contrary, suppose that there exists an exceptional family of elements $\{x_r\}_{r>0}$ for the pair f and g. According to Definition 6.6 of the exceptional family of elements, the vectors $s_r = f(x_r) + \mu_r g(x_r)$, $r > 0$, satisfy the following conditions: $s_r \geq 0$ and $[g(x_r)]^T s_r = 0$. Taking into account that $g(x_r) \geq 0 \; \forall r > 0$, write down the series of relationships

$$[g(x_r)-g(x_1)]^T[f(x_r)-f(x_1)] = [g(x_r)-g(x_1)]^T[s_r - s_1 - \mu_r g(x_r) + \mu_1 g(x_1)] =$$

$$= [g(x_r)-g(x_1)]^T s_r - [g(x_r)-g(x_1)]^T s_1 - [g(x_r)-g(x_1)]^T[\mu_r g(x_r) - \mu_1 g(x_1)] \leq$$

$$\leq -[g(x_r)-g(x_1)]^T \{\mu_r [g(x_r)-g(x_1)] - (\mu_1 - \mu_r) g(x_1)\} =$$

$$= -\mu_r \|g(x_r)-g(x_1)\|^2 + (\mu_1 - \mu_r)[g(x_r)-g(x_1)]^T g(x_1). \tag{6.37}$$

Conditions (6.36), (6.37), and the Cauchy-Buniakowski inequality imply the estimate

$$\varphi(\|g(x_r)-g(x_1)\|) \leq -\mu_r \|g(x_r)-g(x_1)\| + |\mu_1 - \mu_r| \|g(x_1)\|,$$

which contradicts the relationship $\varphi(t) \to +\infty$ as $t \to +\infty$; here $\|g(x_r)-g(x_1)\|$ is taken as t when $r \to +\infty$. Hence, there is no exceptional families of elements for the pair f and g, which implies the solvability of problem (6.35) according to Theorem 6.9 and completes the proof. ■

The following theorems are proven in the same way as Theorems 6.5 and 6.8.

THEOREM 6.11. *Let the mappings $f, g : R^n \to R^n$ satisfy the assumptions of Theorem 6.9, and $C \subset R^n$ be a bounded subset such that, for every $x \notin C$, either $g(x)^T f(x) \geq 0$ holds, or the inequality $g(x) \geq 0$ is not valid. Then problem (6.35) is solvable. Moreover, if for each $x \notin C$, either $g(x)^T f(x) > 0$ is true, or $g(x) \geq 0$ is false, then all the solutions to problem (6.35) belong to the subset C.* ■

THEOREM 6.12. *Let the mappings $f, g : R^n \to R^n$ satisfy assumptions of Theorem 6.9, and $C \subset R^n$ be a nonempty bounded subset such that, for every $x \notin C$, either $g_i(x) f_i(x) > 0$ holds for at least one of $i = 1, \ldots, n$, or the inequality $g(x) \geq 0$ is false. Then problem (6.35) is solvable, and all the solutions belong to the subset C.* ■

Recall that we refer to a closed subset $C \subset R^n$ as *star-shaped* with respect to the vector b, if $x \in C$ implies $[b, x] \subset C$. Then the function $\eta(x)$ that associates any vector $x \neq b$ with the most distant (from b) point of the subset C belonging to the ray starting from b and running through x.

THEOREM 6.13. *Consider the mappings $f, g : R^n \to R^n$ satisfying assumptions of Theorem 6.9, and a nonempty compact subset $C \subset R^n$ which is star-shaped with respect to b. If for every $x \in R^n \setminus \{b\}$ and $y = \eta(x)$ either $g_i(y) f_i(y) > 0$ holds for some $i = 1, \ldots, n$, or $g(y) \geq 0$ is not valid, then problem (6.35) has a solution belonging to the subset C.*

PROOF. First examine the case of continuous function η. Define the mappings $\hat{f}, \hat{g} : R^n \to R^n$ as follows:

$$\hat{f}(x) = \begin{cases} f(x), & \text{if } x \in C; \\ f(\eta(x)), & \text{if } x \notin C; \end{cases} \tag{6.38a}$$

$$\hat{g}(x) = \begin{cases} g(x), & \text{if } x \in C; \\ g(\eta(x)), & \text{if } x \notin C. \end{cases} \tag{6.38b}$$

The pair of mappings $\hat{f}, \hat{g}$ and the subset C clearly satisfy all the conditions of Theorem 6.9. Moreover, outside the subset C, either $\hat{g}_i(x)\hat{f}_i(x) > 0$ holds for at least one number $i = 1, \ldots, n$, or the relationship $\hat{g}(x) \geq 0$ is not valid. Thus we have shown that the pair $\hat{f}$ and $\hat{g}$ has no exceptional families of elements. Hence, Theorem 6.9 implies the existence of a solution $x^* \in C$ to problem (6.35) with respect to the mappings $\hat{f}, \hat{g}$. For each $x \in C$, it follows from (6.38a) and (6.38b) that $\hat{f}(x) = f(x)$, $\hat{g}(x) = g(x)$. Therefore, the vector $x^* \in C$ also solves problem (6.35) with respect to the mappings f and g.

If the function η is not continuous, we can repeat exactly the arguments used when proving Theorem 6.4. Namely, we average the function η with respect to the finite ones over the unit sphere, and introduce the auxiliary continuous mappings f_ε just like in the proof of Theorem 6.4. For thus obtained continuous function η_ε, the existence of a solution of problem (6.35) with respect to the mappings f_ε and g have also been proved above. Now passing to limit as $\varepsilon \to +0$, the continuity of the mappings f and g and closedness of the subset C imply the existence of a solution of problem (6.35) belonging to C which completes the proof. ∎

Now consider a particular case of the implicit complementarity problem with affine mappings f and g. Let $f(x) = Ax + b$, $g(x) = Gx + d$, with A and G being real $n \times n$-matrices, and b, d elements of R^n. Moreover, assume that the matrix G is nonsingular, and that there exists a positive matrix B ($B > 0$) such that

$$A = BG. \tag{6.39}$$

THEOREM 6.14. *Let matrices A, B, G satisfy all the above- mentioned conditions, and at least one of the subsets*

$$C_i = \{x \in R^n \mid Gx + d \geq 0,\ A_i x + b_i \leq 0\}, \quad i = 1, \ldots, n \tag{6.40}$$

be nonempty; here A_i is the i-th row of the matrix A. Then complementarity problem (6.35) with respect to the affine mappings f and g is solvable.

PROOF. First, show that each of the subsets C_i, $i = 1, \dots, n$, defined by (6.40) is compact. On the contrary, suppose that C_i is unbounded for some i, and consider a point $x^0 \in C_i$. Since the subset C_i is polyhedral, there exists a vector $s \neq 0$ with $x^0 + ts \in C_i$ $\forall t \geq 0$. Then (6.40) implies

$$Gs \geq 0, \qquad B_i Gs \leq 0. \tag{6.41}$$

Since the row B_i is positive, (6.41) allows one to conclude that $Gs = 0$. Hence, one obtains $s = 0$ as the matrix G is nonsingular. Therefore, every subset C_i is compact. Now one can apply Theorem 6.11 for the union $C = \bigcup_{i=1}^{n} C_i$. Indeed, for each $y \notin C$ either $Ay + b > 0$ holds, or the inequality $Gy + d \geq 0$ is not valid. Consequently, Theorem 6.11 implies the existence of a solution to problem (6.35) belonging to the subset C which completes the proof. ■

6.6 GENERAL ORDER COMPLEMENTARITY PROBLEM

Let (E, E^*) be a duality of Banach spaces and let $K \subset E$ be a closed convex cone. If E is a vector lattice with respect to the ordering defined by K, and $f_1, f_2, \dots, f_n$ are mappings from E into E^*, we consider the problem: Find $x \in K$ such that

$$\wedge (f_1(x), f_2(x), \dots, f_n(x)) = 0. \tag{6.42}$$

We refer to (6.42) as the *General Order Complementarity Problem.* In Isac (1992), interesting relationships between the General Order and Implicit Complementarity Problems were established. Making use of the latter, one obtains the following existence results.

Let $(R^n, <,>, K)$ be a Hilbert lattice with respect to the ordering defined by K. For $x, y \in K$, it is well-known that $x \wedge y = 0$ if and only if $x^T y = 0$.

THEOREM 6.15. *Let $f_1, f_2, \dots, f_n$ be continuous mappings from R^n into R^n. If the following assumptions are satisfied:*

(i) there exists a unique point $b \in R^n$ such that $f_1(b) = 0$;

(ii) f_1 maps a neighborhood of the point b homeomorphically onto a neighborhood of the origin,

then there exists either a solution of problem (6.42) or an exceptional family of elements for the pair h, f_1 where $h(x) = \wedge (f_2(x), f_3(x), \dots, f_n(x))$ for each $x \in R^n$.

PROOF. Since $(R^n, <,>, R^n_+)$ is a Hilbert lattice, problem (6.42) is tantamount to the implicit complementarity problem (6.27) with $f = h$, $g = f_1$, and $K = R^n_+$. Therefore, the conclusion of the theorem follows from Theorem 6.9, as we can show that the mapping h is continuous. ■

Remark. It is easily verified that the mapping $h = \wedge(f_2, \ldots, f_n)$ has no exceptional family of elements whenever each mapping $f_2, \ldots, f_n$ is without exceptional families of elements.

COROLLARY 6.9. *If $f_2, f_3, \ldots, f_n$ are continuous mappings from R^n into R^n, the mapping f_1 satisfies assumptions (i) and (ii) of Theorem* 6.15, *and none of $f_2, \ldots, f_n$ has exceptional family of elements, then problem* (6.42) *is solvable.* ■

6.7 SEMIDEFINITE COMPLEMENTARITY PROBLEM

In this section, the topological degree theory is applied to study the problem of existence of solutions to the semidefinite complementarity problem (SDCP). A notion of an exceptional family of matrices is introduced, and assertions of the nonstrict alternative type are obtained. Namely, for a continuous mapping, there exists at least one of the following two items: either a solution to the SDCP, or an exceptional family of matrices. Hence, if there is no exceptional families, then at least one solution exists.

6.7.1 PROBLEM SPECIFICATION

We examine the semidefinite complementarity problem (SDCP) in the standard form. Let $R^{n\times n}$ be the Euclidean space of real $n \times n$-matrices with the scalar product

$$\langle A, B\rangle = \mathrm{tr}(A^T B) \qquad \forall A, B \in R^{n\times n}. \tag{6.43}$$

Moreover, let $S^{n\times n}$ be the linear subspace of symmetric real $n \times n$-matrices, and $S^n_+ \subset S^{n\times n}$ be the cone of the positive semidefinite ones; in this section, $A \geq 0$ means $A \in S^n_+$. The interior points of this cone are the positive definite matrices A, i.e. $A \geq 0$ and has the full rank n. We denote that by $A > 0$. Finally, introduce the norm in the space $R^{n\times n}$ in the standard manner: $\|A\| = \langle A, A\rangle^{1/2}$.

Now consider a continuous mapping $F : S^n_+ \to S^{n\times n}$ and define a solution X to the semidefinite complementarity problem as follows:

$$X \geq 0, \quad F(X) \geq 0, \quad \text{and} \quad \langle X, F(X)\rangle = 0. \tag{6.44}$$

It is readily verified that for $A, B \in S^n_+$, the equality $\mathrm{tr}(AB) = 0$ is tantamount to $AB = 0$. Therefore, we can reformulate SDCP as follows: find a matrix $X \geq 0$ such that

$$F(X) \geq 0, \qquad \text{and} \qquad X \cdot F(X) = 0. \tag{6.45}$$

It is well-known that every symmetrical matrix $A \in S^{n\times n}$ admits a complete orthonormal family of eigenvectors $a_1, a_2, \ldots, a_n$. Moreover, its nonzero (necessarily real) eigenvalues $\lambda_1, \lambda_2, \dots, \lambda_k$ $(k \le n)$, together with the respective eigenvectors, form the spectral representation

$$A = \sum_{i=1}^{k} \lambda_i a_i a_i^T; \tag{6.46}$$

here $k = \text{rank}A$ (*cf.* Schatten (1960)). Making use of such spectral representations for $X = A$ and $F(X) = B$, we reduce the equality $X \cdot F(X) = 0$ to the following system of equations:

$$a_i^T b_j = 0, \quad i = 1, 2, \dots, k; \ j = 1, 2, \dots, \ell, \tag{6.47}$$

where $\ell = \text{rank}B$, and $k + \ell \le n$. In particular, if $\text{rank}X = \text{rank}A = n$, we simply have $F(X) = B = 0$.

In this section, we introduce the notions of exceptional and regular exceptional families of matrices and formulate the general (nonstrict) alternative: for each continuous mapping F there exists either a solution to problem (6.44) or a (regular) exceptional family of matrices. Based upon this alternative, we prove both the well-known and new sufficient conditions of existence of solutions to the semidefinite complementarity problems.

Since $S^{n\times n}$ is a finite-dimensional Euclidean space (with $\dim S^{n\times n} = n(n+1)/2$), the Poincaré-Bohl and the Kronecker theorems are applicable to the considered problem.

The section is organized as follows. In Subsection 6.6.2, the definitions of the exceptional and regular exceptional families of matrices are given, and the theorems on the alternative for the semidefinite complementarity problem are proved. Sufficient conditions that guarantee the absence of exceptional families and thus provide for the existence of solutions to the SDCP are presented in Subsection 6.6.3.

6.7.2 THEOREMS OF ALTERNATIVES

Let $F : S^{n\times n} \to S^{n\times n}$ be a continuous mapping and $S_+^n \subset S^{n\times n}$ be the (closed convex) cone of positive semidefinite matrices. Denote by $(S_+^n)^*$ the dual cone (in the subspace $S^{n\times n}$) for S_+^n, i.e.

$$(S_+^n)^* = \{Y \in S^{n\times n} \mid \langle Y, X\rangle \ge 0 \quad \forall X \in S_+^n\}.$$

It is well-known that the cone S_+^n is self-dual, i.e. $(S_+^n)^* = S_+^n$. Therefore, the general complementarity problem: to find a matrix $X \in S_+^n$ such that

$$F(X) \in (S_+^n)^*, \qquad \text{and} \qquad \langle X, F(X)\rangle = 0, \tag{6.48}$$

is equivalent to the semidefinite complementarity problem expressed by (6.44). The cone S_+^n being a closed convex subset, every matrix $Z \in S^{n\times n}$ has the uniquely determined projection onto S_+^n:

$$Z^+ = \text{Pr}_{S_+^n} Z = \arg\min\{\|Z - X\|, \ X \in S_+^n\}. \tag{6.49}$$

Let us introduce the notation $Z^- = Z^+ - Z$ and remark that Z^+ and Z^- depend continuously upon Z.

LEMMA 6.3. *For every symmetric matrix $Z \in S^{n\times n}$, we have: $Z^- \in S^n_+$ and $(Z^-)^T Z^+ = 0$.*

PROOF. Making use of the first order optimality condition for problem (6.49), we write down the inequality

$$\langle Z - Z^+, Z^+ - W\rangle \geq 0 \qquad \forall W \in S^n_+. \tag{6.50}$$

Substituting $W = 0$ and $W = 2Z^+$ into it, we obtain $\langle Z^-, Z^+\rangle = 0$. Now open the brackets in (6.50) and use the latter equality to deduce the inequality

$$\langle Z^-, W\rangle \geq 0 \qquad \forall W \in S^n_+.$$

This means that $Z^- \in (S^n_+)^* = S^n_+$ which implies $Z^- Z^+ = (Z^-)^T Z^+ = 0$ and thus completes the proof. ∎

Lemma 6.3 implies that Z^- is the *normal element to the supporting hyperplane* of the cone S^n_+ at the point Z^+.

Making use of the spectral representation (6.46), it is easy to characterize and construct Z^+ and Z^- explicitly for a given symmetrical matrix Z. In particular, the following result obtains.

LEMMA 6.4. *Let $Z \in S^{n\times n}$ have the spectral representation*

$$Z = \sum_{i=1}^{k} \lambda_i a_i a_i^T - \sum_{j=1}^{\ell} \mu_j b_j b_j^T, \qquad \lambda_i > 0,\ \mu_j > 0,\ k + \ell \leq n.$$

Then

$$Z^+ = \mathrm{Pr}_{S^n_+} Z = \sum_{i=1}^{k} \lambda_i a_i a_i^T, \quad Z^- = \sum_{j=1}^{\ell} \mu_j b_j b_j^T.$$

PROOF. We show that Z^+ satisfies the first order optimality condition (6.50), i.e.

$$\langle Z - Z^+, Z^+ - W\rangle \geq 0 \qquad \forall W \in S^n_+.$$

According to our assumptions,

$$Z - Z^+ = -\sum_{j=1}^{\ell} \mu_j b_j b_j^T.$$

For an arbitrary $W \in S^n_+$, make use of its spectral representation:

$$W = \sum_{i=1}^{m} \nu_i w_i w_i^T, \qquad \nu_i > 0,\ i = 1, \ldots, m \leq n;$$

here w_i, $i = 1, \dots, m$ are the orthonormal subsystem of the eigenvectors of the matrix W corresponding to the positive eigenvalues.

Now use the definition of the scalar product and write

$$\langle Z - Z^+, Z^+ - W\rangle = \mathrm{tr}\left[\left(-\sum_{j=1}^{\ell} \mu_j b_j b_j^T\right)\left(\sum_{i=1}^{k} \lambda_i a_i a_i^T - \sum_{i=1}^{m} \nu_i w_i w_i^T\right)\right] =$$

$$= \mathrm{tr}\left(\sum_{j=1}^{\ell}\sum_{i=1}^{m} \mu_j \nu_i b_j (b_j^T w_i) w_i^T\right) = \sum_{j=1}^{\ell}\sum_{i=1}^{m} \mu_j \nu_i \mathrm{tr}\left(b_j (b_j^T w_i) w_i^T\right) =$$

$$= \sum_{j=1}^{\ell}\sum_{i=1}^{m} \mu_j \nu_i (b_j^T w_i)^2 \geq 0,$$

which establishes Z^+ to solve the projection problem (6.49). As the solution is unique, Lemma 6.4 is proven. ■

DEFINITION 6.6. A set of matrices $\{Z_r\}_{r>0} \subset S_+^n$ is called *an exceptional family (EF)* with respect to the cone S_+^n for the mapping F, if $\|Z_r\| \to +\infty$ as $r \to +\infty$, and for each $r > 0$ there exists a scalar $\mu_r > 0$ such that the matrix $M_r = F(Z_r) + \mu_r Z_r$ has the following properties:

$$M_r \geq 0, \qquad \text{and} \qquad M_r Z_r = 0.$$

In the other words, for each $r > 0$, the matrix M_r is the normal element to the supporting hyperplane of the cone S_+^n at the point Z_r.

THEOREM 6.16. *For each continuous mapping $F : S_+^n \to S^{n\times n}$, there exists either a solution to problem (6.44) or an exceptional family of matrices.*

PROOF. Consider the equation

$$f(Z) = F(Z^+) - Z^- = 0, \tag{6.51}$$

where Z^+ is the projection (6.49), and $Z^- = Z^+ - Z$. It is easy to see that problems (6.49) and (6.51) are tantamount in the following sense. If X solves (6.49), then $Z = X - F(X)$ satisfies equation (6.51). Conversely, if Z is a solution to (6.51), then $X = Z^+$ solves problem (6.49).

In order to investigate problem (6.51) in detail, consider the family of spheres V_r and open balls W_r:

$$V_r = \{Z \in S^{n\times n} \mid \|Z\| = r\},$$

$$W_r = \{Z \in S^{n\times n} \mid \|Z\| < r\}.$$

Now examine the standard homotopy $H(Z,t)$ of the identity mapping $g(Z) = Z$ and the continuous mapping $f(Z)$:

$$H(Z,t) = tZ + (1-t)f(Z), \quad 0 \leq t \leq 1.$$

Apply the Poincaré–Bohl Theorem for $y = 0$, $C = W_r$ and write down $H(Z,t)$ for arbitrary $Z \in \partial W_r = V_r$ and $t \in [0,1]$:

$$H(Z,t) = tZ^+ + (1-t)F(Z^+) - Z^-. \tag{6.52}$$

Two outcomes are possible:

A) There is a number $r > 0$ such that

$$H(Z,t) \neq 0, \quad \forall\, Z \in V_r, \quad \forall\, t \in [0,1].$$

Then the Poincaré–Bohl Theorem implies $\deg(f, W_r, 0) = \deg(g, W_r, 0)$. The degree of the identity mapping is equal to the unity: $\deg(g, W_r, 0) = 1$. Therefore, $\deg(f, W_r, 0) = 1$. Now the Kronecker Theorem implies the existence of solutions to (6.51) within the ball cl W_r, hence problem (6.49) is solvable, too.

B) For every $r > 0$ there exist a matrix $Z_r \in V_r$ and a scalar $t_r \in [0,1)$ such that

$$H(Z_r, t_r) = 0. \tag{6.53}$$

If $t_r = 0$, then Z_r solves (6.51), which implies the solvability of problem (6.49) again. Otherwise, if $t_r > 0$, then it follows from (6.52) and (6.53) that

$$t_r Z_r^+ + (1-t_r)F(Z_r^+) = Z_r^-. \tag{6.54}$$

Dividing both parts of (6.54) by $(1-t_r)$, we obtain

$$F(Z_r^+) + \frac{t_r}{1-t_r} Z_r^+ = \frac{1}{1-t_r} Z_r^-. \tag{6.55}$$

Now set $\mu_r = t_r/(1-t_r)$ and verify that $\{Z_r^+\}$ is an EF for S_+^n and F. Indeed, from (6.55) and Lemma 6.3, the matrix $M_r = F(Z_r^+) + \mu_r Z_r^+$ is the normal element to the supporting hyperplane to the cone S_+^n at the point Z_r^+.

In order to show that the family $\{Z_r^+\}_{r>0}$ is exceptional, one needs to verify that $\|Z_r^+\| \to +\infty$ as $r \to +\infty$. On the contrary, suppose that the family $\{Z_r^+\}_{r>0}$ has a finite accumulation point. On one hand, the equality $\|Z_r^-\| = \sqrt{r^2 - \|Z_r^+\|^2}$ implies that the righthand side of (6.55) comprises an unbounded sequence of matrices. On the other hand, the respective matrices in the lefthand side of (6.55) compose a bounded family due to the continuity of the mapping F. Thus obtained contradiction completes the proof. ∎

DEFINITION 6.7. The exceptional family of matrices $\{Z_r\}$ is referred to as a *regular* one (REF), if

$$\|Z_r\| = r \qquad \forall r > 0.$$

By modifying the proof of Theorem 6.16, we obtain the following result.

THEOREM 6.17. *For each continuous mapping $F : S_+^n \to S^{n\times n}$ there exists either a solution to problem (6.44), or a regular exceptional family of matrices (REF).*

PROOF. Consider equation (6.51). For each $r > 0$ define

$$E_r = \{Z \in S^{n\times n} \mid \|Z^+\| < r\}, \quad m_r = \sup_{Z\in E_r} \|F(Z^+)\|,$$

$$p = 1 + \sqrt{(\max\{r, m_r\})^2 + r^2}.$$

As above, Z^+ is the projection of Z onto the cone S^n_+, defined by (6.49). It is easy to verify that

$$m_r \leq \max_{Z\in \mathrm{cl}\, W_r \cap S^n_+} \|F(Z)\| < +\infty,$$

hence, the scalar value p is finite. Now define

$$D_r = E_r \cap W_p,$$

where W_p is the open ball of radius p, and apply the Poincaré–Bohl Theorem to the set $C = D_r$ and homotopy (6.52). Again, it suffices to consider two cases.

A) There exists a scalar $r > 0$ such that

$$H(Z,t) \neq 0 \qquad \forall\, Z \in \partial D_r, \quad \forall\, t \in [0,1].$$

By an exact repetition of the proof of Theorem 6.16, we can show the existence of a solution to equation (6.51), and therewith, the solvability of problem (6.44).

B) For every $r > 0$ there exist a matrix $Z_r \in \partial D_r$ and a scalar $t_r \in [0,1)$ such that

$$H(Z_r, t_r) = 0.$$

If $t_r = 0$, then the matrix Z_r satisfies equation (6.51), consequently, the matrix $X^r = Z_r^+$ solves (6.44). Otherwise, if $t_r > 0$, then from (6.52)–(6.55) and Lemma 6.3 we deduce that $M_r = F(Z_r^+) + \mu_r Z_r^+$ is the normal element to the supporting hyperplane to the cone S^n_+ at the matrix Z_r^+; here $\mu_r = t_r/(1-t_r)$. In order to show that $\|Z_r^+\| = r$, examine the boundary ∂D_r in detail. It is straightforward that $\partial D_r = \Phi_r \cup U_p$, where

$$\Phi_r = \{Z \in S^{n\times n} \mid \|Z^+\| = r, \ \|Z\| \leq p\} = \partial E_r \cap \mathrm{cl}\ W_p,$$

and

$$U_p = \mathrm{cl}\ E_r \cap V_p.$$

Now verify that $Z_r \notin U_p$. Indeed, equality (6.52), the triangle inequality, and the definitions of E_r and m_r imply the relationships

$$\|Z_r^-\| \leq \max\Big\{\|Z_r^+\|, \|F(Z_r^+)\|\Big\} \leq \max\Big\{r, m_r\Big\}.$$

Making use of the definition of p, we obtain the estimate

$$\|Z_r\|^2 = \|Z_r^+\|^2 + \|Z_r^-\|^2 \leq r^2 + (\max\{r, m_r\})^2 = p^2 - 1,$$

which means that $\|Z_r\| < p$. Thus, it has been proved that $Z_r \in \Phi_r$, hence $\|Z_r^+\| = r$. The exceptional family $\{Z_r^+\}_{r>0}$ has been shown to be regular which completes the proof. ∎

6.7.3 EXISTENCE OF SOLUTIONS

By making use of Theorems 6.16 and 6.17, one can obtain the new proofs of the previously known results, as well as deduce some new sufficient conditions that guarantee the existence of solutions to the semidefinite complementarity problem. Analogues of some of the theorems below for the standard (vector) complementarity problems are well-known (see Harker and Pang (1990), Isac (1992), McLinden (1980)), but for the semidefinite complementarity problem they are all original. We begin with the following result.

THEOREM 6.18. *Let $F : S^n_+ \to S^{n\times n}$ be a continuous mapping and there exist a function $\varphi : [0,+\infty) \to [0,+\infty)$ with the properties*

$$\lim_{t\to+\infty} \varphi(t) = +\infty, \quad \varphi(t) > 0 \text{ for } t > 0, \text{ and}$$

$$\langle X - Y, F(X) - F(Y)\rangle \geq \|X - Y\|\varphi(\|X - Y\|), \ \forall X, Y \in S^n_+. \tag{6.56}$$

Then problem (6.44) *with the mapping F is solvable.*

PROOF. Suppose that there exists an exceptional family of matrices $\{Z_r\}_{r>0} \subset S^n_+$ for the mapping F. According to the definition of exceptional family, the matrices $M_r = F(Z_r) + \mu_r Z_r$ and $M_1 = F(Z_1) + \mu_1 Z_1$ are the normal matrices to the supporing hyperplanes to the cone S^n_+ at the matrices Z_r and Z_1, respectively. Making use of this fact, write down the following series of relationships

$$\begin{aligned}\langle Z_r - Z_1, F(Z_r) - F(Z_1)\rangle &= \langle Z_r - Z_1, M_r - M_1 - \mu_r Z_r + \mu_1 Z_1\rangle = \\ &= \langle Z_r - Z_1, M_r\rangle - \langle Z_r - Z_1, M_1\rangle - \langle Z_r - Z_1, \mu_r Z_r - \mu_1 Z_1\rangle \leq \\ &\leq -\langle Z_r - Z_1, \mu_r(Z_r - Z_1) - (\mu_1 - \mu_r)Z_1\rangle = \\ &= -\mu_r\|Z_r - Z_1\|^2 + (\mu_1 - \mu_r)\langle Z_r - Z_1, Z_1\rangle. \end{aligned} \tag{6.57}$$

By conditions (6.56), (6.57), and the Cauchy-Buniakowski inequality, we finally obtain

$$\varphi(\|Z_r - Z_1\|) \leq -\mu_r\|Z_r - Z_1\| + |\mu_1 - \mu_r|\|Z_1\|,$$

which contradicts the condition $\varphi(t) \to +\infty$ as $t \to +\infty$; here $\|Z_r - Z_1\|$ serves as t when $r \to +\infty$. Thus the cone S^n_+ and mapping F prohibit exceptional families to exist. Therefore, problem (6.44) is solvable, and the theorem is proved. ∎

Remark. If $\varphi(t) = t$, the mapping F satisfying (6.56) is called *strongly monotone*, in contrast to simply *monotone* one for which (6.56) holds with $\varphi(t) = 0$.

THEOREM 6.19. *Consider a continuous mapping $F : S^n_+ \to S^{n\times n}$ and a bounded subset $C \subset S^n_+$ such that for every $X \in S^n_+ \setminus C$ there exists a matrix $Y \in C$ with*

$$\langle X - Y, F(X)\rangle \geq 0. \tag{6.58}$$

Then problem (6.44) *has a solution.*

PROOF. Suppose that there exists an exceptional family of matrices $\{Z_r\} \subset S^n_+$. Starting from some $r > 0$ onward, the matrices Z_r belong to $S^n_+ \setminus C$. According to the theorem assumptions, there exists a matrix $Y_r \in C$ such that the following relationships hold:

$$0 \leq \langle Z_r - Y_r, F(Z_r)\rangle = \langle Z_r - Y_r, M_r - \mu_r Z_r\rangle,$$

with the matrices M_r associated with Z_r in the definition of exceptional family. Rearrange the terms of the latter inequality and obtain

$$\langle Z_r - Y_r, M_r\rangle - \mu_r\|Z_r\|^2 + \mu_r\langle Y_r, Z_r\rangle \leq -\mu_r\|Z_r\|\left(\|Z_r\| - \|Y_r\|\right) < 0,$$

which leads to the contradiction. Therefore, no exceptional family exists, and problem (6.44) is solvable as it follows from Theorem 6.16. ∎

EXAMPLE 6.5. Construct a mapping $F : S^2_+ \to S^{2\times 2}$ as follows. If a matrix $X \in S^2_+$ has a spectral representation

$$X = \mu_1 a_1 a_1^T + \mu_2 a_2 a_2^T,$$

then set

$$F(X) = \lambda_1 a_1 a_1^T + \lambda_2 a_2 a_2^T,$$

where

$$\lambda_1 = \mu_1 + 3\sin\mu_2 - 1, \quad \lambda_2 = \mu_2 + 3\sin\mu_1 - 1.$$

The mapping F is not monotone but it is continuous and satisfies (6.58) over the cone S^2_+ (the unit ball W_1 can be picked up for the set C). By virtue of Theorem 6.19, the semidefinite complementarity problem for this mapping has a solution. In particular, the solution set includes the matrices

$$A = \begin{pmatrix} 1 & 0 \\ 0 & 0 \end{pmatrix}, \quad B = \begin{pmatrix} 0 & 0 \\ 0 & 1 \end{pmatrix}, \text{ and } D = \begin{pmatrix} \xi & 0 \\ 0 & \xi \end{pmatrix},$$

where $0 < \xi < 1$ solves the equation $\xi + 3\sin\xi = 1$. ∎

COROLLARY 6.10. *Let $F : S^n_+ \to S^{n\times n}$ be a continuous mapping and $C \subset S^n_+$ be a bounded subset such that $\langle Y, F(Y)\rangle \geq 0\ \forall Y \in S^n_+ \setminus C$. Then problem (6.44) is solvable. If the strict inequality $\langle Y, F(Y)\rangle > 0$ takes place for each $Y \in S^n_+ \setminus C$, then all the solutions of problem (6.44) belong to the subset C.* ∎

COROLLARY 6.11. *Let $F : S^n_+ \to S^{n\times n}$ be a continuous mapping and $C \subset S^n_+$ be a bounded subset such that $F(Y) \geq 0\ \forall Y \in S^n_+ \setminus C$. Then problem (6.44) is solvable. Moreover, if $0 \in C$ and the stronger condition $F(Y) > 0$ is valid for every $Y \in S^n_+ \setminus C$, then all the solutions of problem (6.44) belong to the subset C.* ∎

Consider further a compact subset $C \subset S^n_+$ which is star-shaped with respect to the origin, i.e. it contains the whole segment $[0, X]$ if a matrix X

belongs to C. Then the function $\eta(X)$ is well-defined which maps $X \in S_+^n$, $X \neq 0$ into the furthermost (from the origin) point of the set C in the ray running through 0 and the matrix X. It is clear that the subset $\Gamma = \{Y = \eta(X) \mid X \in S_+^n, X \neq 0\}$ constitutes the relative boundary of the subset C with respect to the cone S_+^n.

THEOREM 6.20. *Consider a continuous mapping $F : S_+^n \to S^{n\times n}$ and a nonempty subset $C \subset S_+^n$ that is compact and star-shaped with respect to the origin. If the function $\eta(X)$ is continuous, $\eta(X) \neq 0$, and $\langle Y, F(Y)\rangle \geq 0$ for every $Y \in \Gamma$, then problem* (6.44) *has a solution in the subset C.*

PROOF. Define the mapping $G : S_+^n \to S^{n\times n}$ as follows:

$$G(X) = \begin{cases} F(X), & \text{if } X \in C; \\ F(\eta(X)) + \|X - \eta(X)\|X, & \text{if } X \in S_+^n \setminus C. \end{cases} \tag{6.59}$$

The mapping G is clearly continuous over S_+^n and satisfies the following inequality

$$\langle X, G(X)\rangle > 0 \qquad X \in S_+^n \setminus C.$$

The latter provides for the existence of a solution to problem (6.44) with the mapping G over the subset C. However, due to (6.59), the mappings G and F coincide over C. Therefore, problem (6.44) with the mapping F is solvable over the subset C, too. This completes the proof. ■

EXAMPLE 6.6. Construct the mapping $F : S_+^2 \to S^{2\times 2}$ as follows. If a matrix $X \in S_+^2$ has a spectral representation

$$X = \mu_1 a_1 a_1^T + \mu_2 a_2 a_2^T,$$

then set

$$F(X) = \lambda_1 a_1 a_1^T + \lambda_2 a_2 a_2^T,$$

where

$$\lambda_1 = 3\mu_1 - \mu_1\mu_2 - 1, \quad \lambda_2 = 3\mu_2 - \mu_1\mu_2 + 1.$$

Take the set $\{\, X \in S_+^2 \mid \mu_1 + \mu_2 \leq 1\}$ for the subset C. It is easy to see that the latter satisfies the conditions of Theorem 6.20. Indeed, for the relative boundary of the subset C we obtain

$$\langle Y, F(Y)\rangle = \mathrm{tr}\left[\left(\xi a_1 a_1^T + (1-\xi) a_2 a_2^T\right)\left(\left(\xi^2 + 2\xi - 1\right) a_1 a_1^T + \right.\right.$$

$$\left.\left.\left((1-\xi)^2 + 2(1-\xi) + 1\right) a_2 a_2^T\right)\right] = \xi^3 + (1-\xi)^3 + 2\left[\xi^2 + (1-\xi)^2\right] + 1 - 2\xi.$$

Transforming the linear and quadratic terms, we come to the following inequality:

$$2\left[\xi^2 + (1-\xi)^2\right] + 1 - 2\xi = 4\xi^2 - 6\xi + 3 \geq 0 \quad \forall 0 \leq \xi \leq 1.$$

Finally we get

$$\langle Y, F(Y)\rangle \geq \xi^3 + (1-\xi)^3 \geq 0 \qquad \forall 0 \leq \xi \leq 1.$$

Therefore, problem (6.44) has a solution in the simplex C, although the mapping F is clearly non-monotone. For instance, the matrix

$$X = \begin{pmatrix} 1/3 & 0 \\ 0 & 0 \end{pmatrix}$$

solves the problem. ∎

Now we extend Theorem 6.20, giving up the continuity of the function η. In order to do that, first reduce the domain of the function η to the subset $\Omega = V_1 \cap S_+^n$ which is the intersection of the unit sphere around the origin and the cone S_+^n. Furthermore, we assume again that $\eta(X) \neq 0 \ \forall X \in \Omega$, but do not demand the function η to be continuous over Ω. It clearly allows the subset C to have the boundary which contains whole segments of straight rays starting from the origin.

THEOREM 6.21. *Consider a continuous mapping $F : S_+^n \to S^{n\times n}$ and a nonempty compact subset $C \subset S_+^n$ which is star-shaped with respect to the origin and $0 \notin \Gamma$. If $\langle Y, F(Y) \geq 0$ for each matrix Y of the relative boundary Γ, then problem (6.44) has a solution in the subset C.*

PROOF. Consider a function $\zeta : \Omega \to R$ defined by $\eta(X) = \zeta(X)X$ for $X \in \Omega$. Since C is compact, the function ζ is upper semicontinuous over Ω. This implies the inequality

$$\limsup_{k\to\infty} \zeta(X^k) \leq \zeta(X)$$

to hold for each $X \in \Omega$ and every sequence of matrices $\{X^k\} \subset \Omega$ convergent to X. Hence, the Lebesgue subsets $\{X \in \Omega \mid \zeta(X) \geq t\}$ of the function ζ are closed for every $t \in R$. Therefore the function ζ is Lebesgue summable over the subset Ω. Now fix a scalar $\varepsilon > 0$ and consider a family of standard nonnegative finite functions $\{\varphi\}_{X\in V_1}$, each of them being infinitely differentiable, taking positive values over the neighborhood $U_\varepsilon(X) = \{Y \in V_1 \mid \|X - Y\| < \varepsilon\}$ of the matrix X and equal to zero outside it, and

$$\int\limits_{U_\varepsilon(X)} \varphi(Y)dY = 1. \tag{6.60}$$

Furthermore, extend the function ζ to the matrices Y of the subset $V_1 \setminus \Omega$ which belong to the ε-neighborhood of the set Ω, by the rule

$$\zeta(Y) = \zeta\left(\mu \mathrm{Pr}_{S_+^n}(Y)\right), \quad Y \in (V_1 \setminus \Omega) \bigcap \left(\cup_{X\in\Omega} U_\varepsilon(X)\right),$$

where $\mu > 0$ is such a scalar that $\mu \mathrm{Pr}_{S_+^n}(Y) \in \Omega$. Now define $\zeta_\varepsilon : \Omega \to R$ by averaging ζ with respect to the standard finite functions φ as follows:

$$\zeta_\varepsilon(X) = \int\limits_{U_\varepsilon(X)} \zeta(Y)\varphi(Y)dY, \quad \forall X \in \Omega. \tag{6.61}$$

Function ζ_ε is apparently continuous and bounded over Ω and has (like ζ) only positive values. Now consider the compact star-shaped subset $C_\varepsilon \subset S^n_+$ defined by the function $\eta_\varepsilon(X) = \zeta_\varepsilon(X)X$ whose value is the matrix furthermost from the origin, belonging to the subset C_ε and to the ray going from the origin and running through the point $X \in \Omega$. In the other terms,

$$C_\varepsilon = \bigcup_{X\in\Omega} [0, \eta_\varepsilon(X)] .$$

Now we will show that the distance between the boundaries of the subsets C_ε and C is small along with small ε. In order to do that, first prove that for each $x \in \Omega$, there exists a matrix $Y \in \Omega$ in the neighborhood $U_\varepsilon(X)$ such that the matrix $D = \zeta_\varepsilon(X)Y$ belongs to the (relative) boundary Γ of the subset C. On the contrary, suppose that the latter fails, then the closedness of C implies the following alternative: either all the matrices D lie outside the subset C for every $Y \in U_\varepsilon(X)$, or vice versa, they all belong to the interior of C. In the former case, the inequality $\zeta_\varepsilon(X) > \zeta(Y)\ \forall Y \in U_\varepsilon(X)$ holds, and otherwise, the opposite one $\zeta_\varepsilon(X) < \zeta(Y)\ \forall Y \in U_\varepsilon(X)$ takes place. In both cases, average these inequalities by the finite function φ over the neighborhood $U_\varepsilon(X)$ and make use of (6.60) and (6.61), obtaining impossible relationships $\zeta_\varepsilon(X) > \zeta_\varepsilon(X)$ or $\zeta_\varepsilon(X) < \zeta_\varepsilon(X)$, respectively. Thus, our assumption is false, and the needed matrix $Y \in \Omega$ in the neighborhood $U_\varepsilon(X)$ does exist. Also note that

$$\|\eta_\varepsilon(X) - D\| = \zeta_\varepsilon(X)\|X - Y\| \le \zeta_\varepsilon(X)\varepsilon. \tag{6.62}$$

Now fix a ball W large enough so that the subset $K = W \cap S^n_+$ comprises C and all the subsets C_ε for $0 < \varepsilon < \bar{\varepsilon}$, too. The mapping F being continuous, the scalar product $\langle X, F(X)\rangle$ is a uniformly continuous function over K. Therefore, the latter has a continuity module $\omega : K \to R$ with the following properties: the function ω is continuous and non-decreasing,

$$\lim_{\delta\to+0} \omega(\delta) = 0, \tag{6.63}$$

and the inequality

$$|\langle X, F(X)\rangle - \langle Y, F(Y)\rangle| \le \omega(\delta) \tag{6.64}$$

holds for the matrices $X, Y \in K$ such that $\|X - Y\| < \delta$. Now define the continuous mapping F_ε over the cone S^n_+ as follows: $F_\varepsilon(0) = F(0)$, and

$$F_\varepsilon(X) = F(X) + \omega\left(\zeta_\varepsilon\left(\frac{X}{\|X\|}\right)\varepsilon\right)\frac{X}{\rho_\varepsilon^2}, \qquad \forall X \in S^n_+ \setminus \{0\}; \tag{6.65}$$

here $\rho_\varepsilon = \min\{\|X\| \mid X \in \Gamma_\varepsilon\}$, and Γ_ε is the relative boundary of C_ε with respect to the cone S^n_+. It is easy to verify that

$$\langle Y, F_\varepsilon(Y)\rangle = \langle Y, F(Y)\rangle + \omega\left(\zeta_\varepsilon\left(\frac{Y}{\|Y\|}\right)\varepsilon\right)\frac{\|Y\|^2}{\rho_\varepsilon^2} \ge$$

$$\ge \langle Y, F(Y)\rangle + \omega\left(\zeta_\varepsilon\left(\frac{Y}{\|Y\|}\right)\varepsilon\right) \qquad \forall Y = \eta_\varepsilon(X), X \in S^n_+ \setminus \{0\}. \tag{6.66}$$

According to the theorem assumptions, $\langle D, F(D)\rangle \geq 0$ at every point D of the relative boundary Γ. Hence, for arbitrary $X \in \Omega$ and $\varepsilon < \bar{\varepsilon}$, relationships (6.62), (6.63) and (6.66) imply

$$\langle \eta_\varepsilon(X), F_\varepsilon(\eta_\varepsilon(X))\rangle \geq 0.$$

The latter allows one to make use of Theorem 6.20 which provides for the existence of a solution $z_\varepsilon \in C_\varepsilon$ to problem (6.44) with the mapping F_ε. Now as $\varepsilon \to +0$, the distance between the subsets C_ε and C also vanishes. The subset C being closed, every accumulation point of the solution family z_ε belongs to C. Taking that into account, together with the continuity of the mapping F and the function ζ_ε, property (6.64) of the continuity module ω, and definition (6.65) of the function F_ε, pass to the limit as $\varepsilon \to +0$ in the relationships

$$Z_\varepsilon \in S^n_+, \quad F_\varepsilon(Z_\varepsilon) \in S^n_+, \quad \langle Z_\varepsilon, F_\varepsilon(Z_\varepsilon)\rangle = 0.$$

Since the cone S^n_+ is closed, we thus obtain

$$Z \in S^n_+, \quad F(Z) \in S^n_+, \quad \langle Z, F(Z)\rangle = 0$$

for an arbitrary accumulation point Z of the family of solutions $\{Z_\varepsilon\}$ to the auxiliary complemetarity problem. The matrix Z apparently solves (6.44) which completes the proof. ■

In order to deduce new sufficient conditions for the existence of solutions to the semidefinite complementarity problem, we examine some properties of the exceptional family of matrices.

LEMMA 6.5. *For each element of an exceptional family $Z_r \in S^n_+$ the following relationships are true:*

(i) $\langle U, F(Z_r)\rangle \leq 0$ for every non-zero matrix U that belongs to the cone's facet of the minimal dimension containing the element Z_r and satisfies the inequality $\langle U, Z_r\rangle \geq 0$;

(ii) $\langle D, F(Z_r)\rangle \geq 0$ for every (nonzero) feasible direction D from the point Z_r (i.e. not leading outside the cone S^n_+) and such that $\langle D, Z_r\rangle = 0$.

PROOF. Recall that for each matrix $Z_r \in S^n_+$ of the exceptional family, there exists a scalar $\lambda_r > 0$ such that the matrix $M_r = F(Z_r) + \lambda_r Z_r$ is the interior normal element to the supporting hyperplane to the cone S^n_+ at the point Z_r. Consider a direction U that lies in the cone minimal dimension facet containing Z_r. The properties of a supporting hyperplane imply the equality $\langle U, M_r\rangle = 0$ that is tantamount to $\langle U, F(Z_r)\rangle = -\lambda_r\langle U, Z_r\rangle$. Moreover, if $\langle U, Z_r\rangle \geq 0$, then assertion (i) follows from the latter equality.

Again use the properties of the normal element to the supporting hyperplane (*cf.* Lemma 6.3) and obtain inequality $\langle D, M_r\rangle \geq 0$ for each feasible direction D at the point z_r. In additon, if $\langle D, Z_r\rangle = 0$, the the required inequality (ii) follows immediately from the structure of the matrix M_r. The proof is complete. ■

THEOREM 6.22. *Let $F : S^n_+ \to S^{n\times n}$ be a continuous mapping and $C \subset S^n_+$ be a nonempty bounded subset such that for every $X \in S^n_+ \setminus C$ (at least) one of the following assumptions be valid:*

(i) $\langle U, F(X)\rangle > 0$ for a nonzero matrix U that belongs to the cone's minimal dimension facet containing the matrix X, and satisfies $\langle U, X\rangle \geq 0$;

(ii) $\langle D, F(X)\rangle < 0$ for a nonzero feasible direction D at the point X with $\langle D, X\rangle = 0$.

Then problem (6.44) *is solvable. Moreover if assumption (i) holds for $U = X$, all the solutions of* (6.44) *belong to C.*

PROOF. Suppose an exceptional family of matrices $\{Z_r\} \subset S^n_+$ to exist. From some $\bar{r} > 0$ onward, we have $Z_r \in S^n_+ \setminus C$. According to the theorem assumptions, at least one of the conditions (i) or (ii) is valid for the elements Z_r, $r > \bar{r}$ which contradicts the assertion of Lemma 6.5. That denies existence of exceptional families and therefore, establishes the solvability of problem (6.44) (see Theorem 6.16). At last, if either the stronger version of assumption (i), or assumption (ii) take place, we deduce for every $X \in S^n_+ \setminus C$ either $\langle X, F(X)\rangle > 0$, or $F(X) \notin S^n_+$, respectively. Consequently, all the solutions of (6.46) belong to C, which completes the proof. ∎

By making use of the techniques from the proofs of Theorems 6.20–6.21, we obtain the following corollary.

COROLLARY 6.12. *Let $F : S^n_+ \to S^{n\times n}$ be a continuous mapping, and a nonempty subset $C \subset S^n_+$ be compact and star-shaped with respect to the origin with $0 \notin \Gamma$. If for every $X \in S^n_+ \setminus \{0\}$ (at least) one of the following assumptions concerning $Y = \eta(X)$ is valid:*

(i) $\langle U, F(Y)\rangle > 0$ for a non-zero matrix U that lies in the cone's minimal dimension facet containing the matrix Y, and satisfies $\langle U, Y\rangle \geq 0$;

(ii) $\langle D, F(Y)\rangle < 0$ for a non-zero feasible direction D at the point Y such that $\langle D, Y\rangle = 0$,

then problem (6.44) *is solvable. Moreover if (i) holds for $U = Y$, then there exists a solution of* (6.44) *belonging to C.* ∎

6.7.4 CONCLUSION

In conclusion of the chapter, we remark that its key assertions can be formulated as follows. For a continuous mapping, there exists either a solution of the respective complementarity problem or an exceptional family of elements. These theorems open a new research direction in finding out sufficient conditions for the complementarity problem to be solvable. Namely, assumptions that exclude existence of exceptional families guarantee the problem's solvability.

REFERENCES

Bulavsky, V.A, Isac, G., Kalashnikov, V.V. 'Application of topological degree theory to complementarity problems.'- In: *Multilevel Optimization: Algorithms and Applications*, A. Migdalas, P.M. Pardalos and P. Värbrand, eds. Dordrecht-Boston-London: Kluwer Academic Publishers, 1998.

Bulavsky VA, Isac G, Kalashnikov VV. Application of topological degree theory to semi-definite complementarity problems. Optimization, 2001; 49.

Harker PT, Pang J-S. Finite-dimensional variational inequalities and nonlinear complementarity problems: a survey of theory, algorithms and applications. Math. Programming 1990; 48: 161-220.

Isac, G.

1. *Complementarity Problems*. Lecture Notes in Mathematics. Berlin-Heidelberg: Springer-Verlag, 1992.

2. *Topological Methods in Complementarity Theory*. Dordrecht: Kluwer Academic Publishers, 2000.

Kojima M, Shindoh S, Hara S. Interior-point methods for the monotone semidefinite linear complementarity problem in symmetric matrices. SIAM J. Optim. 1997; 7: 86-125.

Lloyd, N.G. *Degree Theory*. Cambridge: Cambridge University Press, 1978.

McLinden, L.'The complementarity problem for maximal monotone multifunctions'. - In: *Variational Inequalities and Complementarity Problems*, R.W. Cottle, F. Giannessi and J.L. Lions, eds. New York: Academic Press, 1980.

Ortega, J., Rheinboldt, W.C. *Iterative Solution of Nonlinear Equations in Several Variables*. New York: Academic Press, 1970.

Schatten, R.*Norm Ideals of Completely Continuous Operators*. Berlin-Göttingen-Heidelberg: Springer-Verlag, 1960.

Smith TE. A solution condition for complementarity problems with an application to spatial price equilibrium. Appl. Math. Comput. 1984; 15: 61-69.

Chapter 7

PSEUDOMONOTONE AND IMPLICIT COMPLEMENTARITY PROBLEMS

In this chapter, we extend our techniques to the case of infinite dimensional complementarity problems. Especial attention is paid to the latter with pseudomonotone operators. The second part of the chapter is devoted to Implicit Complementarity Problems with single-valued and multi-valued mappings.

First, the property of a function defined on an infinite-dimensional spase (Hilbert space) to be without exceptional family of elements (EFE) is investigated. Second, a new notion of exceptional family of elements for a pair of functions involved in the implicit complementarity problem (ICP) is introduced. Based upon this notion and the Leray-Schauder Alternative, a general alternative is obtained which gives more general existence theorems for the implicit complementarity problem. Finally, via the techniques of continuous selections, these existence theorems are extended to the multi-valued implicit complementarity problems (MIPS).

7.1 INTRODUCTION AND PRELIMINARIES

In the previous chapter, we introduced a new method for solving complementarity problems. This method is based upon the notion of *exceptional family of elements* for a continuous function with respect to a closed convex cone in the Euclidean space. This notion (in the form of *an exceptional sequence*) was introduced by Smith (1984) for the Euclidean space R^n and only for the cone R^n_+. However, since our notion is related to the topological degree, it is applicable to any closed convex cone K in the Hilbert space. In this sense, our method is new.

By this method, we studied several problems in complementarity theory and we obtained interesting results: *cf.* Kalashnikov (1995), Isac, Bulasvky and Kalashnikov (1997), Bulavsky, Isac and Kalashnikov (1998), Bulavsky, Isac and Kalashnikov (1999), Isac [1] (1999), Isac [2] (1999), Isac (2000), Isac and Obuchowska (1998), Isac and Zhao (1999), Zhao and Isac (1999), Zhao and Isac (2000). Our method was extended also to variational inequalities by Zhao (1997), Zhao (1998), Zhao and Han (1999), Zhao e.a. (1999). Now, in this chapter we will study the complementarity problem defined by a pseudomonotone operator, using the notion of exceptional family of elements and some *Leray-Schauder-type* alternative. In the previous chapter, we used the topological degree as an instrument. By the Leray-Schauder alternative, we obtain our results with simpler proofs and more evident possibility to generalize the results. In doing that, we follow the recent paper by Isac and Kalashnikov (2001).

The property of a function to be *without exceptional family of elements* (EFE) is a kind of a very general coercivity condition. In Section 7.2, we will show that, for a pseudomonotone operator, the solvability of a complementarity problem is equivalent to the above-mentioned property of the function to be without EFE.

We also use the *strict feasibility* to obtain an existence theorem for complementarity problems generated by pseudomonotone operators and general cones in Hilbert spaces. In Section 7.2 we also study the strict feasibility of a complementarity problem making use of a Leray-Schauder alternative and the notion of δ-exceptonal family of elements (δ-EFE).

Section 7.3 is dedicated to bilevel variational inequality with monotone operators, and Section 7.4 examines the bilevel variational inequality with pseudomonotone mappings. The former is solved by the penalty function approach as in paper Kalashnikov and Kalashnikova (1996), whereas the latter is treated by a regularization method proposed in Kalashnikov and Khan (1999).

Section 7.5 deals with the *implicit complementarity problem (ICP)* that was introduced into the complementarity theory by Bensoussan e.a. (1973), Bensoussan and Lions (1973) as a mathematical tool in the study of some stochastic optimal control problems. The ICP has been studied by several authors Bulavsky, Isac and Kalashnikov (1998), Capuzzo- Dolcetta and Mosco (1980), Isac (1986,1990,1992), Isac, Bulavsky and Kalashnikov (1997), Isac and Goeleven [1],[2] (1993), Mosco (1976,1980), Nikaido (1968), Pang (1981,1982).

Generally, the ICP has been studied via variational or quasi-variational inequalities; *cf.* Chan and Pang (1982), Baiocchi and Capelo (1983), Isac (1990), the fixed point theory; *cf.* Isac (1997,1998), iterative methods, zero-epi mappings, or some algebraic methods. Recently, a new method has been introduced in the study of complementarity problems. This method is based upon the concept of an *exceptional family of elements (EFE)* for a continuous function. In our papers Isac, Bulavsky, Kalashnikov (1997), Bulavsky, Isac, Kalashnikov (1998) we obtained an interesting alternative theorem related to the existence of solutions to *implicit complementarity problems*. One of the assumptions about the function g of the pair (f, g) was, in a certain sense, too restrictive, but this condition was necessary due to the fact that we used the topological degree theory.

Now, in Section 7.5 of this chapter, we will show that by a little modification of the notion of *exceptional family of elements (EFE)* for a pair of functions and replacing the topological degree results by the Leray- Schauder Alternative, we can obtain a general alternative, which implies an existence theorem for implicit complementarity problems. Finally, in Section 7.6, by use of the continuous selection techniques, we extend the foregoing results to the multi-valued implicit complementarity problems (MICP) with lower semi-continuous mappings, and by making use of a compactification approach (*cf.* Avtukhovich, 1999), to the upper semicontinuous mappings.

7.1.1 PRELIMINARIES

Let $(H, \langle\ ,\ \rangle)$ be a Hilbert space and let $X, Y \subset H$ be nonempty subsets of H. We will denote by ∂X the boundary of X, by $\bar{X}$ the closure of X, and by $\mathcal{P}(Y)$ the family of all non-empty subsets of Y. Let $f : X \to \mathcal{P}(Y)$ be a multi-valued mapping. The mapping f is said to be *upper semicontinuous* (u.s.c.) on X if the set $\{x \in X \mid f(x) \subset V\}$ is open in X, whenever V is an open subset in Y.

The multi-valued mapping f is said to be *compact* if $f(X)$ is contained in a compact subset of Y. A subset D of H is called *contractible* if there is a continuous mapping $h : D \times [0, 1] \to D$ such that for all $x \in D$ we have $h(x, 0) = x$ and $h(x, 1) = x_0$ for some $x_0 \in D$. We note that if D is convex, then it is contractible, since for any $x_0 \in D$ the mapping $h(x, t) = tx_0 + (1-t)x$ has the needed property. It is easy to see that a set star-shaped at a point x_0 is also contractible to x_0.

We say that a multi-valued mapping $f : H \to H$ with nonempty values is *completely upper semicontinuous*, if it is upper semicontinuos and for any bounded set $B \subset H$, we have that $f(B) = \cup_{x \in B} f(x)$ is relatively compact. In particular, a mapping $f : H \to H$ is called *completely continuous* if f is continuous and for any bounded set $B \subset H$, $f(B)$ is relatively compact. We say that $f : H \to H$ is a *completely continuous field* if f has a representation of the form $f(x) = x - T(x)$, where $T : H \to H$ is a completely continuous mapping. Obviously, if H is the Euclidean space $(R^n, \langle\ ,\ \rangle)$, then any continuous mapping $f : R^n \to R^n$ is a completely continuous field, since for any $x \in R^n$ we have $f(x) = x - (x - f(x))$.

We say that $K \subset H$ is a *closed pointed convex cone*, if K is a nonempty closed set with the following properties:

(*i*) $K + K \subseteq K$;

(*ii*) $\lambda K \subseteq K$, for all $\lambda \in R_+$;

(*iii*) $K \cap (-K) = \{0\}$.

We say that $K \subset H$ is *closed convex cone* if K is a non-empty closed set with properties *(i)* and *(ii)*. The dual cone K^* of K is defined by

$$K^* = \{y \in K \mid \langle x, y\rangle \geq 0 \text{ for all } x \in K\}.$$

We can show that K^* is a closed convex cone.

Since K is closed and convex, the *projection* operator onto K, denoted by P_K, is well-defined for every $x \in H$; *cf.* Baiocchi and Capelo (1983). We have that for any $x \in H$, $P_K(x)$ is the unique element in K with the property

$$\|x - P_K(x)\| = \min_{y \in K} \|x - y\|.$$

Because K is a closed convex cone, we have that P_K is characterized by the following properties: for every $x \in H$, $P_K(x)$ is the unique element in K such that:

(*i*) $\langle P_K(x) - x, y\rangle \geq 0$, for all $y \in K$;

(*ii*) $\langle P_K(x) - x, P_K(x)\rangle = 0$.

For the proofs the reader is referred to Baiocchi and Capelo (1983).

7.1.2 LERAY-SCHAUDER TYPE ALTERNATIVE

In this chapter, we will use the following Leray-Schauder-type alternatives.

THEOREM 7.1 [NONLINEAR ALTERNATIVE]. *Let $C \subset H$ be a convex set, U a subset open in C and such that $0 \in U$.*

Then each continuous compact mapping $f : \bar{U} \to C$ has at least one of the following two properties:

(*i*) *f has a fixed point;*

(*ii*) *there is an $x_* \in \partial U$ with $x_* = \lambda_* f(x_*)$ for some $\lambda_* \in (0, 1)$.*

PROOF. This result is proved in Dugundji and Granas (1982) using the transversality theory; see also Granas (1976). ■

THEOREM 7.2 [LERAY-SCHAUDER-TYPE ALTERNATIVE]. *Let X be a closed subset of a locally convex space E such that $0 \in intX$ and $f : X \to E$ a compact upper semicontinuous multi-valued mapping with nonempty compact contractible values. If f is fixed-point-free, then it satisfies the following Leray-Schauder condition: there exists $(\lambda_*, x_*) \in (0, 1) \times \partial X$ such that $x_* \in \lambda_* f(x_*)$.*

PROOF. This result is a part of a corollary of the main theorem proved in Ben-el-Mechaiekh and Idzik (1994); see also Granas (1976). A simple proof is also presented in Ben-el-Mechaiekh e.a. (1998). ■

7.2 COMPLEMENTARITY PROBLEMS WITH PSEUDOMONOTONE MAPPINGS

Let $(H, \langle\ ,\ \rangle)$ be a Hilbert space, $K \subset H$ a closed pointed convex cone and $f : H \to H$ a nonlinear mapping. The *nonlinear complementarity problem* defined by f and K is:

$$(\mathrm{NCP}(f, K)) \text{ find } x_* \in K \text{ such that } f(x_*) \in K^* \text{ and } \langle x_*, f(x_*)\rangle = 0.$$

If $f : H \to H$ is a multi-valued mapping, the *multi-valued complementarity problem* generated by f and K is:

$$(\mathrm{MCP}(f, K)) \text{ find } x_* \in K \text{ and } y_* \in f(x_*) \cap K^* \text{ such that } \langle x_*, y_*\rangle = 0.$$

These problems have a plenty of applications and were studied by a number of authors: Cottle e.a. (1992), Isac (1992), Kalashnikov (1995), Hyers, Isac and Rassias (1997).

We will give some existence theorems for the problems NCP(f, K) and MCP(f, K) making use of the notion of exceptional family of elements (EFE), introduced for single-valued mappings in Isac, Bulavsky, Kalashnikov (1997) and for multi-valued mappings in Isac (1999). This notion was systematically used in the study of several kinds of problems in complementarity theory in relation with the topological degree.

DEFINITION 7.1. We say that a family of elements $\{x_r\}_{r>0} \subset K$ is an *exceptional family of elements* (EFE) for a mapping $f : H \to H$, with respect to K, if for every real number $r > 0$ there exists a real number $\mu_r > 0$ such that the vector $u_r = f(x_r) + \mu_r x_r$ satisfies the following conditions:

(*i*) $u_r \in K^*$,
(*ii*) $\langle u_r, x_r\rangle = 0$,
(*iii*) $\|x_r\| \to +\infty$ as $r \to +\infty$.

For a multi-valued mapping, Definition 7.1 must be replaced by the following variant.

DEFINITION 7.2. We say that a family of elements $\{x_r\}_{r>0} \subset K$ is an *exceptional family of elements* (EFE) for a multi-valued mapping $f : H \to H$, with respect to K, if and only if, for every real number $r > 0$ there exists a real number $\mu_r > 0$ and an element $y_r \in f(x_r)$ such that the following conditions are satisfied:

(*i*) $u_r = \mu_r x_r + y_r \in K^*$,
(*ii*) $\langle u_r, x_r\rangle = 0$,
(*iii*) $\|x_r\| \to +\infty$ as $r \to +\infty$.

We say that a multi-valued mapping $f : H \to H$ is a *multi-valued completely upper semicontinuous field*, if f has a representation of the form $f(x) = x - T(x)$ for all $x \in H$, where $T : H \to H$ is a completely upper semicontinuous multi-valued mapping. Now we have the following result.

THEOREM 7.3. *Let $(H, \langle\ ,\ \rangle)$ be a Hilbert space, $K \subset H$ a closed pointed convex cone and $f : H \to H$ a multi-valued completely upper semicontinuous field, with nonempty compact contractible values. Then there exists either a solution to the problem MCP(f, K), or an exceptional family of elements for f with respect to K, in the sense of Definition* 7.2.

PROOF. First, we observe that, because P_K is a continuous operator and for every $x \in H$, $T(x)$ is a compact set, we have that $P_K[x - f(x)] = P_K[T(x)]$ is a multi-valued mapping with nonempty compact contractible values. For the proof of this fact, see the proof of Corollary 1 of Theorem 2 presented in Isac (2000). If the problem MCP(f, K) has a solution, then the proof is finished.

Suppose that the problem MCP(f, K) has no solution. For any positive real number r consider the set $B_r = \{x \in H \mid \|x\| \le r\}$. Obviously, $0 \in \text{int}B_r$. The multi-valued mapping $\Phi(x) = P_K[x - f(x)]$, $x \in H$, is fixed-point-free with respect to any set B_r. Indeed, if there exists $r > 0$ such that Φ has a fixed point x_* in B_r, then $x_* \in P_K[x_* - f(x_*)]$. The latter implies that $x_* \in K$ and there exists $u_* \in f(x_*)$ such that

$$x_* = P_K(x_* - u_*). \tag{7.1}$$

Using (7.1) and applying the properties (i) and (ii) of the projection operator P_K, it is easy to show that (x_*, u_*) is a solution to the problem MCP(f, K), which is a contradiction. Therefore, Φ is fixed-point-free with respect to any set B_r ($r > 0$). Observe that for all $r > 0$, all the assumptions of Theorem 7.2 are satisfied for $X = B_r$ and the multi-valued mapping Φ. Hence, for any $r > 0$, there exists $x_r \in \partial B_r$ and $\lambda_r \in (0, 1)$ such that

$$x_r \in \lambda_r P_K[x_r - f(x_r)]. \tag{7.2}$$

From (7.2) we deduce that there exists $y_r \in f(x_r)$ such that

$$x_r = \lambda_r P_K(x_r - y_r). \tag{7.3}$$

Taking into account (7.3) and again the properties (i) and (ii) of the projection operator P_K, we obtain that $u_r = [(1 - \lambda_r)/\lambda_r]\, x_r + y_r \in K^*$ and $\langle u_r, x_r\rangle = 0$. As $\|x_r\| = r$ for any $r > 0$, we conclude that $\{x_r\}_{r>0}$ is an exceptional family of elements for f with respect to K, in the sense of Definition 7.2, which completes the proof. ∎

COROLLARY 7.1. *Let $(H, \langle\ ,\ \rangle)$ be a Hilbert space, $K \subset H$ a closed pointed convex cone and $f : H \to H$ a completely continuous field. Then, there exists either a solution for the problem* NCP(f, K)*, or an exceptional family of elements for f with respect to K, in the sense of Definition* 7.1. ∎

COROLLARY 7.2. *Let K be a closed pointed convex cone in the Euclidean space R^n, and $f : R^n \to R^n$ a continuous mapping. Then, there exists either a solution to the problem* NCP(f, K), *or an EFE for f with respect to K, in the sense of Definition* 7.1. ■

The following result is a direct consequence of Theorem 7.3.

THEOREM 7.4. *Let $(H, \langle\ ,\ \rangle)$ be a Hilbert space, $K \subset H$ a closed pointed convex cone and $f : H \to H$ a multi-valued, completely upper semicontinuous field, with nonempty compact contractible values. If f is without EFE in the sense of Definition* 2.2, *then the problem* MCP(f, K), *has a solution.* ■

The notion of pseudomonotone operator was introduced in Karamardian (1976). Until now, this notion has been studied by several authors Cottle and Yao (1992), Hadjisavvas and Schaible (1993,1996), Karamardian and Schaible (1990), Schaible and Yao (1995), Yao (1994). In the cited papers, we can find a number of existence theorems for complementarity problems generated by monotone or pseudomonotone operators in R^n or in Hilbert spaces. Because Yao (1994) considered variational inequalities defined by multi-valued pseudomonotone operators, it is also natural to consider complementarity problems generated by multi-valued pseudomonotone operators. In this section, we will show that, if a complementarity problem generated by a pseudomonotone multi-valued mapping has a solution, then the mapping is without exceptional family of elements.

We recall that a mapping $f : H \to H$ is said to be *pseudomonotone* on K, if for every pair of distinct points $x, y \in K$ we have $\langle x - y, f(y)\rangle \geq 0$ implies $\langle x - y, f(x)\rangle \geq 0$. Obviously, a monotone mapping is pseudomonotone, whereas the converse is not true: *cf.* Karamardian (1996).

Suppose that $f : H \to H$ is a multi-valued mapping. We say that f is *pseudomonotone* on K if, for any distinct points $x, y \in K$ and arbitrary $u \in f(x)$ and $w \in f(y)$, the inequality $\langle x - y, w\rangle \geq 0$ implies $\langle x - y, u\rangle \geq 0$.

Now, we have the following result.

THEOREM 7.5. *Let $(H, \langle\ ,\ \rangle)$ be a Hilbert space and $K \subset H$ a closed pointed convex cone. Suppose that $f : H \to H$ is a multi-valued pseudomonotone mapping on K. If the problem* MCP(f, K) *has a solution, then f is without exceptional families of elements with respect to K in the sense of Definition* 7.2.

PROOF. Suppose that MCP(f, K) has a solution (x_*, y_*), i.e. $x_* \in K$, $y_* \in f(x_*) \cap K^*$, and $\langle x_*, y_*\rangle = 0$, which is equivalent to the following variational inequality

$$\langle x - x_*, y_*\rangle \geq 0, \quad \text{for all } x \in K. \tag{7.4}$$

As f is pseudomonotone then (7.4) implies

$$\langle x - x_*, u\rangle \geq 0 \text{ for any } x \in K \text{ and } u \in f(x). \tag{7.5}$$

Suppose that f has an exceptional family of elements $\{x_r\}_{r>0} \subset K$. In this case, for any $r > 0$ there exist $\mu_r > 0$ and $w_r \in f(x_r)$ such that

(*i*) $u_r = \mu_r x_r + w_r \in K^*$,
(*ii*) $\langle x_r, u_r\rangle = 0$, and
(*iii*) $\|x_r\| \to +\infty$ as $r \to +\infty$.

Now choose x_r such that $\|x_*\| < \|x_r\|$. Making use of (7.5) we obtain

$$0 \leq \langle x_r - x_*, w_r\rangle = \langle x_r - x_*, u_r - \mu_r x_r\rangle =$$

$$= \langle x_r, u_r\rangle - \langle x_*, u_r\rangle - \mu_r\|x_r\|^2 + \mu_r\langle x_*, x_r\rangle \leq$$

$$\leq -\mu_r\|x_r\|^2 + \mu_r\|x_*\|\|x_r\| = \mu_r\|x_r\|\left(\|x_*\| - \|x_r\|\right) < 0,$$

which is a contradiction. Therefore, f is without exceptional family of elements, with respect to K, in the sense of Definition 7.2, and the proof is complete. ∎

COROLLARY 7.3. *Let $(H, \langle\ ,\ \rangle)$ be a Hilbert space, $K \subset H$ a closed pointed convex cone and $f : H \to H$ a multi-valued pseudomonotone mapping with respect to K. Suppose that f is a completely upper semicontinuous field with nonempty compact contractible values. Then, the problem* MCP(f, K) *has a solution if and only if, f is without exceptional families of elements with respect to K, in the sense of Definition* 7.2.

PROOF. The corollary is a consequence of Theorems 7.4 and 7.5. ∎

COROLLARY 7.4 *Let $(H, \langle\ ,\ \rangle)$ be a Hilbert space, $K \subset H$ a closed pointed convex cone and $f : H \to H$ a completely continuous field which is pseudomonotone with respect to K. Then, the problem* NCP(f, K) *has a solution if and only if, f is without exceptional families of elements with respect to K, in the sense of Definition* 7.1. ∎

COROLLARY 7.5. *Let $(R^n, \langle\ ,\ \rangle)$ be the Euclidean space, and let $K \subset R^n$ be a closed pointed convex cone. If $f : R^n \to R^n$ is a continuous pseudomonotone operator with respect to K, then the problem* NCP(f, K) *has a solution if and only if, f is without exceptional families of elements, with respect to K, in the sense of Definition* 7.1. ∎

7.2.1 STRICT FEASIBILITY, SOLVABILITY, AND EXCEPTIONAL FAMILIES OF ELEMENTS

In this subsection, we will establish some interesting relationships between strict feasibility, solvability and the notion of exceptional family of elements. Let $(H, \langle\ ,\ \rangle)$ be a Hilbert space, $K \subset H$ a closed pointed convex cone and $f : H \to H$ a multi-valued mapping. Consider the multi-valued complementarity problem MCP(f, K). Now we define the *strict dual* of K as the set

$$\widehat{K^*} = \{y \in H \mid \langle y, x\rangle > 0, \text{ for all } x \in K \setminus \{0\}\}$$

(here y is considered as an element of the topological dual of H). For some cones, the strict dual can be empty.

We say that the cone K is *well-based* if and only if, there exists a bounded closed convex set $B \subset K$ such that $0 \notin B$ and $K = \cup_{\lambda \geq 0} \lambda B$.

It is well-known that a closed pointed convex cone $K \subset H$ is well-based if and only if, there exists a continuous linear functional $\varphi : H \to R$ and a constant $c > 0$ such that $c\|x\| \leq \varphi(x)$ for all $x \in K$. Obviously, in this case $\varphi \in \widehat{K^*}$, hence, $\widehat{K^*}$ is nonempty. When $\widehat{K^*}$ is nonempty, we say that the problem NCP(f, K) is *strictly feasible* if there exists an element $x_0 \in K$ such that $f(x_0) \in \widehat{K^*}$.

The following result can be easily deduced from Lemma 3.2 of Cottle and Yao (1992); in fact, it has been already obtained in the proof of Theorem 3.3 in Cottle and Yao (1992) using the classical coercivity condition for multivalued operators: see Yao (1994). But with the purely methodical purpose, we show below how this well-known result can be obtained using the notion of exceptional family of elements (EFE).

THEOREM 7.6. *Let $(H, \langle\ ,\ \rangle)$ be a Hilbert space, $K \subset H$ a closed, pointed, well-based convex cone and $f : H \to H$ a multi-valued completely upper semi-continuous field. If f is pseudomonotone with respect to K and there exists an element $x_0 \in K$ such that $f(x_0) \cap \widehat{K^*} \neq \emptyset$, then the problem* MCP$(f, K)$ *has a solution.*

PROOF. We show that f is without EFE. Assume the opposite, i.e. there exists $\{x_r\}_{r>0} \subset K$ such that:

(L1) for every $r > 0$, there exists $\mu_r > 0$ and an element $y_r \in f(x_r)$ such that $u_r = \mu_r x_r + y_r \in K^*$ and $\langle x_r, u_r \rangle = 0$;

(L2) $\|x_r\| \to +\infty$ as $r \to +\infty$.

Pick up an element $y_0 \in f(x_0) \cap \widehat{K^*}$. Consider $\langle y_0, x_r - x_0 \rangle$ and demonstrate that it is positive for all $r > r_0 > 0$. It is well-known: *cf.* Hyers, Isac and Rassias (1997), Chapter 1, – that since K is well-based, there exists a $\varphi \in H^*$ and a constant $c > 0$ such that $c\|x\| \leq \langle \varphi, x \rangle$, for all $x \in K$. Moreover, the subset

$$D = \{x \in K \mid \langle \varphi, x \rangle = 1\}$$

is weakly compact in H (it follows from K being well-based and H reflexive, that K is weakly locally compact). Now, making some transformations, one gets

$$\langle y_0, x_r - x_0 \rangle = \langle y_0, x_r \rangle - \langle y_0, x_0 \rangle =$$

$$= \langle y_0, \frac{x_r}{\langle \varphi, x_r \rangle} \rangle \cdot \langle \varphi, x_r \rangle - \langle y_0, x_0 \rangle.$$

Since $\frac{x_r}{\langle \varphi, x_r \rangle} \in D$, then

$$\langle y_0, \frac{x_r}{\langle \varphi, x_r \rangle} \rangle \geq \varepsilon > 0,$$

where $\varepsilon = \min_{x \in D}\langle y_0, x\rangle > 0$ due to the weak compactness of D, continuity of $\langle y_0, x\rangle$ in the weak topology and the property $\langle y_0, x\rangle > 0$ for all $x \in K \setminus \{0\}$ as $y_0 \in \widehat{K^*}$. Therefore, we have

$$\langle y_0, x_r - x_0\rangle \geq \varepsilon\langle \varphi, x_r\rangle - \langle y_0, x_0\rangle \geq$$

$$\geq \varepsilon c\|x_r\| - \langle y_0, x_0\rangle > 0 \quad \text{for all } r > r_0,$$

where $r_0 > 0$ is such that

$$\|x_r\| > \frac{\langle y_0, x_0\rangle}{\varepsilon c} \quad \forall r > r_0.$$

Now the pseudomonotonicity of f implies

$$\langle y_r, x_r - x_0\rangle \geq 0 \quad \text{for all } r > r_0 \text{ and any } y_r \in f(x_r).$$

From the last inequality and (L1), we obtain (for the y_r from (L1))

$$0 \leq \langle y_r, x_r - x_0\rangle = \langle u_r - \mu_r x_r, x_r - x_0\rangle =$$

$$= \langle u_r, x_r\rangle - \mu_r\langle x_r, x_r\rangle - \langle u_r, x_0\rangle + \mu_r\langle x_r, x_0\rangle \leq$$

$$\leq -\mu_r\|x_r\|^2 + \mu_r\|x_r\|\|x_0\| = -\mu_r\|x_r\|\left(\|x_r\| - \|x_0\|\right),$$

for all $r > r_0$. Since $\|x_r\| \to +\infty$ as $r \to +\infty$, there exists $r_1 > 0$ such that $\|x_r\| > \|x_0\|$ for all $r > r_1$. Hence, for $r > \max\{r_0, r_1\}$ we have

$$0 \leq -\mu_r\|x_r\|\left(\|x_r\| - \|x_0\|\right) < 0.$$

This contradiction proves that f is without exceptional families of elements, and the assertion of the theorem is a consequence of Theorem 2.3. ∎

In Theorem 7.6, the strict feasibility is essential. Now we will show that the notion of exceptional family of elements can be also used to study the strict feasibility of the problem NCP(f, R^n_+). First we introduce the following definition. Let $(R^n, \langle,\rangle)$ be the Euclidean space.

DEFINITION 7.3. Let $\delta > 0$ be an arbitrary real number and $f : R^n \to R^n$ a continuous function. We say that $\{x^r\}_{r>0} \subset R^n_+$ is a *δ-exceptional family of elements (δ-EFE)* for f with respect to R^n_+, if and only if for every $r > 0$ there exists $t_r \in (0,1)$ such that

(*i*) $\|x^r\| \to +\infty$ as $r \to +\infty$,
(*ii*) $f_i(x^r) + \mu_r x^r_i = \delta$, for some $\mu_r > 0$, if $x^r_i > 0$,
(*iii*) $f_i(x^r) \geq \delta$, if $x^r_i = 0$.

THEOREM 7.7. *Let $f : R^n \to R^n$ be a continuous function. Then, either the problem* NCP(f, R^n_+) *is strictly feasible, or for any $\delta > 0$, the function f has a δ-exceptional family of elements with respect to R^n_+.*

PROOF. Let δ be an arbitrary positive real number. If the problem NCP(f, R^n_+) is strictly feasible, the proof is finished. Suppose that the problem NCP(f, R^n_+) is not strictly feasible. Consider the function

$$\Psi_\delta(x) = ((\Psi_\delta)_i(x))_{i=1}^n$$

defined for any $x \in R^n$ by

$$(\Psi_\delta)_i(x) = \sqrt{x_i^2 - f_i(x) + \delta + \sqrt{[f_i(x) - \delta]^2}}, \quad i = 1, 2, \ldots, n.$$

The function Ψ_δ is continuous and $\Psi_\delta(R^n_+) \subseteq R^n_+$. We apply the nonlinear alternative (Theorem 7.1) to the function Ψ_δ, considering $C = R^n_+$ and

$$U = U_r = \{x \in R^n_+ \mid \|x\| < r\}$$

with an arbitrary $r > 0$.

The function Ψ_δ is fixed-point-free on every set

$$\bar{U}_r = \{x \in R^n_+ \mid \|x\| \leq r\}.$$

Indeed, if for some $r > 0$ there exists $x^r \in \bar{U}_r$ such that $\Psi_\delta(x^r) = x^r$, then we can show that the problem NCP(f, R^n_+) is strictly feasible, which is a contradiction. Hence, Ψ_δ is fixed-point-free on any $\bar{U}_r$, $r > 0$. In this case, applying Theorem 7.1, we obtain for any $r > 0$ an element $x^r \in R^n_+$, such that $\|x^r\| = r$ and a real number $t_r \in (0, 1)$ such that $x^r = t_r \Psi_\delta(x^r)$. Now recalling the definition of Ψ_δ we have for every $i = 1, 2, \ldots, n$,

$$x_i^r = t_r \left[\sqrt{(x_i^r)^2 - f_i(x^r) + \delta + \sqrt{[f_i(x^r) - \delta]^2}} \right]. \tag{7.6}$$

If $x_i^r = 0$, then from (7.6) we deduce that

$$-f_i(x^r) + \delta + \sqrt{[f_i(x^r) - \delta]^2} = 0,$$

which implies $f_i(x^r) \geq \delta$.

Suppose that $x_i^r \neq 0$, i.e. $x_i^r > 0$. In this case, it follows from (7.6) that

$$(1/t_r)x_i^r - x_i^r = -f_i(x^r) + \delta + \sqrt{[f_i(x^r) - \delta]^2},$$

which implies

$$[(1 - t_r)/t_r]\, x_i^r + f_i(x^r) - \delta = \sqrt{[f_i(x^r) - \delta]^2}. \tag{7.7}$$

If we denote in (7.7) by $u = f_i(x^r) - \delta$, then we obtain

$$[(1 - t_r)/t_r]\, x_i^r + u = \sqrt{u^2},$$

and finally,

$$u = -\left[(1 - t_r)/(2t_r)\right] x_i^r.$$

Hence, we have the equality

$$f_i(x^r) + \mu_r x_i^r = \delta,$$

where $\mu_r = [(1 - t_r)/(2t_r)] > 0$.

Therefore, $\{x^r\}_{r>0}$ is a δ-exceptional family of elements for f with respect to R^n_+, and the proof is complete. ∎

Remark. A similar result as Theorem 7.7 was recently communicated to the first author of this book by D.B. Zhao and Li. Their result is based on the notion of μ-antifeasible family of elements, which is different to our definition, and on a very long proof based upon the topological degree.

Considering the results presented in this section, it is interesting to know if there exist other classes of mappings, different to that of pseudomonotone operators, with the property that the complementarity problem has a solution if and only if the mapping is without exceptional families of elements.

7.3 SOLVING BILEVEL VARIATIONAL INEQUALITIES WITH MONOTONE MAPPINGS

As it was shown in Chapter 4, the Stackelberg model is reduced to solving a mathematical program with a variational inequality or a (nonlinear) complementarity problem as a constraint. Therefore, the first part of this section examines an approach to solution of problems of that kind. The latter consists in a variational reformulation of the optimization criterion and looking for a solution of thus obtained variational inequality among the points satisfying the initial variational constraints. A penalty function perturbation is applied to get a sequence of approximate solutions converging to that of the original problem with monotone operators.

However, this method is not efficient enough to regularize variational inequalities with non-coercive pseudomonotone operators. The main reason for its deficiency is the fact that an arbitrary perturbation of a pseudomonotone operator may not be pseudomonotone. In the second part of this section, we study the variational inequalities with non-coercive pseudomonotone operators. Instead of the classical regularization method to process the problem, we use a different approach which is based on the theory of linear compact operators.

7.3.1 PROBLEM SPECIFICATION

The problem of solving a mathematical program with variational inequalities or complementarity conditions as constraints arises quite frequently in the analysis of physical and socio-economic systems, see, e.g., Harker and Choi (1991), Outrata (1994), among others. The paper Harker and Choi (1991) presents an exterior-point penalty method based on M.J. Smith's optimization formulation of the finite-dimensional variational inequality problem Smith (1984).

Another approach to solving the above-mentioned problem consists, on the contrary, in a variational reformulation of the optimization criterion and looking for a solution of thus obtained variational inequality among the points satisfying the initial variational inequality constraints. In Subsection 7.5.2 of this section, we examine conditions under which the set of the feasible points is non-empty, and compare the conditions with those established previously in Harker and Pang (1990). Subsection 7.5.3 describes a penalty function method solving the two-level problem after having reduced it to a single variational inequality with a penalty parameter.

7.3.2 EXISTENCE THEOREM

Let X be a nonempty closed convex subset of R^n and G a continuous mapping from X into R^n. Suppose that G is pseudomonotone with respect to X, i.e.

$$(x-y)^{\mathrm{T}}G(y)\geq 0 \quad \text{implies} \quad (x-y)^{\mathrm{T}}G(x)\geq 0 \quad \forall x,y\in X, \tag{7.8}$$

and that there exists a vector $x^0 \in X$ such that

$$G(x^0)\in \operatorname{int}\,(0^+X)^*, \tag{7.9}$$

where int$(\cdot)$ denotes the interior of the set. Here 0^+X is the *recession cone* of the set X, i.e. the set of all directions $s \in R^n$ such that $X + s \subset X$; at last, C^* is the dual cone of $C \subset R^n$, i.e.

$$C^* = \{y\in R^n \mid y^{\mathrm{T}}x\geq 0 \qquad \forall x\in C\}. \tag{7.10}$$

Hence, condition (7.9) implies that the vector $G(x^0)$ lies within the interior of the dual to the recession cone of the set X.

Under these assumptions, the following result obtains.

PROPOSITION 7.1. *The variational inequality problem: to find a vector* $z \in X$ *such that*

$$(x-z)^{\mathrm{T}}G(z)\geq 0 \qquad \forall x\in X, \tag{7.11}$$

has a nonempty compact convex solution set.

PROOF. It is well-known (Karamardian, 1976) that the pseudomonotonicity (7.8) and continuity of the mapping G imply the convexity of the problem (7.11) solution set

$$Z = \{z\in X \mid (x-z)^{\mathrm{T}}G(z)\geq 0 \qquad \forall x\in X\}, \tag{7.12}$$

if the latter is nonempty. Now we show the existence of at least one solution to this problem. In order to do that, we use the following fact from Proposition

6.2: if there exists a nonempty bounded subset D of X such that for every $x \in X \setminus D$ there is a $y \in D$ with

$$(x - y)^{\mathrm{T}} G(x) > 0, \tag{7.13}$$

then problem (7.11) has a solution. Moreover, the solution set (7.12) is bounded because $Z \subset D$. Now, we construct the set D as follows:

$$D = \left\{ x \in X \mid (x - x^0)^{\mathrm{T}} G(x^0) \leq 0 \right\}. \tag{7.14}$$

The set D is clearly nonempty as it contains the point x^0. Now we show that D is bounded, even if X is not so. On the contrary, suppose that there exists an unbounded sequence $\{x^k\} \subseteq D$ such that $\|x^k - x^0\| \to +\infty$ when $k \to \infty$. Without lack of generality, assume that $x^k \neq x^0,\ k = 1, 2, \ldots$, and consider the inequality

$$\frac{(x^k - x^0)^{\mathrm{T}} G(x^0)}{\|x^k - x^0\|} \leq 0, \qquad k = 1, 2, \ldots, \tag{7.15}$$

which follows from definition (7.14) of the set D. Again not affecting generality, accept that the normed sequence $(x^k - x^0)/\|x^k - x^0\|$ converges to a vector $s \in R^n$, $\|s\| = 1$. It is well-known (Rockafellar, 1970, Theorem 8.2) that $s \in 0^+ X$. From (7.15), we deduce the limit relationship

$$s^{\mathrm{T}} G(x^0) \ \leq \ 0. \tag{7.16}$$

Since $0^+ X \neq \{0\}$ (as X is unbounded and convex), we have $0 \in \partial(0^+ X)^*$, hence $G(x^0) \neq 0$. Now it is easy to see that inequality (7.16) contradicts assumption (7.9). Indeed, the inclusion $G(x^0) \in \operatorname{int}(0^+ X)^*$ implies that $s^{\mathrm{T}} G(x^0) > 0$ for any $s \in 0^+ X, s \neq 0$. The contradiction proves the boundedness of the set D, and the statement of Proposition 7.1 therewith. Indeed, for a given $x \in X \setminus D$, one can pick $y = x^0 \in D$ with the inequality $(x - y)^{\mathrm{T}} G(y) > 0$ taking place. The latter, jointly with the pseudomonotonicity of G, implies the required condition (7.13) and completes the proof therewith. ∎

Now suppose further that the problem (7.11) solutions set Z contains more than one element, and consider the following variational inequality problem: to find a vector $z^* \in Z$ such that

$$(z - z^*)^{\mathrm{T}} F(z^*) \geq 0 \qquad \text{for all} \quad z \in Z. \tag{7.17}$$

Here, the mapping $F : X \to R^n$ is continuous and strictly monotone over X; i.e.

$$(x - y)^{\mathrm{T}} [F(x) - F(y)] > 0 \qquad \forall x, y \in X, x \neq y. \tag{7.18}$$

In this case, the compactness and convexity of the set Z guarantee (Proposition 6.2) the existence of a unique (due to the strict monotonicity of F) solution z of problem (7.17). We refer to problem (7.11), (7.12), (7.17) as the *bilevel variational inequality* (BVI). In the next subsection, we present a penalty function algorithm solving the BVI without explicit description of the set Z.

7.3.3 PENALIZATION APPROACH

Fix a positive parameter ε and consider the following parametrized variational inequality problem: Find a vector $x_\varepsilon \in X$ such that

$$(x - x_\varepsilon)^{\mathrm{T}}[G(x_\varepsilon) + \varepsilon F(x_\varepsilon)] \geq 0 \qquad \text{for all} \quad x \in X. \tag{7.19}$$

If we assume that the mapping G is monotone over X, i.e.

$$(x - y)^{\mathrm{T}}[G(x) - G(y)] \geq 0 \quad \forall x, y \in X, \tag{7.20}$$

and keep on all the above assumptions regarding G, F and Z, then the following result obtains.

PROPOSITION 7.2. *For each $\varepsilon > 0$ sufficiently small, problem* (7.19) *has a unique solution x_ε. Moreover, x_ε converge to the solution z^* of* BVI (7.11), (7.12), (7.17) *as $\varepsilon \to 0$.*

PROOF. Since G is monotone and F is strictly monotone, the mapping $\Phi_\varepsilon = G + \varepsilon F$ is strictly monotone on X for any $\varepsilon > 0$. It is also clear that if x^0 satisfies (7.9) then the following inclusion holds

$$\Phi_\varepsilon(x^0) = G(x^0) + \varepsilon F(x^0) \in \text{int}\ (0^+ X)^*, \tag{7.21}$$

if $\varepsilon > 0$ is small enough. Hence, Proposition 7.1 implies the first assertion of Proposition 7.2; namely, for every $\varepsilon > 0$ satisfying (7.21), variational inequality (7.19) has a unique solution x_ε .

From the continuity of F and G, it follows that each (finite) limit point $\bar{x}$ of the generalized sequence $Q = \{x_\varepsilon\}$ of solutions to problem (7.19) solves variational inequality (7.11); that is, $\bar{x} \in Z$. Now we prove that the point $\bar{x}$ solves problem (7.17), too. In order to do that, we use the following relationships valid for any $z \in Z$ due to (7.12), (7.19) and (7.20):

$$(z - x_\varepsilon)^{\mathrm{T}}[G(z) - G(x_\varepsilon)] \geq 0, \tag{7.22}$$

$$(z - x_\varepsilon)^{\mathrm{T}} G(z) \leq 0, \tag{7.23}$$

$$(z - x_\varepsilon)^{\mathrm{T}} G(x_\varepsilon) \geq -\varepsilon (z - x_\varepsilon)^{\mathrm{T}} F(x_\varepsilon). \tag{7.24}$$

Subtracting (7.24) from (7.23) and using (7.22), we obtain the following series of inequalities

$$0 \leq (z - x_\varepsilon)^{\mathrm{T}}[G(z) - G(x_\varepsilon)] \leq \varepsilon (z - x_\varepsilon)^{\mathrm{T}} F(x_\varepsilon). \tag{7.25}$$

From (7.25) we have $(z - x_\varepsilon)^{\mathrm{T}} F(x_\varepsilon) \geq 0$ for all $\varepsilon > 0$ and $z \in Z$. Since F is continuous, the following limit relationship holds: $(z - \bar{x})^{\mathrm{T}} F(\bar{x}) \geq 0$ for each $z \in Z$, which means that $\bar{x}$ solves (7.17).

Thus we have proved that every limit point of the generalized sequence Q solves bilevel variational inequality (7.11), (7.12), (7.17). Hence, Q can have at most one limit point. To complete the proof of Proposition 7.2, it suffices to establish that the set Q is bounded, and consequently, the limit point exists.

On the contrary, suppose that there exists an unbouded sequence $\{x_{\varepsilon_k}\}$ of solutions to parametrized problem (7.19) where $\varepsilon_k \to 0$ as $k \to \infty$. Without loss of generality, suppose that $x_{\varepsilon_k} \neq x^0$ for each k, and $\frac{(x_{\varepsilon_k} - x^0)}{\|x_{\varepsilon_k} - x^0\|} \to s \in R^n$, $\|s\| = 1$; here x^0 is the vector from condition (7.11). As $\|x_{\varepsilon_k} - x^0\| \to +\infty$, we get $s \in 0^+X$ (*cf.* Rockafellar, 1970). As the mappings G and F are monotone, the following inequalities take place for all $k = 1, 2, \ldots$:

$$(x_{\varepsilon_k} - x^0)^{\mathrm{T}}[G(x_{\varepsilon_k}) + \varepsilon_k F(x_{\varepsilon_k})] \leq 0, \tag{7.26}$$

$$(x_{\varepsilon_k} - x^0)^{\mathrm{T}}[G(x^0) + \varepsilon_k F(x^0)] < 0. \tag{7.27}$$

Dividing inequality (7.27) by $\|x_{\varepsilon_k} - x^0\|$ we obtain

$$\frac{(x_{\varepsilon_k} - x^0)^{\mathrm{T}}}{\|x_{\varepsilon_k} - x^0\|} \cdot [G(x^0) + \varepsilon_k F(x^0)] < 0, \qquad k = 1, 2, \ldots, \tag{7.28}$$

which yields (as $\varepsilon_k \to 0$) the limit inequality $s^{\mathrm{T}} G(x^0) \leq 0$. Since $s \neq 0$, the latter inequality contradicts assumption (7.9). This contradiction shows the set Q to be bounded which completes the proof of Proposition 7.2. ■

EXAMPLE 7.1. Let $\Omega \subseteq R^m$, $\Lambda \subseteq R^n$ be subsets of finite-dimensional Euclidean spaces and $f : \Omega \times \Lambda \to R$, $g : \Omega \times \Lambda \to R^n$ be continuous mappings. Consider the following mathematical program with variational inequality constraint:

$$\min_{(u,v) \in \Omega \times \Lambda} f(u, v), \tag{7.29}$$

subject to

$$g(u, v)^T (w - v) \geq 0, \qquad \forall w \in \Lambda. \tag{7.30}$$

If the function f is continuously differentiable, then problem (7.29)–(7.30) is obviously tantamount to BVI (7.11), (7.12), (7.17) with the gradient mapping $f'(z)$ used as $F(z)$ and $G(u, v) = [0; g(u, v)]$; here $z = (u, v) \in \Omega \times \Lambda$. As an example, examine the case when

$$f(u, v) = (u - v - 1)^2 + (v - 2)^2; \quad g(u, v) = uv; \quad \Omega = \Lambda = R^1_+. \tag{7.31}$$

Then it is readily verified that $z^* = (1; 0)$ and

$$\Phi_\varepsilon(u, v) = \left[\varepsilon(2u - 2v - 2); uv + \varepsilon(-2u + 4v - 2)\right]. \tag{7.32}$$

Now solve the variational inequality: find $(u_\varepsilon, v_\varepsilon) \in R^2_+$ such that

$$\Phi_\varepsilon(u_\varepsilon, v_\varepsilon)^T \left[(u, v) - (u_\varepsilon, v_\varepsilon)\right] \geq 0 \quad \forall (u, v) \in R^2_+, \tag{7.33}$$

and obtain

$$u_\varepsilon = v_\varepsilon + 1; \quad v_\varepsilon = -\frac{1}{2} - \varepsilon + \sqrt{(\frac{1}{2} + \varepsilon)^2 + 4\varepsilon}. \tag{7.34}$$

Clearly $(u_\varepsilon, v_\varepsilon) \to z^*$ when $\varepsilon \to 0$.

Unfortunately, it is not always the case, because the mapping $[0; g(u, v)]$ is not usually monotone with respect to (u, v), even if $g(u, v)$ is such with respect to v for each fixed u. ■

7.4 A REGULARIZATION APPROACH TO VARIATIONAL INEQUALITIES WITH PSEUDOMONOTONE MAPPINGS

In the previous section, we assumed that mappings F and G were both (at least) monotone, in order to establish the convergence of solutions of the perturbed problems to the solution of the initial one. In this section, we apply another approach to regularize variational inequality with pseudomonotone mappings.

From now on, unless specifically stated $\mathcal{X}$ denotes a real Banach space (possibly nonreflexive), $\mathcal{X}^*$ the topological dual of $\mathcal{X}$, $\langle \cdot, \cdot \rangle$ the associated pairing and $\|\cdot\|$ norm in $\mathcal{X}$, $\Omega \subseteq \mathcal{X}$ nonempty, closed, convex and $\mathcal{F} : \mathcal{D}(\mathcal{F}) \subseteq \mathcal{X} \longrightarrow \mathcal{X}^*$ with $\Omega \subseteq \mathcal{D}(\mathcal{F})$. The symbols "$\longrightarrow$" and "$\rightharpoonup$" denote the strong and the weak convergence respectively.

This section is concerned with the following variational inequality: Find an $x \in \Omega$ such that

$$\langle \mathcal{F}x, z - x \rangle \geq 0; \quad \forall z \in \Omega. \tag{7.35}$$

We denote by $\mathcal{S}(\mathcal{F}, \Omega)$ the set of all solutions to the above variational inequality problem.

In recent years the theory of variational inequalities has emerged as an important branch of applied and industrial mathematics. This theory provides us with a convenient mathematical apparatus for uniformly studying a wide range of problems arising in diverse fields as fluid flows through porous media, structural mechanics, elasticity, economics, local and global optimization etc.; see, for instance, the books by Baiocchi and Capelo (1984) and Kinderlehrer and Stampacchia (1980).

A great number of existence results for problem (7.35) are available when either domain Ω is bounded or the operator $\mathcal{F}$ is coercive. However variational formulations of many engineering and economics models lead to inequalities like (7.35) with unbounded domains and non-coercive operators. These problems are generally ill-posed in the sense of Tikhonov and Arsenin (1977). In order to make them well-posed various regularization methods have been designed in recent past; see, for instance, a recent survey article by Nashed and Liu (1996). Central idea of these methods is to regularize (7.35) by supplying an arbitrary mapping $\mathcal{G}$ which together with $\mathcal{F}$ provides stability. So far these methods have been explored mainly for monotone mappings where solvability of regularized problems is highly dependent on the fact that for $\varepsilon_n > 0$, $\mathcal{F}+\varepsilon_n\mathcal{G}$ is monotone provided that $\mathcal{F}$ and $\mathcal{G}$ are monotone.

Much recently an enlargement of the class of monotone mappings, has obtained great attention. It has been observed, for example by Isac (1993) that the existing classical regularization methods are not efficient enough to handle (7.35) with $\mathcal{F}$ being non-coercive and pseudomonotone. Only one of the reasons for this deficiency is the fact that an arbitrary perturbation of a pseudomonotone mapping may not be pseudomonotone. In the present section we study regularization methods for (7.35) with $\mathcal{F}$ non-coercive and pseudomonotone. In order to overcome the aforesaid difficulties we use a different approach based on the theory of linear compact operators in an arbitrary Hilbert space. As a historical remark we would like to mention that this approach is similar to the singular perturbation arguments of the so called elliptic super regularization introduced by Browder and Ton (1968) to prove the existence of solutions to functional equations with coercive operators.

The rest of the section is organised as follows. Subsection 7.6.2 is devoted to the preliminaries. In Subsection 7.6.3, we consider the regularization procedure. After proving the solvability of regularized problem we establish the weak convergence of the sequence of regularized solutions. The final subsection contains the concluding remarks.

7.4.1 PRELIMINARIES

Below we briefly set forth some basic definitions and results which we use here. For more details, we refer the reader to Zeidler (1993), Yao (1994).

Let $\mathcal{B}$ denote a real reflexive Banach space, $\mathcal{B}^*$ the topological dual of $\mathcal{B}$, $\langle\cdot,\cdot\rangle$ the associated pairing and $\|\cdot\|$ norm in $\mathcal{B}$, $\mathcal{K} \subseteq \mathcal{B}$ a nonempty closed and convex subset. We recall the following definitions.

DEFINITION 7.4. Let $\mathcal{A} : \mathcal{D}(\mathcal{A}) \subseteq \mathcal{B} \longrightarrow \mathcal{B}^*$ and $x, z \in \mathcal{D}(\mathcal{A})$ be arbitrary.

1. The mapping $\mathcal{A}$ is called monotone, if

$$\langle \mathcal{A}x - \mathcal{A}z, x - z\rangle \geq 0.$$

2. The mapping $\mathcal{A}$ is called strictly monotone, if

$$\langle \mathcal{A}x - \mathcal{A}z, x - z\rangle > 0, \quad x \neq z.$$

3. The mapping $\mathcal{A}$ is called strongly monotone, if

$$\langle \mathcal{A}x - \mathcal{A}z, x - z\rangle \geq m\|x - z\|^2, m > 0.$$

DEFINITION 7.5.

1. The mapping $\mathcal{A}$ is called *demicontinuous* at a point $x \in \mathcal{D}(\mathcal{A})$ if for any sequence $x_n \longrightarrow x$ with $x_n \in \mathcal{D}(\mathcal{A})$, we have $\mathcal{A}x_n \rightharpoonup \mathcal{A}x$. The mapping $\mathcal{A}$ is called demicontinuous, if it is demicontinuous at every point $x \in \mathcal{D}(\mathcal{A})$.

2. The mapping $\mathcal{A}$ is called *hemicontinuous* at a point $x_0 \in \mathcal{D}(\mathcal{A})$, if for any vector x such that $x_0 + tx \in \mathcal{D}(\mathcal{A})$ for $0 \leq t \leq \alpha(\alpha = \alpha(x) > 0$ and for any sequence $t_n \longrightarrow 0$ as $n \longrightarrow \infty(0 < t_n \leq \alpha)$, we have $\mathcal{A}(x_0 + t_n x) \rightharpoonup \mathcal{A}x_0$. The mapping $\mathcal{A}$ is called hemicontinuous, if it is hemicontinuous at every point $x \in \mathcal{D}(\mathcal{A})$.

We need the following result which is often referred to as the *Minty's lemma* and can be found in Bakushinskii and Goncharskii (1994).

THEOREM 7.8. *Let $\mathcal{A} : \mathcal{D}(\mathcal{A}) \equiv \mathcal{B} \longrightarrow \mathcal{B}^*$ be a single-valued monotone and hemicontinuous mapping and $f \in \mathcal{B}^*$ be an arbitrary element. Then the condition that $x \in \mathcal{K}$ such that*

$$\langle \mathcal{A}x - f, z - x\rangle \geq 0; \quad \forall z \in \mathcal{K}, \tag{7.36}$$

is equivalent to that x is a solution to the following system of variational inequalities

$$\langle \mathcal{A}z - f, z - x\rangle \geq 0; \quad \forall z \in \mathcal{K}. \tag{7.37}$$

If in addition either $\mathcal{K}$ is bounded or $\mathcal{A}$ is coercive, (i.e $\langle \mathcal{A}x, x\rangle/\|x\| \longrightarrow +\infty$ as $\|x\| \longrightarrow +\infty$), then (7.35) is solvable. If coercivity of $\mathcal{A}$ is replaced by a stronger assumption, namely: $\mathcal{A}$ is strongly monotone, then the solution to (7.35) is unique. ■

Remark. It is worth mentioning that under the assumptions of the above theorem, the solutions set of (7.35) is closed and convex. Roughly speaking, coercivity condition implies the existence of a ball $\mathcal{B}_R[0]$ with radius $R > 0$ such that no point outside $\mathcal{B}_R[0]$ can be a candidate for solution.

In the following definition we recall the concept of pseudomonotone mapping.

DEFINITION 7.6. Let $\mathcal{A} : \mathcal{D}(\mathcal{A}) \subseteq \mathcal{B} \longrightarrow \mathcal{B}^*$ and $x, z \in \mathcal{D}(\mathcal{A})$ be arbitrary. The mapping $\mathcal{A}$ is called *pseudomonotone*, if

$$\langle \mathcal{A}x, z - x\rangle \geq 0 \Rightarrow \langle \mathcal{A}z, z - x\rangle \geq 0.$$

Remark. The concept of pseudomonotonicity in the above definition was introduced by Karamardian (1976). It is easy to see that a monotone operator is pseudomonotone but not conversely. Statement of Theorem 7.8 remains true if we replace monotonicity by pseudomonotonicity, see Yao (1994). At this point we would like to mention that there is another notion of pseudomonotonicity introduced by Brézis (see Zeidler, 1993) which has no known relation with the above definition.

7.4.2 REGULARIZATION

Throughout this subsection, $\mathcal{H}$ denotes a real Hilbert space and $\mathcal{H}^*$ be its topologocal dual. Since $\mathcal{H}$ and $\mathcal{H}^*$ are isomorphic, we will identify these spaces in a usual way and consider no distinction between the scalar product in $\mathcal{H}$ and the duality pairing between $\mathcal{H}$ and $\mathcal{H}^*$ (both denoted by $\langle\cdot,\cdot\rangle_{\mathcal{H}}$). We denote by $\|\cdot\|_{\mathcal{H}}$ the norm in $\mathcal{H}$. Let σ be a single-valued mapping from $\mathcal{H}$ into $\mathcal{X}$, σ^* being the dual mapping of σ from $\mathcal{X}^*$ into $\mathcal{H}$ and $\mathcal{G} : \mathcal{H} \longrightarrow \mathcal{H}$.

Let $\{\varepsilon_n\}$ for $n \in \mathcal{N}$ be a sequence of positive reals which is strictly decreasing and converging towards zero.

We consider the following regularized variational inequality: Find an $x_{\varepsilon_n} \in \sigma^{-1}(\Omega \cap \sigma(\mathcal{H})) \subset \mathcal{H}$ such that

$$\langle \sigma^*\mathcal{F}\sigma x_{\varepsilon_n} + \varepsilon_n \mathcal{G}x_{\varepsilon_n}, z - x_{\varepsilon_n}\rangle_{\mathcal{H}} \geq 0; \quad \forall z \in \sigma^{-1}(\Omega \cap \sigma(\mathcal{H})), \tag{7.38}$$

where $\varepsilon_n > 0$. In order to justify the formulation of regularized problem we need the following results.

LEMMA 7.1. *Let σ be bounded with $\sigma(\mathcal{H}) \subseteq \mathcal{D}(\mathcal{F})$ and $\mathcal{F} : \mathcal{D}(\mathcal{F}) \subseteq \mathcal{X} \longrightarrow \mathcal{X}^*$ be pseudomonotone. Then $\mathcal{T} = \sigma^*\mathcal{F}\sigma$ is the well-defined pseudomonotone mapping from the whole of $\mathcal{H}$ to $\mathcal{H}$.*

PROOF. For arbitrary $x, z \in \mathcal{H}$ we have $\sigma(x), \sigma(z) \in \mathcal{D}(\mathcal{F})$ and hence $\mathcal{T}x$ and $\mathcal{T}z$ are well-defined. Now we establish the pseudomonotonicity of $\mathcal{T}$. Indeed, for $x, z \in \mathcal{H}$ and $\langle \mathcal{T}x, z - x\rangle_{\mathcal{H}} \geq 0$ we have

$$\langle \sigma^*\mathcal{F}\sigma x, z - x\rangle_{\mathcal{H}} \geq 0$$

$$\Rightarrow \qquad \langle \mathcal{F}\sigma x, \sigma z - \sigma x\rangle \geq 0$$

$$\Rightarrow \qquad \langle \mathcal{F}\sigma z, \sigma z - \sigma x\rangle \geq 0$$

as $\mathcal{F}$ is pseudomonotone and $\sigma x, \sigma z \in \mathcal{D}(\mathcal{F})$. This further implies that

$$\langle \sigma^*\mathcal{F}\sigma z, z - x\rangle_{\mathcal{H}} \geq 0$$

$$\Rightarrow \qquad \langle \mathcal{T}z, z - x\rangle_{\mathcal{H}} \geq 0.$$

This completes the proof. ■

The following lemma plays an important role in our main results (see Ton, 1969).

LEMMA 7.2. *Let σ be linear and continuous, then $\sigma^{-1}(\Omega \cap \sigma(\mathcal{H}))$ is a closed and convex subset of $\mathcal{H}$.*

PROOF. Let us assume that $\Omega \cap \sigma(\mathcal{H})$ is nonempty in $\mathcal{X}$ (otherwise the statement is trivially true). As σ is one-to-one mapping, $\sigma^{-1}(\Omega \cap \sigma(\mathcal{H}))$ is a subset in $\mathcal{H}$. Convexity of this set follows as a consequence of the convexity of Ω and linearity of the single-valued mapping σ.

To prove the closedness, let $x_n \in \sigma^{-1}(\Omega \cap \sigma(\mathcal{H}))$ with $x_n \longrightarrow x$. We claim that $x \in \sigma^{-1}(\Omega \cap \sigma(\mathcal{H}))$. Indeed, $\sigma x_n \in \Omega$, which implies $\sigma x \in \Omega$ as σ is continuous and Ω is closed. Also, we have $\sigma^{-1}(\sigma x) = x$ and this together with the previous argument yields $x \in \sigma^{-1}(\Omega \cap \sigma(\mathcal{H}))$. The proof is complete. ■

Now we are in a position to discuss the solvability of (7.38).

THEOREM 7.9.*Let the mapping $\mathcal{F} : \mathcal{D}(\mathcal{F}) \subseteq \mathcal{X} \longrightarrow \mathcal{X}^*$ be pseudomonotone, demicontinuous and satisfy $\langle \mathcal{F}x, x\rangle \geq 0$, $\forall x \in \mathcal{D}(\mathcal{F})$; the mapping $\sigma : \mathcal{H} \longrightarrow \mathcal{X}$ be linear and compact with the dual $\sigma^* : \mathcal{X}^* \longrightarrow \mathcal{H}$, and also satisfy the condition that $\sigma(\mathcal{H}) \subseteq \mathcal{D}(\mathcal{F})$, $\mathcal{G} : \mathcal{H} \longrightarrow \mathcal{H}$ be strongly monotone and hemicontinuous. Then for every $\varepsilon_n > 0$, (7.38) is solvable.*

PROOF. First we claim that $(\varepsilon_n\mathcal{G} + \sigma^*\mathcal{F}\sigma)$ is coercive. Indeed, for any $x \in \mathcal{H}$ we have

$$\begin{aligned}
&\langle (\ \varepsilon_n \mathcal{G} + \sigma^* \mathcal{F} \sigma) x, x \rangle_{\mathcal{H}} \\
&= \varepsilon_n \langle \mathcal{G} x, x \rangle_{\mathcal{H}} + \langle \sigma^* \mathcal{F} \sigma x, x \rangle_{\mathcal{H}} \\
&= \varepsilon_n \langle \mathcal{G} x, x \rangle_{\mathcal{H}} + \langle \mathcal{F} \sigma x, \sigma x \rangle \\
&\geq \varepsilon_n \left(m \|x\|^2 - \|\mathcal{G}(0)\| \, \|x\| \right)
\end{aligned}$$

which proves the assertion. Now by virtue of coercivity of the operator $(\varepsilon_n \mathcal{G} + \sigma^* \mathcal{F} \sigma)$ there is no loss of generality to assume that the set $\sigma^{-1}(\Omega \cap \sigma(\mathcal{H}))$ is bounded.

In order to prove the main assertion, we define a nonlinear mapping Λ from $\sigma^{-1}(\Omega \cap \sigma(\mathcal{H}))$ into $\sigma^{-1}(\Omega \cap \sigma(\mathcal{H}))$ as follows. Let $\omega \in \sigma^{-1}(\Omega \cap \sigma(\mathcal{H}))$ be arbitrary, then $\Lambda \omega = \{\omega^* \mid \omega^* \in \sigma^{-1}(\Omega \cap \sigma(\mathcal{H}))\}$ with

$$\langle \varepsilon_n \mathcal{G} \omega^* + \sigma^* \mathcal{F} \sigma \omega, z - \omega^* \rangle_{\mathcal{H}} \geq 0; \quad \forall z \in \sigma^{-1}(\Omega \cap \sigma(\mathcal{H}))\}. \tag{7.39}$$

We observe the following properties of Λ:

1. For all $x \in \sigma^{-1}(\Omega \cap \sigma(\mathcal{H}))$, Λx is a singleton set, as it follows from Theorem 7.8.

2. The mapping Λ is continuous.

   Let $x_n \longrightarrow x$ with $x_n^* = \Lambda x_n$ and $x^* = \Lambda x$. Setting $z = 1/2(x_n^* + x^*)$ in variational inequality (7.38) with x_n and x as fixed points, we get

   $$\langle \varepsilon_n \mathcal{G} x_n^* + \sigma^* \mathcal{F} \sigma x_n, 1/2(x_n^* + x^*) - x_n^* \rangle \geq 0$$

   and

   $$\langle \varepsilon_n \mathcal{G} x^* + \sigma^* \mathcal{F} \sigma x, 1/2(x_n^* + x^*) - x^* \rangle \geq 0.$$

   Combining the two inequalities above, we get

   $$\varepsilon_n/2 \langle \mathcal{G} x_n^* - \mathcal{G} x^*, x^* - x_n^* \rangle + 1/2 \langle \sigma^* \mathcal{F} \sigma x_n - \sigma^* \mathcal{F} \sigma x, x^* - x_n^* \rangle \geq 0.$$

   Using the strong monotonicity of $\mathcal{G}$, we obtain

   $$m\varepsilon_n/2 \|x_n^* - x^*\|^2 \ \leq \ (1/2) \|x_n^* - x^*\| \, \|\sigma^* \mathcal{F} \sigma x_n - \sigma^* \mathcal{F} \sigma x\|$$

   which further implies

   $$m\varepsilon_n \|\Lambda x_n - \Lambda x\| \ \leq \ \|\sigma^* \mathcal{F} \sigma x_n - \sigma^* \mathcal{F} \sigma x\|.$$

   Now the continuity of $\Lambda(\cdot)$ follows from the continuity of $\sigma^* \mathcal{F} \sigma$ from weak topology to the strong topology.

3. Λ is compact.

   Let $\|x_n\| \leq M$, then from the reflexivity, there exists $x \in \sigma^{-1}(\Omega \cap \sigma(\mathcal{H}))$, such that $x_{n_j} \rightharpoonup x$, and the previous argument yields $\Lambda x_{n_j} \longrightarrow \Lambda x$.

Now the Schauder theorem confirms the existence of a fixed point for $\Lambda(\cdot)$, which results in the solvability of (7.38). ■

In order to study the stability properties of (7.38) we consider the following problem: Find an $x \in \sigma^{-1}(\Omega \cap \sigma(\mathcal{H}))$ such that

$$\langle \sigma^* \mathcal{F} \sigma x, z - x \rangle \geq 0; \qquad \forall z \in \sigma^{-1}(\Omega \cap \sigma(\mathcal{H})). \tag{7.40}$$

We denote the set of all solutions to (7.38) by $\mathcal{S} = \mathcal{S}(\sigma^* \mathcal{F} \sigma, \sigma^{-1}(\Omega \cap \sigma(\mathcal{H})))$, and $\Pi = \Omega \cap \sigma(\mathcal{H}))$.

THEOREM 7.10. *In addition to the hypotheses of Theorem 7.9, assume the following:*

1. *The set $\Omega \cap \sigma(\mathcal{H})$ is dense in Ω;*
2. *The set $\mathcal{S}(\sigma^* \mathcal{F} \sigma, \sigma^{-1}(\Omega \cap \sigma(\mathcal{H})))$ is nonempty.*

Then:

(I) The sequence $\{x_{\varepsilon_n}\}_{n=1}^{\infty} \subset \mathcal{H}$ is uniformly bounded; here $x_{\varepsilon_n} \in \bar{\mathcal{S}}_{\varepsilon_n}(\sigma^* \mathcal{F} \sigma, \sigma^{-1}(\Omega \cap \sigma(\mathcal{H})))$ is chosen arbitrarily.

(II) The sequence $\{x_{\varepsilon_n}\}_{n=1}^{\infty} \subset \mathcal{H}$ converges strongly to $x^* \in \mathcal{S}(\sigma^* \mathcal{F} \sigma, \sigma^{-1}(\Pi))$. Moreover x^* can be distinguished by the property that it is the unique solution to the problem: Find an $x \in \mathcal{S}(\sigma^* \mathcal{F} \sigma, \sigma^{-1}(\Omega \cap \sigma(\mathcal{H}))$ such that

$$\langle \mathcal{G} x, z - x \rangle_{\mathcal{H}} \geq 0; \quad \forall z \in \mathcal{S}(\sigma^* \mathcal{F} \sigma, \sigma^{-1}(\Omega \cap \sigma(\mathcal{H})). \tag{7.41}$$

(III) The σ image of x^* is a solution to the original $\mathcal{VIP}$ (7.35), that is

$$\sigma x^* \in \mathcal{S}(\mathcal{F}, \Omega).$$

PROOF. (*I*) As the assumptions of Theorem 7.9 hold, there exists an $x_{\varepsilon_n} \in \sigma^{-1}(\Omega \cap \sigma(\mathcal{H}))$ such that

$$\langle \varepsilon_n \mathcal{G} x_{\varepsilon_n}, z - x_{\varepsilon_n} \rangle_{\mathcal{H}} \geq \langle \sigma^* \mathcal{F} \sigma x_{\varepsilon_n}, x_{\varepsilon_n} - z \rangle_{\mathcal{H}}; \quad \forall z \in \sigma^{-1}(\Omega \cap \sigma(\mathcal{H})). \tag{7.42}$$

Let $x \in \sigma^{-1}(\Omega \cap \sigma(\mathcal{H}))$ be an element which satisfies (7.40). By arranging $z = x_{\varepsilon_n}$ in (7.40), and using the pseudomonotonicity of the operator $\sigma^* \mathcal{F} \sigma$ we obtain

$$\langle \sigma^* \mathcal{F} \sigma x_{\varepsilon_n}, x_{\varepsilon_n} - x \rangle_{\mathcal{H}} \geq 0. \tag{7.43}$$

Now by substituting $z = x$ in (7.41) and combining it with (7.43), we obtain

$$\varepsilon_n \langle \mathcal{G} x_{\varepsilon_n}, x - x_{\varepsilon_n} \rangle_{\mathcal{H}} \geq 0.$$

The latter combined with the fact that $\varepsilon_n > 0$ implies

$$\langle \mathcal{G} x_{\varepsilon_n}, x - x_{\varepsilon_n} \rangle_{\mathcal{H}} \geq 0. \tag{7.44}$$

By virtue of the strong monotonicity of the mapping $\mathcal{G}$ and the above inequality, we conclude that

$$m\|x - x_{\varepsilon_n}\|^2 \leq \|\mathcal{G}x\| \, \|x - x_{\varepsilon_n}\|$$

which finally implies

$$\|x_{\varepsilon_n}\| \leq \mathcal{M}$$

where $\mathcal{M}$ is a positive constant independent of ε_n.

(II) In view of the previous part and the reflexivity of Hilbert spaces, $\{x_{\varepsilon_n}\}_{n=1}^{\infty}$ is weakly compact. Therefore one can select a subsequence denoted by $\{x_{\kappa_n}\}_{n=1}^{\infty}$ converging weakly to some element $x^* \in \mathcal{H}$. Furthermore $\sigma^{-1}(\Omega \cap \sigma(\mathcal{H}))$ being closed is weakly closed by virtue of Mazur's Theorem and consequently $x^* \in \sigma^{-1}(\Omega \cap \sigma(\mathcal{H}))$. Now replacing x_{ε_n} in (7.44) by x_{κ_n}, we get

$$\langle \sigma^* \mathcal{F} \sigma x_{\kappa_n} + \kappa_n \mathcal{G} x_{\kappa_n}, z - x_{\kappa_n} \rangle_{\mathcal{H}} \geq 0; \quad \forall z \in \sigma^{-1}(\Omega \cap \sigma(\mathcal{H})).$$

This implies

$$\langle \sigma^* \mathcal{F} \sigma x_{\kappa_n}, z - x_{\kappa_n} \rangle \geq \kappa_n \langle \mathcal{G} x_{\kappa_n}, x_{\kappa_n} - z \rangle; \quad \forall z \in \sigma^{-1}(\Omega \cap \sigma(\mathcal{H})).$$
$$\geq -\kappa_n \|\mathcal{G} x_{\kappa_n}\| \, \|x_{\kappa_n} - z\|.$$

Passing this to the limit $n \longrightarrow \infty$, using the compactness of the mapping $\sigma^* \mathcal{F} \sigma$, the local boundedness of monotone operators and the fact that $\kappa_n \longrightarrow 0$, we conclude

$$\langle \mathcal{F} \sigma x^*, \sigma z - \sigma x^* \rangle \geq 0; \quad \forall \sigma z \in \Omega \cap \sigma(\mathcal{H}).$$

In the other words, setting $u^* = \sigma x^*$ we have

$$\langle \mathcal{F} u^*, u - u^* \rangle \geq 0; \quad \forall u \in \Omega \cap \sigma(\mathcal{H}). \tag{7.45}$$

Now we prove that the whole sequence also converges to the point x^* in the strong topology. Let us denote by $\bar{x}$ the unique solution to (7.41). We claim that $\bar{x} = x^*$. Indeed, as an analogue of (7.44) we have

$$\langle \mathcal{G} x_{\kappa_n}, x - x_{\kappa_n} \rangle \geq 0 \quad \forall x \in S,$$

which implies (due to the monotonicity of $\mathcal{G}$) that

$$\langle \mathcal{G} x, x - x_{\kappa_n} \rangle \geq 0, \quad \forall x \in S.$$

Passing the above inequality to the limit $n \longrightarrow \infty$, we obtain

$$\langle \mathcal{G} x, x - x^* \rangle \geq 0, \quad \forall x \in S.$$

From the Minty's lemma, the above inequality can also be written as

$$\langle \mathcal{G} x^*, x - x^* \rangle \geq 0, \quad \forall x \in S,$$

in particular,

$$\langle \mathcal{G} x^*, \bar{x} - x^* \rangle \geq 0.$$

If we set $z = x^*$ in (7.43), we obtain

$$\langle \mathcal{G}\bar{x}, x^* - \bar{x} \rangle \geq 0.$$

Now if we sum up the above two inequalities, we immediately get

$$\langle \mathcal{G}\bar{x} - \mathcal{G}x^*, x^* - \bar{x} \rangle \geq 0$$

which in view of the strong monotonicity of the operator $\mathcal{G}$ confirms that $\bar{x} = x^*$. Since x^* is the unique limit of any weakly convergent subsequence of $\{x_{\varepsilon_n}\}_{n=1}^{\infty}$, we conclude that the whole sequence $\{x_{\varepsilon_n}\}_{n=1}^{\infty}$ converges weakly to x^*. Finally, for the strong convergence we again use the fact that for the mapping $\mathcal{G}$, we have

$$\langle \mathcal{G}x_{\varepsilon_n} - \mathcal{G}x^*, x_{\varepsilon_n} - x^* \rangle \geq m\|x_{\varepsilon_n} - x^*\|^2.$$

This further implies

$$\langle \mathcal{G}x^*, x^* - x_{\varepsilon_n} \rangle \geq m\|x_{\varepsilon_n} - x^*\|^2.$$

Passing the above inequality to the limit $n \longrightarrow \infty$, we get

$$\|x_{\varepsilon_n} - x^*\|^2 \longrightarrow 0.$$

The strong convergence of the sequence now follows as a consequence of the weak convergence and the norm convergence.

(III) Now we claim that the above inequality is indeed valid for all $u \in \Omega$. Let us begin with an arbitrary $\overline{u} \in \Omega$. Since $\Omega \cap \sigma(\mathcal{H})$ is dense in Ω, there exist $\{u_k\}_{k=1}^{\infty} \subset \Omega \cap \sigma(\mathcal{H})$ such that $u_k \longrightarrow \overline{u}$ as $k \longrightarrow \infty$. Since $\mathcal{F}$ is demicontinuous, we have $\mathcal{F}u_k \rightharpoonup \mathcal{F}\overline{u}$ as $k \longrightarrow \infty$. However applying Theorem 7.8 to (7.45) we get the Minty's formulation of variational inequality which is as follows:

$$\langle \mathcal{F}u, u - u^* \rangle \geq 0; \qquad \forall u \in \Omega \cap \sigma(\mathcal{H}).$$

Setting $u = u_k$ we get

$$\langle \mathcal{F}u_k, u_k - u^* \rangle \geq 0.$$

Letting $k \longrightarrow \infty$ and recalling that

$$\mathcal{F}u_k \rightharpoonup \mathcal{F}\overline{u}, \; (u_k - u^*) \longrightarrow (\overline{u} - u^*)$$

we get

$$\langle \mathcal{F}\overline{u}, \overline{u} - u^* \rangle \geq 0; \qquad \text{for } \overline{u} \in \Omega.$$

Repeating the use of the Minty's formulation of variational inequality, we conclude that

$$\langle \mathcal{F}u^*, \overline{u} - u^* \rangle \geq 0.$$

Since $\overline{u} \in \Omega$ is an arbitrary element, we get the desired result. The proof is complete. ■

7.4.3 CONCLUDING REMARKS

We conclude this section with some remarks that highlight differences between existing regularization methods and our approach.

The traditional methods to regularize for example, a variational inequality with monotone non-coercive mapping comprised of the following steps:

- Solvability of the regularized problem for certain ε_n.
- Boundedness of the sequence of regularized solutions.
- Weak and strong convergences of the sequence of the regularized solutions to the solution of original problem.

The solvability of the regularized problem highly depends on the fact that the sum of two monotone mappings is again a monotone mapping. This further helps to establish that the sum of a coercive monotone mapping and a monotone mapping is a coercive monotone mapping. In the case of pseudomonotone operators these results are not available. In order to have a result of this kind we are bound to impose extra condition namely, the mapping is positive on the diagonal.

For monotone operators boundedness of the sequence of regularized solutions is equivalent to the nonvoidness of the solution set of the original problem. This equivalence and also the convergence to the solution of the original problem highly relies on the Minty's formulation of the regularized variaionlal inequality. For pseudomonotone mappings the Minty's formulation is not available for the regularized variational inequality: *cf.* (7.36). In the absence of Minty's formulation, strong compactness arguments are needed to attain the weak convergence. In our setting we have acomplished this with the help of the linear compact mapping σ. In suitable environment such mappings always exist as it was proved by Browder and Ton (1968). As we are doing convergence analysis in the topology of Hilbert spaces, the original Banach space might be non-reflexive.

7.5 INFINITE DIMENSIONAL IMPLICIT COMPLEMENTARITY PROBLEM

Let $(H, \langle\ ,\ \rangle)$ be a Hilbert space and $K \subset H$ a closed convex cone. Given two mappings $f, g : H \to H$, the *implicit complementarity problem* defined by the (ordered) pair of mappings (f, g) and the cone K is

$$\mathrm{ICP}(f,g,K) : \begin{cases} \text{find } x_* \in H \text{ such that} \\ f(x_*) \in K^*,\ g(x_*) \in K, \\ \text{and } \langle g(x_*), f(x_*)\rangle = 0. \end{cases}$$

Many of the well-known existence theorems for the problem $\mathrm{ICP}(f,g,K)$ demand that the mappings f and g be subject to some strong restrictions. For example, Isac (1985,1987) requires f to be strongly monotone and Lipschitz continuous with respect to g (in other words, f must be g-strongly monotone and g-Lipschitz continuous).

For some conditions that assure existence of solutions to this problem, the reader is referred to Capuzzo-Dolcetta and Mosco (1980), Isac (1986,1990,1992), Isac e.a. (1997), Isac and Goeleven [1] (1993), Isac and Goeleven [2] (1993), Mosco (1976,1980), Nikaido (1968), Pang (1981,1982), Hyers e.a. (1997), Chan and Pang (1982).

In this section, we will study the existence of solution to the problem ICP(f, g, K) by making use of the Leray-Schauder type alternatives.

7.5.1 EXISTENCE ALTERNATIVES

First, consider the Euclidean space $(R^n, \langle\ ,\ \rangle)$ and continuous mappings $f, g : R^n \to R^n$. Let $K \subset R^n$ be a closed convex cone. Before examining the problem ICP(f, g, K), we recall from Isac, Bulavsky, Kalashnikov (1997), Bulavsky, Isac, Kalashnikov (1998), the following notion.

DEFINITION 7.7. We say that a family of elements $\{x_r\}_{r>0} \subset R^n$ is an *exceptional family of elements (EFE)* for the pair (f, g) with respect to the cone K, if the following conditions are satisfied:

(1) $\|x_r\| \to +\infty$ as $r \to +\infty$;

(2) $g(x_r) \in K$ for any $r > 0$;

(3) for every $r > 0$, there exists $\mu_r > 0$ such that $s_r = f(x_r) + \mu_r g(x_r) \in K^*$ and $\langle g(x_r), s_r\rangle = 0$.

Using this notion in our paper Bulavsky, Isac, Kalashnikov (1998), the following result was proved.

THEOREM 7.11 [BULAVSKY, ISAC and KALASHNIKOV (1998)]. *Let $f, g : R^n \to R^n$ be continuous mappings. If the following assumptions are satisfied:*

(1) *the equation $g(x) = 0$ has a unique solution $b \in R^n$;*

(2) *g maps homeomorphically a neighborhood of the element b onto a neighborhood of the origin,*

then there exists either a solution to the problem ICP(f, g, K), *or an exceptional family of elements for the couple (f, g) (in the sense of Definition* 7.7.) ∎

The proof of this result, given in Bulavsky, Isac, Kalashnikov (1998), is based upon the topological degree theory, and assumptions (1) and (2) were imposed by this method of proof. As a consequence of Theorem 7.11 we have the fact that, if a couple of continuous mappings (f, g) is without exceptional families of elements, then there exists a solution to the problem ICP(f, g, K). By this way, several existence theorems for implicit complementarity problems are proved in Isac, Bulavsky and Kalashnikov (1997) and Bulavsky, Isac and Kalashnikov (1998).

Now we will show that by a little modification of the notion of exceptional family of elements introduced in Definition 7.7 and replacing the topological degree theory by a Leray-Schauder type alternative, we can obtain a more general result than Theorem 7.11 by a much simpler proof.

Let $(H, \langle\ ,\ \rangle)$ be a Hilbert space, $K \subset H$ a closed pointed convex cone and $f, g : H \to H$ completely continuous fields. We introduce the following notion.

DEFINITION 7.8. We say that a family of elements $\{x_r\}_{r>0} \subset H$ is an *exceptional family of elements (EFE)* for the pair (f, g) of completely continuous fields, with respect to K, if the following conditions are satisfied:

(1) $\|x_r\| \to +\infty$ as $r \to +\infty$;

(2) for any $r > 0$, there exists $\mu_r > 0$ such that $s_r = \mu_r x_r + f(x_r) \in K^*$, $v_r = \mu_r x_r + g(x_r) \in K$ and $\langle v_r, s_r \rangle = 0$.

We have the following result.

THEOREM 7.12. *Let $(H, \langle\ ,\ \rangle)$ be a Hilbert space, $K \subset H$ a closed convex cone and $f, g : H \to H$ completely continuous fields such that $f(x) = x - T(x)$ and $g(x) = x - S(x)$, where $T, S : H \to H$ are completely continuous mappings. Then there exists either a solution to the problem* ICP(f, g, K), *or an exceptional family of elements $\{x_r\}_{r>0}$ for the pair (f, g). Moreover, if $S(K) \subseteq K$, we have that the problem* ICP(f, g, K) *has either a solution in K, or an exceptional family of elements in the sense of Definition* 7.8 *such that$\{x_r\}_{r>0} \subset K$.*

PROOF. Using the properties (i) and (ii) of the projection operator P_K, it is easy to show that the problem ICP(f, g, K) has a solution if and only if the equation

$$g(x) = P_K\left[g(x) - f(x)\right] \tag{7.46}$$

has a solution in H. Consider the mapping

$$\Phi(x) = x - g(x) + P_K\left[g(x) - f(x)\right]; \quad x \in H, \tag{7.47}$$

and observe that equation (7.46) has a solution if and only if the mapping Φ has a fixed point in H. From representations of f and g we have

$$\Phi(x) = x - g(x) + P_K\left[g(x) - f(x)\right] = S(x) + P_K\left[-S(x) + T(x)\right].$$

We note that the mapping Φ is completely continuous. Now if the mapping Φ does have a fixed point then the problem ICP(f, g, K) has a solution, and the proof is finished.

Otherwise, let Φ have no fixed point in the space H. For any $r > 0$, consider the set $U_r = B_r = \{x \in H \mid \|x\| < r\}$. For any $r > 0$, the restriction of Φ·onto the subset U_r is a continuous compact mapping without fixed points. Applying Theorem 7.1 to the restriction of the mapping Φ and the sets $C = H$ and $U = U_r$, we obtain that for all $r > 0$, there exists $x_r \in \partial U_r = \{x \in H \mid \|x\| = r\}$, that is with $\|x_r\| = r$, and $\lambda_r \in (0, 1)$ such that

$$x_r = \lambda_r\left[S(x_r) + P_K\left[T(x_r) - S(x_k)\right]\right]. \tag{7.48}$$

From (7.48) we deduce

$$\frac{1}{\lambda_r}x_r - S(x_r) = P_K\left[T(x_r) - S(x_k)\right],$$

which implies, using the properties (i) and (ii) of P_K,

$$\begin{cases} \langle \lambda_r^{-1}x_r - T(x_r), y\rangle \geq 0 \text{ for all } y \in K, \\ \text{and} \\ \langle \lambda_r^{-1}x_r - T(x_r), \lambda_r^{-1}x_r - S(x_r)\rangle = 0. \end{cases}$$

Therefore we have that $\lambda_r^{-1}x_r - T(x_r) \in K^*$ and $\lambda_r^{-1}x_r - S(x_r) \in K$. Now we obtain

$$s_r = \mu_r x_r + f(x_r) = \left(\frac{1}{\lambda_r} - 1\right)x_r + x_r - T(x_r) \in K^*$$

and

$$v_r = \mu_r x_r + g(x_r) = \left(\frac{1}{\lambda_r} - 1\right)x_r + x_r - S(x_r) \in K,$$

(where $\mu_r = \lambda_r^{-1} - 1 > 0$ for any $r > 0$).

Now it is also easy to see that $\langle v_r, s_r\rangle = 0$ for any $r > 0$. According to Definition 7.8, the family of elements $\{x_r\}_{r>0}$ is exceptional, and the first assertion of the theorem is proved.

If $S(K) \subseteq K$, then also $\Phi(K) \subseteq K$. We apply the nonlinear alternative (Theorem 7.1) to the case when $C = K$, and for any $r > 0$ we consider $U = U_r(K) = \{x \in K \mid \|x\| < r\}$. If the problem ICP$(f, g, K)$ has a solution in K then the proof is finished. Otherwise, as in the first part of the proof, we construct $\{x_r\}$ where $x_r \in K$ for each $r > 0$. This finishes the proof of the second assertion of the theorem, which completes the proof as a whole. ■

COROLLARY 7.6. *Let $(H, \langle\ ,\ \rangle)$ be a Hilbert space, $K \subset H$ a closed convex cone and $f, g : H \to H$ completely continuous fields. If the pair (f, g) is without exceptional families of elements (in the sense of Definition* 7.8 *), then the problem* ICP(f, g, K) *has a solution.* ■

COROLLARY 7.7. *Let $(R^n, \langle\ ,\ \rangle)$ be the Euclidean space, $K \subset R^n$ a closed convex cone and $f, g : R^n \to R^n$ continuous mappings.If the pair (f, g) is without exceptional families of elements, then the problem* ICP(f, g, K) *has a solution.* ■

7.5.2 PAIRS OF MAPPINGS WITHOUT EXCEPTIONAL FAMILIES OF ELEMENTS

In view of Corollary 7.6, we want to know when a pair of completely continuous fields is without exceptional families of elements (EFEs) in the sense of Definition 7.8. We give now an example of such a pair of mappings.

Let $(H, \langle\ ,\ \rangle)$ be a Hilbert space, $K \subset H$ a closed pointed convex cone and $f, g : H \to H$ completely continuous fields such that $f(x) = x - T(x)$ and $g(x) = x - S(x)$ for all $x \in H$.

THEOREM 7.13. *If $S(K) \subseteq K$ and there exists $\rho > 0$ such that for any $x \in K$, $\|x\| > \rho$, there exists $y \in K$ such that*

$$\begin{cases} \langle g(x) - y, f(x) \rangle \geq 0, \text{ and} \\ \langle g(x) - y, x \rangle > 0, \end{cases} \tag{7.49}$$

then the problem ICP(f, g, K) *has a solution in* K.

PROOF. It suffices to show that the pair (f, g) has no exceptional family of elements in K. Suppose the contrary, i.e. there exists an exceptional family of elements $\{x_r\}_{r>0} \subset K$ for the pair (f, g). For any $r > 0$ such that $\|x_r\| > \rho$ we have an element $y_r \in K$ satisfying the following inequalities

$$\begin{cases} \langle g(x_r) - y_r, f(x_r) \rangle \geq 0, \text{ and} \\ \langle g(x_r) - y_r, x_r \rangle > 0. \end{cases} \tag{7.50}$$

Since, according to Definition 7.8,

$$s_r = \mu_r x_r + f(x_r) \in K^*, \; v_r = \mu_r x_r + g(x_r) \in K, \quad \text{and } \langle v_r, s_r \rangle = 0,$$

we have

$$0 \leq \langle g(x_r) - y_r, f(x_r) \rangle = \langle v_r - \mu_r x_r - y_r, s_r - \mu_r x_r \rangle =$$

$$= \langle v_r, s_r \rangle - \langle \mu_r x_r, s_r \rangle - \langle y_r, s_r \rangle - \langle v_r, \mu_r x_r \rangle + \mu_r^2 \|x_r\|^2 + \langle y_r, \mu_r x_r \rangle \leq$$

$$\leq -\langle v_r, \mu_r x_r \rangle + \mu_r^2 \|x_r\|^2 + \langle y_r, \mu_r x_r \rangle =$$

$$= -\langle \mu_r x_r + g(x_r), \mu_r x_r \rangle + \mu_r^2 \|x_r\|^2 + \langle y_r, \mu_r x_r \rangle =$$

$$-\mu_r^2 \|x_r\|^2 - \langle g(x_r), \mu_r x_r \rangle + \mu_r^2 \|x_r\|^2 + \langle y_r, \mu_r x_r \rangle =$$

$$-\langle g(x_r), \mu_r x_r \rangle + \langle y_r, \mu_r x_r \rangle = -\mu_r \left[\langle g(x_r) - y_r, x_r \rangle \right] < 0,$$

which is a contradiction.

Hence, the pair (f, g) is without exceptional families of elements in K, and by Corollary 7.6 of Theorem 7.12 the problem ICP(f, g, K) has a solution. Moreover, due to the condition $S(K) \subseteq K$, there is a solution in K. The proof is complete. ∎

Remark. We refer to condition (7.49) as condition (θ_g), due to the following. In Isac (2000), for a given function $f : H \to H$, we introduced the property below:

$$\begin{cases} \text{there exists } \rho > 0 \text{ such that for each} \\ x \in K \text{ with } \|x\| > \rho, \text{ there exists} \\ y \in K \text{ with } \|y\| < \|x\| \text{ such that} \\ \langle x - y, f(x) \rangle \geq 0, \end{cases}$$

which we called condition (θ). For a pair of mappings (f, g), the condition (θ_g) is an extention of condition (θ). Indeed, if $g(x) \equiv x$ for all $x \in H$, then

$$\langle g(x) - y, x\rangle = \langle x - y, x\rangle = \langle x, x\rangle - \langle y, x\rangle \geq$$
$$\geq \|x\|^2 - \|y\|\|x\| = \|x\|\left(\|x\| - \|y\|\right).$$

Hence, if $\|y\| < \|x\|$, then we have

$$\langle g(x) - y, x\rangle = \langle x - y, x\rangle > 0.$$

In Isac (2000), we presented several existence theorems based on condition (θ), which is a strict generalization of the Karamardian's condition introduced in Karamardian (1971). By specializing condition (θ_g) we can show that several classes of pairs of mappings are without exceptional families of elements.

PROPOSITION 7.3. *Let $f, g : H \to H$ be two mappings. If there exists $\rho > 0$ such that for any $x \in K$ with $\|x\| > \rho$, there exists $y \in K$ such that*

(1) $\langle g(x) - y, f(x)\rangle \geq 0$,

(2) $\langle g(x), x\rangle \geq \alpha\|x\|^2, \quad \alpha \in R_+ \setminus \{0\}$, *and*

(3) $\|y\| < \alpha\|x\|$,

then the pair (f, g) satisfies the condition (θ_g).

PROOF. The assertion of the proposition is implied by the definition of condition (θ_g) and the following fact:

$$\langle g(x) - y, x\rangle = \langle g(x), x\rangle - \langle y, x\rangle \geq$$

$$\alpha\|x\|^2 - \|y\| \cdot \|x\| = \|x\|\left(\alpha\|x\| - \|y\|\right) > 0. \qquad \blacksquare$$

DEFINITION 7.9. We say that the pair (f, g) satisfies the Karamardian type condition, if one can find a bounded subset $D \subset K$ such that for all $x \in K \setminus D$, there exists a $y \in D$ such that $\langle g(x) - y, f(x)\rangle \geq 0$ and $\langle g(x), x\rangle \geq \alpha\|x\|^2$, for some $\alpha > 0$.

THEOREM 7.14. *If the pair (f, g) satisfies the Karamardian type condition, then it also satisfies the condition (θ_g).*

PROOF. We can suppose $0 < \alpha < 1$. Because D is bounded, there exists $\rho_0 > 0$ such that $D \subset \{x \in K \mid \|x\| \leq \rho_0\}$. In this case, for every $x \in K$ with $\|x\| > \rho$, where $\rho > \frac{\rho_0}{\alpha}$, there exists $y \in D$ such that

$$\langle g(x) - y, f(x)\rangle \geq 0, \qquad \langle g(x), x\rangle \geq \alpha\|x\|^2, \quad \text{and} \quad \|y\| \leq \alpha\|x\|.$$

Hence, all the assumptions of Proposition 7.3 are satisfied with $\alpha = 1$, and the assertion of the theorem is a consequence of Proposition 7.3. $\blacksquare$

7.5.3 CONTINUOUS SELECTIONS APPROACH TO MULTI-VALUED IMPLICIT COMPLEMENTARITY PROBLEM

In this subsection, we consider the *multi-valued implicit complementarity problem* (MICP) specified as follows. Given two multi-valued mappings

$f, g : R^n \to 2^{R^n}$, the MICP(f, g, K) defined by the (ordered) pair of mappings (f, g) and the cone K is: find $x_* \in R^n$ such that there exist $u \in f(x_*)$ and $v \in g(x_*)$ satifying the following conditions:

$$u \in K^*, \qquad v \in K, \qquad \text{and} \quad \langle u, v\rangle = 0. \tag{7.51}$$

To establish the existence of solutions to problem (7.51), we can use the following approach: select single-valued continuous branches (selections) from the multi-valued mappings f and g, and use Theorem 7.12 immediately. In order to select the continuous selections, one need imposing certain conditions on the continuity of the underlying multi-valued mappings.

Introduce the binary operation of addition of subsets (the Minkowski sum):

$$A + C = \{a + c \mid a \in A,\ c \in C\}.$$

DEFINITION 7.10. We say that the multi-valued mapping $f : X \to 2^Y$ is *(i) lower semicontinuous* (l.s.c.) at the point x, if for an arbitrary $\varepsilon > 0$ there exists a $\delta > 0$ such that $\|x - y\| < \delta$ implies $f(x) \subset f(y) + \varepsilon B$, where B is the unit ball centered at the origin. The mapping is called lower semicontinuous, if it is such at every point $x \in X$.

(ii) upper semicontinuous (u.s.c) at the point x, if for an arbitrary $\varepsilon > 0$ there exists a $\delta > 0$ such that $\|x - y\| < \delta$ implies $f(y) \subset f(x) + \varepsilon B$, where B is the unit ball centered at the origin. The mapping is called upper semicontinuous, if it is such at every point $x \in X$.

For the lower semicontinuous mappings, the problem of continuous selections was succesfully solved by Michael (1956).

DEFINITION 7.11. A single-valued continuous mapping $\varphi : X \to Y$ is referred to as a *single-valued continuous selection (branch)* of the multi-valued mapping $f : X \to 2^Y$, if $\varphi(x) \in f(x)$ for every $x \in X$. A topological vector space X is called *paracompact* if from every open cover of it, one can extract a locally finite subcover.

THEOREM 7.15 [Michael (1956)]. *Let $f : X \to 2^Y$ be a multi-valued lower semicontinuous mapping of the paracompact X to the Banach space Y, where $f(x)$ is a nonempty closed convex subset of Y for each $x \in X$. Then there exists a contionuous selection $\varphi : X \to Y$.* ■

As we are dealing with the finite-dimensional Euclidean space R^n, we can rewrite the convex-valued theorem by Michael for R^n. To do that, recall that by the Stone theorem, every metric space is paracompact.

THEOREM 7.16. *Let $f : R^n \to 2^{R^m}$ be a multi-valued lower semicontinuous mapping of (the paracompact) R^n to (the Banach space) R^m, where $f(x)$ is a nonempty closed convex subset of R^m for each $x \in R^n$. Then there exists a contionuous selection $\varphi : R^n \to R^m$.* ■

Now we extend the existence results to the MICP using the theorem by E. Michael on the continuous selection for the lower semicontinuous mapping.

DEFINITION 7.12. We say that a family of elements $\{x_r\}_{r>0} \subset R^n$ is an *exceptional family of elements* for the pair (f,g) of multi-valued mappings $f,g : R^n \to 2^{R^n}$, with respect to the cone $K \subset R^n$, if the following conditions are satisfied:

(1) $\|x_r\| \to +\infty$ as $r \to +\infty$;

(2) for every $r > 0$, there exist a real number $\mu_r > 0$ and elements $y_r \in f(x_r)$, $z_r \in g(x_r)$ such that $s_r = \mu_r x_r + y_r \in K^*$, $v_r = \mu_r x_r + z_r \in K$, and $\langle s_r, v_r \rangle = 0$.

The following result extends Theorem 7.12 of this section to the lower semicontinuous mappings.

THEOREM 7.17. *Let $K \subset R^n$ be a closed convex cone with the vertex at the origin, and $f,g : R^n \to 2^{R^n}$ lower semicontinuous multi-valued mappings with nonempty convex closed values. Then there exists either a solution to the problem MICP(f,g,K), or an exceptional family of elements in the sense of Definition* 7.9. *Moreover, if $K - g(K) \subseteq K$, then the problem* MICP(f,g,K) *has either a solution in K, or an exceptional family of elements $\{x_r\}_{r>0} \subset K$.*

PROOF. We use Theorem 7.16 and obtain the existence of two continuous selections $\varphi, \psi : R^n \to R^n$ from the multi-valued mappings f,g, respectively. Now it is easy to see that the pair of single-valued continuous functions (φ, ψ) satisfy all the conditions of Theorem 7.9. Hence, there exists either a solution to the problem ICP(φ, ψ, K) that evidently solves the initial multi-valued problem MICP(f,g,K), too, or an exceptional family of elements $\{x_r\}_{r>0}$ for the pair (φ, ψ) which also is clearly an exceptional family of elements the initial pair (f,g) according to Definition 7.12. Moreover, $K - g(K) \subseteq K$ implies $K - \psi(K) \subseteq K$, and Theorem 7.12 allows us to conclude that the ICP(φ, ψ, K), and hence, the MICP(f,g,K), has either a solution in K, or an exceptional family of elements $\{x_r\}_{r>0} \subset K$ in the sense of Definition 7.12. The proof is complete. ∎

Remark. The foregoing result looks quite nice since no compactness of the values of the multi-valued mappings has been required. To illustrate this fact, consider the following examples.

EXAMPLE 7.2. Let $K = R_+ \subset R^1$ and the lower semicontinuous mappings f,g are given by the following formulas:

$$f(x) = \begin{cases} [2, \infty), & \text{for} \quad -\infty < x \le 0; \\ [1-x, \infty), & \text{for} \quad 0 < x \le 1; \\ [x-1, \infty), & \text{for} \quad 1 < x < +\infty; \end{cases}$$

$$g(x) = \begin{cases} [-1, -0.5], & \text{for} \quad -\infty < x \le -1; \\ [x, x+0.5], & \text{for} \quad -1 < x < +\infty. \end{cases}$$

It is easy to see that all the points of the interval $[-0.5, 0]$ and the isolated point $x_* = 1$ solve the MICP(f, g, K). Indeed, as $K^* = K = R_+$ here, for any point $x \in [-0.5, 0]$ one can choose $2 \in f(x) \cap K^*$, $0 \in g(x) \cap K$, and $\langle 2, 0 \rangle = 0$. Similarly, we choose $0 \in f(1) \cap K^*$, $1 \in g(1) \cap K$, and $\langle 0, 1 \rangle = 0$. It is also evident that there is no exceptional families of elements for the pair (f, g) with respect to the cone K in the sense of Definition 7.12. Indeed, if $x_r < 0$ then $g(x_r) \subset (-K) \setminus \{0\}$, hence $v_r = \mu_r x_r + z_r \in (-K) \setminus \{0\}$, i.e. $v_r \notin K$ for any $\mu_r > 0$ and $z_r \in g(x_r)$. On the other hand, if $x_r > 0$ then $s_r = \mu_r x_r + y_r > 0$ and $v_r = \mu_r x_r + z_r > 0$, which implies that $\langle s_r, v_r \rangle > 0$, for every $\mu_r > 0$, $y_r \in f(x_r)$ and $z_r \in g(x_r)$. ∎

EXAMPLE 7.3. Consider again $K = R_+ \subset R^1$ and the lower semicontinuous mappings f, g, given by the following formulas:

$$f(x) = \begin{cases} (-\infty, -2], & \text{for } -\infty < x \le 0; \\ (-\infty, x - 1], & \text{for } 0 < x < 1; \\ (-\infty, 1 - 2x], & \text{for } 1 \le x < +\infty; \end{cases}$$

$$g(x) = \begin{cases} [-1, -0.5], & \text{for } -\infty < x \le -1; \\ [x, x + 0.5], & \text{for } -1 < x < +\infty. \end{cases}$$

It is clear that no solution to the MICP(f, g, K) exists whereas the family $\{x_r = r\}_{r>0}$ is an exceptional family of elements, in the sense of Definition 7.12. Indeed, $\|x_r\| = r \to +\infty$ as $r \to \infty$, and having put

$$\mu_r = \begin{cases} (1 - r)/r, & \text{for } 0 < r < 1; \\ (2r - 1)/r, & \text{for } r \ge 1; \end{cases}$$

$$z_r = r + 0.5, \qquad \text{and} \quad y_r = \begin{cases} r - 1, & \text{for } 0 < r < 1; \\ 1 - 2r, & \text{for } r \ge 1, \end{cases}$$

we obtain

$$v_r = \mu_r x_r + z_r = \begin{cases} 1 - r + r + 0.5 = 1.5, & \text{for } 0 < r < 1; \\ 2r - 1 + r + 0.5 = r - 0.5, & \text{for } r \ge 1; \end{cases}$$

$$s_r = \mu_r x_r + y_r = 0 \quad \text{for all} \quad r > 0.$$

Hence, $s_r = 0 \in K^*$, $0 < v_r \in K$ and $\langle s_r, v_r \rangle = 0$ for all $r > 0$. ∎

7.6 COMPACTIFICATION APPROACH TO MICP WITH UPPER SEMICONTINUOUS MAPPINGS

Unfortunately, for the upper semicontinuous multi-valued mappings, no results concerning selection of continuous single-valued selections are known.

Moreover, Silin (1988) presented an example of an upper semicontinuous multi-valued mapping, every single-valued selection of which is non-continuous over a subset of full measure.

However, now we will show that the MICP(f, g, K) with a multi-valued mapping f and a single-valued one-to-one correspondence g is equivalent to a (multi-valued) variational inequality. It is well-known (Aubin, 1988) that the multi-valued variational inequality with upper semicontinuous mapping is solvable on every compact domain. Therefore, making use of the compactification approach applied previously to the standard complementarity problem (see Avtukhovich, 1999), we will extend the foregoing existence results onto the particular case of the multi-valued implicit complementarity problem MIPC(f, g, K) with upper semicontinuous mapping f.

DEFINITION 7.13. Consider a multi-valued mapping $F : M \to 2^{R^n}$. We say that an element $x_* \in M$ solves the *variational inequality VI*(F, M) if there exists a vector $u \in F(x_*)$ such that

$$\langle u, x - x_* \rangle \geq 0 \quad \text{for all} \quad x \in M. \tag{7.52}$$

PROPOSITION 7.4. *Let $f : R^n \to 2^{R^n}$ be a multi-valued mapping and $g : R^n \to C \subset R^n$ a one-to-one correspondence with $K \subseteq C \cap g(R^n)$. Then the* MICP$(f, g, K)$ *defined by* (7.51) *is solvable simultaneously with the* VI(F,K) *defined by* (7.52), *where $F : K \to 2^{R^n}$ is given by*

$$F(y) = \left(f \circ g^{-1}\right)(y) \quad \forall y \in K.$$

PROOF. Let $x_* \in R^n$ solve the MICP(f, g, K). It means that $g(x_*) \in K$ and there exists $u \in f(x_*) \cap K^*$ such that

$$\langle u, z - g(x_*) \rangle \geq 0 \qquad \forall z \in K. \tag{7.53}$$

Taking $y_* = g(x_*)$ we have that

$$u \in \left(f \circ g^{-1}\right)(y_*) = F(y_*),$$

and (7.53) implies that

$$\langle u, z - y_* \rangle \geq 0 \qquad \forall z \in K, \tag{7.54}$$

which means that y_* solves the VI(F, K).

Conversely, if an element $y_* \in K$ solves VI(F, K), that is if $\exists u \in F(y_*)$ satisfying (7.54), then we put $x_* = g^{-1}(y_*)$ and conclude that

$$u \in \left(f \circ g^{-1}\right)(y_*) = f(x_*). \tag{7.55}$$

Now we show that also $u \in K^*$ and

$$\langle u, g(x_*) \rangle = 0. \tag{7.56}$$

Indeed, if $y_* = 0$ then (7.56) trivially holds, as well as (7.54) implies

$$\langle u, z \rangle \geq 0 \tag{7.57}$$

for all $z \in K$, i.e. $u \in K^*$. Otherwise, if $y_* \neq 0$ then by substituting $z = 0$ and $z = 2y_*$ into (7.54) we deduce (7.56) which again combined with (7.54) implies (7.57) for each $z \in K$ that means $u \in K^*$. Thus, we have shown that $u \in f(x_*) \cap K^*$, $g(x_*) \in K$ and (7.56) holds which means that x_* solves the MICP(f, g, K) and completes the proof. ∎

It is well-known that the VI(F, K) is equivalent to the multi-valued standard complementarity problem CP(F, K): Find an $x \in K$ such that there exists a vector $u \in F(x)$ with the properties

$$u \in K^* \qquad \text{and} \quad \langle u, x \rangle = 0.$$

Based upon the concept of exceptional family of elements for the (multi-valued) standard complementarity problem (*cf.* Bulavsky, Isac, Kalashnikov,1998,1999), we introduce a corresponding notion of exceptional family of elements for the MICP(f, g, K).

DEFINITION 7.14. We say that a family of elements $\{x_r\}_{r>0} \subset R^n$ is an *exceptional family of elements* (EFE) for the pair (f, g) of a multi-valued mapping $f : R^n \to 2^{R^n}$ and a one-to-one correspondence $g : R^n \to C \subset R^n$, $K \subseteq C \cap g(R^n)$, with respect to the cone K, if the following conditions are satisfied:

(1) $\|g(x_r)\| \to +\infty$ as $r \to +\infty$;

(2) for every $r > 0$, $z_r = g(x_r) \in K$, and there exist a real number $\mu_r > 0$ and an element $y_r \in f(x_r)$ such that $s_r = \mu_r g(x_r) + y_r \in K^*$, and $\langle s_r, g(x_r) \rangle = 0$. If $r = 0, 1, 2, \ldots$, we refer to the EFE as a *discrete* one.

Consider an upper semicontinuous multi-valued mapping $f : R^n \to 2^{R^n}$ with nonempty convex closed values and a single-valued continuous one-to-one correspondence $g : R^n \to C$ such that $K \subseteq C \cap g(R^n)$. As shown above, the multi-valued implicit complementarity problem MICP(f, g, K) can be reduced to the multi-valued variational inequality VI(F, K) with the same domain and the upper semicontinuous mapping $F = f \circ g^{-1}$. The latter problem is always solvable if its domain is a convex compact and the function F values are nonempty convex compacts in R^n: *cf.* Aubin (1988). Therefore, in order to obtain existence results for the MICP(f, g, K) we need compactifying it. In doing that, we will follow mainly the techniques of compactification proposed recently by Avtukhovich (1999) for the standard complementarity problems.

Denote by $K_{\pi/2}$ the subset $K_{\pi/2} = K \cap B_{\pi/2}$ where

$$B_{\pi/2} = B(0, \pi/2) = \{p \in R^n \mid \|p\| \leq \pi/2\}$$

is the ball with radius $\pi/2$ and centre at the origin which is obviously convex and compact. Denote also $\widehat{K}_{\pi/2} = K \cap \operatorname{int} B_{\pi/2}$ and $B = B(0, 1)$. Define a continuous mapping $\sigma_i : R^n \to B(0,\pi/2)$ as follows: for any vector $\hat{q} \in R^n \neq 0$ set

$$\sigma_i(\hat{q}) = \frac{\hat{q}}{\|\hat{q}\|}\arctan(\|\hat{q}\|) \in B(0,\pi/2),$$

and $\sigma_i(0) = 0$. If we restrict σ_i to the cone K, we obtain $\sigma_i : K \to K_{\pi/2}$ with

$$\sigma_i(q) = \frac{q}{\|q\|}\arctan\left(\|q\|\right) \in K_{\pi/2}$$

for all $q \in K$. Depending upon the domain of the mapping σ_i, its inverse mapping is either $\sigma_i^{-1} : B(0,\pi/2) \to R^n$ or $\sigma_i^{-1} : K_{\pi/2} \to K$.

Now we need defining a multi-valued mapping $\psi : K_{\pi/2} \to 2^B$. The vectors of $K_{\pi/2}$ may lie either in the interior or on the boundary of this subset. Hence, consider the two possible cases.

If a vector p is in the interior of the subset $K_{\pi/2}$, i.e. $\|p\| < \pi/2$, then define the needed mapping as follows:

$$\psi(p) = \frac{2}{\pi}\sigma_i \circ F\left(\sigma_i^{-1}(p)\right). \tag{7.58}$$

Now in order to define ψ for vectors p on the (relative) boundary of $K_{\pi/2}$, i.e. if $\|p\| = \pi/2$, we define three multi-valued mappings $\chi, \xi : \widehat{K}_{\pi/2} \to 2^B$ and $\tilde{\varphi} : \widehat{K}_{\pi/2} \to 2^B$ as follows. For any vector $z \in \widehat{K}_{\pi/2}$, denote by

$$\chi\left(\sigma_i^{-1}(z)\right) =$$

$$= \left\{s \in K^* \mid \|s\| = 1, \exists \gamma, \mu > 0 : \ s \in \gamma F\left(\sigma_i^{-1}(z)\right) + \mu\sigma_i^{-1}(z), \langle s, \sigma_i^{-1}(z)\rangle = 0\right\},$$

$$\xi\left(\sigma_i^{-1}(z)\right) =$$

$$= \{0, \text{if } \exists\{z_r\}_{r>0} \subset K : \sigma_i(z_r) \to z, F(z_r) \cap \{-\mu z_r \mid \mu > 0\} \neq \emptyset \forall r > 0\},$$

$$\tilde{\varphi}(z) = \operatorname{conv}\left\{\frac{2}{\pi}\left(\sigma_i \circ F \circ \sigma_i^{-1}\right)(z) \cup \left(\chi \circ \sigma_i^{-1}\right)(z) \cup \left(\xi \circ \sigma_i^{-1}\right)(z)\right\}.$$

Let $\Gamma = \{(z,s) \mid s \in \tilde{\varphi}(z)\}$ denote the graph of the mapping $\tilde{\varphi}$, and $\bar{\Gamma}$ - its closure. Now we can define the mapping ψ for the vectors $p \in K_{\pi/2}$ with $\|p\| = \pi/2$ as follows:

$$\psi(p) = \left\{s \mid (p,s) \in \bar{\Gamma}\right\}. \tag{7.59}$$

Finally, making use of (7.58) and (7.59), one can define the multi-valued mapping $\psi : K_{\pi/2} \to 2^B$ as follows:

$$\psi(p) = \begin{cases} \frac{2}{\pi}\left(\sigma_i \circ F \circ \sigma_i^{-1}\right)(p), & \text{for } p \in K_{\pi/2},\ \|p\| < \pi/2, \\ \left\{s \mid (p,s) \in \bar{\Gamma}\right\}, & \text{for } p \in K_{\pi/2},\ \|p\| = \pi/2. \end{cases} \tag{7.60}$$

PROPOSITION 7.5. *The mapping ψ is closed, and its values are nonempty, compact and convex.*

PROOF. First, by Nikaido (1968), Theorem 4.6, the continuity of σ_i and upper semicontinuity of F imply the upper semicontinuity of the multi-valued mapping $\sigma_i \circ F$. The latter mapping has nonempty convex compact values, hence it is closed: *cf.* Nikaido (1968) or Avtukhovich (1999, Lemma 5.1.1), and then (7.58) implies that the mapping ψ is closed and has nonempty compact convex values at the points $p \in K_{\pi/2}$ such that $\|p\| < \pi/2$. Second, the same properties of ψ at the points $p \in K_{\pi/2}$ such that $\|p\| = \pi/2$, is deduced from the closedness of the graph of the convex hull of mappings χ, ξ and $\frac{2}{\pi}\sigma_i \circ F$ which have convex compact values (see (7.59)). The proof is complete. ■

Remark. Note that values of the mappings χ and ξ may be empty sets as well.

Below, we will use the following assumption.

M1. Consider a sequence of elements $\{\hat{z}_m\}_{m=1}^{\infty} \subset K_{\pi/2}$ such that $\hat{z}_m \to q$ as $m \to \infty$ with $\|q\| = \pi/2$. Consider also a sequence $\{s_m\}_{m=1}^{\infty}$ of elements $s_m \in \frac{2}{\pi}\sigma_i \circ F\left(\sigma_i^{-1}(\hat{z}_m)\right)$ such that $s_m \to s$ as $m \to \infty$. If for each m there exists a $\hat{q}_m \in K_{\pi/2}$ such that $\langle s_m, \hat{q}_m - \hat{z}_m \rangle < 0$, then there exists a vector $\hat{q} \in K_{\pi/2}$ with $\langle s, \hat{q} - q \rangle < 0$.

Remark. Assumption **M1** means that if for every element $\hat{z}_m = \sigma_i(z_m) \in K_{\pi/2}$ of the images of a sequence $\{z_m\}_{m=1}^{\infty}$ in the cone K there exists a vector s_m from the corresponding value of the mapping $\psi(\hat{z}_m)$ such that the pair $(s_m, \hat{z}_m)$ **does not** solve the variational inequality $\mathrm{VI}(\psi, K_{\pi/2})$, then the limits of the images $\sigma_i(z_m)$ and the corresponding vectors s_m, i.e., the pair $(s, \hat{z})$ **does not** solve the $\mathrm{VI}(\psi, K_{\pi/2})$, either.

Now we are in a position to prove the following theorem of equivalence of the $\mathrm{MICP}(f, g, K)$ and the $\mathrm{VI}(\psi, K_{\pi/2})$. Before doing this, recall that a multi-valued mapping $\varphi : R^n \to 2^{R^n}$ is called *closed* if its graph $\Gamma = \{(x, s) | s \in \varphi(x)\}$ is a closed subset of R^n.

THEOREM 7.18. *The following three assertions are valid.*

(1) *A vector $p \in R^n$ and a vector $u \in f(p)$ solve the* $\mathrm{MICP}(f, g, K)$ *if and only if the vector $q = (\sigma_i \circ g)(p)$ has $\|q\| < \pi/2$ and solves the* $\mathrm{VI}(\psi, K_{\pi/2})$ *together with the corresponding vector $v = \frac{2}{\pi}\sigma_i(u)$.*

(2) *If $\{z_r\}_{r>0}$ is an exceptional family of elements for the pair (f, g) with respect to the cone K, in the sense of Definition* 7.14, *then there exists a vector $q \in K_{\pi/2}$ with $\|q\| = \pi/2$ that solves the* $\mathrm{VI}(\psi, K_{\pi/2})$.

(3) *Suppose that*

- *assumption M1 is valid;*
- *there exists a vector $q \in K_{\pi/2}$ with $\|q\| = \pi/2$ that solves the* $\mathrm{VI}(\psi, K_{\pi/2})$, *and*
- *there is no vector $q \in K_{\pi/2}$ with $\|q\| < \pi/2$ that solves the* $\mathrm{VI}(\psi, K_{\pi/2})$.

Then there exists a discrete exceptional family of elements for the pair (f, g) *with respect to the cone* K, *in the sense of Definition* 7.14.

PROOF. Proposition 7.5 implies that the MICP(f, g, K) with the pair (f, g) satisfying the conditions of the theorem, is equivalent to the variational inequality VI(F, K) with $F = f \circ g^{-1}$. In its turn, the latter variational inequality is equivalent to the (multi-valued) standard complementarity problem CP(F, K), which allows one to apply directly Avtukhovich (1999), Theorem 6.2.1, proved for the standard CP(F, K) and obtain all the needed assertions. However, we will prove all the assertions of the theorem explicitly.

PROOF OF ASSERTION (1). Consider the necessary part first.

(A). Let the vector p and the respective vector $u \in f(p)$ solve the MICP(f, g, K), i.e. $g(p) \in K$, $u \in f(p) \cap K^*$, and $\langle u, g(p) \rangle = 0$. As $u \in K^*$ then for every $z \in K$, the inequality $\langle u, z \rangle \geq 0 = \langle u, g(p) \rangle$ holds. The inclusion $K_{\pi/2} \subset K$ implies that the the latter inequality is valid for all vectors $z \in K_{\pi/2}$:

$$\langle u, z \rangle \geq 0 = \langle u, q \rangle, \quad \text{where} \quad q = (\sigma_i \circ g)(p). \tag{7.61}$$

It is obvious that $\|q\| < \pi/2$. As $p = \left(g^{-1} \circ \sigma_i^{-1}\right)(q)$ then $f(p) = \left(F \circ \sigma_i^{-1}\right)(q)$. Therefore, if $u = 0 \in f(p)$ then $v = \frac{2}{\pi}\sigma_i(u) = 0$ trivially solves the VI$(\psi, K_{\pi/2})$. Otherwise, if $u \neq 0$, multiply both parts of inequality (7.61) by $\frac{2}{\pi}\frac{1}{\|u\|}\arctan(\|u\|)$ and set $v = \frac{2}{\pi}\frac{u}{\|u\|}\arctan(\|u\|)$ which yields

$$\langle v, z \rangle \geq \langle v, q \rangle \quad \text{for all} \quad z \in K_{\pi/2}.$$

Since $v \in \frac{2}{\pi}\left(\sigma_i \circ F \circ \sigma_i^{-1}\right)(q) = \psi(q)$, the latter inequality implies that the vector $q = (\sigma_i \circ g)(p) \in K_{\pi/2}$, together with the vector $v \in \psi(q)$, solves the VI$(\psi, K_{\pi/2})$.

(B). Conversely, let a vector $q \in K_{\pi/2}$ with $\|q\| < \pi/2$ solve the VI$(\psi, K_{\pi/2})$, together with a corresponding vector $v \in \psi(q)$. This means that $\langle v - z, q \rangle \geq 0$ for all $z \in K_{\pi/2}$. As the mapping σ_i is obviously one-to-one correspondence between K and $K_{\pi/2}$, and g, in its turn, is also one-to-one correspondence, there exists a unique vector $p \in R^n$ such that $q = (\sigma_i \circ g)(p)$. We assume that $g(p) \neq 0$, otherwise p is a trivial solution of the MICP(f, g, K). Indeed, if $g(p) = 0$ then $q = 0$ hence $v \in \psi(0) \cap K^*$. Below we demonstrate that $v \in K^*$ implies the existence of a $u \in f(p) \cap K^*$ which means that this p with $g(p) = 0$ solves the MICP(f, g, K),

If $v = 0 \in \psi(q)$ then it is trivial that $u = 0 = \sigma_i^{-1}(v) \in f(p)$ solves the MICP(f, g, K) together with p. Now let $v \neq 0$. Since $v \in \psi(q) = \frac{2}{\pi}\left(\sigma_i \circ f \circ g^{-1} \circ \sigma_i^{-1}\right)(q)$ and the mapping σ_i is continuous, there exists a vector $u \in f(p)$ such that

$$v = \frac{2}{\pi}\frac{u}{\|u\|}\arctan(\|u\|) = \frac{2}{\pi}\sigma_i(u). \tag{7.62}$$

Denote by $K_\varepsilon = K \cap B(0, \varepsilon)$ where $B(0, \varepsilon)$ is the ball of radius ε centered at the origin. Select $\varepsilon > 0$ such that $q + K_\varepsilon \subseteq K_{\pi/2}$ (which can be done due to

$\|q\| < \pi/2$), then for any vector $z \in K_\varepsilon$ one has $q + z \in K_{\pi/2}$. Since the vector q (together with v) solves the VI$(\psi, K_{\pi/2})$, we get $\langle v, q+z\rangle \geq \langle v, q\rangle$, hence $\langle v, z\rangle \geq 0$ for all $z \in K_\varepsilon$. As $K_\varepsilon \subset K$, the property (ii) of the cone implies $\lambda z \in K$ for an arbitrary $z \in K_\varepsilon$ and each $\lambda \geq 0$. Thus we get $\langle v, z\rangle \geq 0$ for all $z \in K$ which means that $v \in K^*$, but as K^* also is a cone, definition (7.62) of u implies $u \in K^*$.

Since $q \in K_{\pi/2}$ solves the VI$(\psi, K_{\pi/2})$ then the inequality

$$\langle v, \lambda q - q\rangle = (\lambda - 1)\langle v, q\rangle \geq 0$$

is valid for all vectors $\lambda q \in K_{\pi/2}$. As $\|q\| < \pi/2$, the inequality can hold for all $\lambda > 0$ only if $\langle v, q\rangle = 0$ which implies $\langle u, g(p)\rangle = 0$ because

$$q = (\sigma_i \circ g)(p) = \frac{g(p)}{\|g(p)\|} \arctan\left(\|g(p)\|\right) \text{ and } v = \frac{2}{\pi}\sigma_i(u).$$

Thus, given a solution q of the VI$(\psi, K_{\pi/2})$ with the corresponding vector $v \in \psi$, we have constructed the vector $p = \left(g^{-1} \circ \sigma_i^{-1}\right)(q)$ and the corresponding vector $u = \sigma_i^{-1}(\pi/2v)$ such that $g(p) \in K$, $u \in f(u) \cap K^*$ and $\langle u, g(p)\rangle = 0$, i.e. that p and $u \in f(p)$ solve the MICP(f, g, K). Assertion (1) of the theorem is proved.

PROOF OF ASSERTION (2). Let $\{x_r\}_{r>0}$ be an exceptional family of elements for the pair (f, g) with respect to the cone K in the sense of Definition 7.14, i.e. $z_r = g(x_r) \in K$ and $\|g(x_r)\| \to +\infty$ as $r \to +\infty$. Moreover, for each $r > 0$ there exist $\mu_r > 0$ and $y_r \in f(x_r)$ such that

$$s_r = f(x_r) + \mu_r g(x_r) \in K^* \quad \text{and} \quad \langle s_r, g(x_r)\rangle = 0. \tag{7.63}$$

Denote by $\hat{z}_r = (\sigma_i \circ g)(x_r) = \sigma_i(z_r) \in K_{\pi/2}$. Since $K_{\pi/2}$ is a convex compact, then one can extract a subsequence $\{\hat{z}_{r_m}\}_{m=1}^{\infty} \subset K_{\pi/2}$ such that there exists

$$\lim_{m\to\infty} \hat{z}_{r_m} = \lim_{m\to\infty} (\sigma_i \circ g)(x_{r_m}) = \lim_{m\to\infty} \frac{g(x_{r_m})}{\|g(x_{r_m})\|} \arctan\left(\|g(x_{r_m})\|\right) = q; \tag{7.64}$$

it is clear that $\|q\| = \pi/2$.

If $s_r = 0$ then the subset $F(z_r) \cap \{-\mu_r z_r | \mu_r > 0\} \neq \emptyset$. According to the definition of the mapping ξ it means that $s_r \in \xi\left(\sigma_i^{-1}(\hat{z}_r)\right)$.

Otherwise, if $s_r \neq 0$, then $\hat{s}_r = \frac{s_r}{\|s_r\|} \in \chi\left(\sigma_i^{-1}(\hat{z}_r)\right)$. Indeed, $\hat{s}_r \in K^*$, $\|\hat{s}_r\| = 1$, and choose $\hat{\gamma} = \frac{1}{\|s_r\|} > 0$ for γ and $\hat{\mu}_r = \frac{\mu_r}{\|s_r\|} > 0$ for μ from the definition of the mapping χ. Now it is easy to verify that $\hat{s}_r \in \hat{\gamma}_r F(z_r) + \mu_r z_r$ and $\langle \hat{s}_r, z_r\rangle = 0$.

As the family $\{\hat{s}_r\}_{r>0}$ belongs to the unit sphere, one can extract a subsequence $\{\hat{s}_{r_m}\}_{m=1}^{\infty}$ such that

$$\lim_{m\to\infty} \hat{s}_{r_m} = \frac{s}{\|s\|}. \tag{7.65}$$

By multiplying both sides of the equality $\langle s_{r_m}, z_{r_m}\rangle = 0$ by the scalars

$$\frac{1}{\|s_{r_m}\|}\frac{1}{\|z_{r_m}\|}\arctan\left(\|z_{r_m}\|\right),$$

we come to the equality $\langle \hat{s}_{r_m}, \hat{z}_{r_m} \rangle = 0$ the limit of which gives $\langle \frac{s}{\|s\|}, q \rangle = 0$ due to (7.64) and (7.65). Since the cone K^* is closed we have $\frac{s}{\|s\|} \in K^*$. The inequalty $\langle \frac{s}{\|s\|}, z \rangle \geq 0 = \langle \frac{s}{\|s\|}, q \rangle$ takes place for every $z \in K$, in particular, for all $z \in K_{\pi/2}$. Hence

$$\langle \frac{s}{\|s\|}, z - q \rangle \geq 0$$

is valid for each $z \in K_{\pi/2}$.

Finally, note that

$$\hat{s}_r \in \text{conv} \left\{ \chi\left(\sigma_i^{-1}(\hat{z}_r)\right) \cup \xi\left(\sigma_i^{-1}(\hat{z}_r)\right) \right\},$$

i.e. $\hat{s}_r \in \psi\left(\sigma_i^{-1}(\hat{z}_r)\right)$, and as the mapping ψ is closed, we obtain $\frac{s}{\|s\|} \in \psi(q)$. Thus we have proved that the vector q determined in (7.64) has $\|q\| = \pi/2$ and solves the VI$(\psi, K_{\pi/2})$, together with the corresponding vector $\frac{s}{\|s\|} \in \psi(q)$. The proof of assertion (2) of the theorem is finished.

PROOF OF ASSERTION (3). Conversely, let $q \in K_{\pi/2}$, $\|q\| = \pi/2$ solve the VI$(\psi, K_{\pi/2})$, i.e. there exists a corresponding vector $s \in \psi(q)$ such that for each $\hat{z} \in K_{\pi/2}$, the inequality $\langle s, \hat{z} - q \rangle \geq 0$ holds true. Let $\{z_m\}_{m=1}^{\infty} \subset K$ be a sequence of points such that the sequence $\hat{z}_m = \sigma_i(z_m)$ converges to q. As the mapping ψ is closed at q with $\|q\| = \pi/2$, there exists a sequence $\{s_m\}_{m=1}^{\infty}$ such that $s_m \in \psi(\hat{z}_m)$ and $\lim_{m\to\infty} s_m = s \in \psi(q)$.

The sequence $\{z_m\}_{m=1}^{\infty}$ can be divided into two subsequences (without affecting generality, we will denote them by the same symbols as the original one) for which, according to the definition of mapping ψ (see (7.60)), one and only one of the two following outcomes holds:

- the subset $\text{conv}\left\{\chi\left(\sigma_i^{-1}(\hat{z}_m)\right) \cup \xi\left(\sigma_i^{-1}(\hat{z}_m)\right)\right\} = \emptyset$;
- the subset $\text{conv}\left\{\chi\left(\sigma_i^{-1}(\hat{z}_m)\right) \cup \xi\left(\sigma_i^{-1}(\hat{z}_m)\right)\right\} \neq \emptyset$.

In case of the first outcome, for each member of the subsequence $\{z_m\}$, one has $s_m \in \psi(\hat{z}_m)$, $\|\hat{z}_m\| = \|\sigma_i(z_m)\| < \pi/2$ and hence, from the third assumption of the Theorem's part (3), $\hat{z}_m$ cannot be a solution of the VI$(\psi, K_{\pi/2})$, i.e. there exists a vector $\hat{q}_m \in K_{\pi/2}$ such that $\langle s_m, \hat{q}_m - \hat{z}_m \rangle < 0$. Since $s_m \to s$ as $m \to \infty$, assumption M1 guarantees the existence of such a vector $\hat{q} \in K_{\pi/2}$ that $\langle s, \hat{q} - q \rangle < 0$ which contradicts q being a solution of the VI$(\psi, K_{\pi/2})$. Thus obtained contradiction shows that the first outcome is impossible.

In case of the second outcome, one can explicitly construct a discrete exceptional family of elements. Indeed, as

$$s_m \in \text{conv} \left\{ \chi\left(\sigma_i^{-1}(\hat{z}_m)\right) \cup \xi\left(\sigma_i^{-1}(\hat{z}_m)\right) \right\},$$

we must take into account the structure of the mappings χ and ξ. Again, without loss of generality, we can restrict ourselves to examining only two possible variants: $s_m \in \chi\left(\sigma_i^{-1}(\hat{z}_m)\right) \forall m$ or $s_m \in \xi\left(\sigma_i^{-1}(\hat{z}_m)\right) \forall m$.

Let $s_m \in \chi\left(\sigma_i^{-1}(\hat{z}_m)\right)$. According to the definition of the mapping χ, we have $s_m \in K^*$, $\|s_m\| = 1$, and there exist positive values γ_m, μ_m such that

$$s_m \in \gamma_m F\left(\sigma_i^{-1}(\hat{z}_m)\right) + \mu_m \sigma_i^{-1}(\hat{z}_m), \tag{7.66}$$

and

$$\langle s_m, \sigma_i^{-1}(\hat{z}_m)\rangle = 0. \tag{7.67}$$

Actually (7.66)–(7.67) defines a (discrete) exceptional family of elements $\{x_m\}_{m=1}^{\infty}$ as follows: substitute $g(x_m) = z_m = \sigma_i^{-1}(\hat{z}_m)$, divide both sides of (7.66) and (7.67) by γ_m, recall that $F = f \circ g^{-1}$, take into account that $\hat{s}_m = \frac{1}{\gamma_m} s_m \in K^*$ and $\hat{\mu}_m = \frac{\mu_m}{\gamma_m} > 0$ and obtain the needed properties of the family $\{x_m\}$:

$$\hat{s}_m \in [f(x_m) + \hat{\mu}_m g(x_m)] \cap K^* \quad \text{and} \quad \langle \hat{s}_m, g(x_m)\rangle = 0,$$

required in part 2) of Definition 7.14 of the exceptional family of elements. Part 1) of the definition, that is $\|g(x_m)\| \to +\infty$ as $m \to \infty$ is guaranteed by the structure of the mapping σ_i and the property $\lim_{m\to\infty} \sigma_i(g(x_m)) = q$ where $\|q\| = \pi/2$.

At last, consider the second variant where

$$s_m = 0 \in \xi\left(\sigma_i^{-1}(\hat{z}_m)\right) = \xi(z_m) \quad \forall m.$$

According to the definition of the mapping ξ, for the sequence $\{z_m\}_{m=1}^{\infty} \subset K$ such that $\sigma_i(z_m) \to q$ (with $\|q\| = \pi/2$ which implies that $\|z_m\| \to \infty$ as $m \to \infty$), one can choose $\mu_m > 0$ so that $s_m = 0 \in F(z_m) + \mu_m z_m$. Clearly $s_m = 0 \in K^*$ and $\langle s_m, z_m\rangle = 0$, which allows to conclude that the sequence $x_m = g^{-1}(z_m)$ is a discrete exceptional family of elements for the MICP(f, g, K).

Thus we have shown that if the VI$(\psi, K_{\pi/2})$ has a solution q at the (relative) boundary of the subset $K_{\pi/2}$, i.e. if $\|q\| = \pi/2$, and does not have ones at the interior of $K_{\pi/2}$, then there exists (a corresponding) exceptional family of elements for the MICP(f, g, K). The proof of assertion (3) and hence, of the theorem as a whole, is complete. ∎

Now as a corollary of Theorem 7.18, we extend the theorem of alternative to the case of upper semicontinuous mappings.

THEOREM 7.19 *Consider an upper semicontinuous multi-valued mapping $f : R^n \to 2^{R^n}$ with nonempty closed convex values and a single-valued continuous one-to-one correspondence $g : R^n \to C$ such that $K \subseteq C \cap g(R^n)$. Moreover, let assumption* M1 *be valid. Then there exists either a solution to the* MICP(f, g, K), *or a discrete exceptional family of elements, in the sense of Definition* 7.14.

PROOF. Construct the mapping $\psi : K_{\pi/2} \to 2^B$ by the formula (7.60). The multi-valued mapping ψ is an upper semicontinuous mapping of the convex compact $K_{\pi/2}$ with nonempty compact convex values in B (*cf.* Proposition 7.3). Hence, the variational inequality VI$(\psi, K_{\pi/2})$ is solvable: *cf.* Aubin (1988) or Avtukhovich (1999, Theorem 5.2.3).

If the norm of the vector from $K_{\pi/2}$ that solves VI$(\psi, K_{\pi/2})$ is less than $\pi/2$, then Theorem 7.18 implies, by assertion (1), that there is a vector $p \in R^n$

and an element $u \in f(p)$ that solve the MICP(f, g, K), and the first alternative is valid.

Otherwise, if the norm of the vector from $K_{\pi/2}$ that solves VI$(\psi, K_{\pi/2})$ is equal to $\pi/2$ and there is no other solutions with the norm less than $\pi/2$, then Theorem 7.18, by assertion (3) implies that there exists a discrete exceptional family of elements for the pair (f, g) with respect to the cone K, in the sense of Definition 7.14. The proof is complete. ∎

Remark. Again, as in the remark about the lower semicontinuous mappings, we see that the values of the multi-valued upper semicontinuous mappings involved need not being compact. To illustrate this fact, consider the following example.

EXAMPLE 7.4. Let $K = R_+ \subset R^1$, the upper semicontinuous mapping f be defined by

$$f(x) = \begin{cases} \left[\frac{1}{x^2}, \infty\right), & \text{for } \; -\infty < x < 0; \\ [0, \infty), & \text{for } \; x = 0; \\ \left[\frac{1}{x^2} - 1, \infty\right), & \text{for } \; 0 < x \le 1; \\ [x^2 - 1, \infty), & \text{for } \; x > 1. \end{cases}$$

and

$$g(x) = \begin{cases} \arctan(x) + \pi/2 - 1, & \text{for } \; x < 0; \\ x + \pi/2 - 1, & \text{for } \; x \ge 0. \end{cases}$$

The point $x_* = -\tan\left(\frac{\pi}{2} - 1\right)$ is obviously the unique solution of the MICP(f, g, K). It is also clear that there is no exceptional family of elements for the pair (f, g) with respect to the cone K, in the sense of Definition 7.14. Indeed, condition $\|g(x_r)\| \to \infty$ as $r \to \infty$ implies that $x_r > 0$ for all $r \ge r_0$ for some $r_0 > 0$. However in this case, $g(x_r) = x_r + \pi/2 - 1 > 0$ and $s_r = \mu_r g(x_r) + y_r > 0$ for all $\mu_r > 0$ and $y_r \in f(x_r)$. Therefore $\langle s_r, g(x_r)\rangle > 0$ which prohibits $\{x_r\}_{r>0}$ being an exceptional family of elements in the sense of Definition 7.14.

On the other hand, if we put $h(x) = -f(x)$, then we obtain the existence not only of the unique solution $x_* = -\tan\left(\frac{\pi}{2} - 1\right)$ of the MICP(h, g, K) but also of an exceptional family of elements in the sense of Definition 7.11. Indeed, $\{x_r = r\}_{r>0}$ is the exceptional family of elements, as $g(x_r) = r + \pi/2 - 1 > 0$ and $s_r = \mu_r g(x_r) + y_r = 0$ (where $y_r \in h(x_r)$) if we set

$$y_r = \begin{cases} 1 - \frac{1}{r^2}, & \text{for } \; 0 < r \le 1; \\ 1 - r^2, & \text{for } \; r > 1; \end{cases}$$

and

$$\mu_r = \begin{cases} \frac{1/r^2 - 1}{r + \pi/2 - 1}, & \text{for } \; 0 < r \le 1; \\ \frac{r^2 - 1}{r + \pi/2 - 1}, & \text{for } \; r > 1. \end{cases}$$

Hence, $g(x_r) \in K$, $s_r = 0 \in K^*$ and $\langle s_r, g(x_r)\rangle = 0$ for all $r > 0$, which completes the proof. ∎

Remark. The latter example in which there exist simultaneously the solution and the exceptional family of elements does not contradict Theorem 7.19, as the alternative of this theorem is non-strict in general case. It would be interesting to establish if the alternatives of Theorems 7.17 and 7.19 are strict in some special cases, as, for example, when the mappings f and/or g are pseudomonotone or maximal monotone ones: *cf.* Isac and Kalashnikov (2001).

Remark. The results of Theorem 7.18 and Theorem 7.19 allow one to interpret the exceptional family of elements as a "solution at the infinity", analogous to the concept of the infinite singular point in Complex Analysis.

REFERENCES

Aubin, J.-P. *Nonlinear Analysis and Its Economic Applications.* Moscow: Mir, 1988 (*translation into Russian*).

Avtukhovich, EV. "Existence of solution for multi-valued complimentarity problem", Working Paper, Computer Center of the Russian Academy of Sciences, Moscow, 1999 (*in Russian*).

Baiocchi, C., and Capelo, A. *Variational Inequalities and Quasi-Variational Inequalities: Applications to Free Boundary Value Problems.* New York: Wiley and Sons, 1984.

Bakushinskii, A.B., and Goncharskii, A.V. *Iterative Methods for Ill-Posed Problems.* Dordrecht: Kluwer Academic Publishers, 1994.

Ben-el-Mechaiekh H, Idzik A. A Leray-Schauder type theorem for approximable maps. Proc. Amer. Math. Soc. 1994; 122:105-109.

Ben-el-Mechaiekh H, Chebbi S, Florenzano M. A Leray-Schauder type theorem for approximable maps: a simple proof. Proc.Amer. Math. Soc. 1998;126:2345-2349.

Bensoussan A, Lions J-L. Nouvelle formulation de problèmes de contrôle impulsionnel et applications. C.R.Acad.Sci., Paris, 1973; Sér.A–B 276:1189-1192.

Bensoussan A, Gourset M., Lions J-L. Contrôle impulsionnel et inéquations quasi-variationnelles stationaires. C.R.Acad.Sci., Paris, 1973; Sér. A–B 276:1279-1284.

Browder FE. Existence and approximation of solution of nonlinear variational inequalities. Proc. National Acad. Sci., USA, 1966; 56:1080-1086.

Browder FE, and Ton BA. Nonlinear functional equations in Banach spaces and elliptic super regularization. Math. Zeitschr. 1968; 105:177-195.

Bulavsky, VA, Isac, G, Kalashnikov, VV.

1. 'Application of topological degree theory to complementarity problems'.- In: *Multilevel Optimization: Algorithms and Applications*, A. Migdalas, P.M. Pardalos and P. Värbrand, eds. Dordrecht-Boston-London: Kluwer Academic Publishers, 1998.

2. Application of topological degree theory to semidefinite complementarity problems. Proceedings of the International Conference on Operations Research OR'98; 1998 August 31-September 03; Zurich- Berlin-Heidelberg: Springer-Verlag, 1999.

Bulavsky VA, Kalashnikov VV.

1. One-parametric drive method to study equilibrium. Ekonomika i Matematicheskie Metody (*in Russian*) 1994; 30:129-138.

2. Equilibria in generalized Cournot and Stackelberg models. Ekonomika i Matematicheskie Metody (*in Russian*) 1995; 31:164-176.

3. 'An alternative model of spatial competition'.- In: *Operations research and decision aid methodologies in traffic and transportation management*, M. Labbé, G. Laporte, K. Tanczos, P. Toint, eds. NATO ASI Series, Series F: Computer and Systems Science, Vol. 166, Berlin-

Heidelberg: Springer-Verlag, 1998.

Capuzzo-Dolcetta, J. and Mosco, U. 'Implicit complementarity problems and quasi-variational inequalities'. In: *Variational Inequalities and Complementarity Problems, Theory and Applications*, R.W. Cottle, F. Giannessi and J.L. Lions, eds. New York-London: John Wiley and Sons, 1980.

Chan D, Pang J-S. The generalized quasi-variational problem. Math. Oper. Res. 1982; 7:211-222.

Cottle, R.W., Pang, J.S., and Stone, R.E.. *The Linear Complementarity Problem.* New York: Academic Press, 1992.

Cottle RW, Yao JC. Pseudomonotone complementarity problems in Hilbert space. J. Optim. Theory Appl. 1992; 75:281-295.

Dugundji, J., Granas, A. *Fixed Point Theory.* Warszawa: PWN – Polish Scientific Publishers, 1982.

Eaves BC. On the basic theorem of complementarity. Math. Programming 1971; 1:68-75.

Granas A. Sur la méthode de continuité de Poincaré. C.R. Acad.Sci., Paris, France, 1976; 282:978-985.

Hadjisavvas N, Schaible S.

1. On strong pseudomonotonicity, (semi)strict quasimonotonicity. J. Optim. Theory Appl. 1993; 79:139-156

2.Quasimonotone variational inequalities in Banach spaces. J. Optim. Theory Appl. 1996; 90:95-111.

Harker PT, Pang JS. Finite-dimensional variational inequalities and nonlinear complementarity problems: a survey of theory, algorithms and applications. Math. Programming 1990; 48: 161-220.

Harker PT, Choi SC. A penalty function approach for mathematical programs with variational inequality constraints. Information and Decision Technologies 1991; 17:41-50.

Hyers, D.H., Isac, G., and Rassias, T.M. *Topics in Nonlinear Analysis and Applications*. Singapore: World Scientific, 1997.

Isac G

1. On the implicit complementarity problem in Hilbert spaces. Bull. Austral. Math.Soc. 1985;32:251-260.

2. Complementarity problem and coincidence equations on convex cones. Boll.Un.Mat.Ital. 1986; B(6) 5:925-943.

3. Fixed point theory and complementarity problems in Hilbert spaces. Bull. Austral. Math. Soc. 1987;36:295-310.

4. Fixed point theory, coincidence equations on convex cones and complementarity problem. Contemporary Mathematics 1988; 72:139-155.

5. A special variational inequality and the implicit complementarity problem. J. Fac. Sci. Univ. Tokyo, Sect.IA, Math. 1990;37:109-127.

6. *Complementarity Problems*, Lecture Notes in Mathematics. Berlin-Heidelberg: Springer Verlag, Vol. 1528, 1992.

7. Tikhonov regularization and the complementarity problem in Hilbert spaces. J. Math. Analysis Appl. 1993; 174:53-66.

8. Exceptional families of elements for k-fields in Hilbert spaces and complementarity theory. Proceedings of the International Conference on Optimization Techniques and Applications (ICOTA'98); 1998 July 1-3; Perth, Australia.

9. A generalization of Karamardian's condition in complementarity theory. Nonlinear Analysis Forum 1999; 4:49-63.

10. On the solvability of multivalued complementarity problem: A topological method. Proceedings of the Fourth European Workshop on Fuzzy Decision Analysis and Recognition Technology (EFDAN'99); 1999 June 14-15; Dortmund, Germany (R. Felix, ed.), 1999.

11. Exceptional family of elements, feasibility and complementarity. J. Optim.Theory Appl. 2000;104:577-588.

12. 'Exceptional family of elements, feasibility, solvability and continuous paths of ε-solutions for nonlinear complementarity problems'. -In: *Approximation and Complexity in Numerical Optimization: Continuous and Discrete Problems*, P.M. Pardalos, ed., 2000.

13. *Topological Methods in Complementarity Theory*. Dordrecht: Kluwer Academic Publishers, 2000.

Isac G, Bulavsky VA, Kalashnikov VV. Exceptional families, topological degree and complementarity problems. J. Global Opt. 1997; 10:207-225.

Isac G, Carbone A. Exceptional families of elements for continuous functions. Some applications to complementarity theory. J. Global Opt. 1999;15:181-196.

Isac G, Goeleven D.

1. Existence theorems for the implicit complementarity problems. Intern. J. Math. and Math. Sci. 1993; 16:67-74.

2. The implicit general order complementarity problem: models and iterative methods. Ann. Oper. Res. 1993; 44:63-92.

Isac G, Kalashnikov VV. Exceptional families of elements, Leray- Schauder alternatives, pseudomonotone operators and complementarity. J. Optim. Theory Appl. 2001; 109: 69-83.

Isac G, Obuchowska WT. Functions without exceptional families of elements and complementarity problems. J.Optim.Theory Appl. 1998; 99:147-163.

Isac, G., and Zhao, YB.

1. "Exceptional Families of Elements and the Solvability of Variational Inequalities for Unbounded Sets in Infinite Dimensional Hilbert Spaces", Preprint, University of Beijing, 1999.

2. Exceptional families of elements and the solvability of variational inequalities for unbounded sets in infinite dimensional Hilbert spaces. J. Math. Anal. Appl. 2000; 246:544-556.

Kalashnikov, VV. "Complementarity Problem and Generalized Oligopoly Models", Habilitation Thesis, Central Economics and Mathematics Institute, Moscow, Russia, 1995 (in Russian).

Kalashnikov, VV. "Fixed Point Existence Theorems Based upon Topological Degree Theory", Working Paper, Moscow, Central Economics and Mathematics Institute (*in Russian*),1995.

Kalashnikov VV and Kalashnikova NI. Solving two-level variational inequality. J.Global Optim. 1996; 8:289-294.

Kalashnikov VV and Khan AA. A regularization approach for variational inequalities with pseudomonotone operators. Operations Research Proceedings 1999, Selected papers of the Symposium on Operations Research (SOR'99); 1999 September 1-3; Magdeburg; Germany (K. Inderfurth e.a., eds.). Berlin- Heidelberg: Springer-Verlag, 2000.

Karamardian S.

1. Generalized complementarity problem. J. Optim.Theory Appl. 1971; 8:161-168.

2. Complementarity problems over cones with monotone and pseudomonotone maps. J. Optim. Theory Appl. 1976; 18:445-454.

Karamardian S, Schaible S. Seven kinds of monotone maps. J. Optim. Theory Appl. 1990; 66:37-46.

Kinderlehrer, D., and Stampacchia, G. *An Introduction to Variational Inequalities and Their Applications*. New York: Academic Press, 1980.

Michael E. Selected selection theorems. Amer. Math. Monthly 1956; 63:233-238.

Mosco, U.

1. *Implicit Variational Problems and Quasi-Variational Inequalities*, Lecture Notes in Math., No. 543, Springer-Verlag, 1976.

2. 'On some nonlinear quasi-variational inequalities and implicit complementarity problems in stochastic control theory'. In: *Variational Inequalities and Complementarity Problems, Theory and Applications*, R.W. Cottle, F. Giannessi and J.L. Lions, eds. New York: John Wiley and Sons, 1980.

Nashed, M.Z., and Liu, F. 'On nonlinear ill-posed problems II: monotone operator equations and monotone variational inequalities'. In: *Theory and Applications of Nonlinear Operators of Accretive and Monotone Type*, A.A. Karsatos, ed. Lecture Notes in Pure and Applied Mathematics, Vol.178, New York: Marcel Dekker, Inc., 1996.

Nikaido, H. *Convex Structures and Mathematical Economics*. New York: Acad. Press, 1968.

Outrata JV.

1. On necessary optimality conditions for Stackelberg problems. J. Optim. Theory Appl. 1993; 76:305-320.

2. On optimization problems with variational inequality constraints. SIAM Journal on Optimization 1994; 4:340-357.

Pang, J.-S.

1. 'The implicit complementarity problem'.- In: *Nonlinear Programming IV*, O.L. Mangasarian, R.R. Meyer and S.M. Robinson, eds. New York-London: Academic Press, 1981.

2. On the convergence of a basic iterative method for the implicit complementarity problem. J.Optim.Theory Appl. 1982; 37:149-162.

Rockafellar, R.T. *Convex Analysis*. Princeton, NJ: Prinston University Press, 1970.

Schaible, S. 'Generalized monotonicity - concepts and uses'. In: *Variational Inequalities and Network Equilibrium Problems*, F. Giannessi and A. Maugeri, eds. New York: Plenum Publishing Corporation, 1995.

Schaible S, Yao JC. On the equivalence of nonlinear complementarity problems and least-elements problems. Math. Programming 1995; 70:191-200.

Sherali HD, Soyster AL and Murphy FH. Stackelberg-Nash-Cournot equilibria: characterizations and computations. Oper. Res. 1983; 31: 171-186.

Silin DB. Some properties of upper semicontinuous multi-valued mappings. Proc. of Math. Inst. of the Russian Academy of Science (Steklov Institute) 1998; 185: 222-235 *(in Russian)*.

Smith MJ. A descent algorithm for solving monotone variational inequalities and monotone complementarity problems. J. Optim. Theory Appl. 1984; 44:485-496.

Smith TE. A solution condition for complementarity problem with an application to spatial price equilibrium. Appl. Math. Comp. 1984; 15:61-69.

Tikhonov, A.N., and Arsenin, V.Y. *Solution of Ill-Posed Problems*. New York: J.Wiley & Sons, 1977.

Ton BA. Nonlinear operators on convex subsets of Banach spaces. Math. Ann. 1969; 181:35-44.

Yao JC. Multivalued variational inequalities with K-pseudomonotone operators. J. Optim.Theory Appl. 1994; 83:391-403.

Zeidler, E. *Nonlinear Functional Analysis and its Applications*, II/B. New York: Springer-Verlag, 1993.

Zhao YB.

1. Exceptional family and finite-dimensional variational inequality over polyhedral convex

set. Appl. Math. Comp. 1997; 87:111-126.

2. "Existence Theory and Applications for Finite-Dimensional Variational Inequality and Complementarity Problems", Ph.D. Thesis, Institute of Applied Mathematics, Academia Sinica, Beijing, China, 1998.

Zhao YB, Han JY. Exceptional family of elements for a variational inequality problem and its applications. J. Global Opt. 1999; 14: 313-330.

Zhao YB, Han JY, Qi HD. Exceptional family and existence theorems for variational inequality problems. J. Optim. Theory Appl. 1999;101: 475-495.

Zhao, Y.B., and Isac, G.

1. "Properties of a Multivalued Mapping and Existence of Central Path for Some Nonmonotone Complementarity Problems", Preprint, University of Beijing, 1999.

2. Quasi-P_* and $P(\tau, \alpha, \beta)$-maps, exceptional family of elements and complementarity problems. J. Optim. Theory Appl. 2000; 105:213-231.

3. Properties of a multivalued mapping and existence of central path for some non-monotone complementarity problems, *(to appear in: SIAM J. Control Opt.).*

PART III.

NUMERICAL METHODS FOR SOLVING COMPLEMENTARITY PROBLEMS

Chapter 8

COMPLEMENTARITY PIVOT METHODS

This chapter deals with pivot methods that constituted the first and important instrument in solving Linear Complementarity Problems. Developed under the impact of the simplex method, they revealed many crucial differences between the linear programming and complementarity problems.

In Section 8.2, we describe the Lemke type algorithms following mainly the paper by Eaves (1971). Section 8.3 is dedicated to two interesting extensions of the Lemke's algorithm based upon the projective transformations and developed by Todd (1976). The idea of block pivot algorithm described in Section 8.4 follows an original pigeonholing approach by Kostreva (1978). At last, Section 8.5 presents a summary of conditions under which the complementary pivot algorithm solves Linear Complementarity Problem. This section follows mainly Chapter 2.3 by Murty (1988).

8.1 INTRODUCTION

We describe here complementary pivot methods that can be used to solve *complementarity problems (CP)* for a continuous mapping $f : R^n \to R^n$. The standard CP finds a vector $x \in R^n$ such that

$$x \geq 0, \quad f(x) \geq 0, \quad \text{and} \quad x^T f(x) = 0. \tag{8.1}$$

To start with, we assume the mapping f to be affine, i.e. $f(y) = My + q$ where M is an $n \times n$-matrix, and $q \in R^n$.

Then problem (8.1) can be rewritten in the form

$$Ix - My = q, \quad x \geq 0, \; y \geq 0, \tag{8.2}$$

$$\text{and} \quad x^T y = 0, \tag{8.3}$$

where I is identity. We will also refer to system (8.2) as $q \mid M$. If the system has a solution, then $q \mid M$ is called *feasible* (otherwise *infeasible*). If (x, y) is a solution such that $x^T y = 0$ (either x_i or y_i is zero for $i = 1, \ldots, n$) then we call (x, y) a *complementary solution* of $q \mid M$. By $d|q|\ M$ we denote the system

$$Ix - My - dz = q, \quad x \geq 0,\ y \geq 0,\ z \geq 0,$$

where $d \in R^{n \times 1}$ and z is a scalar variable. If (x, y, z) is a solution such that $x \perp y$, then, again, we call it a complementary solution of $d|q|M$. This system is important in the search of complementary solutions of $q|M$.

In the next section, we describe Lemke's pivot algorithm and the class of matrices $\mathcal{L}$ such that Lemke's algorithm processes (i.e. solves or shows no solution exists) linear complementarity problems for $M \in \mathcal{L}$.

8.2 LEMKE'S PIVOTING ALGORITHM

As we describe it the Lemke's algorithm generates a sequence of adjacent complementary bases. Here we define adjacency and complementary bases. First, we introduce some notation.

8.2.1 NOTATION

(1) Let $R^{n \times m}$ denote the set of $n \times m$-matrices with real coefficients. Let $i = 1, \ldots, m$, $A \in R^{m \times n}$, and $B \in R^{n \times m}$. Then by A^i and B_i we denote the i-th row of A and the i-th column of B. Let $\beta = (\beta_1, \ldots, \beta_\ell)$ where $\beta_i \in \{1, \ldots, m\}$ for $i = 1, \ldots, \ell$; define A^β and B_β by

$$A^\beta = \begin{pmatrix} A^{\beta_1} \\ \vdots \\ A^{\beta_\ell} \end{pmatrix}, \qquad B_\beta = (B_{\beta_1}, \ldots, B_{\beta_\ell}).$$

Let $\alpha \subset \{1, \ldots, m\}$; by A^α and B_α we denote A^γ and B_γ where $\gamma = (\gamma_1, \ldots, \gamma_\ell)$, $\{\gamma_1, \ldots, \gamma_\ell\} = \alpha_i$, and $\gamma_1 < \gamma_2 < \cdots < \gamma_\ell$.

(2) By $A \geq 0$ and $A > 0$ we denote that each entry of A is nonnegative and positive, respectively. By $A \succeq 0$ [$A \succ 0$] we denote that each row of A is *lexico-nonnegative* [*lexico-positive*]. This means that the first nonzero entry of the row in question is nonnegative [positive].

(3) Let $A \in R^{n \times m}$ and $B \in R^{n \times \ell}$ then by $A \succeq 0$, $B \succeq 0$, and $A \perp B$ we denote, in addition to the fact that the rows of A and B are lexico-nonnegative, that for $i = 1, \ldots, n$ either A^i or B^i is zero. Such A and B are then said to be *complementary*.

(4) By $e = e_n$ we denote the vector $(1, \ldots, 1)^T \in R^{1 \times n}$. By a *P-matrix* (respectively, *P_0-matrix*) we mean a matrix with positive (respectively, nonnegative) principal minors. A matrix A is referred to as one of *bounded type* if it possesses one and hence all of the following properties:

(i) $\{x \geq 0 \mid Ax \leq 0\} = \emptyset$,

(ii) for any b, $\{x \geq 0 \mid Ax \leq b\}$ is bounded,

(iii) there is a $\lambda \geq 0$ such that $\lambda A > 0$.

8.2.2 LEXICO-POSITIVE INEQUALITY SYSTEMS

To conveniently obtain arbitrarily large and small "perturbations" of q we follow Eaves (1971) and often use the *lexico-system*

$$IX - MY = Q, \quad X \succeq 0, \, Y \succeq 0, \quad X \perp Y,$$

where X, Y, and Q are matrices. Results are readily translated to the systems where X, Y, and Q are column vectors. First, we state two properties of the lexico-system

$$LX = Q, \qquad X \succeq 0,$$

where $L \in R^{m \times n}$ and $Q \in R^{m \times \ell}$ (and hence $X \in R^{n \times \ell}$). These properties are merely analogues of those of the inequality system: $Lx = q$, $x \geq 0$ where $q \in R^{m \times 1}$. The first lemma is the alternative principle.

LEMMA 8.1. *The system $LX = Q$, $X \succeq 0$ has a solution if and only if there does not exist a $\lambda \in R^{1 \times m}$ such that $\lambda L \geq 0$ and $\lambda Q \prec 0$.* ■

If $L\bar{X} = Q$ then we say that $\bar{X}$ *uses* column i of L if $\bar{X}^i \neq 0$.

LEMMA 8.2. *Let $\bar{X}$ be a solution of $LX = Q$, $X \succeq 0$. Then there is a solution which uses a linearly independent subset of the columns used by $\bar{X}$.* ■

Suppose that $LX = Q$ and $X \succeq 0$. Notice that there is a $t_0 > 0$ such that $LX[t] = Q[t]$ and $X[t] \succeq 0$ for $0 \leq t \leq t_0$ where $A[t] = A_1 + A_2 t + \cdots + A_\ell t^{\ell-1}$ and $A \in R^{m \times \ell}$. The use of lexicography for perturbations was introduced in Dantzig e.a. (1955).

8.2.3 THE LINEAR COMPLEMENTARITY PROBLEMS $Q|M$ AND $d|Q|M$

By $Q|M$ we denote the system

$$IX - MY = Q, \qquad X \succeq 0, \quad Y \succeq 0,$$

where $I \in R^{n \times n}$ (identity), $M \in R^{n \times n}$, and $Q \in R^{n \times \ell}$. If the system has a solution, then $Q \mid M$ is called *feasible* (otherwise *infeasible*). If (X, Y) is a solution such that $X \perp Y$ (either X^i or Y^i is zero for $i = 1, \ldots, n$) then we call (X, Y) a *complementary solution* of $Q \mid M$.

By $d|Q|M$ we denote the system

$$IX - MY - dZ = Q, \qquad X \succeq 0, \, Y \succeq 0, \, Z \succeq 0,$$

where $d \in R^{n\times 1}$. If (X, Y, Z) is a solution such that $X \perp Y$, then, again, we call it a *complementary solution* of $d|Q|M$. Lemke (1965) showed that this system is important in the search of complementary solutions of $Q|M$.

8.2.4 COMPLEMENTARY BASES AND ADJACENCY

(1) If $\beta \subset \{1,\ldots,2n+1\}$, $|\beta| = n$, $(I,-M,-d)_\beta$ has an inverse, $(I,-M,-d)_\beta^{-1} q \geq 0$, and $\{i, i+n\} \not\subset \beta$ for $i = 1,\ldots,n$, then we shall call β a *complementary basis* of $d|q|M$. If, in addition, $2n+1 \notin \beta$ then we call β a complementary basis of $q|M$. Of course, a complementary basis induces a complementary solution. Note that $(I,-M,-d)$ is $n \times (2n+1)$ and that the index $2n+1$ corresponds to the column $-d$.

(2) Let β be a complementary basis of $d|q|M$. If $2n+1 \notin \beta$ then define $\bar{\beta} = \{2n+1\}$. If $2n+1 \in \beta$, then there is exactly one $i \in \{1,\ldots,n\}$ such that $\{i, i+n\} \cap \beta = \emptyset$; in this case, define $\bar{\beta} = \{i, i+n\}$. We can now define *adjacency*.

(3) Two complementary bases α_i and β of $d|q|M$ are defined to be *adjacent* if

$$\alpha_i = \beta \cup \{i\} \setminus \{j\}$$

where $i \in \bar{\beta}$ and $j \in \beta$, or equivalently, if

$$\beta = \alpha_i \cup \{j\} \setminus \{i\}$$

where $j \in \bar{\alpha}_i$ and $i \in \alpha_i$.

For example, suppose that $n = 4$ and that $\alpha_i = \{1,2,3,9\}$, $\beta = \{9,2,3,8\}$ and $\gamma = \{5,2,3,8\}$ are complementary bases. Then $\bar{\alpha}_i = \{4,8\}$, $\bar{\beta} = \{1,5\}$ and $\bar{\gamma} = \{9\}$; also α_i and β are adjacent and β and γ are adjacent.

8.2.5 DEGENERACY

If a solution of $Q|M$ or $d|Q|M$ uses at least n linearly independent columns then we say the solution is *nondegenerate*. If each complementary solution of $Q|M$ (respectively, $d|Q|M$) is nondegenerate then we say that $Q|M$ (respectively, $d|Q|M$) is *nondegenerate*.

The following facts of degeneracy take place. For any system $d|Q|M$, the system $d|(Q,I)|M$ is nondegenerate. Any complementary basis of $d|(Q,I)|M$ remains such with respect to $d|Q|M$; if two complementary bases of $d|(Q,I)|M$ are adjacent, they remain adjacent with respect to $d|Q|M$. If $d|Q|M$ is nondegenerate, then $d|Q|M$ have the same complementary basis and adjacency structure. If $d|Q|M$ is nondegenerate, then each complementary basis is adjacent to exactly 0,1, or 2 complemenatry bases; in fact, if β is a complementary basis and $i \in \bar{\beta}$ then β is adjacent to at most one complementary basis of form $\beta \cup \{i\} \setminus \{j\}$ where $j \in \beta$. If $d|Q|M$ is nondegenerate and $(d, Q) \succ 0$, then there is exactly one basis of form $\{1,\ldots,n,2n+1\}$. If $Q|M$ is nondegenerate, then complementary solutions and complementary bases are in one-to-one correspondence.

The next lemmas are important in proofs that there are an odd number of solutions.

LEMMA 8.3. *If $Q|M$ is nondegenerate then there is a $d > 0$ such that $d|Q|M$ is nondegenerate.* ∎

LEMMA 8.4. *Suppose*

$$\begin{pmatrix} 0 & Q \\ 1 & 0 \end{pmatrix} \Bigg| \begin{pmatrix} M & c \\ a & b \end{pmatrix}$$

is nondegenerate. Then there is a $d > 0$ such that

$$\begin{pmatrix} d \\ 0 \end{pmatrix} \Bigg| \begin{pmatrix} 0 & Q \\ 1 & 0 \end{pmatrix} \Bigg| \begin{pmatrix} M & c \\ a & b \end{pmatrix}$$

is nondegenerate ($M \in R^{n\times n}$, $Q \in R^{n\times \ell}$, $d \in R^{n\times 1}$, $a \in R^{1\times n}$, $b \in R^1$, $1 \in R^1$, and $c \in R^{n\times 1}$). ∎

8.2.6 DESCRIPTION OF LEMKE'S ALGORITHM

The following procedure describes Lemke's algorithm.

(1) **ALGORITHM.** Given the system $d|q|M$ where $(d,q) \succ 0$, apply to the system $d|(q,I)|M$ the following steps:

(i) Let $\beta^1 = \beta^0$ be the complementary basis of form $\{1,\ldots,n,2n+1\}\backslash\{j\}$ and let $k = 1$.

(ii) If there is a complementary basis α_i different from β^{k-1} and adjacent to β^k, then let $\beta^{k+1} = \alpha_i$. Otherwise, stop.

(iii) Increase k by 1 and return to step (ii). ∎

For example if

$$d = \begin{pmatrix} 1 \\ 0 \end{pmatrix}, \quad q = \begin{pmatrix} -1 \\ 1 \end{pmatrix}, \quad M = \begin{pmatrix} -2 & 1 \\ -1 & 0 \end{pmatrix},$$

then the algorithm generates $\beta^1 = \{2,5\}$, $\beta^2 = \{3,5\}$, and $\beta^3 = \{3,4\}$.

THEOREM 8.1. *The algorithm can be executed. It generates a unique finite sequence $\beta^1,\ldots,\beta^\ell$ of distinct complementary bases of $d|q|M$. The terminal basis $\gamma = \beta^\ell$ will have at least one of the following two properties:*

(a) $2n+1 \notin \gamma$.

(b) There is an $i \in \bar{\gamma}$ such that $(\gamma \cup \{i\}) \cap \{n+1,\ldots,2n\} \neq \emptyset$ and such that there is no complementary basis of form $\gamma \cup \{i\} \setminus \{j\}$ where $j \in \gamma$. ∎

If (a) is the terminal condition then we have a complementary basis of $q_i^{(m)}$ and hence a complementary solution of $q_i^{(m)}$. If (b) is the terminal condition then a *secondary ray* of $d|q|M$ is generated. The next subsection discusses secondary rays.

Proof of convergence of this algorithm (and that of Lemke (1965)) is not based on any "monotone evolution" but rests upon the fact that the generated complementary bases cannot repeat and that there are only finitely many complementary bases.

8.2.7 SECONDARY RAYS AND THE BASIC THEOREM

Here we define a secondary ray following Eaves (1971). The issue is that for a broad and interesting class of matrices M the existence of a secondary ray for $d|Q|M$ implies that $Q|M$ is infeasible. Hence if $Q|M$ is feasible and M is of this class, Lemke's algorithm will compute a complementary solution.

Assume that $(d, Q) \succ 0$ and consider the ray

$$\left\{ \begin{pmatrix} X \\ Y \\ Z \end{pmatrix} + \theta \begin{pmatrix} x, & 0 \\ y, & 0 \\ z, & 0 \end{pmatrix} \mid \theta \geq 0 \right\}$$

of the set of solutions of $d|Q|M$ where

1. $IX - MY - dZ = Q$,
2. $Ix - My - dz = 0$,
3. $X, Y, Z, x, y, z \succeq 0$,
4. $X \perp Y, X \perp y, x \perp Y, x \perp y$,
5. $(y, z) \neq 0$,
6. $(Y, y) \neq 0$,

$(X, Y, Q \in R^{n\times\ell}; Z \in R^{1\times\ell}; x, y \in R^{n\times 1}; z \in R^1)$. We call such a ray a *secondary ray*, and we shall say that (X, Y, Z, x, y, z) generates the ray.

The following theorem of Lemke is our basic tool and used repeatedly.

THEOREM 8.2. *Assume $(d, Q) \succ 0$. If Lemke's algorithm is given $d|Q|M$ then it generates either a complementary solution of $Q|M$ or a secondary ray of $d|Q|M$ with $Z \succ 0$. If $d|Q|M$ is nondegenerate and has no secondary rays with $Z \succ 0$ or $z > 0$, the $Q|M$ has an odd number of complementary solutions.* ∎

The first assertion follows from Theorem 8.1; proof of the second assertion relies on the facts of degeneracy and the fact that a graph has an even number of odd elements. However, a detailed proof is supplied in Eaves (1969).

8.2.8 THE CLASS $\mathcal{L}$

Here a class of matrices which we call $\mathcal{L}$ is defined. It is stated that Lemke's algorithm will proceed these matrices. We show that $\mathcal{L}$ includes P-matrices, certain P_0-matrices called *adequate*, copositive plus matrices, and matrices for bimatrix games (Nash equilibrium points). Further it is stated that if $Q|M$

is feasible and nondegenerate and if $M \in \mathcal{L}$ then $Q|M$ has an odd number of complementary solutions.

By definition $\mathcal{L}$ is the intersection of two classes each of which has interest in its own right.

Class $\mathcal{L}$: $\mathcal{L} = \mathcal{L}_1 \cap \mathcal{L}_2$.

Class $\mathcal{L}_1$: $M \in R^{n\times n}$ is in $\mathcal{L}_1$ if for every $y \in R^{n\times 1}$, $0 \neq y \geq 0$, there is an $i = 1, \ldots, n$ such that $y^i > 0$ and $(My)^i \geq 0$.

Class $\mathcal{L}_2$: $M \in R^{n\times n}$ is in $\mathcal{L}_2$ if for every $y \in R^{n\times 1}$, $0 \neq y \geq 0$, such that $My \geq 0$ and $y^T My = 0$, there are diagonal matrices $\Lambda \geq 0$ and $\Omega \geq 0$ such that $\Omega y \neq 0$ and $(\Lambda M + M^T \Omega)y = 0$.

In "$\mathcal{L}_1$" we refer to i as a *defining index* for M and y. In "$\mathcal{L}_2$" matrices Λ and Ω may depend on y as well as M. We say a matrix is in $\mathcal{L}_*$ if it is in $\mathcal{L}_1$ and $\mathcal{L}_2$ but $\mathcal{L}_2$ always by default, that is, $0 \neq y \geq 0$, $My \geq 0$, and $y^T My = 0$ is impossible. Hence:

Class $\mathcal{L}_*$: $M \in R^{n\times n}$ is in $\mathcal{L}_*$ if for every $y \in R^{n\times 1}$, $0 \neq y \geq 0$, there is an $i = 1, \ldots, n$ such that $y^i > 0$ and $(My)^i > 0$.

The next lemma exhibits a simple fact about $\mathcal{L}_*$ and $\mathcal{L}_1$.

LEMMA 8.5. *The following three statements are equivalent:*

(i) $M \in \mathcal{L}_*$.

(ii) *For every $q \geq 0$, $q_i^{(m)}$ has a unique complementary solution.*

(iii) *For every $Q \succeq 0$, $Q|M$ has a unique complementary solution;*

and the following three statements are equivalent:

(i) $M \in \mathcal{L}_1$.

(ii) *For every $q > 0$, $q_i^{(m)}$ has a unique complementary solution.*

(iii) *For every $Q \succ 0$, $Q|M$ has a unique complementary solution.* ∎

The next theorems show that Lemke's algorithm will process $Q|M$ if $M \in \mathcal{L}$ or $M \in \mathcal{L}_*$. Further, if $Q|M$ is feasible, nondegenerate, and $M \in \mathcal{L}$, then it is shown that $Q|M$ has an odd number of complementary solutions. The same result holds if $Q|M$ is nondegenerate and $M \in \mathcal{L}_*$.

THEOREM 8.3. *Assume $(d, Q) \succ 0$ and $M \in \mathcal{L}$. If Lemke's algorithm is applied to $d|Q|M$, then either a complementary solution of $Q|M$ is computed or multipliers showing $Q|M$ infeasible are computed. If in addition $Q|M$ is feasible and nondegenerate then $Q|M$ has an odd number of complementary solutions.* ∎

THEOREM 8.4. *Assume $(d, Q) \succ 0$ and $M \in \mathcal{L}_*$. If Lemke's algorithm is applied to $d|Q|M$, then a complementary solution of $Q|M$ is generated. If $Q|M$ satisfies the nondegeneracy condition then $Q|M$ has an odd number of solutions.* ■

The next lemma suggests that Lemke's algorithm (when applied to $d|Q|M$) will not always generate a complementary solution to $q_i^{(m)}$ unless $M \in \mathcal{L}_*$.

LEMMA 8.6. *Problem $d|q|M$ has no secondary ray for all $d > 0 > q$ if and only if $M \in \mathcal{L}_*$.* ■

Various subclasses of $\mathcal{L}$ are now described.

Lemke defined copositive plus matrices and showed that his algorithm would process them.

DEFINITION 8.1. A matrix M is *copositive plus* if both
(i) $y \geq 0$ implies $y^T MY \geq 0$,
(ii) $y \geq 0$, $y^T My = 0$ imply $(M + M^T)y = 0$.

To prove that these matrices are in $\mathcal{L}$ we merely let Ω and Λ be I. A strictly copositive matrix $(0 \neq y \geq 0 \Longrightarrow y^T My > 0)$ is in $\mathcal{L}_*$.
If M and N are in $\mathcal{L}$ then so are
(i) PMP^T (principal rearrangement),
(ii) $\begin{pmatrix} M & 0 \\ 0 & N \end{pmatrix}$ (direct summation),
(iii) $\Lambda M \Omega$ (positive scaling),

where P is a permutation matrix and Λ and Ω are diagonal matrices with positive diagonals. The copositive plus class is closed under (i) and (ii) but not (iii).

Danzig and Cottle (1967) defined a class which included a matrix M if and only if every principal submatrix $\bar{M}$ of M had the property that $0 \neq \bar{Z} \geq 0$ and $\bar{M}\bar{Z} \leq 0$ is impossible. It is easy to see that this class and $\mathcal{L}_*$ are one and the same. They showed that Lemke's algorithm would generate a complementary solution for $q_i^{(m)}$ for any q if $M \in \mathcal{L}_*$. They also proved the next lemma.

LEMMA 8.7. *P-matrices are in $\mathcal{L}_*$.* ■

This lemma follows from the next one proved in Fiedler and Ptak (1962).

LEMMA 8.8. *$M \in R^{n \times n}$ is a P-matrix if and only if for every $y \neq 0$ there is an i such that $(My)^i y^i > 0$.* ■

In an analysis of an engineering problem related to the linear complementarity problem, Ingleton (1966) defined a class of matrices called *adequate.* He showed that if M is adequate and q is a linear combination of columns of M then $q_i^{(m)}$ has a complementary solution.

DEFINITION 8.2. A matrix $M \in R^{n \times n}$ is defined to be *adequate* if it is P_0 and if for every $\alpha \subset \{1, \ldots, n\}$, $\det(M^\alpha_\alpha) = 0$ implies that the rows of M^α are linearly dependent and the columns of M_α are linearly dependent.

Under weaker condition than that, Eaves (1971) showed M is in $\mathcal{L}$.

DEFINITION 8.3. $M \in R^{n \times n}$ is defined to be *row adequate* if M is a P_0-matrix and for every $\alpha \subset \{1, \ldots, n\}$, $\det(M^\alpha_\alpha) = 0$ implies that the rows of M^α are linearly dependent.

LEMMA 8.9. *M is row adequate if and only if for any y, $(y^T M)_i y^i \leq 0$ for $i = 1, \ldots, n$ implies $y^T M = 0$.* ■

THEOREM 8.5. *If M is row adequate then it is in $\mathcal{L}$.* ■

Cottle (1968) noted that if R is copositive plus and if T is skew-symmetric then $R + T$ is copositive plus. In particular, if R and S are copositive plus then so is

$$\begin{pmatrix} S & A^T \\ -A & R \end{pmatrix}.$$

Class $\mathcal{L}$ is not closed under the latter operation; however, copositive plus is not closed under the operation discussed in the following lemmas.

LEMMA 8.10. *If $R \in \mathcal{L}$, $S \in \mathcal{L}$, $P > 0$, and $N < 0$ then*

$$\begin{pmatrix} R & P \\ N & S \end{pmatrix} \quad \text{and} \quad \begin{pmatrix} S & N \\ P & R \end{pmatrix}$$

are in $\mathcal{L}$. ■

LEMMA 8.11. *If $R \in \mathcal{L}_*$ and $S \in \mathcal{L}_*$, $P \geq 0$, and Q arbitrary then*

$$\begin{pmatrix} R & P \\ Q & S \end{pmatrix} \quad \text{and} \quad \begin{pmatrix} S & Q \\ P & R \end{pmatrix}$$

are in $\mathcal{L}_*$. ■

8.2.9 THE CLASS P_0

It is now well known that M is a P-matrix if and only if $q|M$ has a unique solution for every q. A number of researchers contributed to this fact, see Samuelson e.a. (1958), Ingleton (1966), Cottle (1968), Parsons (1967), Cottle and Danzig (1968), and Murty (1968). Lemke (1967) showed that if M is positive semidefinite and if $q|M$ has a nondegenerate complementary solution then it is unique. Eaves (1971) showed that if M is a P_0-matrix and $q|M$ has a nondegenerate complementary solution then it is unique. However, this condition is not sufficient, that is, $q|M$ may have a unique solution for every q for which $q|M$ is nondegenerate and yet M is not a P_0-matrix.

THEOREM 8.6. *If $M \in P_0$ and $Q|M$ has a nondegenerate solution then it is unique.* ■

Lemke's algorithm will not process all matrices in P_0. That is, $d|Q|M$ where $M \in P_0$ might have a secondary ray even though $Q|M$ is feasible, for example, let

$$d = \begin{pmatrix} 1 \\ 1 \end{pmatrix}, \quad Q = \begin{pmatrix} -1 \\ 0 \end{pmatrix}, \quad M = \begin{pmatrix} 0 & 1 \\ 0 & 0 \end{pmatrix}.$$

Clearly (X, Y, Z, x, y, z) generates a secondary ray where

$$X = \begin{pmatrix} 0 \\ 1 \end{pmatrix}, \quad Z = 2, \quad y = \begin{pmatrix} 1 \\ 0 \end{pmatrix}$$

and $Y, x, z = 0$, and clearly $(\bar{X}, \bar{Y}, \bar{Z})$ is a complementary solution where

$$\bar{X} = \begin{pmatrix} 0 \\ 0 \end{pmatrix}, \quad \bar{Y} = \begin{pmatrix} 0 \\ 1 \end{pmatrix}, \quad \bar{Z} = \begin{pmatrix} 0 \\ 0 \end{pmatrix}.$$

8.3 EXTENSIONS OF LEMKE'S ALGORITHM

As it was demonstrated above, Lemke's algorithm for the linear complementarity problem fails when a desired pivot is not blocked. A projective transformation proposed by Todd (1976) overcomes this difficulty. The transformation is performed computationally by adjoining a new row to a schema of the problem and pivoting on the element in this row and the unit constant column. Two new algorithms result, and some conditions for their success are well-grounded.

For convenience purpose, rewrite the linear complementarity problem as follows: Find $n \times 1$ vectors w, z satisfying

$$w = Mz + q, \quad w \geq 0, \quad z \geq 0, \tag{8.4}$$

$$w^T z = 0, \tag{8.5}$$

for given $n \times n$ matrix M and $n \times 1$ vector q.

Lemke's method introduces an artificial variable z_0 to obtain

$$w = Mz + q + ez_0, \quad w, z, z_0 \geq 0, \tag{8.6}$$

$$w^T z = 0, \tag{8.7}$$

where (e, q) has lexicographically positive rows (generally, we take $e > 0$, e.g., the vector of ones). Pivots are made in the linear system (8.6) maintaining (8.7) from an easily accessible *primary ray* until:

(a) z_0 leaves the basis; a solution to (8.4)–(8.5) is obtained;

or

(b) a desired pivot is unblocked; a *secondary* ray is encountered.

In Todd (1973), the Lemke's algorithm is extended by adjoining a simple bounding constraint to truncate (8.6). The additional power of the resulting algorithms was demonstrated in Todd (1973); roughly, the occurence of case (b) in Lemke's algorithm is turned to advantage by permitting the search of more almost-complementary paths.

In Todd (1976), an alternative way of dealing with unboundedness is investigated. A projective transformation is applied to the polyhedron of feasible solutions to (8.4) or (8.6). This results in a polytope of which part corresponds to the original polyhedron and the rest to the behavior of the original polyhedron *past infinity*; in fact, to the behavior of the facets in the opposite orthant. The advantage lies in obtaining a bounded system, so that more almost-complementary paths can be searched.

8.3.1 PROJECTIVE TRANSFORMATIONS

For d a $1 \times n$ vector, let D, D^+, and D^- denote the set of $u \in R^n$, with du equal to, greater than, or less than 1, respectively. The mapping $\phi_d : R^n \setminus D \to R^n \setminus D$ is defined by

$$\phi_d(u) = [1/(du-1)]u.$$

For given $m \times n$ matrix A and $m \times 1$ vector b, denote by P the set

$$P = \{u \in R^n \mid v = b - Au \geq 0,\ u \geq 0\},$$

by Q the set

$$Q = \{u' \in R^n \mid v' = -b - (A - bd)u' \geq 0,\ u' \geq 0\},$$

and by R the set

$$R = \{u' \in R^n \mid v' = -b - (A - bd)u' \leq 0,\ u' \leq 0\}.$$

THEOREM 8.7. *There is a one-to-one correspondence between $P \cap D^+$ and $Q \cap D^+$ and between $P \cap D^-$ and $R \cap D^-$. The facets $u_i = 0$ and $u'_i = 0$ and the facets $v_j = 0$ and $v'_j = 0$ correspond for $1 \leq i \leq n$, $1 \leq j \leq m$.*

PROOF. The correspondence in either direction is given by ϕ_d. Note that

$$d\phi_d(u) - 1 = 1/(du - 1),$$

showing that

$$u' = \phi_d(u) \Leftrightarrow u = \phi_d(u').$$

It is now easy to check the theorem by direct substitution. ∎

Remark. The following comments are pertinent:

(i) If $b \not\geq 0$, then $0 \notin P$; then, there is $d > 0$ with $P \subseteq D^+$.

(ii) For any $d \leq 0$, $P \subseteq D^-$.

(iii) If P is nonempty and bounded, then under the conditions of (i), $Q \subseteq D^+$; and, under the conditions of (ii), $R \subseteq D^-$.

8.3.2 ALGORITHMS

In this subsection, we describe two algorithms for the LCP presented in Todd (1976). The first is applicable only when $q < 0$, but is somewhat simpler than the second. However, the second algorithm is a direct extension of Lemke's, and so is at least as powerful as Lemke's. In fact, a simple example shows that it is more powerful.

For the first algorithm, we assume $q < 0$. We then take $d > 0$ sufficiently large, so that $qd + M$ has at least one negative row. In the context of the last subsection, we will be applying ϕ_d to the feasible region given by (5.4); our choice of d assures that this region lies in D^+.

The transformation is performed by adjoining a row with associated free basic variable f to obtain

$$\begin{cases} w = Mz + q \cdot 1, & w \geq 0, \quad z \geq 0, \\ f = dz + (-1) \cdot 1, & f \text{ free.} \end{cases} \tag{8.8}$$

Now, pivot on the -1, to make 1 "basic" and f nonbasic:

$$\begin{cases} w = (qd + M)z + (-q) \cdot f, & w \geq 0, \quad z \geq 0, \\ 1 = dz + (-1) \cdot f, & f \text{ free.} \end{cases} \tag{8.9}$$

Now, relabel w, z, f, and 1 by $x, y, 1$, and g to obtain

$$\begin{cases} x = (qd + M)y + (-q) \cdot 1, & x \geq 0, \quad y \geq 0, \\ g = dy + (-1) \cdot 1, & g \text{ free.} \end{cases} \tag{8.10}$$

Algorithm A produces a sequence of solutions to the system (8.10). These solutions correspond either to solutions of (8.8), or to solutions of (8.8) with q replaced by 0 or $-q$. Denote the corresponding systems (8.8.0) and (8.8.$-q$), respectively.

LEMMA 8.12. *Suppose that the linear system* (8.10) *is bounded. Then, there are one-to-one correspondences between:*

(a) *basic solutions to* (8.10) *with g positive and basic solutions to* (8.8);

(b) *solutions to* (8.10) *with g zero and solutions to* (8.8.0) *with f zero;*

(c) *basic solutions to* (8.10) *with g negative and basic solutions to* (8.8.$-q$).

In each case, $w_i = 0$ if and only if $x_i = 0$, and $z_i = 0$ if and only if $y_i = 0$. ∎

The first and the last parts of the lemma show that, in almost all cases, pivots in (8.10) can be accomplished by pivots in (8.8) with the basic variables either all nonnegative or all nonpositive. However, if a solution to (8.10) with g zero must be generated, the system (8.8) is inadequate; one would have to pivot on a zero element. The projective transformation allows these "vertices at infinity" of (8.8) to be passed through without difficulty. This behavior is illustrated in the following example.

EXAMPLE 8.1. Take

$$M = \begin{pmatrix} 0 & 3 \\ 2 & -1 \end{pmatrix}, \quad q = \begin{pmatrix} -2 \\ -1 \end{pmatrix}, \quad d = (1, 2).$$

Algorithm A generates a sequence of basic solutions to (8.10); g is always a basic variable. For simplicity of exposition, we suppose that (8.10) is nondegenerate, so that the n basic variables other than g are strictly positive. In the contrary case, one must replace x and y by matrices, g by a row vector, and the constants $-q$ and -1 by the matrix $(-q, I)$ and the vector $(-1, 0, \ldots, 0)$. X and Y are then required to be lexicographically nonnegative.

Assuming nondegeneracy, a *basic* solution to (8.10) has $n+1$ basic variables; one is g, and the rest are strictly positive. A solution is *complementary* if

$$x^T y = 0,$$

and *i-almost complementary* if

$$x^T y = x_i y_i.$$

The graph G_i, for $1 \le i \le n$, has for nodes all i-almost complementary basic solutions to (8.10), two such solutions are adjacent if their midpoint is also i-almost complementary. Note that, by Lemma 8.12, the structure of each G_i is the same for any d for which (8.10) is bounded. Choosing d so that $qd + M$ has a negative row is the simplest way to guarantee this.

PROPOSITION 8.1. *For $1 \le i \le n$, a node G_i has degree of 1 or 2 according to whether the corresponding solution is complementary or not.*

PROOF. If (x, y) is complementary, exactly one of x_i, y_i is nonbasic. Introducing this into the basis gives the unique i-almost complementary basic solution adjacent to (x, y). If (x, y) is not complementary,

$$x^T y = x_i y_i > 0.$$

There is then precisely one index j such that x_j and y_j are nonbasic. Introducing these into the basis gives the only two adjacent i-almost complementary basic solutions. ■

DEFINITION 8.4. By *search* G_i from (x, y) to find (x', y') we mean trace the unique path in G_i from the complementary solution (x, y) to the different complementary solution (x', y'). This is performed by a sequence of pivots in (8.10). Thus, by (x, y) we denote the whole tableau corresponding to that basic solution.

We now state the following algorithm.

ALGORITHM A.[M.TODD]

STEP 0. Set $(x,y) = (-q,0)$, the initial solution to (8.10). Set $i = 1$.

STEP 1. Search G_i from (x,y) to find (x',y') with associated value g' for g. If $g' > 0$, go to Step 4. Otherwise, replace (x,y) by (x',y'). If $i < n$, go to Step 3.

STEP 2. If $g' = -1$, stop with an indication of failure. Otherwise, set $i = 0$.

STEP 3. Increase i by 1, and go to Step 1.

STEP 4. In the tableau corresponding to (x',y'), write $w, z, 1$, and f for x, y, g, and 1. A solution to (8.9) results, with f nonbasic and 1 "basic". Now, pivot f into and 1 out of the basis to give a solution to (8.8) with $w^T z = 0$, hence a solution to the LCP.

We now develop an algorithm for the general case when $q \not< 0$. Consider the system (8.6), where

$$e = (1, 1, \ldots, 1)^T.$$

Exactly as in Lemke's algorithm, we first introduce z_0 to replace w_i, where

$$q_i = \min_{1 \leq j \leq n} q_j.$$

We may assume without loss in generality, that $q_i < 0$, otherwise $w = q$, $z = 0$ solves the LCP; $i = 1$; and $q_1 < q_j$, $1 < j \leq n$, otherwise use $e = (\frac{1}{2}, 1, \ldots, 1)^T$.

It is from this solution that a transformation $\phi_{(d_0,d)}$ is made. We choose $d_0 > 0$, so that $1 + q_1 d_0 < 0$, and choose $d > 0$, so that $-M^1 + q_1 d < 0$, where M^j, $1 \leq j \leq n$, denotes the jth row of M. We now are in the context of the previous Remark (ii). As before, the transformation is performed by adding a row with associated free basic variable f to obtain

$$\begin{pmatrix} z_0 \\ w_2 \\ \vdots \\ w_n \end{pmatrix} = e w_1 + \begin{pmatrix} -M^1 \\ M^2 - M^1 \\ \vdots \\ M^n - M^1 \end{pmatrix} z + \begin{pmatrix} -q_1 \\ q_2 - q_1 \\ \vdots \\ q_n - q_1 \end{pmatrix} \cdot 1,$$

$$f = d_0 w_1 + dz + (+1) \cdot 1, \tag{8.11}$$

$$w \geq 0, \quad z \geq 0, \quad z_0 \geq 0, \quad f \text{ free}.$$

Pivot on the $+1$, to make 1 "basic" and f nonbasic:

$$\begin{pmatrix} z_0 \\ w_2 \\ \vdots \\ w_n \end{pmatrix} = \begin{pmatrix} 1 + q_1 d_0 \\ 1 - (q_2 - q_1) d_0 \\ \vdots \\ 1 - (q_n - q_1) d_0 \end{pmatrix} w_1 + \begin{pmatrix} -M^1 + q_1 d \\ M^2 - M^1 - (q_2 - q_1) d \\ \vdots \\ M^n - M^1 - (q_n - q_1) d \end{pmatrix} z + \begin{pmatrix} -q_1 \\ q_2 - q_1 \\ \vdots \\ q_n - q_1 \end{pmatrix} f,$$

$$1 = -d_0 w_1 - dz + (+1) f, \tag{8.12}$$

$$w \geq 0, \quad z \geq 0, \quad z_0 \geq 0, \quad f \text{ free}.$$

Now, relabel w, z, z_0, f, and 1 by $x, y, y_0, 1$, and g to obtain:

$$\begin{pmatrix} y_0 \\ x_2 \\ \vdots \\ x_n \end{pmatrix} = \begin{pmatrix} 1 + q_1 d_0 \\ 1 - (q_2 - q_1) d_0 \\ \vdots \\ 1 - (q_n - q_1) d_0 \end{pmatrix} x_1 + \begin{pmatrix} -M^1 + q_1 d \\ M^2 - M^1 - (q_2 - q_1) d \\ \vdots \\ M^n - M^1 - (q_n - q_1) d \end{pmatrix} y + \begin{pmatrix} -q_1 \\ q_2 - q_1 \\ \vdots \\ q_n - q_1 \end{pmatrix} \cdot 1,$$

$$g = -d_0 x_1 - dy + (+1) \cdot 1, \tag{8.13}$$

$$x \geq 0, \quad y \geq 0, \quad y_0 \geq 0, \quad g \text{ free.}$$

Algorithm B produces a sequence of solutions to the system (8.13). Denote by (8.11.0) and (8.11.-q) the system (8.11) with q replaced by 0 and $-q$, respectively. The proof of the following result is similar to that of Lemma 8.12. Note that f is positive in any solution to (8.11).

LEMMA 8.13. *Suppose that the linear system* (8.13) *is bounded. Then, there are one-to-one correspondences between:*

(a) *basic solutions to* (8.13) *with g positive and basic solutions to* (8.11);

(b) *solutions to* (8.13) *with g zero and solutions to* (8.11.0) *with f zero;*

(c) *basic solutions to* (8.13) *with g negative and basic solutions to* (8.11.-q).

In each case, $w_i = 0$ if and only if $x_i = 0$, and $z_i = 0$ if and only if $y_i = 0$. ■

Again, we assume that (8.13) is nondegenerate, so that in any basic solution the n basic variables other than g take strictly positive values. A solution to (8.13) is *complementary* if $y_0 = 0$ and $x^T y = 0$, 0-*almost complementary* if $x^T y = 0$, and i-*almost complementary* if $y_0 = 0$ and $x^T y = x_i y_i$. Define G_i, $0 \leq i \leq n$, in the obvious way. Again, it is easy to show the following proposition.

PROPOSITION 8.2. *For $0 \leq i \leq n$, a node of G_i has degree 1 or 2 according to whether the corresponding solution is complementary or not.* ■

Searching G_i, $1 \leq i \leq n$, has the same meaning as before: but note that our initial solution is no longer complementary, and cycling in G_0 can occur. We can now state the following algorithm.

ALGORITHM B[M.TODD]

STEP 0. Search G_0 from the initial solution

$$y_0 = -q_1, \qquad x = q - q_1 e,$$

first introducing y_1, until either of the following situations occurs: (a) a solution is obtained for which g takes the value +1; then cycling has occured; stop with an indication of failure; or (b) y_0 leaves the basis, and a complementary solution (x, y) is obtained; drop the current column for y_0; if the value of g is positive, go to Step 4; otherwise, set $i = 1$ and $(x^0, y^0) = (x, y)$.

STEP 1. Search G_i from the current solution (x, y) to find (x', y'), with associated value g' for g. If $g' > 0$, go to Step 4. Otherwise, replace (x, y) with (x', y'). If $i < n$, go to Step 3.

STEP 2. If $(x, y) = (x^0, y^0)$, stop with an indication of failure. Otherwise, set $i = 0$.

STEP 3. Increase i by 1, and go to Step 1.

STEP 4. In the tableau corresponding to (x', y'), write $w, z, 1$, and f for x, y, g, and 1. A solution to (8.12) results, with f nonbasic and 1 "basic". Now, pivot f into and 1 out of the basis to give a solution to (8.11) with $w^T z = 0$ and $z_0 = 0$, hence a solution to the LCP.

EXAMPLE 8.2. We use Algorithms A and B to solve the LCP with

$$M = \begin{pmatrix} 0 & 3 \\ 2 & -1 \end{pmatrix}, \qquad q = \begin{pmatrix} -2 \\ -1 \end{pmatrix}.$$

It can be shown that Lemke's algorithm fails on this problem. For algorithm A, take $d = (1, 2)$. We obtain the tableau

		z_1	z_2	1
w_1	=	0	3	−2
w_2	=	2	−1	−1
f	=	1	2	−1*

Pivoting in the lower right corner gives

		y_1	y_2	1
x_1	=	−2*	−1	2
x_2	=	1	−3	1
g	=	1	2	−1

where we have replaced $w, z, 1$, and f by x, y, g, and 1. We now search G_1:

		x_1	y_2	1
y_1	=	−1/2	−1/2	1
x_2	=	−1/2	−7/2*	2
g	=	−1/2	3/2	0

obtaining a complementary solution. However, the value of g is zero. We now search G_2:

		x_1	x_2	1
y_1	=	−3/7	1/7	5/7
y_2	=	−1/7	−2/7	4/7
g	=	−5/7	−3/7	6/7*

obtaining a complementary solution with g positive. Replace x, y, g, and 1 by $w, z, 1$, and f, and pivot in lower right corner:

		w_1	w_2	1
z_1	=	1/6	1/2	5/6
z_2	=	1/3	0	2/3
f	=	5/6	1/2	7/6

Hence, $z_1 = 5/6, z_2 = 2/3, w = 0$ solves the LCP.

We now apply Algorithm B. Adding an artificial variable z_0, we obtain

		z_0	z_1	z_2	1
w_1	=	1^*	0	3	−2
w_2	=	1	2	−1	−1

Pivoting z_0 in and w_1 out, we obtain the first two rows of the tableau:

		w_1	z_1	z_2	1
z_0	=	1	0	−3	2
w_2	=	1	2	−4	1
f	=	1	1	1	1^*

where last row $(d_0, d, 1)$ satisfies our conditions. Pivoting in the lower right corner gives:

		x_1	y_1	y_2	1
y_0	=	−1	-2^*	−5	2
x_2	=	0	1	−5	1
g	=	−1	−1	−1	1

where we have replaced $w, z, z_0, 1$, and f by x, y, y_0, g, and 1. We now search G_0, introducing y_1:

		x_1	y_0	y_2	1
y_1	=	$-1/2^*$	−1/2	−5/2	1
x_2	=	−1/2	−1/2	−15/2	2
g	=	−1/2	1/2	3/2	0

(8.14)

obtaining a complementary solution with g zero. We eliminate the column for y_0, record this solution, and search G_1:

		y_1	y_2	1
x_1	=	−2	−5	2
x_2	=	1	-5^*	1
g	=	1	4	−1

obtaining a complementary solution with g negative. We now search G_2:

		y_1	x_2	1
x_1	=	-3^*	1	1
y_2	=	1/5	−1/5	1/5
g	=	9/5	−4/5	−1/5

obtaining a complementary solution with g negative. We check that this is not solution (8.14), then search G_1:

		x_1	x_2	1
y_1	=	−1/3	1/3	1/3
y_2	=	−1/15	−2/15	4/15
g	=	−3/5	−1/5	$2/5^*$

obtaining a complementary solution with g positive. Replace x, y, g, and 1 by $w, z, 1$, and f, and pivot in lower right corner:

		w_1	w_2	1
z_1	=	1/6	1/2	5/6
z_2	=	1/3	0	2/3
f	=	3/2	1/2	5/2

Hence, $z_1 = 5/6$, $z_2 = 2/3$, $w = 0$ solves the LCP. ∎

Before studying the algorithms in more detail, we note that the projection transformations can be used to advantage in linear programming also. Given a nonempty feasible region

$$\{x \in R^n \mid Ax \geq b, x \geq 0\}$$

with $b > 0$, most standard techniques would require artificial variables, unless the problem happened to be dual feasible. If cx is to be minimized, then its optimal value must be nonnegative if finite; for, if x is feasible, so is λx for any $\lambda > 1$. Let us assume that the optimal value is positive; by scaling, if necessary, we can assume that $cx > 1$ for all feasible x. Then, we can proceed as follows:

$$s = Ax - b \cdot 1, \quad s \geq 0, \quad x \geq 0,$$
$$(\min)\ u = cx + (-1) \cdot 1, \quad u \quad \text{free.} \tag{8.15}$$

Now, pivot 1 into the basis and u out:

$$s = (A - bc)x + b \cdot u, \quad s \geq 0, \quad x \geq 0,$$
$$1 = cx + (-1) \cdot u, \quad u \quad \text{free.} \tag{8.16}$$

Relabeling, we obtain

$$s = (A - bc)y + b \cdot 1, \quad t \geq 0, \quad y \geq 0,$$

$$(\max)\ v = cy + (-1) \cdot 1, \quad v \ \text{ free.} \tag{8.17}$$

Under our assumptions, if (8.17) is solved, its optimal value will be positive. Relabeling t, y, v, and 1 with s, x, 1, and u, and pivoting u into and 1 out of the basis will give an optimal tableau for (8.15). Note that an initial basic feasible solution is available for (8.17).

8.3.3 PROPERTIES OF THE ALGORITHMS

Recall first that both algorithms are independent of the choice of d or (d_0, d). This follows from Lemmas 8.12 and 8.13.

Secondly, both algorithms must terminate. Since the number of pairs of basic solutions to (8.10) or (8.13) is finite, the algorithms must terminate unless an edge of some G_i is repeated in the same direction. In Algorithm A, simple combinatorial arguments confirm that the first edge repeated must lead from the initial solution to its successor in G_1. From (8.10), the initial solution is the only one with $g = -1$, since $d > 0$. Step 2 prohibits cycling in this cage. Similar arguments confirm that cycling in Algorithm B is precluded by the stopping of Steps 0(a) and 2.

The stopping rules in Step 2 of Algorithm B are very simple to apply, since the initial solutions are simple to recognize. The same is not true for Step 2 in Algorithm B. The solution (x^0, y^0) can be made more easily recognizable: we may either add another artificial row the tableau with positive coefficients for all nonbasic variables; or, if g is negative, we may go to Step 4, obtaining a nonpositive complementary solution from which Algorithm A can be applied. Todd (1973) indicates, however, that the extra work in either approach is not worthwhile.

Since both algorithms terminate, one's concern is with conditions for which they terminate successfully in Step 4. We investigate first Algorithm A.

A direct comparison of Algorithm A with Lemke's algorithm is difficult; below we compare sufficient conditions. If the first column of M is positive, however, searching G_1 in (8.10) is equivalent after the first pivot to Lemke's Scheme II (described in Lemke (1968)), introducing first z_1. The value of g becomes positive after the first pivot and stays positive unless Lemke's Scheme II fails in a secondary ray. Thus, in this case, Algorithm A is at least as powerful as Lemke's Scheme II.

To compare Algorithm A with Lemke's algorithm (more precisely, his Scheme I in Lemke (1968)), recall that a secondary ray of (8.6)–(8.7) is a ray

$$\{(w, z, z_0) + \theta(v, u, u_0) \mid \theta \geq 0\}$$

whose every element satisfies (8.6)–(8.7) with

$$(v, u, u_0) \neq 0, \quad (z, u) \neq 0, \quad (z_0, u_0) \neq 0.$$

THEOREM 8.8. *Assume that $q < 0$, and e in (8.6) is $-q$. Then,*

(a) *there is a solution to (8.10) with $x^T y = 0$ and $g = 0$ if and only if (8.6)–(8.7) has a secondary ray with $u_0 = 0$;*

(b) *there is a solution to (8.10) with $x^T y = 0$, and $g < 0$ if and only if (8.4)–(8.5), with $-q$ replacing q, has more than one solution.* ∎

One can use Theorem 8.8. to show that Algorithm A will process the LCP [i.e. find a solution if (8.4) is consistent] for the most general classes of matrices M for which it is known that Lemke's algorithm will process the LCP. Recall the definistions of the classes $\mathcal{L}$ and $\mathcal{L}(-q)$ as follows (these classes were introduced in Eaves (1971) and Garcia (1973), respectively):

$M \in \mathcal{L}$ if and only if

(a) $0 \neq y \geq 0 \quad \Rightarrow \exists j$ with $y_j > 0$ and $M^j y \geq 0$,

(b) $0 \neq y \geq 0, My \geq 0, y^T My = 0 \Rightarrow \exists$ a diagonal matrix $\Lambda \geq 0, \Omega \geq 0$

$$\text{with } \Omega \neq 0 \text{ and } (\Lambda M + M^T\Omega)y = 0; \tag{8.18}$$

and

$M \in \mathcal{L}(-q)$ if and only if

(a) $w = Mz \geq 0,\ 0 \neq z \geq 0,\ w^T z = 0 \Rightarrow \exists x$
with $y = -M^T x,\ 0 \leq x \leq z,\ 0 \leq y \leq w$, and $x \neq 0$,
(b) the LCP with $-q$ replacing q has a unique solution. (8.19)

COROLLARY 8.1. *If $M \in \mathcal{L}$ or $M \in \mathcal{L}(-q)$, Algorithm A will find a solution to (8.4)–(8.5) unless (8.4) is inconsistent.* ∎

For M in $\mathcal{L}, \mathcal{L}(-q)$ or copositive plus, etc., it is generally easier to show that Algorithm A processes the LCP than to show that Lemke's does. Theorem 8.8, however, is much stronger, allowing new results for Lemke's algorithm to apply also to Algorithm A.

The next paragraph gives another class of LCP's which Algorithm A will solve; Lemke's algorithm may fail on these problems. We will use the following result of Fiedler and Ptak (1962).

THEOREM 8.9. *If a square matrix M has nonpositive off-diagonal entries and there is a $v > 0$ with $vM > 0$, then M is invertible and M^{-1} is nonnegative.* ∎

THEOREM 8.10. *If $q < 0$ and there is a $v > 0$ such that $qvM - vqM$ has positive off-diagonal entries, and if further M is invertible with $M^{-1}q \leq 0$, then Algorithm A will solve (8.4)–(8.5).* ∎

Theorem 8.10 implies, for example, that Algorithm A will solve the LCP with

$$M = \begin{pmatrix} -2 & 4 \\ 5 & -7 \end{pmatrix}, \quad q = (-2, -1)^T,$$

for which Lemke's algorithm fails. In fact, this M satisfies the conditions of Theorem 8.10 with $v = (2, 1)$ for *any* $q < 0$. It is difficult to characterize such M; they must be positive off their diagonals, with negative diagonals, but these conditions are not sufficient. If, further, $\exists v > 0$ with $vM < 0$, then the first condition of Theorem 8.10 is trivial, but $M^{-1}q \geq 0$ for all $q < 0$; furthermore, (8.4) is infeasible.

The proof of Theorem 8.10 is similar to that of Theorem 4 of Todd (1973); in that case, $M^{-1}q \leq 0$ was necessary for feasibility of (8.4). That is not the case here; if

$$M = \begin{pmatrix} -12 & -1 & 13 & -1 \\ 13 & -11 & 13 & -1 \\ 13 & -1 & -12 & -1 \\ 13 & -1 & 13 & -11 \end{pmatrix}, \quad q = (-1, -1, -1, -1)^T,$$

then the first condition of Theorem 8.10 is satisfied with

$$v = (1, 1, 1, 1),$$

but the second not. Algorithm A fails, yet

$$w = (0, 25, 0, 25)^T, \quad z = (1, 0, 1, 0)^T$$

solves the LCP.

We turn now to Algorithm B.

LEMMA 8.14. *Algorithm B is at least as strong as Lemke's algorithm.*

PROOF. Each basic solution encountered in Lemke's algorithm until termination gives a corresponding solution to (8.13), by Lemma 8.13. Thus, Step 0 of Algorithm B exactly duplicates Lemke's algorithm until the value of g becomes nonpositive. At this stage (if it is reached), Lemke's algorithm will have terminated in a secondary ray, while Algorithm B still offers a possibility of success. ■

Now conditions for the success of Algorithm B are given.

THEOREM 8.11. *Assume that* $0 > q_1 < q_j$ *for* $2 \leq j \leq n$, *and* $\forall y$ *with* $0 \neq y \geq 0$ *and* $y_1 = 0$, $\exists j$ *with* $y_j > 0$ *and*

$$q_j M^1 y \geq q_1 M^j y.$$

Then, Algorithm B does not terminate in Step 0(a). The conclusion also follows if $q < 0$. ■

The dependence of the condition of Theorem 8.11 on q is critical. Indeed, it can be shown that, if M satisfies the condition for all q, M is nonnegative (and even some nonnegative matrices fail the condition for some q); even if M only satisfies the condition for all $q \leq 0$, M satisfies condition (a) of (8.18) (and some matrices in L fail the condition for some q).

To guarantee success of Algorithm B, we must also eliminate termination in Step 2.

THEOREM 8.12. *If $q \neq 0$ and $\exists v \geq 0$ with $vM < 0$, then any complementary solution to* (8.13) *has $g > 0$ if* (8.4) *is feasible.* ∎

The combination of Theorems 8.11 and 8.12 shows for example that Algorith B will solve the LCP with

$$M = \begin{pmatrix} -3 & 1 & 1 \\ 1 & -3 & 1 \\ 1 & 1 & -3 \end{pmatrix}, \quad q = (-1, 4, 4)^T,$$

for which Lemke's algorithm fails.

8.4 BLOCK-PIVOT METHODS

These algorithms presented by Kostreva (1978) differ from most methods previously applied to CPs because they utilize the concept of a block-pivot or multiple exchange of basic and nonbasic variables. An example of a method which permits the exchange of two or more variables simultaneously was that of Chandrasekaran (1970). Chandrasekaran's method was applied to affine functions $f(x) = Mx + q$ in which the coefficient matrix was a Z-matrix (i.e. a square matrix with nonpositive off-diagonal elements). Tamir (1974) extended the method to certain nonlinear functions known as Z-functions and M-functions by incorporating nonlinear Jacobi and Gauss-Seidel iterative processes. The convergence proof in Chandrasekaran (1970), Tamir (1974) depends on a monotonic increase in the number of positive basic variables. Algorithms by Kostreva (1978) do not require f to be a Z-function or an M-function and do not necessarily increase in monotonic fashion the number of positive basic variables. Only complementary points are considered, while the decision as to how to proceed is based entirely on the sign pattern of the n current basic variables.

8.4.1 PIGEONHOLING

For a concise presentation we consider the following. Given $f : R^n \to R^n$, a continuous function, the *complementarity problem associated with* f is:

(CPf) Find $x \in R^n$ such that for each index $i \in N = \{1, 2, \ldots, n\}$

$$x_i \geq 0, \tag{8.20}$$

$$f_i(x) \geq 0, \tag{8.21}$$

$$x_i \cdot f_i(x) = 0. \tag{8.22}$$

Equation (8.22) in (CPf) is known as the *complementarity condition*. Since any solution to (CPf) satisfies (8.20) and (8.21), equation (8.22) is here equivalent to $x^T f(x) = 0$.

DEFINITION 8.5. Suppose $x \in R^n$ satisfies the complementarity condition (8.22) of (CPf). Define the *complementary point associated with* x by $s = x + f(x)$.

By (8.22) of (CPf), there exists an index set $I_x \subseteq N$ such that

$$f_i(x) = 0 \quad \text{for} \quad i \in I_x, \qquad x_i = 0 \quad \text{for} \quad i \in N \setminus I_x.$$

This yields the following about s, the complementary point associated with x:

$$s_i = \begin{cases} x_i & \text{for } i \in I_x, \\ f_i(x) & \text{for } i \in N \setminus I_x \end{cases}$$

We say that the *index set* I_x *leads to the complementary point* s.

The vector s is well described as a *basic vector* (in linear programming terminology), since it contains all the *dependent* or *basic* variables. Those variables not in s, the *nonbasic* variables, are understood to be equal to zero. In order to recover x and $f(x)$ from s, one needs to know I_x.

DEFINITION 8.6. If $y = f(x)$, (CPf) is *nondegenerate* if for each $x \in R^n$, at most n of the $2n$ variables (y, x) simultaneously vanish.

LEMMA 8.15. *If* (CPf) *is nondegenerate, and* $S = \{s \mid s$ *is a complementary point for* (CPf)$\}$, *then* S *does not intersect orthant boundaries in* R^n.

PROOF. Follows from Definitions 8.5 and 8.6. ■

DEFINITION 8.7. Let S denote the set of all complementary points for a nondegenerate (CPf). The set S is said to be *pigeonholed* if each orthant in R^n contains one and only one complementary point.

The above phrase takes its derivation from the well-known pigeonhole principle:

LEMMA 8.16. *If* m *objects are distributed over* m *places in such a way that no place receives more than one object, then each place receives exactly one object.* ■

The class of functions known as P-functions was introduced by Moré and Rheinboldt (1973). Use of the definition of P-functions can be made in connection with Lemma 8.16 to produce an existence–uniqueness result for (CPf). But first recall the definition of a P-function.

DEFINITION 8.8. If $f : R^n \to R^n$ is a continuous function, then f is a *P-function on a set* S if for all $x, y \in S$ with $x \neq y$, there exists an index $i = i(x, y)$ such that

$$(x_i - y_i)\,[f_i(x) - f_i(y)] > 0.$$

The following theorem contains the uniqueness result necessary for our study of the complementary points for (CPf).

THEOREM 8.13. *Let $f : R^n \to R^n$ be a P-function on R^n. Then*

(i) *Each orthant $\mathcal{O} \subseteq R^n$ contains at most one complementary point.*

(ii) *For each complementary point s, there is a unique x such that s is the complementary point associated with x.*

(iii) *If (CPf) is nondegenerate, at most one index set leads to a given complementary point.*

PROOF. In (i) the assumption of two such points contradicts Definition 8.8. Part (ii) is treated similarly. To show (iii), use Lemma 8.15 and Definition 8.5. ■

Using Theorem 8.13, one may deduce that several classes of functions which have appeared in the literature have pigeonholed complementary points. The identification of such classes is summarized in the following theorem.

THEOREM 8.14. *Let $f : R^n \to R^n$ be a P-function on R^n, and let (CPf) be nondegenerate. Then S, the set of complementary points for (CPf), is pigeonholed if f belongs to any one of the following subclasses of the class of P-functions:*

(i) *Affine functions $f(x) = Mx + q$, where the square matrix M has all principal minors positive.*

(ii) *The set of functions which are continuously differentiable and have a positively bounded Jacobian matrix [i.e., there exists a number δ, with $0 < \delta < 1$, such that for every $x \in R^n$ every principal minor of the Jacobian matrix $J_f(x)$ lies between δ and δ^{-1}].*

(iii) *Uniform P-functions on R^n [i.e., there exists a constant $c > 0$ such that for all $x, y \in R^n$ there is an index $i = i(x, y)$ such that*

$$(x_i - y_i)\,[f_i(x) - f_i(y)] \geq c\|x - y\|^2].$$

(iv) *Nondegenerate P-functions [i.e., functions such that for each subset $I^{(k)}$ of N, the function $f^{(k)} : R^n \to R^n$ defined by*

$$f_i^{(k)}(x) = \begin{cases} f_i(x) & \text{for} \quad i \in I^{(k)}, \\ x_i & \text{for} \quad i \in N \setminus I^{(k)} \end{cases}$$

is a homeomorphism from R^n onto R^n].

PROOF. Theorem 8.13 asserts that there is at most one complementary point s in each orthant. Once the existence of 2^n complementary points is demonstrated, the proof follows from Lemma 8.16.

For cases (i) and (iv) the complementary points are given by

$$s^{(k)} = \left[f^{(k)}\right]^{-1}(0) + f\left[\left[f^{(k)}\right]^{-1}(0)\right],$$

where superscript $k = 1, 2, \ldots, 2^n$ numbers the subsets of N.

Functions with property (ii) retain the property after principal pivots. Thus one may perform a principal pivot coresponding to any subset of N. In the non-degenerate (CPf), setting the nonbasic variables to zero yields a unique complementary point. Hence, in (ii) there are exactly 2^n complementary points.

The functions of (iii) have exactly 2^n complementary points. This is obtained most easily by applying Corollary 3.5 of J. Moré (1974) to each orthant of R^n. ■

8.4.2 PERMUTATIONS

The algorithms we discuss are designed to make use of the geometric properties of certain P-functions, namely the pigeonholing of complementarity points.

DEFINITION 8.9. A *direct algorithm* for (CPf) is a method for recursively determining a sequence $\{I^{(k)}\}_{k=1}^{m}$ of index sets together with some way of determining at each step whether or not $I^{(k)}$ leads to the desired complementary point. Such algorithms are *successful* if the desired index set is found and no index set is considered more than once. A *block-pivot direct algorithm* permits two successive index sets $I^{(t)}$ and $I^{(t+1)}$ to differ by more than one element. Any direct algorithm which is not a block-pivot is called a *single-pivot direct algorithm.*

A few preliminaries from group theory lead to a family of block-pivot direct algorithms.

DEFINITION 8.10. Let $J = \{1, 2, \ldots, 2^n\}$. Denote by T the set of all permutations on J. The set T forms a group under the operation of composition of functions, $\circ$. Then $(T, \circ)$ is called the symmetric group of degree 2^n. If $j \in J$ and $\theta \in T$, then the *orbit of the number j under θ* consists of the numbers $j, \theta(j), \theta^2(j), \ldots, \theta^{\ell-1}(j)$, where ℓ is the smallest positive integer such that $\theta^{\ell}(j) = j$. The integer ℓ is the *length of the orbit.* By a *cycle of θ* we mean the ordered set $\left(j, \theta(j), \theta^2(j), \ldots, \theta^{\ell-1}(j)\right)$.

One of the main facts about permutations is expressed in the following result (Herstein (1964)).

LEMMA 8.17. *Every permutation can be uniquely expressed as a product of disjoint cycles.* ■

THEOREM 8.15. *Let $f(x)$ be a P-function on R^n. Let* (CPf) *be nondegenerate and have its set of complementary points S pigeonholed. Then for each $\theta \in T$, θ yields a successful direct algorithm (of the block-pivot type) for* (CPf). ■

These algorithms can be implemented using another complementary point (and hence another index set) as a starting point. Since the starting point $s = 0 + f(0)$ requires no system to be solved, it is generally used. If another starting point $\hat{s}$ is desired, the corresponding index set $\hat{I}$ and the orthant R^n_+ should be given the same number α.

The algorithm easily adapts itself to solve the generalized complementarity problem studied in Habetler and Price (1971) [i.e. find $x \in K$, a solid cone, such that $f(x) \in K^*$, the polar of K and $x^T f(x) = 0$] for the special case of $K = \mathcal{O} = K^*$, an orthant of R^n. To see this simply give $\emptyset$, the empty set, and $\mathcal{O}$, the desired orthant, the same number α.

A case of special interest is to let θ be the identity map. One way to implement this is given in the following rule:

$$I^{(k+1)} = \{i | \ s_i^{(k)} < 0\}. \tag{8.23}$$

Notice that if $s^{(t)} > 0$, then $I^{(t+1)} = \emptyset$, so that R^n_+ and $\emptyset$ do have the same number α.

The direct algorithm (8.23) serves as a heuristic choice of permutation. To see why, consider the following example: let $f(x) = (f_1(x_1), f_2(x_2), \dots, f_n(x_n))^T$ be a diagonal mapping with positively bounded (diagonal) Jacobian matrix. Then (8.23) converges in one step, as the sign pattern of $f(0)$ automatically identifies the index set I^*. One may argue that the more "diagonally dominant" $f(x)$ is, the more likely (8.23) is to converge on the first step.

8.4.3 DISCUSSION

Perhaps the most obvious point is that in order to use a direct algorithm one must deal with the system of equations corresponding to the index set $I^{(k)}$ at the k-th step. Many methods are available, and the selection of the best method to handle the equations will depend very much on the application. One should refer to Ortega and Rheinboldt (1970) for the most comprehensive collection of traditional techniques and to Eaves (1976) for material on newer methods. The solution of complementary problems is made simple by the direct-algorithm approach, since readily available and robust software already exists for solving systems of equations. A program with as few as 50–100 user generated statements can be used to solve a complementarity problem.

One should note that highly accurate solutions to the intermediate systems are not required in order to have these algorithms operate successfully. All that is needed is that the approximating vector $\hat{s}^{(k)}$ have the same pattern of signs as the kth-step complenetary point $s^{(k)}$. This allows for a relaxed stopping rule for these intermediate systems.

The structure of the intermediate systems is also important. A reduction in the dimensionality generally aids all methods used for solving systems of equations. This may be accomplished on the kth-step system

$$0 = \begin{cases} f_i(x), & i \in I^{(k)}, \\ x_i, & i \in N \setminus I^{(k)}, \end{cases}$$

by directly substituting the value 0 into the f_i's. Then $f_i(x)$ depends only on variables which are indexed by the current index set $I^{(k)}$.

One somewhat disappointing feature of the direct algorithms of this section is that they break down in the case of degenerate (CPf). In the degenerate case, pigeonholing does not occur and the permutation θ_j is not defined at certain elements of J. One may rationalize that such occurences are very rare, and if they do arise they must be dealt with by a perturbation techniques.

It is not difficult to see that certain permutations θ when composed with certain θ_f's will lead to very long (and hence very expensive) cycles in the permutation $g = \theta \circ \theta_f$. Until further study is made of the question of how to choose a θ, Kostreva (1978) recommends limiting the number m of Definition 8.9 to be less than $4n$. Practical experience to date indicates that if I^* is not found in $4n$ steps, it will take many more steps than this. In such a case, either another permutation θ should be generated or perhaps another direct algorithm from Habetler and Kostreva (1978) or Kostreva (1976) should be tried on the problem.

8.5 SUMMARY: CONDITIONS UNDER WHICH THE COMPLEMENTARITY PIVOT ALGORITHM WORKS

This section follows mainly the very informative book by Murty (1988). We recall here several classes of matrices that are useful in the study of the LCP.

Let $M = (m_{ij})$ be a square matrix of order n. It is said to be a

COPOSITIVE MATRIX if $y^T My \geq 0$ for all $y \geq 0$.

STRICTLY COPOSITIVE MATRIX if $y^T My > 0$ for all $y \geq 0$, $y \neq 0$.

COPOSITIVE PLUS MATRIX if it is a copositive matrix and whenever $y \geq 0$, and satisfies $y^T My = 0$, we have $y^T(M + M^T)y = 0$.

P-**MATRIX** if all its principal subdeterminants are positive.

Q-**MATRIX** if the LCP(q, M) has a solution for every $q \in R^n$.

NEGATIVE DEFINITE MATRIX if $y^T My < 0$ for all $y \neq 0$.

NEGATIVE SEMIDEFINITE MATRIX if $y^T My \leq 0$ for all $y \in R^n$.

Z-**MATRIX** if $m_{ij} \leq 0$ for all $i \neq j$.

PRINCIPALLY NONDEGENERATE MATRIX if all its principal determinants are non-zero.

PRINCIPALLY DEGENERATE MATRIX if at least one of its principal subdeterminants is zero.

$\mathcal{L}_1$-**MATRIX** if for every $y \geq 0$, $y \in R^n$, there is an i such that $y_i > 0$ and $M_{i.}y \geq 0$. If M is an L_1-matrix, an i like it is called a *defining index* for M and y. These matrices are also called *semimonotone*.

$\mathcal{L}_2$-**MATRIX** if for every $y \geq 0$, $y \in R^n$, such that $My \geq 0$ and $y^T My = 0$, there are diagonal matrices, $\Lambda \geq 0$, $\Omega \geq 0$ such that $\Omega y \neq 0$ and $(\Lambda M + M^T \Omega)y = 0$. An equivalent definition is that for each $z \geq 0$, satisfying $w = Mz \geq 0$ and $w^T z = 0$; there exists a $\hat{z} \geq 0$ satisfying $\hat{w} = -(\hat{z}^T M)^T$, $w \geq \hat{w} \geq 0$, $z \geq \hat{z} \geq 0$.

$\mathcal{L}$-**MATRIX** if it is both an $\mathcal{L}_1$-matrix and $\mathcal{L}_2$-matrix.

$\mathcal{L}_*$-**MATRIX** if for every $y \geq 0$, $y \in R^n$, there is an i such that $y_i > 0$ and $M_{i.}y > 0$. If M is an $\mathcal{L}_*$-matrix, an i like it is called a *defining index* for M and y.

P_0-**MATRIX** if all its principal subdeterminants are ≥ 0.

ROW ADEQUATE MATRIX if it is a P_0-matrix and whenever the principal subdeterminant corresponding to some subset $J \subset \{1, \dots, n\}$ is zero, then the set of row vectors of M corresponding to J, $\{M_{i.} : i \in J\}$ is linearly dependent.

COLUMN ADEQUATE MATRIX if it is a P_0-matrix and whenever the principal subdeterminant corresponding to some subset $J \subset \{1, \dots, n\}$ is zero, then the set of column vectors of M corresponding to J, $\{M_{.j} : j \in J\}$ is linearly dependent.

ADEQUATE MATRIX if it is both row and column adequate.

Examples of principally degenerate matrices are $\begin{pmatrix} 0 & 4 \\ 0 & -10 \end{pmatrix}$, $\begin{pmatrix} 1 & 1 \\ 1 & 1 \end{pmatrix}$. Examples of principally nondegenerate matrices are $\begin{pmatrix} -1 & 0 \\ 0 & -1 \end{pmatrix}$, $\begin{pmatrix} 3 & 1 \\ 0 & -2 \end{pmatrix}$. The notation C_0-*matrix* is used to denote copositive matrices, and the notation C_+-*matrix* is used to denote copositive plus matrices.

It is well-known that every positive semidefinite matrix is also a copositive plus matrix. Also, the square matrix M is negative definite or negative semidefinite, if and only if $-M$ is positive definite or positive semidefinite, respectively.

8.5.1 RESULTS ON LCPs ASSOCIATED WITH COPOSITIVE PLUS MATRICES

THEOREM 8.16. *If M is a copositive plus matrix and the system of constraints* (8.4) *has a feasible solution, then the LCP* (8.4)–(8.5) *has a solution and the complementary pivot algorithm will terminate with the complementary feasible basis. Conversly, when M is a copositive plus matrix, if the comple-*

mentary pivot algorithm applied on (8.4)–(8.5) *terminates in ray termination, the system of constraints* (8.4) *must be infeasible.* ■

COROLLARY 8.2. *If M is strict copositive the complementary pivot algorithm applied on* (8.4)–(8.5) *terminates with a solution of the LCP.* ■

Thus all strict copositive matrices are Q-matrices. Also, if $M = (m_{ij}) \geq 0$ and $m_{ii} > 0$ for all i, M is strict copositive and hence a Q-matrix.

Later on we state that if M is a P-matrix, the complementarity pivot algorithm terminates with a complementary feasible solution when applied on the LCP(q, M). When the complementary pivot algorithm is applied on a LCP in which the matrix M is not a copositive plus or a P-matrix, it is still possible that the algorithm terminates with a complementary feasible basis for the problem. However, in this general case it is also possible that the algorithm stops with ray termination even if a solution to the LCP exists.

PROCESSING A LCP(q, M)

An algorithm for solving LCPs is said to *process* a particular LCP(q, M) for given q and M, if the algorithm is guaranteed to either determine that the LCP(q, M) has no solution, or find a solution for it, after a finite amount of computational effort.

Suppose M is a copositive plus matrix, and consider the LCP(q, M), for given q. When the complementary pivot algorithm is applied on this LCP(q, M), either it finds a solution; or it ends up in ray termination which implies that this LCP has no solution by Theorem 8.16. Hence, the complementary pivot alsorithm processes the LCP(q, M) whenever M is a copositive plus matrix.

8.5.2 RESULTS ON LCPs ASSOCIATES WITH $\mathcal{L}$- AND $\mathcal{L}_*$-MATRICES

Here we state that the complementary pivot algorithm will process the LCP(q, M) whenever M is an $\mathcal{L}$- or $\mathcal{L}_*$-matrix. The results in this section are from Eaves (1971), they extend the results presented in the previous subsection considerably.

LEMMA 8.18. *If M is an $\mathcal{L}_1$-matrix , the LCP(q, M) has a unique solution for all $q > 0$, and conversely.* ■

LEMMA 8.19. *If M is an $\mathcal{L}_*$-matrix, the LCP(q, M) has a unique solution for all $q \geq 0$, and conversely.* ■

THEOREM 8.17. *The complementary pivot algorithm processes the LCP(q, M) if M is an $\mathcal{L}$-matrix.* ■

THEOREM 8.18. *If M is an $\mathcal{L}_*$-matrix, when the complementary pivot algorithm is applied on the LCP(q, M), it terminates with a complementary feasible solution.* ■

Theorems 8.17 and 8.18 make it possible for us to conclude that the complementary pivot algorithm processes the LCP(q, M) for a much larger class of matrices M than the copositive plus class proved in Theorem 8.16. We will now state several results establishing that a variety of matrices are in fact $\mathcal{L}$- or $\mathcal{L}_*$-matrices. By virtue of Theorems 8.17 and 8.18, this establishes that the complementary pivot method processes the LCP(q, M) whenever M is a matrix of one of these types.

All copositive plus matrices are $\mathcal{L}$-matrices. This follows because when M is copositive plus, $y \geq 0$ implies $y^T My \geq 0$, and if y is such that $y \geq 0, y^T My = 0$ then $(M + M^T)y = 0$, hence M satisfies the definition of being an $\mathcal{L}$-matrix by taking the diagonal matrices Λ and Ω to be both I (the unit matrix). A strictly copositive matrix is clearly an $\mathcal{L}_*$-matrix. From the definitions, it can be verified that PMP^T (obtained by principal rearrangement of M), $\Lambda M \Omega$ (obtained by positive row and column scaling of M) are $\mathcal{L}$-matrices if M is, whenever P is a permutation matrix and Λ, Ω are diagonal matrices with positive diagonal elements. Copositive plus matrices M satisfy the property that PMP^T is also copositive plus whenever P is a permutation matrix, but if M is copositive plus, $\Lambda M \Omega$ may not be copositive when Λ, Ω are diagonal matrices with positive diagonal entries. Also if M, N are L-matrices, so is $\begin{pmatrix} M & 0 \\ 0 & N \end{pmatrix}$.

8.5.3 A VARIANT OF THE COMPLEMENTARITY PIVOT ALGORITHM

In the versions of complementarity pivot algorithm discussed so far, we have chosen the original column vector associated with the artificial variable z_0 to be $-e_n$. Given a column vector $d \in R^n$ satifying $d > 0$, clearly we can choose the original column vector associated with z_0 to be $-d$ instead of $-e_n$ in the complementary pivot algorithm. If this is done, the original tableau turns out to be:

w	z	z_0	
I	$-M$	$-d$	q

$$w \geq 0, \quad z \geq 0, \quad z_0 \geq 0. \tag{8.24}$$

If $q \geq 0$, $(w = q, z = 0)$ is a solution of the LCP(q, M) and we are done. So assume $q \not\geq 0$. Determine t to satisfy $(q_t/d_t) = \min\{(q_i/d_i) \mid i = 1, \ldots, n\}$. Ties for t can be broken arbitrarily. It can be verified that if a pivot step is performed in (8.24), with the column vector of z_0 as the pivot column, and the tth row as the pivot row; the right hand side constants vector becomes nonnegative after the pivot step. So $(w_1, \ldots, w_{t-1}, z_0, w_{t+1}, \ldots, w_n)$ is a feasible basic vector for (8.24). It is an almost complementarity feasible vector as defined earlier. Choose z_t as the entering variable into this initial almost complementary feasible basic vector $(w_1, \ldots, w_{t-1}, z_0, w_{t+1}, \ldots, w_n)$, and continue by choosing entering variables using the complementary pivot rule as before.

We will now illustrate this variant of the complementary pivot algorithm using a numerical example by Kostreva (1976).

EXAMPLE 8.3. Consider the LCP(q, M), where

$$M = \begin{pmatrix} -1.5 & 2 \\ -4 & 4 \end{pmatrix}, \qquad q = \begin{pmatrix} -5 \\ 17 \end{pmatrix}.$$

Let $d = (5, 16)^T$. We will apply the complementarity pivot algorithm on this LCP, using $-d$ as the original column of the artificial variable z_0.

Basic variables	w_1	w_2	z_1	z_2	z_0	q	
w_1	1	0	1.5	−2	−5*	−5	$t = 1$
w_2	0	1	4	−4	−16	17	
z_0	−1/5	0	−3/10	2/5	1	1	
w_2	−16/5	1	−8/10	12/5	0	33	

The entering variable is z_1. The updated column vector of z_1 in the canonical tableau with respect to the basic vector (z_0, w_2) is nonpositive. So the algorithm ends up in ray termination. ■

EXAMPLE 8.4. Consider the LCP(q, M) discussed in Example 8.3. Let $d = e_2 = (1, 1)^T$. We will apply the complementarity pivot algorithm on this LCP with $-e_2$ as the original column of the artificial variable z_0.

Basic variables	w_1	w_2	z_1	z_2	z_0	q	
w_1	1	0	1.5	−2	−1*	−5	$t = 1$
w_2	0	1	4	−4	−1	17	
z_0	−1	0	−3/2	2	1	5	
w_2	−1	1	5/2*	−2	0	22	
z_0	−8/5	3/5	0	4/5*	1	91/5	
z_1	−2/5	2/5	1	−4/5	0	44/5	
z_2	−2	3/4	0	1	5/4	91/4	
z_1	−2	1	1	0	1	27	

Now we have terminated with a complementary feasible basic vector, and the corresponding solution of the LCP is $w = 0$, $z = (z_1, z_2) = (27, \frac{91}{4})$. ■

These examples taken from Kostreva (1976) illustrate the fact that, given general LCP(q, M), the complementarity pivot algorithm applied on it with a given positive vector d may end up in ray termination; and yet when it is run with a different positive vector d it may terminate with a solution of the LCP. The question of how to find a good vector d seems to be a hard problem, for which no answer is known. There are LCPs which are known to have solutions, and yet when the complementary pivot algorithm is applied on them with any positive vector d, it always ends up in ray termination.

If M is a copositive plus matrix, and if the complementary pivot algorithm with any positive vector d ends up in ray termination when applied on the LCP(q, M), then it can be proved that the LCP(q, M) has no solution (in fact it can be proved that "$w - Mz = q$" does not even have a nonnegative solution). Thus any LCP(q, M) where M is a copositive plus matrix, will be processed by the complementary pivot algorithm with any positive vector d.

REFERENCES

Chandrasekaran R. A special case of the complementarity pivot problem. Opsearch 1970; 7: 263-268.

Cottle RW.

1. Nonlinear programs with positively bounded Jacobians. SIAM J. Appl.Math. 1966; 14:147-158.

2. On a problem in linear inequalities. J. London Math. Soc. 1968; 43:378-384.

Cottle RW, and Dantzig GB. Complementarity pivot theory of mathematical programming. Linear Algebra and Its Applications. 1968; 1:103-125.

Eaves, B.C.

1. "The linear complementarity problem in mathematical programming," Research Report 69-4, Operations Research House, Stanford University, Stanford, California, 1969.

2. The linear complementarity problem. Management Science. 1971; 17:612-634.

3. On quadratic programming. Management Science. 1971; 17: 698-711.

4. 'A short course in solving equations with *PL* homotopies.'- In: *Nonlinear Programming*, Vol. IX, SIAM-AMS Proceedings, R.W. Cottle and C.E. Lemke, eds. American Mathematical Society, Providence, 1976.

Fiedler M, and Ptak V. On matrices with non-positive off-diagonal elements and positive principal minors. Czechoslovak Mathematical Journal. 1962; 12:382-400.

Garcia CB. Some classes of matrices in linear complementarity theory. Math. Programming. 1973; 5:299-310.

Habetler CJ, and Kostreva MM. On a direct algorithm for nonlinear complementarity problems. SIAM J. Control Optim. 1978; 16: 504-511.

Habetler CJ, and Price AL. Existence theory for generalized nonlinear conplementarity problems. J. Optim. Theory Appl. 1971; 8:161-168.

Herstein, I.N., *Topics in Algebra*. New York: Blaisdell, 1964.

Kostreva, M.M.

1. "Direct algorithms for complementarity problems," Ph.D. thesis, Rensselaer Polytechnic Institute, Troy, N.Y., 1976.

2. Block pivot methods for solving the complementarity problem. Linear Algebra and its Applications. 1978; 21: 207-215.

Moré J. Coercivity conditions in nonlinear complementarity problems. SIAM Review. 1974; 16:1-15.

Moré J, and Rheinboldt W. On P- and S-functions and related classes of n-dimensional nonlinear mappings. Linear Algebra and Applications. 1973; 6:45-68.

Ingleton AW. A problem in linear inequalities. Proc. London Math. Soc., Third Series. 1966; 16: 519-536.

Lemke CE.

1. Bimatrix equilibrium points and mathematical programming. Management Science. 1965; 7:681-689.

2. "On complementarity pivot theory," RPI Math Report No. 76, Rensselaer Polytechnic Institute, Troy, New York, 1967.

3. 'On Complementarity Pivot Theory'.- In: *Mathematics of the Decision Sciences*, G.B. Dantzig and A.F. Veinott, Jr., eds., American Mathematical Society, Providence, Rhode Island, 1968.

Murty, K.G.

1. "On the number of solutions to the complementarity quadratic programming problem," Doctoral Dissertation, Engineering Science, University of California, Berkeley, 1968.

2. *Linear Complementarity, Linear and Nonlinear Programming*. Berlin: Heldermann Verlag, 1988.

Nash J. Non-cooperative games. Annals of Mathematics. 1951; 54:286-295.

Ortega, J.M, and Rheinboldt, W.C. *Iterative Solution of Nonlinear Equations in Several Variables*. New York: Academic Press, 1970.

Tamir A. Minimality and complementarity properties associated with Z-functions and M-functions. Math. Programming. 1974; 7:17-31.

Todd, M.J.

1. "Dual families of linear programs," Technical Report No. 197, Department of Operations Research, Cornell University, 1973.

2. Extensions of Lemke's algorithm for the linear complementarity problem. J. Optim. Theory Appl. 1976; 20:397-416.

Chapter 9

SCARF TYPE ALGORITHMS

Citing Scarf (1973), one of the major triumphs of mathematical economics has been the proof of the existence of a solution for the neoclassic model of economic equilibrium. When cast in a mathematical form the general equilibrium model becomes a system of simultaneous equations and inequalities so complex that the existence of a solution can be guaranteed only by an appeal to fixed point theorems rather than by more elementary and constructively oriented techniques.

For certain models, it is not only important to know that there exists at least one solution, but one would like to show the existence of a continuum of solutions. A particular example comes from general equilibrium theory with price rigidities, where a continuum of solutions on the unit cube as a polytope is shown to exist in Herings (1998). It is therefore important to have generally applicable tools that guarantee the existence of a continuum of zero points to a certain system of equaitions.

One of the powerful tools to prove the existing results has been recently elaborated in a series of works, which consists in designing a simplicial variable dimension algorithm on a polytope. This type of algorithm was initiated by Scarf (1967), simplicial homotopy methods were developed by Eaves (1972). The simplicial restart variable dimension algorithm was introduced by van der Laan and Talman (1979) to compute a fixed point of a continuous function from the unit simplex into itself. Such an algorithm generates a unique sequence of simplices of varying dimension in a simplicial subdivision of the set and connects the arbitrarily chosen starting point with an approximate solution. For other recent developments, we refer to Talman and Yamamoto (1989), Yamamoto (1993), Brown e.a. (1996), DeMarzo and Eaves (1996), Yang (1996, 1999), and van der Laan e.a. (1998). Allgower and George (1990), Todd (1976), and Yang (1999) provide comprehensive treatments of simplicial algorithms.

This chapter follows the paper by Herings, Talman and Yang (1999) which constructs a new simplicial algorithm operating on a triangulation of the polytope and generating a piecewise linear path of points connecting two distinct faces of the polytope. Each point on the path is an approximate zero point. As the mesh size of the triangulation goes to zero, the path converges to a connected (continuous) set of zero points linking the two distinct faces. As a consequence, these results generalize Browder's fixed point theorem (1960) and the earlier results by Herings e.a. (1996) on the n-dimensional unit cube.

Section 9.2 states the problem, and in Section 9.3 three sufficient conditions are cited for the existence of a connected set of zero points of an upper semicontinuous multi-valued mapping over an arbitrary polytope which link together two distinct faces of the polytope. In Section 9.4, the algorithm is described and its convergence proof is presented. Section 9.5 analyzes the accuracy of the approximation of zero points and proves the existence theorems. In Section 9.6, Browder's and MasColell's theorems are derived as special cases, and an economic application is given. Section 9.7 provides some concluding comments.

9.1 INTRODUCTION

Consider the following problem: Given a multi-valued mapping $\varphi : P \rightarrow R^n$, with P and arbitrary polytope, what reasonable conditons can guarantee the existence of a continuum of solutions x to the system

$$0 \in \varphi(x).$$

The Herings, Talman and Yang (1999) approach is to show that there is a connected subset of solutions that links together at least two distinct points, thereby guaranteeing the continuum. The cited paper shows that any upper semicontinuous multi-valued mapping with some mild (boundary) conditions will have a connected set of zero points linking together two distinct faces of the polytope P.

It is well known that under certain conditions a multi-valued mapping defined on a polytope has a solution to the variable inequality problem. Herings, Talman and Yang (1999) generalize the variational inequality (VI) problem and define a parametric VI problem and show that under similar conditions a multi-valued mapping defined on a polytope P has a connected set of solutions to the parametric VI problem, called parametrized stationary points. The set of parametrized stationary points connects two distinct faces of P. With respect to some given nonzero vector c, on one of these faces, denoted by F^-, the value of $c^T x$ is minimized for $x \in P$, while on the other face, denoted by F^+, the value of $c^T x$ is maximized for $x \in P$. A special case occurs when both F^- and F^+ are vertices of P and the set of parametrized stationary points contains both these vertices. Under the three different conditions the set of parametrized stationary points is a connected set of zero points linking the two distinct faces F^- and F^+ of P.

In the paper by Herings, Talman and Yang (1999) a simplicial algorithm is proposed that generates within a simplicial subdivision of P a finite sequence of simplices of varying dimension. This sequence connects two different simplices, one simplex lying in the face F^- of P and the other lying in the face F^+ of P. The sequence of simplices connecting these two simplices is generated by the algorithm by making a sequence of semi-lexicographic pivot steps in a linear system of equations. In case the face F^- is not a vertex of P, the algorithm starts by finding a suitable simplex in F^-. Next it generates a sequence of simplices of varying dimension in P. It is possible that the algorithm returns to F^-. Then it generates a third simplex in F^0 from where a sequence of adjacent simplices in P is generated. It is shown that the algorithm eventually finds a simplex in F^- from which a sequence of adjacent simplices reaching F^+ is generated.

Induced by the sequence of adjacent simplices, the algorithm yields a piecewise linear path of parametrized stationary points of a piecewise linear approximation of the underlying mapping. When the mesh size of the simplicial subdivision of P goes to zero the sequence (or at least subsequence) of piecewise linear paths converges to a connected set of parametrized stationary points of the original mapping. This set has points in common with both faces F^- and F^+.

9.2 PROBLEM SPECIFICATION

Denote by I_m the set of positive integers $\{1, \ldots, m\}$ and consider an arbitrary full-dimensional polytope P that has the following representation as a (compact) polyhedron:

$$P = \left\{x \in R^n \mid \left(a^i\right)^T x \leq b_i, \ \forall i \in I_m\right\}. \tag{9.1}$$

For each subset I of I_m, define

$$F(I) = \left\{x \in P \mid \left(a^i\right)^T x = b_i, \ \forall i \in I\right\}. \tag{9.2}$$

Then $F(I)$ is called a *face* of P unless it is empty. Note that $F(\emptyset) = P$. Let

$$\mathcal{I} = \{I \subset I_m \mid F(I) \text{ is a face of } P\}. \tag{9.3}$$

A face F of the set $F(I)$ of dimension one lower than the dimension of $F(I)$ is called a *facet* of $F(I)$. The polytope P is said to be *simple* if the dimension of any face $F(I)$ of P is equal to $n - |I|$. Throughout the chapter, whenever we use a polytope P, it is assumed that P is simple and that its representation as a polyhedron has no redundant constraints.

The following simple assertions are valid.

(i) For each face F of P, the set $I \in \mathcal{I}$ with $F = F(I)$ is unique and given by the set $\left\{i \in I_m \mid \left(a^i\right)^T x = b_i, \ \forall x \in F\right\}$.

(ii) The set $F(I)$ is a vertex of P if and only if $I \in \mathcal{I}$ with $|I| = n$.

(iii) If $I \in \mathcal{I}$, then $I \setminus \{i\} \in \mathcal{I}$ for any $i \in I$.

(iv) For some $I \in \mathcal{I}$, F is a facet of $F(I)$ if and only if $F = F(I \cup \{i\})$ for some $i \notin I$ with $I \cup \{i\} \in \mathcal{I}$.

(v) For any $I \in \mathcal{I}$, the vectors a^i with $i \in I$ are linearly independent.

Let c be an arbitrary nonzero vector in R^n. Then F^+ will denote the face of P such that for each $x^+ \in F^+$ it holds that $c^T x^+ = \max_{x \in P} c^T x$, and F^- will denote the face of P such that for each $x^- \in F^-$ it holds that $c^T x^- = \min_{x \in P} c^T x$. Let $t^+ = c^T x^+$ for $x^+ \in F^+$ and $t^- = c^T x^-$ for $x^- \in F^-$. Since P is full-dimensional, it follows that $t^- < t^+$ and therefore $F^- \cap F^+ = \emptyset$. We define

$$I^+ = \left\{ i \in I_m \mid \left(a^i\right)^T x = b_i, \ \forall x \in F^+ \right\}, \tag{9.4}$$

$$I^- = \left\{ i \in I_m \mid \left(a^i\right)^T x = b_i, \ \forall x \in F^- \right\}. \tag{9.5}$$

So $F^+ = F(I^+)$ and $F^- = F(I^-)$.

We need some further notation. For each $I \in \mathcal{I}$, we define

$$A(I) = \left\{ y \in R^n \mid y = \sum_{i \in I} \mu_i a^i + \beta c, \mu_i \geq 0, \forall i \in I, \text{and} \beta \in R \right\}, \tag{9.6}$$

$$A_0(I) = \left\{ y \in R^n \mid y = \sum_{i \in I} \mu_i a^i, \ \mu_i \geq 0, \ \forall i \in I \right\}, \tag{9.7}$$

$$A^*(I) = \left\{ x \in R^n \mid x^T y \leq 0, \ \forall y \in A(I) \right\}, \tag{9.8}$$

$$A_0^*(I) = \left\{ x \in R^n \mid x^T y \leq 0, \ \forall y \in A_0(I) \right\}. \tag{9.9}$$

Note that $A(\emptyset) = \{y \in R^n \mid y = \beta c, \ \beta \in R\}$, $A_0(\emptyset) = \{0^n\}$, $A^*(\emptyset) = \{x \in R^n \mid c^T x = 0\}$, and $A_0^*(\emptyset) = R^n$. Moreover, for any $I \in \mathcal{I}$ we have that $A_0(I) \subset A(I)$, $A^*(I) \subset A_0^*(I)$, $A^*(I) \cap A(I) = \{0^n\}$, and $A_0^*(I) \cap A_0(I) = \{0^n\}$.

Let $\varphi : P \to R^n$ be a multi-valued mapping that satisfies the following assumption.

R1. The mapping $\varphi : P \to R^n$ values are nonempty compact convex subsets.

For an arbitrary (single-valued) function $f : P \to R^n$, the *stationary point* (or *variational inequality*) *problem* on the polytope P is to find a point $x^* \in P$ such that

$$\left(x^* - x\right)^T f\left(x^*\right) \geq 0, \quad \forall x \in P. \tag{9.10}$$

Such a point x^* is called a *stationary point* of f on P. It is well known that a continuous function on a convex compact set has a stationary point: *cf.* Eaves (1971), Hartman and Stampacchia (1966).

Below we introduce a stationary point of a mapping φ on the polytope P with respect to the vector c.

DEFINITION 9.1. A point $x^* \in P$ is a *stationary point* of a mapping φ on the polytope P *with respect to the nonzero vector* c if and only if there exists a $f \in \varphi(x^*)$ such that $(x^* - x)^T f \geq 0$ for all $x \in P$ satisfying $c^T x = c^T x^*$.

We call the problem of finding a stationary point with respect to a nonzero vector c a *parametric variational inequality problem* (PVIP). A solution to it is referred to as a *parametrized stationary point.* Now we have the following assertion which is an immediate consequence of the Linear Programming Theory.

LEMMA 9.1. *A point x^* is a parametrized stationary point of the mapping φ on P with respect to c if and only if, for some $I \in \mathcal{I}$, it holds that $x^* \in F(I)$ and $\varphi(x^*) \cap A(I) \neq \emptyset$.* ■

A point $x^* \in P$ is called a *zero point* of the mapping φ if and only if $0^n \in \varphi(x^*)$ where 0^n denotes the n-vector of zeros. Any zero point of φ is clearly a parametrized stationary point. The main purpose of the following sections of the chapter is to present conditions on φ that guarantee the existence and allow to calculate a continuum of zero points of a mapping φ on P such that it contains points of both F^- and F^+.

9.3 EXISTENCE CONDITIONS

This result is on the existence of parametrized stationary points.

THEOREM 9.1. *Let $\varphi : P \to R^n$ be an arbitrary multi-valued mapping satisfying assumption* R1 *and let a vector $c \in R^n \setminus \{0^n\}$ be given. Then there exists a connected set C of parametrized stationary points of φ on P with respect to c, such that $C \cap F^+ \neq \emptyset$ and $C \cap F^- \neq \emptyset$.* ■

The theorem makes clear that there exists a continuum of parametrized stationary points with a special topological structure. The set of parametrized stationary points has a connected subset that links two faces F^- and F^+.

In order to obtain the existence of a continuum of zero points, one need imposing certain conditions on the mapping φ. The following three results list sufficient conditions for the existence of a continuum of zero points of φ.

THEOREM 9.2. *Let $\varphi : P \to R^n$ be an arbitrary multi-valued mapping satisfying assumption* R1 *and let a vector $c \in R^n \setminus \{0^n\}$ be given. If for each $x \in F(I)$, $I \in \mathcal{I}$, the inclusion is valid*

$$\varphi(x) \cap A(I) \subset \{0^n\}, \tag{9.11}$$

then there exists a connected set C of zero points of φ such that $C \cap F^+ \neq \emptyset$ and $C \cap F^- \neq \emptyset$. ■

Condition (9.11) in the theorem states that for any x in the face $F(I)$ of P, the image $\varphi(x)$ may not have nonzero elements in common with $A(I)$.

THEOREM 9.3. *Let $\varphi : P \to R^n$ be an arbitrary multi-valued mapping satisfying assumption* R1 *and let a vector $c \in R^n \setminus \{0^n\}$ be given. If for any $x \in F(I)$, $I \in \mathcal{I}$, it holds that*

$$(i) \quad \varphi(x) \cap A_0^*(I) \cap A(I) \subset \{0^n\}; \tag{9.12}$$

$$(ii) \quad \varphi(x) \cap A_0^*(I) \neq \emptyset, \tag{9.13}$$

then there exists a connected set C of zero points of φ such that $C \cap F^+ \neq \emptyset$ and $C \cap F^- \neq \emptyset$. ■

Conditions (9.12) and (9.13) in the theorem imply that for any x in the face $F(I)$ of P, the image $\varphi(x)$ may not have nonzero elements in common with $A_0^*(I) \cap A(I)$ and that at least on element of $\varphi(x)$ lies in $A_0^*(I)$.

THEOREM 9.4. *Let $\varphi : P \to R^n$ be an arbitrary multi-valued mapping satisfying assumption* R1 *and let a vector $c \in R^n \setminus \{0^n\}$ be given. If for each $x \in F(I)$, $I \in \mathcal{I}$, it holds that*

$$\varphi(x) \cap A^*(I) \neq \emptyset, \tag{9.14}$$

then there exists a connected set C of zero points of φ such that $C \cap F^+ \neq \emptyset$ and $C \cap F^- \neq \emptyset$. ■

Condition (9.14) in the theorem says that for any x in the face $F(I)$ of P, at least one element of $\varphi(x)$ lies in $A^*(I)$.

The three theorems state different conditions under which a continuum of zero points can be shown to exist. All these conditions are by no means trivial ones: for instance, they prevent the mapping φ to be constant with a compact convex value set not containing zero. Moreover, there is a logical order in these sufficient conditions. Theorem 9.2 demands a weak condition but such that holds for all elements of φ. Theorem 9.4 requires a strong condition but holding for only one element in $\varphi(x)$. Theorem 9.3 is in between: a very weak condition for all elements in $\varphi(x)$ together with a rather weak condition for some element in $\varphi(x)$.

The following result claims that for the special case where φ is a (single-valued) function f, Theorem 9.2 is the strongest and Theorem 9.4 the weakest. Note that assumption T1 implies that the function f is continuous.

THEOREM 9.5. *Let P be any polytope and let $c \in R^n \setminus \{0^n\}$. The collection of functions satisfying the conditions of Theorem* 9.2 *contains the collection of functions satisfying the conditions of Theorem* 9.3, *which contains the collection of functions satisfying the conditions of Theorem* 9.4.

PROOF. Suppose that a function $f : P \to R^n$ satisfies the conditions of Theorem 9.4. Take any $x \in F(I)$, $I \in \mathcal{I}$, so $f(x) \in A^*(I)$. Since $A^*(I) \subset A_0^*(I)$ and $A(I) \cap A^*(I) = \{0^n\}$, one obtains that $f(x) \in A_0^*(I)$ and $f(x)$ not in $A(I)$ unless $f(x) = 0^n$. Hence the conditions of Theorem 9.3 are satisfied. Suppose now that a function satisfies the conditions of Theorem 9.3. Again pick up an arbitrary point $x \in F(I)$, $I \in \mathcal{I}$, so $f(x) \in A_0^*(I)$ and $f(x)$ not in $A(I) \cap A_0^*(I)$ unless $f(x) = 0^n$. Therefore, $f(x)$ not in $A(I)$ unless $f(x) = 0^n$ which completes the proof. ■

Another remarkable feature of the framework proposed by Herings, Talman and Yang (1999) is that all the theorems above are proved in a constructive manner, as it is demonstrated below.

9.4 THE SCARF TYPE ALGORITHM

In this section, an algorithm on the polytope P will be described that generates a piecewise linear path of parametrized stationary points, with respect to a given nonzero vector c, of a piecewise linear approximation of the mapping φ. The piecewise linear approximation is constructed with respect to some simplicial subdivision of the set P. In this and the next sections, it is assumed that the faces F^+ and F^- are vertices of P, denoted by x^+ and x^-, respectively. In that case the piecewise linear path generated by the algorithm contains both x^- and x^+ and can be traced by a sequence of semi-lexicographic pivot steps in a system of linear equations. The general case is discussed in Section 9.6.

For a nonnegative integer t, a *t-dimensional simplex* or *t-simplex*, denoted by σ, is defined as the convex hull of $t+1$ affinely independent points $x^1, \ldots, x^{t+1}$ in R^n. One often writes $\sigma = \sigma\left(x^1, \ldots, x^{t+1}\right)$ and calls $x^1, \ldots, x^{t+1}$ the vertices of σ. A $(t-1)$-simplex being the convex hull of t vertices of σ is said to be a *facet* of σ. The facet $\tau\left(x^1, \ldots, x^{i-1}, x^{i+1}, \ldots, x^{t+1}\right)$ is called the *facet* of $\sigma\left(x^1, \ldots, x^{t+1}\right)$ *opposite to the vertex* x^i. For k, $0 \leq k \leq t$, a k-simplex being the convex hull of $k+1$ vertices of σ is referred to as a *k-face* or *face* of σ. A finite collection $\mathcal{T}$ of n-simplices is a *triangulation* of the polytope P if

(i) P is the union of all the simplices in $\mathcal{T}$;

(ii) The intersection of any two simplices of $\mathcal{T}$ is either the empty set or a common face of both.

Let $\mathcal{T}$ be any triangulation of P. Then every face $F(I)$ of P is subdivided into t-simplices where $t = n - |I|$. For example one can take the V-triangulation of Talman and Yamamoto (1989). Since $\mathcal{T}$ is finite and P is compact, every facet τ of an $(n-|I|)$-simplex σ in $F(I)$ either lies in the boundary of $F(I)$ and is only a facet of σ or is a facet of exactly one other $(n-|I|)$-simplex in $F(I)$. Let f be a *simplicial approximation* of φ with respect to $\mathcal{T}$. This means that $f(x) = \varphi(x)$ at each vertex of $\mathcal{T}$ and f is affine on each simplex of $\mathcal{T}$.

A row vector is *lexico-positive* if it is a nonzero vector and its first nonzero entry is positive. A matrix is said to be *lexico-positive* if all its rows are lexico-positive. A matrix is called *semilexico-positive* if each its row except possibly for the last one is lexico-positive.

DEFINITION 9.2. Let $\tau(x^1,\ldots,x^t)$ be a $(t-1)$- simplex in $F(I)$ where $I \in \mathcal{I}$ with $I = \{i_{t+1},\ldots,i_n\}$, $t = n - |I|$. The $(n+1)\times(n+1)$-matrix

$$A_{\tau,I} = \begin{pmatrix} 1 & \cdots & 1 & 0 & \cdots & 0 & 0 \\ -f(x^1) & \cdots & -f(x^t) & a^{i_{t+1}} & \cdots & a^{i_n} & c \end{pmatrix}$$

is the *label matrix* of τ with respect to I. The simplex τ is *I-complete* if $A_{\tau,I}^{-1}$ exists and is semilexico-positive, i.e. the first nonzero entry in every row, except possibly for the last one, is positive.

Notice that if for an I-complete simplex τ one changes the ordering of the first n columns of matrix $A_{\tau,I}$, the inverse of the resulting matrix still exists and is semilexico-positive. Clearly, if, for some $I \in \mathcal{I}$, a $(t-1)$-simplex $\tau(x^1,\ldots,x^t)$ is an I-complete facet of a simplex $\sigma\left(x^1,\ldots,x^{t+1}\right)$ in some face $F(I)$, then the system of $n+1$ linear equations with $n+2$ variables

$$\sum_{j=1}^{t+1}\lambda_j\begin{pmatrix}1\\-f(x^j)\end{pmatrix} + \sum_{i\in I}\mu_i\begin{pmatrix}0\\a^i\end{pmatrix} + \beta\begin{pmatrix}0\\c\end{pmatrix} = \begin{pmatrix}1\\0\end{pmatrix} \tag{9.15}$$

has a solution $(\lambda,\mu,\beta) = (\lambda_1,\ldots,\lambda_{t+1},(\mu_i)_{i\in I},\beta)$ satisfying $\lambda_j \geq 0$ for $j \in I_{t+1}$ and $\mu_i \geq 0$ for $i \in I$, with $\lambda_{t+1} = 0$. Let x be defined by $x = \sum_{j=1}^{t+1}\lambda_j x^j$ at a solution (λ,μ,β) of (9.15), then x lies in σ and is a parametrized stationary point of f with respect to c.

The following result is a special case of Theorem 2.6 in Fujishige and Yang (1998) that also has been proved in a constructive way.

THEOREM 9.6. *Consider any polytope Q given by*

$$Q = \left\{x \in R^n \mid \left(c^i\right)^T x \leq d_i,\ i \in I_n \text{ and } \left(c^0\right)^T x = d_0\right\}.$$

Assume that Q is an $(n-1)$-dimensional simple polytope with no redundant constraints. For any $g \in R^n$, there exists a unique subset $I = \{j_1,\ldots,j_{n-1}\}$ of I_n with $|I| = n-1$ such that the following matrix

$$B^{-1} = \begin{pmatrix} 1 & 0 & 0 & \cdots & 0 & 0 \\ g & c^{j_1} & c^{j_2} & \cdots & c^{j_{n-1}} & c^0 \end{pmatrix}^{-1}$$

exists and is semilexico-positive. ∎

The complete proof of the algorithm convergence can be found in Herings, Talman and Yang (1999) and is based upon the following auxiliary results. First, it is shown that each of the two 0-simplices $\{x^-\}$ and $\{x^+\}$ is I-complete in the face $F(I)$ for a unique index set $I \in \mathcal{I}$ containing $n-1$ indices.

LEMMA 9.2. *Let $x^1 = x^-$ $[x^+]$ and $\tau = \{x^1\}$. Then there exists a unique subset $I = \{j_1,\ldots,j_{n-1})$ of I^- $[I^+]$ with $|I| = n-1$ such that τ is an I-complete 0-simplex in $F(I)$.* ∎

The next two lemmas are important for the algorithm to be well defined. We also provide their proofs to see the nature of the algorithm steps.

LEMMA 9.3. *Let σ be a t-simplex in $F(I)$ where $I \in \mathcal{I}$, $t = n - |I|$ and $I = \{i_{t+1}, \ldots, i_n\}$. If σ has an I-complete facet τ, then exactly one of the following two cases occurs:*

(1) The simplex σ is an $\bar{I}$-complete simplex in $F\left(\bar{I}\right)$ where $\bar{I} = I \setminus \{i\}$ for precisely one index $i \in I$;

(2) The simplex σ has exactly one other I-complete facet $\bar{\tau}$.

PROOF. Let x^{t+1} be the vertex of σ opposite to τ, and let $y = A_{\tau,I}^{-1}\left(1, -f\left(x^{t+1}\right)^T\right)^T$. Notice that $y \neq 0$. Let $K = \{i \in I_n \mid y_i > 0\}$. Since $A_{\tau,I}y = \left(1, -f\left(x^{t+1}\right)^T\right)^T$, we have $\sum_{i=1}^t y_i = 1$. This implies that K is nonempty.

Consider the ratio vectors $(1/y_j)\left(A_{\tau,I}^{-1}\right)_j$ for all $j \in K$. Choose $k \in K$ such that the k-th ratio vector is the minimum in the lexicographic order over all such ratio vectors. Since $A_{\tau,I}^{-1}$ is regular, k is uniquely determined. Now consider the following two cases (1) and (2).

(1) If $k \in I_n \setminus I_t$, then let $\ell = i_k$ and $\bar{I} = I \setminus \{\ell\}$. Clearly, $\bar{I} \in \mathcal{I}$ and σ is in $F\left(\bar{I}\right)$. Let B be the matrix obtained from $A_{\tau,I}$ by replacing its k-th column by $\left(1, -f\left(x^{t+1}\right)^T\right)^T$. It follows from Lemma 4.5 (Herings, Talman and Yang, 1999) that B^{-1} exists and is semilexico-positive. By reordering the columns of B we get $A_{\sigma,\bar{I}}$ whose inverse exists and is semilexico-positive. So σ is $\bar{I}$-complete.

(2) If $k \in I_t$, then let $\bar{\tau}$ be the facet of σ opposite to the vertex x_k. Using Lemma 4.5 (Herings, Talman and Yang, 1999), it follows from the choice of k that $A_{\bar{\tau},I}^{-1}$ exists and is semilexico-positive. Hence $\bar{\tau}$ is an I-complete $(t-1)$-simplex in $F(I)$.

It also follows immediately from Lemma 4.5 (Herings, Talman and Yang, 1999) that if any column other than the k-column is replaced, then the inverse of the resulting matrix is not semi-lexicopositive. The proof is complete. ■

In the next Lemma 9.4, the analogous case is considered that consists in making a lexicographic pivot step with a column $\left(1, \left(a^i\right)^T\right)^T$.

LEMMA 9.4. *Let σ be an I-complete $(t-1)$- simplex in $F(I)$ where $I \in \mathcal{I}$, $t = n - |I|$ and $I = \{i_{t+1}, \ldots, i_n\}$. If σ is in $F(\bar{I})$ and $\bar{I} \neq I^-$ or $\bar{I} \neq I^+$, where $\bar{I} = I \cup \{\ell\} \in \mathcal{I}$ for some $\ell \in I_n \setminus I$, then exactly one of the following two cases occurs:*

(1) There exists a unique set $J \in \mathcal{I}$ with $|J| = |I|$ and $J \neq I$ so that σ is in $F(J)$ and is J-complete;

(2) There exists exactly one facet τ of σ which is in $F(\bar{I})$ and is $\bar{I}$-complete.

PROOF. Let $\ell \in I$, $x = \left(0, \left(a^\ell\right)^T\right)^T$ and $y = A^{-1}_{\sigma,I}x$. Note that $y \neq 0$. Let $K = \{i \in I_n \mid y_i > 0\}$. Note that $A_{\sigma,I}y = \left(0, \left(a^\ell\right)^T\right)^T$. We need considering the following two cases (i) and (ii):

Case (i). If there exists an index $j \in I_t$ such that $y_j < 0$, then there exists $i \in I_t$ such that $y_i > 0$ since $\sum_{i=1}^t y_i = 0$. Hence K is nonempty.

Case (ii). Suppose that $y_i = 0$ for all $i \in I_t$; then we have

$$a^\ell = \sum_{i=t+1}^{n} y_i a^{j_{i-t}} + y_{n+1}c. \tag{9.16}$$

If $y_i \leq 0$ for all $i = t+1, \ldots, n$, then (9.16) is impossible since $\bar{I}$ is neither equal to I^- nor equal to I^+ (Lemma 4.7 by Herings, Talman and Yang, 1999). Hence there exists at least one index $i \in I_n \setminus I_t$ such that $y_i > 0$, and again K is nonempty.

Consider the ratio vectors $(1/y_j)\left(A^{-1}_{\sigma,I}\right)_j$ for all $j \in K$. Choose $k \in K$ such that the k-th ratio vector is the minimum in the lexicographic order over all such ratio vectors. Since $A^{-1}_{\sigma,I}$ is regular, k is uniquely determined. Now consider the following two cases (1) and (2).

(1) If $k \in I_n \setminus I_t$, then let $p = i_k$ and $J = I \cup \{\ell\} \setminus \{p\}$. Clearly, $J \in \mathcal{I}$, $J \neq I$, $|J| = |I|$ and σ is in $F(J)$. Let B be the matrix obtained from $A_{\sigma,I}$ by replacing its k-th column by x. It follows from Lemma 4.5 (Herings, Talman and Yang, 1999) that B^{-1} exists and is semilexico-positive. It is clear that $A_{\sigma,J} = B$. So σ is a J-complete $(t-1)$-simplex in $F(J)$.

(2) If $k \in I_t$ then let τ be the facet of σ opposite to the vertex x^k. Clearly, τ is a $(t-2)$-simplex in $F(\bar{I})$. Let B be the matrix obtained from $A_{\sigma,I}$ by replacing its k-th column by x. It follows from Lemma 4.5 (Herings, Talman and Yang, 1999) that B^{-1} exists and is semilexico-positive. So τ is an $\bar{I}$-complete $(t-2)$-simplex in $F(\bar{I})$.

Again it follows from Lemma 4.5 (Herings, Talman and Yang, 1999) that if any other column is replaced, then the new matrix is no longer semilexico-positive, which completes the proof. ∎

Now the Scarf type algorithm can be described. A graph $G = (\mathcal{V}, \mathcal{A})$ is constructed where $\mathcal{V}$ denotes the set of nodes and $\mathcal{A}$ denotes the set of edges. Each I-complete $(n-|I|-1)$-simplex is a *node* in $\mathcal{V}$. An I-complete $(n-|I|-1)$-simplex τ^1 in $F(I)$ and a J-complete $(n-|J|-1)$-simplex τ^2 in $F(J)$ are said to be *adjacent* complete simplices if $I = J = L$ and τ^1 and τ^2 are both facets of an $(n-|L|)$-simplex σ in $F(L)$, or τ^1 is a facet of τ^2 and τ^2 is an $(n-|I|)$-simplex in $F(I)$, or τ^2 is a facet of τ^1 and τ^1 is an $(n-|J|)$-simplex in $F(J)$. Two adjacent complete simplices τ^1 and τ^2 are connected by an *edge* $e = \{\tau^1, \tau^2\} \in \mathcal{A}$. The *degree of a node* τ in G is defined to be the number of nodes connected with it and denoted by $\deg(\tau)$. A *path* in G from node $\tau^0 = \{x^-\}$ to node τ^ℓ is defined as a sequence of the form $(\tau^0, e_1, \tau^1, \ldots, e_\ell, \tau^\ell)$ where $\tau^0, \tau^1, \ldots, \tau^\ell$ are nodes and $e_1, \ldots, e_\ell$ are edges such that $e_i = \{\tau^{i-1}, \tau^i\}$ for $i \in I_\ell$. A path is *simple* if all its nodes and edges are different.

THEOREM 9.7. *Let T be a triangulation of P. Starting at the vertex x^-, the algorithm generates a finite sequence of adjacent J-complete simplices for varying $J \in \mathcal{I}$ which leads to the vertex x^+.*

PROOF. By Lemma 9.3, $\{x^-\}$ is an I-complete 0-simplex in $F(I)$ for some unique set $I \in \mathcal{I}$ with $|I| = n - 1$. As $\{x^-\}$ lies in the boundary of $F(I)$, there exists a unique 1-simplex σ in $F(I)$ having $\{x^-\}$ as its facet. By Lemma 9.3, either σ is an $\bar{I}$-complete simplex in $F(\bar{I})$ where $\bar{I} = I \setminus \{i\}$ for some unique $i \in I$, or σ has exactly one other I-complete facet $\bar{\tau}$. Hence there exists a unique adjacent complete simplex to $\{x^-\}$. That is, $\deg(\{x^-\}) = 1$. Similarly, by using Lemmas 9.3 and 9.4, one can prove that $\deg(\{x^+\}) = 1$.

In all other cases, we prove that if τ is an I-complete $(n - |I| - 1)$-simplex in $F(I)$ for some $I \in \mathcal{I}$, τ has exactly two adjacent complete simplices. There are two possibilities: either τ lies in the interior of $F(I)$ or τ lies in the boundary of $F(I)$. If τ lies in the interior of $F(I)$, then τ is a facet of exactly two complete $(n - |I|)$-simplices in $F(I)$ according to Lemma 9.4. If τ lies in the boundary of $F(I)$, then there exists exactly one $(n - |I|)$-simplex σ in $F(I)$ having τ as its facet. By Lemma 9.4 either σ is an $\bar{I}$-complete $(n - |\bar{I}| - 1)$- simplex in $F(\bar{I})$ for a unique $\bar{I} \in \mathcal{I}$ with $|\bar{I}| = |I| - 1$ and has no other I-complete facets, or σ has exactly one other I-complete facet. This yields one adjacent complete simplex to τ. On the other hand, as τ lies in the boundary of $F(I)$, then τ lies in $F(\tilde{I})$ for some unique set $\tilde{I} \in \mathcal{I}$ with $|\tilde{I}| = |I| + 1$. By Lemma 9.4, either τ is J-complete for some unique set $J \in \mathcal{I}$ with $|J| = |I|$ and $J \neq I$, or τ has exactly one $\tilde{I}$-complete facet. In the former case, τ lies in $F(\tilde{I})$ and hence there exists exactly one simplex $\bar{\sigma}$ in $F(\tilde{I})$ having τ as its facet. It follows again from Lemma 9.4 that there exists exactly one other complete simplex adjacent to τ. This allows one to conclude that τ has exactly two adjacent complete simplices. In other words, we have $\deg(\tau) = 2$.

As shown above, the degree of each node in the graph $G = (\mathcal{V}, \mathcal{A})$ is at most two. Exactly two nodes have degree equal to one. Since the number of simplices in P is finite, the number of nodes in G is finite, too. Since $\deg(\{x^-\}) = 1$, it easy to see that there exists a simple finite path starting from $\{x^-\}$. The end node of this path must be a node of degree one and different from x^-. The only possibility is that τ coincides with $\{x^+\}$ which completes the proof of the theorem. ∎

It is evident that every simplex on the path between $\{x^-\}$ and $\{x^+\}$ examined by the above theorem contains a parametrized stationary point of the piecewise linear approximation f of φ with respect to $\mathcal{T}$. By setting the straight line segments between the parametrized stationary points of any two adjacent simplices, we obtain a piecewise linear path of parametrized stationary points of f connecting the points x^- and x^+.

COROLLARY 9.1. *Let $\mathcal{T}$ be any triangulation of P. Then with respect to the vector $c \in R^n \setminus \{0^n\}$ there is a piecewise linear path of parametrized stationary points of the piecewise linear approximation f of φ with respect to $\mathcal{T}$ and this path connects x^- and x^+.* ■

9.5 CONSTRUCTIVE PROOFS OF THE EXISTENCE THEOREMS

In this section it is still assumed that x^- and x^+ are unique. First, it will be argued that the points lying on the path given in Corollary 9.1 indeed all correspond to approximate parametrized stationary points of the mapping φ. To show that, a sequence of triangulations $\mathcal{T}^r$ with mesh size converging to zero is taken. According to Corollary 9.1 this yields, for every natural $r \in \mathcal{N}$, a continuous piecewise linear function $\pi^r : [0,1] \to P$ with image set $\pi^r([0,1])$ connecting x^- and x^+. It will be shown that if q^r is an arbitrary point in $\pi^r([0,1])$ and the sequence $(q^r)_{r\in\mathcal{N}}$ converges to q, then q is a parametrized stationary point of $\varphi(q)$ with respect to c. Under the conditions of Theorems 9.2, 9.3 and 9.4, the piecewise linear approximation can be chosen in such a way that q is a zero point of φ. Furthermore, it is stated in Theorem 9.10 that there exists a connected set of zero points of φ containing both x^- and x^+.

THEOREM 9.8. *Let $\varphi : P \to R^n$ be a mapping satisfying Assumption* T1 *and let $c \in R^n \setminus \{0^n\}$. For $r \in \mathcal{N}$, let $\mathcal{T}^r$ be a triangulation of P with mesh size smaller than $1/r$ and let $f^r : P \to R^n$ be a piecewise linear approximation of φ with respect to $\mathcal{T}^r$ and let $\pi^r : [0,1] \to P$ be the corresponding continuous function with image set connecting x^- and x^+. Let $(q^r)_{r\in\mathcal{N}}$ be an arbitrary convergent sequence of points in P with limit q^* where $q^r \in \pi^r([0,1])$. Then q^* is a parametrized stationary point of φ with respect to c.*

PROOF. Let $(\lambda_1^r, \ldots, \lambda_{n+1}^r, x^{1,r}, \ldots, x^{n+1,r}, s^{1,r}, \ldots, s^{n+1,r})_{r\in\mathcal{N}}$ be a sequence of points in $R_+^{n+1} \times \prod_{i=1}^{n+1} P \times \prod_{i=1}^{n+1} R^n$ satisfying $\sum_{j=1}^{n+1} \lambda_j^r = 1$, $\sigma^r(x^{1,r}, \ldots, x^{n+1,r})$ be a simplex of $\mathcal{T}^r$, $q^r = \sum_{j=1}^{n+1} \lambda_j^r x^{j,r} \in \pi^r([0,1])$, and $s^{j,r} \in f^r(x^{j,r})$. Note that it may happen that $\lambda_j^r = 0$ for some $j \in I_{n+1}$. By definition, $f^r(q^r) = \sum_{j=1}^{n+1} \lambda_j^r s^{j,r}$. Define $s^r = f^r(q^r)$, then $s^r = \beta^r c + \sum_{j\in I^r} \mu_j^r a^j$, for some $\beta^r \in R$, for some $\mu_j^r \geq 0$, $\forall j \in I^r$, and for some $I^r \in \mathcal{I}$ such that σ^r lies in $F(I^r)$. As $\bigcup_{q\in P} \varphi(q)$ is bounded, the sequence defined above belongs to a compact set, and without loss of generality it can be assumed to converge to an element $(\lambda_1^*, \ldots, \lambda_{n+1}^*, x^{1,*}, \ldots, x^{n+1,*}, s^{1,*}, \ldots, s^{n+1,*})$. Define $s^* = \sum_{j=1}^{n+1} \lambda_j^* s^{j,*}$. It is evident that $s^r \to s^*$ as $r \to \infty$. Since for every

$r \in \mathcal{N}$ the mesh size of $\mathcal{T}^r$ is smaller that $1/r$, it holds for every $j \in I_{n+1}$ that $x^{j,*} = q^*$. Making use of the upper semicontinuity of φ we have that $s^{j,*} \in \varphi(q^*)$ for every $j \in I_{n+1}$. Moreover, $\beta^r \to \beta^*$, as $r \to \infty$, for some β^*, and without loss of generality $I^r = I^*$ for all r large enough for some $I^* \in \mathcal{I}$, and $\mu_j^r \to \mu_j^*$ for all $j \in I^*$ for some $\mu_j^* \geq 0$, as $r \to \infty$. As φ is convex-valued, $\sum_{j=1}^{n+1} \lambda_j^* = 1$ and $\lambda_j^* \geq 0$, $\forall j \in I_{n+1}$, then $s^* \in \varphi(q^*)$. At last, $q^* \in F(I^*)$ and $s^* = \beta^* c + \sum_{j \in I^*} \mu_j^* a^j \in A(I^*)$. Hence, according to Lemma 9.1, q^* is a parametrized stationary point of φ with respect to c which completes the proof. ■

In order to give a constructive proof of Theorems 9.2, 9.3 and 9.4 the piecewise linear approximation f of φ with respect to a triangulation $\mathcal{T}$ should be chosen as follows (such an approximation will be referred to as a *proper* one). In case of Theorem 9.2, any piecewise linear approximation of φ with respect to $\mathcal{T}$ can be chosen. Next consider the case of Theorem 9.3. If a point x in the (relative) interior of a face $F(I)$ is a vertex of a simplex of the triangulation, it implies that at least one element of $\varphi(x)$ belongs to the set $A_0^*(I)$, and this element is assigned as the value of the piecewise linear approximation at x. In case of Theorem 9.4 a common element of the sets $A^*(I)$ and $\varphi(x)$ is assigned to a vertex x of a simplex if x belongs to the (relative) interior of $F(I)$.

THEOREM 9.9. *Let $\varphi : P \to R^n$ be a mapping satisfying the conditions of one of Theorems 9.2, 9.3 or 9.4. For $r \in \mathcal{N}$, let $\mathcal{T}^r$ be a triangulation of P with mesh size smaller than $1/r$ and let $f^r : P \to R^n$ be a proper piecewise linear approximation of φ with respect to $\mathcal{T}^r$. Let $(q^r)_{r \in \mathcal{N}}$ be an arbitrary convergent sequence of points in P with limit q^* such that $q^r \in \pi^r([0,1])$ for any $r \in \mathcal{N}$. Then q^* is a zero point of φ.*

PROOF. First consider Theorem 9.2. Following the proof of Theorem 9.8 the limit point $s^* \in \varphi(q^*)$ is an element of $A(I^*)$ whereas q^* is an element of $F(I^*)$. The latter property implies that $s^* = 0^n$, hence q^* is a zero point of φ.

Next consider Theorem 9.3. Consider again the convergent sequence of simplices σ^r in $F(I^*)$ constructed when proving Theorem 9.8. Then the vertex $x^{j,r}$ of σ^r belongs to some face $F\left(I^{j,r}\right)$ of P with $I^* \subset I^{j,r}$. Therefore, one has that $A_0^*\left(I^{j,r}\right) \subset A_0^*\left(I^*\right)$, and hence $s^r \in A_0^*\left(I^*\right)$ for all r due to the properness of f^r. Consequently, also $s^* \in A_0^*\left(I^*\right)$ whence $s^* \in A_0^*\left(I^*\right) \cap A(I^*) \subset \{0^n\}$, i.e. $s^* = 0^n$ and q^* is a zero point of φ.

In case of Theorem 9.4, a similar agrument as for the previous case allows one to conclude that $s^* \in A^*\left(I^*\right) \cap A(I^*)$. Since the latter intersection can contain only zero-vector, one obtains again $s^* = 0^n$ which implies that q^* is a zero point of φ and completes the proof of the theorem. ■

From Theorem 9.9 the next results follow immediately.

COROLLARY 9.2. *Let $\varphi : P \to R^n$ be a mapping satisfying the conditions of one of Theorems 9.2, 9.3 or 9.4. For $r \in \mathcal{N}$, let $\mathcal{T}^r$ be a triangulation of P with mesh size smaller than $1/r$ and let $f^r : P \to R^n$ be a proper piecewise* *linear approximation of φ with respect to T^r. Then for every $\varepsilon > 0$ there exists an $r_\varepsilon \in \mathcal{N}$ such that for all $r \geq r_\varepsilon$, it holds $\|f^r(q^r)\|_\infty < \varepsilon$ whenever $q^r \in \pi^r([0,1])$.*

PROOF. Suppose that there exists a sequence $(q^r, f^r(q^r))_{r\in\mathcal{N}}$ with $q^r \in \pi^r([0,1])$ and $\|f^r(q^r)\|_\infty \geq \varepsilon$ for every $r \in \mathcal{N}$. Since P and $\bigcup_{q\in P}\varphi(q)$ are compact sets, there exists a convergent subsequence $(q^r_k, f^r_k(q^r_k))_{k\in\mathcal{N}}$ with a limit say (q^*, s^*), where $\|s^*\|_\infty \geq \varepsilon > 0$. Repeating the proof of Theorem 9.8 one can show that $s^* = 0$, yielding a contradiction which completes the proof. ■

Making use of Theorems 9.7 and 9.9 it can be shown that there exists a connected set C in P such that $x^- \in C$, $x^+ \in C$, and $0^n \in \varphi(q)$, $\forall q \in C$. Hence, there is a continuum of zero points of φ being approximated by the algorithm of Section 9.4. The latter is confirmed by the theorem below proven in Herings, Talman and Yang (1999).

THEOREM 9.10. *Let $\varphi : P \to R^n$ be a mapping satisfying the conditions of one of Theorems 9.2, 9.3 or 9.4. Then there exists a connected set C of points in P such that $x^- \in C$, $x^+ \in C$, and $0^n \in \varphi(q)$, $\forall q \in C$.* ■

Similarly, one can show that in case of Theorem 9.1, there exists a connected set C of parametrized stationary points in P with respect to c such that $x^- \in C$ and $x^+ \in C$. At last, it is worthy to mention that the algorithm proposed can be adapted for computing a continuum of parametrized stationary points or zero points of φ on P in case the faces F^- and F^+ are not vertices of P (*cf.* Herings, Talman and Yang, 1999).

9.6 FIXED POINT THEOREMS

In this section we follow Herings, Talman and Yang (1999) in showing that the fundamental fixed points theorems of Browder (1960) and MasColell (1974) can be derived from Theorem 9.4. Browder proved the continuous function fixed point theorem, and MasColell extended the result to the case of upper semicontinuous mapping.

THEOREM 9.11. *Let P be an n-dimensional polytope and let $\varphi : P\times[0,1] \to P$ be any mapping satisfying Assumption* T1. *Then the set*

$$D = \{(x,t) \in P \times [0,1] \mid x \in \varphi(x,t)\} \tag{9.17}$$

contains a connected set C such that

$$C \cap (P \times \{0\}) \neq \emptyset \quad \textit{and} \quad C \cap (P \times \{1\}) \neq \emptyset. \tag{9.18}$$

PROOF. One can rewrite the set $P \times [0,1]$ as

$$W = \Big\{(x,t) \in R^{n+1} \mid \left((a^i)^T, 0\right)\left(x^T, t\right)^T \leq b_i, \quad \forall i \in I_m,$$

$$(0, \ldots, 0, -1) \left(x^T, t\right)^T \leq 0,$$
$$\left.(0, \ldots, 0, 1) \left(x^T, t\right)^T \leq 1\right\}. \tag{9.19}$$

Let $c = (0, \ldots, 0, 1) \in R^{n+1}$. Obviously W is simple and has no redundant constraints. Moreover, $F^- = \{(x,t) \in W \mid t = 0\}$ and $F^+ = \{(x,t) \in W \mid t = 1\}$.

Construct the mapping $\psi : W \to R^{n+1}$ as

$$\psi(x,t) = (\varphi(x,t) - \{x\}) \times \{0\}. \tag{9.20}$$

We will show that for any $I \in \mathcal{I}$ and any $(x,t) \in F(I)$, we have $\psi(x,t) \subset A^*(I)$. For $I = \emptyset \in \mathcal{I}$, clearly $\psi(x,t) \subset A^*(\emptyset)$ since $c^T z = 0$ for any $z \in \psi(x,t)$. Now take any nonempty set I from $\mathcal{I}$. One has to consider the following two cases:

(1) In case $I \subset I_m$, take any (x,t) in the face $F(I)$ of W and an arbitrary $z \in \psi(x,t)$. The following equalities are valid

$$\left(a^i\right)^T x = b_i, \ \forall i \in I; \quad z = \left((u-x)^T, 0\right)^T \tag{9.21}$$

for some $u \in P$. Take any $y \in A(I)$, that is,

$$y = \sum_{i \in I} \lambda_i \left((a^i)^T, 0\right)^T + \beta c \tag{9.22}$$

for some $\lambda_i \geq 0$ and $\beta \in R$. Thus, one has

$$s^T y = \sum_{i \in I} \lambda_i (u-x)^T a^i. \tag{9.23}$$

As $u \in P$ it holds $(a^i)^T u \leq b_i$ for all $i \in I_m$. On the other hand, as $(a^i)^T u = b_i$ for all $i \in I$, the inequality is valid $s^T y \leq 0$ which means that $\psi(x,t) \subset A^*(I)$.

(2) If $I \not\subset I_m$, either $m+1$ or $m+2$ belongs to I. Suppose that $m+2$ is in I, i.e. $t = 1$ (the case of $m+1 \in I$ is treated by the similar argument). Take any (x,t) in the face $F(I)$ of W and any $z \in \psi(x,t)$. One clearly has

$$(a^i)^T x = b_i, \ \forall i \in I \setminus \{m+2\}; \ t = 1; \ z = \left((u-x)^T, 0\right)^T \tag{9.24}$$

for some $u \in P$. Pick up an arbitrary $y \in A(I)$, that is

$$y = \sum_{i \in I \setminus \{m+2\}} \lambda_i \left((a^i)^T, 0\right)^T + \beta c \tag{9.25}$$

for some $\lambda_i \geq 0$ and $\beta \in R$. Thus, one has

$$s^T y = \sum_{i \in I \setminus \{m+2\}} \lambda_i (u-x)^T a^i. \tag{9.26}$$

As $u \in P$, we have $(a^i)^T u \leq b_i$ for all $i \in I_m$. Since $(a^i)^T u = b_i$ for all $i \in I \setminus \{m+2\}$, it follows that $s^T y \leq 0$ which again implies that $\psi(x,t) \subset A^*(I)$. By Theorem 9.4 there exists a connected set C in W such that

$$0^{n+1} \in \psi(x,t), \ \forall (x,t) \in C; \ F^- \cap C \neq \emptyset; \ F^+ \cap C \neq \emptyset. \tag{9.27}$$

The latter clearly implies that $x \in \varphi(x,t)$ for each $(x,t) \in C$ and relationships (9.18) are valid which completes the proof. ∎

Theorem 9.4 can be also used to demonstrate the existence of a continuum of equilibria in a pure exchange economy with general price rigidity; *cf.* Schalk and Talman (1999) for details. Price vectors in such an economy with n commodities are restricted to an n-dimensional simple polytope P without redundant constraints in the interior of R^n_+. Each vector a^i, $i \in I_m$, determining a facet of P is normalized to have length 1. Let c be any element of the positive orthant R^n_{++} with length equal to one. Define $Q = \{q \mid (a^i)^T q \leq b_i + 1,\ \forall i \in I_m\}$. Clearly, Q is a polytope containing P in its interior. For any $q \in Q$, let $p(q)$ be the orthogonal projection of q onto P and let $I(q)$ be such that $p(q)$ is in the interior of the face $F(I(q))$ of P. Then there exist uniquely nonnegative numbers μ_i, $i \in I(q)$ such that $p = p(q) + \sum_{i \in I(q)} \mu_i a^i$.

At $q \in Q$ define the price vector by $p(q)$ and a continuous rationing scheme $(r^i, d_i(q))$, $i \in I_m$, such that $r^i = a^i - (c^T a^i)c$, $d_i(q) = 0$ if $\mu_i = \mu_i(q) = 1$ and $d_i(q) = \infty$ if $\mu_i = \mu_i(q) = 0$ or i not in $I(q)$. This rationing scheme determines the constraints on the net supply of the consumers. Given a utility function u^h and an initial endowment w^h, consumer $h \in H$ maximizes his utility function $u^h(x)$ subject to his budget constraint given by $[p(q)]^T x \leq [p(q)]^T w^h$ and rationing constraints $(r^i)^T(x - w^h) \leq d_i(q)$, $i \in I_m$. The solutions set $x^h(q)$ yields the constraint excess demand set $z^h(q) = x^h(q) - w^h$ of consumer $h \in H$ at q. Adding up these sets over all consumers in H gives the constraint excess demand $z(q)$ induced by q. Under certain economic conditions, the mapping z satisfies Assumption T1 and $[p(q)]^T z = 0$ for all $z \in z(q)$, $q \in Q$.

A constraint equilibrium is obtained if $q \in P$ is such that $0^n \in z(q)$. At such an equilibrium for any $i \in I_m$ the i-th rationing scheme is *active*, i.e. $d_i(q)$ is finite only if the i-th constraint of the prices is *binding*, i.e. $[p(q)]^T a^i = b_i$.

Now let F^- $[F^+]$ be the face of Q on which $c^T x$ is minimized [maximized] on Q. These two faces induce trivial constrained equilibria where either all demand or all supply is completely rationed away.

To see that there is a *connected set of constrained equilibria* linking the F^- and F^+, define the mapping φ on Q by $\varphi(q) = \{y \mid y = z - (c^T z)c,\ z \in z(p)\}$, $q \in Q$. We show that $\varphi(q) \subset A^*(I)$ when q lies in the interior of the face $F(I)$ of Q. Indeed, $c^T y = 0$ for any $y \in \varphi(q)$. Moreover, for any $y \in \varphi(q)$ and $i \in I(q)$ one has $(a^i)^T y = (a^i)^T z - ((a^i)^T c)\, c^T z = (r^i)^T z \leq 0$ for some $z \in z(q)$. Therefore, any $y \in \varphi(q)$ is an element of $A^*(I)$, and hence $\varphi(q) \subset A^*(I)$ if $q \in F(I)$. According to Theorem 9.4 there exists a connected set C in Q containing both F^- and F^+ such that every point $q \in C$ is a zero point of φ, i.e. $0^n \in \varphi(q)$. For such a q there is a $z \in z(q)$ satisfying $z - (c^T z)c = 0^n$. Because $[p(q)]^T z = 0$, one obtains that $(c^T z)[p(q)]^T c = 0$. Hence, $c^T z = 0$, since $[p(q)]^T c > 0$. These implies that $z = 0^n$ and therefore $0^n \in z(q)$ inducing a constrained equilibrium. Consequently there exists a connected set of constrained equilibria linking the two faces F^- and F^+. ∎

9.7 COMMENTS

Whenever a mathematical model of some phenomenon is constructed, for instance in engineering or economics, the first question to ask is whether a solution to the model exists. A very powerful tool that is used to this end is Brouwer's fixed point theorem (*cf.* Brouwer (1912)). When the model is not a system of equations but a system of multi-valued mappings, Kakutani's fixed point theorem (1941) is invoked. An alternative to fixed point theorems consists of using intersection theorems on polytopes, with the KKM theorem of Knaster, Kuratowski and Mazurkiewicz (1929) perhaps the most prominent example. It is well-known that there exists a close relatioship between fixed point theorems and intersection theorems. Yet another alternative consists of results that claim the existence of solutions to variational inequality problems, the existence of stationary points, or the existence of zero points.

The results presented in this chapter follow the paper by Herings, Talman and Yang (1999) that generalizes earlier results of Browder (1960), MasColell (1974), and of Herings, Talman and Yang (1996). In case of Browder's theorem the polytope is the cartesian product of a polytope of one dimension less and the unit interval $[0, 1]$, while c is the unit vector with the one on the last position. MasColell's result is an extension of Browder's theorem to the case of multi- valued mappings. Both Browder and MasColell proved their results via a rather sophisticated machinery. As the Herings, Talman and Yang (1999) approach is constructive, it is possible to obtain an alternative but constructive proof of the above results.

Intersection theorems with a continuum of intersection points can be found in Freidenfelds (1974) and Herings and Talman (1998). Although Freidenfeld's intersection theorem typically has a continuum of intersection points, this is not necessarily the case. Freidenfeld's result generalizes the intersection theorem by Scarf (1967). Herings and Talman's (1998) results generalize a number of intersection theorems on the unit simplex, including the Scarf's result and the KKM-theorem, to intersection theorems on the unit cube and show the existence of a continuum of intersection points. It should be mentioned that compared with a large amount of existence results for a single fixed or zero point, existence theorems for a continuum of fixed or zero points are quire rare. It is also worth pointing out that the results in Herings, Talman and Yang (1999) presented in this chapter for the polytope can be extended to the nonempty convex compact sets by using the subgradient set of a convex function.

REFERENCES

Allgower, E.L. and Georg, K. *Numerical Continuation Methods: An Introduction.* Berlin: Springer-Verlag, 1990.

Brouwer LEJ. Über Abbildung von Mannigfaltigkeit. Mathematische Annalen. 1912; 71: 97-115.

Browder FE. On continuity of fixed points under deformation of continuous mapping. Summa Brasiliensis Mathematicae. 1960; 4: 183-191.

Brown DJ, DeMarzo PM and Eaves BC. Computing zeros of sections of vector bundles using homotopies and relocalization. Mathematics of Operations Research. 1996; 21: 26-43.

DeMarzo PM and Eaves BC. Computing equilibria of GEI by relocalization on a Grassmann manifold. Journal of Mathematical Economics. 1996; 26: 479-497.

Eaves BC.

1. On the basic theory of complementarity. Math. Programming. 1971; 1: 68-75.

2. Homotopies for computation of fixed points. Math. Programming. 1972; 3: 1-22.

Freidenfelds J. A set intersection theorem and applications. Math. Programming. 1974; 7: 199-211.

Fujishige S, and Yang Z. A lexicographic algebraic theorem and its applications. Linear Algebra and its Applications. 1998; 279: 75-91.

Hartman P, and Stampacchia G. On some nonlinear elliptic differential functional equations. Acta Mathematica. 1966; 115: 271-310.

Herings PJJ. On the existence of a continuum of constrained equilibria. Journal of Mathematical Economics. 1998; 30: 257-273.

Herings PJJ, and Talman AJJ. Intersection theorems with a continuum of intersection points. J. Optim. Theory Appl. 1998; 96: 311-335.

Herings PJJ, Talman AJJ, and Yang Z.

1. The computation of a continuum of constrained equilibria. Mathematics of Operations Research. 1996; 21: 675-696.

2. Variational Inequality Problems with a Continuum of Solutions: Existence and Computation. Discussion Paper No. 9972, CentER, Tilburg University, Tilburg, The Netherlands, 1999.

Kakutani S. A generalization of Brouwer's fixed point theorem. Duke Mathematical Journal. 1941; 8: 457-459.

Knaster B, Kuratowski C, and Mazurkiewicz C. Ein Beweis des Fixpunktsatzes für n-dimensionale Simplexe. Fundamenta Matematicae. 1929; 14: 132-137.

van der Laan G and Talman AJJ. A restart algorithm for computing fixed points without an extra dimension. Math. Programming. 1979; 17: 74-84.

van der Laan G, Talman AJJ, and Yang Z. Existence and approximation of robust solutions of variational inequality problems over polytopes. SIAM J. Control Optim. 1998; 37: 333-352.

MasColell A. A note on a theorem of F. Browder. Math. Programming. 1974; 6: 229-233.

Scarf H. The approximation of fixed points by a continuous mapping. SIAM J. Appl. Math. 1967; 15: 1328-1343.

Scarf, H., Hansen, T. *The Computation of Economic Equilibria*. New Haven-London: Yale University Press, 1973.

Schalk S and Talman AJJ. "Rationing Equilibia under General Price Rigidities". Mimeo, Tilburg University, Tilburg, The Netherlands, 1999.

Talman AJJ and Yamamoto Y. A simplicial algorithm for stationary point problems on polytopes. Math. Oper. Res. 1989; 14: 383-399.

Todd, M.J. *The Computation of Fixed Points and Applications*. Lecture Notes in Economics and Mathematical Systems 124. Berlin: Springer-Verlag, 1976.

Yamamoto Y. A path-following procedure to find a proper equilibrium of finite games. Intern. J. Game Theory. 1993; 22: 249-259.

Yang Z.

1. A simplicial algorithm for computing robust stationary points of a continuous function on the unit simplex. SIAM J. Control Optim. 1996; 34: 191-206.

2. *Computing Equilibria and Fixed Points*. Boston: Kluwer Academic Publishers, 1999.

Chapter 10

NEWTON-LIKE METHODS

This chapter deals with the problem of finding out an equilibrium in an oligopolistic market model where several subjects supply a single homogeneous product in a non-cooperative manner. The problem is reduced to a nonlinear equation some terms of which are determined by solving nonlinear complementarity problems. An algorithm is presented that combines the Newton method steps with the dichotomy techniques. Under certain assumptions, the algorithm is shown to be convergent at the quadratic rate. Finally, the algorithm is extended to the case of nonlinear production costs, and its linear convergence is demonstrated.

10.1 PROBLEM SPECIFICATION

Consider n firms producing a homogeneous product and a single extraneous supplier of the same item. Let q_i denote the i-th firm's output, whereas $Q \geq 0$ be the volume of the extraneous supply. Then the total product volume G is given by $G = Q + \sum q_i$.

First, we list the assumptions needed as a base to describe and justify the problem and the algorithm to solve it.

S1. Suppose that each firm has a linear cost function $f_i(q_i) = c_i q_i,\ c_i > 0, i = 1, \ldots, n$.

S2. Assume that the inverse demand function $p = p(G)$ is twice differentiable, its first derivative takes negative values whereas its second derivative is positive for all $G > 0$. Moreover, suppose that $p(G) \to 0$ as $G \to +\infty$, the second derivative p'' is Lipschitz continuous over each segment $[a, b],\ b > a > 0$, and the function $p(G)G$ is concave with respect to G.

In framework of the classical Cournot model, each participant chooses its output value $q_i \geq 0$ maximizing its profit function $\mu_i(G, q_i) = q_i p(G) - f_i(q_i)$ under a conjecture that all the other producers do not change their outputs. Therefore, in terms of the first order optimality conditions, an *equilibrium* of the Cournot model is determined as a vector $(G, q_1, \ldots, q_n) \in R_+^{n+1}$ with $G > 0$ satisfying the balance equality $G = Q + \sum q_i$, and such that for each i the value $q_i \geq 0$ solves the following complementarity problem

$$\varphi_i(G, q_i) = c_i - q_i p'(G) - p(G) \geq 0, \tag{10.1}$$

and

$$q_i \varphi_i(G, q_i) = 0. \tag{10.2}$$

In Chapters 4 and 5 of this book, we examined a more general situation when each participant uses its own conjecture concerning the market reaction to its output volume changing. In this case, the factor before $(-p'(G))$ in the term $\varphi_i(G, q_i)$ need not be equal to q_i as in (10.1) but represents some function of variables G and q_i. In this chapter, we study the case when that function is linear with respect to G and q_i. More precisely, we determine the *extended equilibrium* as a vector $(G, q_1, \ldots, q_n)$ with $G > 0$ satisfying the balance equality

$$G = Q + \sum q_i, \tag{10.3}$$

and such that for each $i = 1, \ldots, n$ the value $q_i \geq 0$ solves the complementarity problem

$$\varphi_i(G, q_i) = c_i - \alpha_i q_i p'(G) - \sigma_i G p'(G) - p(G) \geq 0, \tag{10.4}$$

and

$$q_i \varphi_i(G, q_i) = 0. \tag{10.5}$$

We showed in Chapter 4 that under all the above assumptions concerning the function $p(G)$ and the parameters α_i, σ_i satifying the relationships $\alpha_i > 0$, $\sigma_i \geq 0$, $\alpha_i + \sigma_i \leq 1$, a solution to problem (10.3)–(10.5) exists uniquely.

In recent years, a lot of papers appeared in which, in addition to the Cournot firms, there is one particular agent supplying the same product and acting in a special manner. Namely, it sets production levels in an optimal fashion by explicitly considering the reaction of the other firms to its output variations. In contrast to this, the other agents attempt to maximize their individual profits under the Cournot assumption that the remaining firms will hold their outputs at the existing levels. The special agent is sometimes called a *leader* or a *Stackelberg* firm as it was Stackelberg (1934) who originally proposed such a strategy of behaviour. A numerical method solving that kind of problems through the use of the piecewise linear approximation of the Cournot-firms' joint reaction curve is devised by Sherali, Soyster and Murphy (1983). In order to obtain the approximation, they solve mathematical programs with one constraint depending upon a parameter. The parameter value is picked up so that the corresponding Lagrange multiplier should equal zero. Harker and Choi (1991) also consider the model of the market with a leader and reduce it to a bilevel optimization problem on a polyhedron. The latter problem is solved then by an exterior penalty function method using the simplicial decomposition algorithm.

In Chapter 5, we generalized the Stackelberg model and embedded it into the above frameworks. In order to do it, one need defining the influence quotients for the leaders, as was done, for example, in Section 5.1, in terms of the first order optimality conditions.

In the paper by Bulavsky and Kalashnikov (1994), the authors presented an algorithm to compute the optimal extraneous supply Q by the leader, where his expected profit function $\mu_1(Q)$ is maximized (here $i = 1$ is the number assigned to the leader). However, to implement this algorithm, one has to compute at each step the equilibrium for the remaining agents (followers), i.e. to solve the problem (10.3)–(10.5), given a value $Q \geq 0$. Since the latter procedure is used a great number of times, we had to develop an algorithm fast enough to do it.

In contrast to the papers by Ruffin (1971), Okuguchi (1973), Roberts and Sonnenschein (1976), Szidarovszky and Yakowitz (1977), Sherali e.a. (1983), Novshek (1985), Cohen (1984), Nagurney and Zhao (1991), Nagurney e.a. (1992), in which various methods of mathematical programming and complementarity theory were used to solve problem (10.3)–(10.5), this section examines a method of iterations by the parameter G. The idea of that approach was proposed by Marcotte (1984,1985,1988), developed by Harker and Pang (1988,1990), and can be desribed as follows. For each value $G > 0$, agent i chooses the optimal (according to its conjecture concerning the market reaction) output volume $q_i = q_i(G)$ as a unique (due to the properties of the function p and the positivity of value α_i) solution to problem (10.4)–(10.5). Under the above assumptions, the solution is clearly given by the explicit formula

$$q_i(G) = \max\left\{0, \frac{c_i - p(G) - \sigma_i G p'(G)}{\alpha_i p'(G)}\right\}, \quad i = 1, \ldots, n. \tag{10.6}$$

Thus problem (10.3)–(10.5) reduces to finding a fixed point of the mapping $\sum q_i(G) + Q$, i.e. a solution G^* of the scalar equation

$$F(G) = G - Q - \sum_{i \in I(G)} q_i(G) = 0, \tag{10.7}$$

where $I(G) = \{i \mid q_i(G) > 0\}$. In contrast to Marcotte (1984,1985,1988), we are going to apply Newton-type steps to solve (10.7). Conditions imposed here on the function $p(G)$ provide for the existence of the first derivative of the function $q_i(G)$ at the points where it takes positive values. Besides, it is readily verified that after having vanished, the function $q_i(G)$ never changes its value with G growing up. Therefore, the latter function is piecewise smooth, and so is the gap function $F(G)$. Moreover, under a certain additional assumption about $p(G)$, the function $q_i(G)$ is concave over the intersection of the intervals $M_i = \{G > 0 \mid 0 < q_i(G) < G\}$ and $(G^*, +\infty)$. Solution G^* of equation (10.7) clearly belongs to each interval M_i, $i \in I(G^*)$. Hence, the gap function $F(G)$ is convex over the intervals into which the half-line $[G^*, +\infty)$ is divided by the upper bounds of sets M_i, $i \in I(G^*)$. However, it is not so over the whole half-line $[G^*, +\infty)$. Due to that, one cannot use the pure Newton method to solve equation (10.7).

Taking into account all the above information, in Section 10.3 we describe an algorithm solving equation (10.7) after having examined some properties of the individual response function $q_i(G)$ in Section 10.2. The algorithm combines the Newton-like steps with the dichotomy. In Section 10.4, we show the algorithm to be convergent at the quadratic rate. Section 10.5 is dedicated to the extending of the algorithm to the case where the marginal cost of production increases. Finally, in Section 10.6, we present an optimal step accuracy control for bilevel optimization algorithms and its application to the Newton-like algorithm for solving nonlinear comlementarity problems.

10.2 INDIVIDUAL RESPONSE FUNCTIONS

In this section, we study the *individual response function* $q_i(G)$, the value of which is the optimal response of the i-th producer to the price $p(G)$ of the product unit.

LEMMA 10.1. *Under assumptions* S1–S2, *the set* $M_i = \{G > 0 \mid 0 < q_i(G) < G\}$ *is an interval* (H_i, G_i) *of the positive real half-axis.*

PROOF. It is readily verified that $M_i = A_i \cap B_i$ where

$$A_i = \{G > 0 \mid c_i - p(G) - \sigma_i p'(G)G < 0\},$$

$$B_i = \{G > 0 \mid c_i - p(G) - (\alpha_i + \sigma_i)p'(G)G > 0\}.$$

Since $0 \le \sigma_i < 1$, the function $p(G) + \sigma_i p'(G)G = (1 - \sigma_i)p(G) + \sigma_i [p(G)G]'$ is monotone decreasing and vanishing as $G \to +\infty$. Hence, for each $i = 1, \ldots, n$ there exists a $G_i > 0$ such that $q_i(G) = 0$ for $G \ge G_i$, and $q_i(G) > 0$ at $0 < G < G_i$. Therefore, the set A_i coincides with the interval $(0, G_i)$. On the other hand, by a similar argument, we obtain that $B_i = (H_i, +\infty)$ where $H_i > 0$ is a (unique) solution of the equation $c_i = p(G) + (\alpha_i + \sigma_i)p'(G)G$. At last, since $q_i(G_i) = 0$, then G_i belongs to B_i, hence $G_i \ge H_i$. Thus we have demonstrated that $M_i = (H_i, G_i)$ which completes the proof. ■

As mentioned above, assumptions concerning $p(G)$ provide for the differentiability of the individual response functions $q_i(G)$ over the interval where $q_i(G) > 0$. Hence the latter functions are piecewise smooth which implies the same property of the *gap function* $F(G)$.

S3. In addition, assume that the function $\psi(G) = \dfrac{p''(G)}{[p'(G)]^2 G}$ does not increase as G grows.

We note that the latter condition is not too restrictive. In particular, it holds for the inverse demand function $p(G) = AG^{-\gamma}$, $A > 0$, $0 < \gamma \le 1$, widely used in applications. As it was established in Chapter 4, the following relationships are true under the above assumptions:

$$F(G) > 0 \qquad \text{for} \qquad G > G^*, \tag{10.8}$$

and

$$\sum_{i \in I(G)} q_i'(G) < 1 \qquad \text{for} \qquad G \geq G^*; \tag{10.9}$$

here G^* is the solution of equation (10.7). Also we need the following property of the individual response functions.

LEMMA 10.2. *Under assumptions* S1–S3, *each individual response function* $q_i(G)$ *is concave over the interval* $[G^*, G_i)$, $i = 1, \ldots, n$.

PROOF. Being monotone, the function $\psi(G)$ is differentiable almost everywhere over the interval $[G^*, G_i]$. Therefore, the function $\dfrac{p''(G)}{p'(G)}$ is also differentiable almost everywhere (the latter property can also be deduced from the Lipschitz continuity of p''). Differentiating the individual response function $q_i(G)$ and using (10.6), we obtain the equality

$$q_i'(G) = -\frac{1+\sigma_i}{\alpha_i} - q_i(G)\frac{p''(G)}{p'(G)} - \frac{\sigma_i}{\alpha_i} G \frac{p''(G)}{p'(G)}, \tag{10.10}$$

$i = 1, \ldots, n$. Hence, the second derivative $q_i''(G)$ exists at the same points where the derivative $\left(\dfrac{p''(G)}{p'(G)}\right)'$ does. Omitting the variable G for the brevity purpose, write down the following explicit expression for the second derivative

$$q_i''(G) = -q_i \left[\left(\frac{p''}{p'}\right)' - \left(\frac{p''}{p'}\right)^2\right] - \frac{\sigma_i}{\alpha_i} G \left[\left(\frac{p''}{p'}\right)' - \left(\frac{p''}{p'}\right)^2\right] + \frac{1}{\alpha_i}\frac{p''}{p'} =$$

$$= -\left(\frac{q_i}{G} + \frac{\sigma_i}{\alpha_i}\right)\left\{G\left(\frac{p''}{p'}\right)' - G\left(\frac{p''}{p'}\right)^2 - \frac{p''}{p'}\right\} + \left(\frac{1}{\alpha_i} - \frac{\sigma_i}{\alpha_i} - \frac{q_i}{G}\right)\frac{p''}{p'}. \tag{10.11}$$

Now we show that $q_i''(G) \leq 0$. Indeed, after transforming and integrating the term in the curled braces in (10.11), we get

$$\left\{G\left(\frac{p''}{p'}\right)' - G\left(\frac{p''}{p'}\right)^2 - \frac{p''}{p'}\right\} = G\frac{p''}{p'}\left\{\left(\frac{p''}{p'}\right)'\frac{p'}{p''} - \frac{p''}{p'} - \frac{1}{G}\right\} =$$

$$= G\frac{p''}{p'}\frac{d}{dG}\left(\ln\frac{|p''|}{|p'|} - \ln|p'| - \ln G\right) = G\frac{p''}{p'}\frac{d}{dG}\left[\ln\frac{|p''|}{(p')^2 G}\right].$$

Since $p' < 0$, $p'' > 0$, and $\psi(G)$ is non-increasing, the first term in (10.11) is nonpositive, whereas the sign of the second term is opposite to that of the term in parentheses. The latter can be estimated as follows

$$\frac{1}{\alpha_i} - \frac{\sigma_i}{\alpha_i} - \frac{q_i}{G} \geq 1 - \frac{q_i}{G} \geq 0$$

due to the properties of α_i, σ_i and inequality (10.8) holding for $G \in [G^*, G_i)$. Thus, the second derivative $q_i''(G)$ has been shown to be nonpositive over the interval $[G^*, G_i)$ which implies the function q_i concavity and completes the proof. ∎

EXAMPLE 10.1. We illustrate Lemmas 10.1 and 10.2 with a concrete inverse demand function. Namely, as a sample of p, choose the following power function widely used in applications (*cf.* Sherali e.a. 1983)

$$p(G) = AG^{-\gamma},\ A > 0,\ 0 < \gamma \leq 1. \tag{10.12}$$

Make use of the explicit formula (10.6) and obtain

$$q_i(G) = \max\left\{0, \frac{c_i - AG^{-\gamma} + \sigma_i \gamma AG^{-\gamma}}{-\alpha_i \gamma AG^{-\gamma-1}}\right\}, \quad i = 1, \dots, n.$$

Hence, it follows immediately that the subset $M_i = \{G > 0 : 0 < q_i(G) < G\}$ from Lemma 10.1 is indeed an interval (G_i^-, G_i^+) with

$$G_i^- = \frac{A^{1/\gamma}(1 - \gamma\alpha_i - \gamma\sigma_i)^{1/\gamma}}{c_i^{1/\gamma}}, \quad G_i^+ = \frac{A^{1/\gamma}(1 - \gamma\sigma_i)^{1/\gamma}}{c_i^{1/\gamma}}.$$

Moreover, the function $q_i = q_i(G)$ over that interval takes the form

$$q_i(G) = \tau G - \nu G^{1+\gamma}, \quad \tau = \frac{1 - \gamma\sigma_i}{\gamma\alpha_i}, \quad \nu = \frac{c_i}{\gamma\alpha_i A}. \tag{10.13}$$

Making use of (10.13), it is easy to verify that the function q_i is concave over the interval (G^*, G_i^+), which confirms the statement of the lemma. ∎

10.3 DESCRIPTION OF ALGORITHM

The algorithm consists in the stepwise updating the left and right bounds G^- and G^+, respectively, of the interval containing the solution G^* of equation (10.7). As mentioned above, the method makes use of procedures of two types: Newton-like steps and the dichotomy techniques.

PRELIMINARY STEP. At the preliminary stage, we put $G^- := Q \geq 0$ and look for G^+ using the standard scheme of doubling the interval length. Namely, after fixing an arbitrary $\Delta > 0$, accept $G^0 := G^- + \Delta$ and calculate the gap $F(G^0)$. If $|F(G^0)| \leq \varepsilon$ where $\varepsilon > 0$ is the terminal precision parameter, the procedure is stopped after having taken $G^* := G^0$. If $F(G^0) < -\varepsilon$, then put $G^- := G^0$, $\Delta := 2\Delta$, and go to the beginning of the preliminary stage. Further, in case of $F(G^0) > \varepsilon$ and $\sum q_i(G^0) > 0$, we assign $G^+ := G^0$ and go to the Principal Step of the algorithm. Finally, if $F(G^0) > \varepsilon$ and $\sum q_i(G^0) = 0$, we set $\Delta := \Delta/2$ and return to the beginning of the preliminary step having taken $G^0 := G^0 + \Delta$. It is clear that after a finite number of preliminary steps, the value G^* is covered by the interval $[G^-, G^+]$.

PRINCIPAL STEP. Construct the sets $I(G^+) \subseteq I(G^-)$ (where $I(G) = \{1 \leq i \leq n \mid q_i(G) > 0\}$) and set $k := 0$. From here onward, we add a unit to the current value of k before executing either of Procedures 10.1, 10.2, or 10.3.

PROCEDURE 10.1. Realize a Newton-type step by the formula

$$G^k = G^+ - \frac{G^+ - Q - \sum_{I(G^+)} q_i(G^+)}{1 - \sum_{I(G^+)} q_i'(G^+ - 0)} \tag{10.14}$$

and calculate the gap function value at the point G^k. If $|F(G^k)| \leq \varepsilon$, then assume $G^* := G^k$ and stop the algorithm. If $F(G^k) > \varepsilon$, update $G^+ := G^k$, $I(G^+) := I(G^k)$ and return to the beginning of Procedure 10.1. Finally, if $F(G^k) < -\varepsilon$, then choose $G^- := G^k$, $I(G^-) := I(G^k)$ and go to Procedure 10.2.

PROCEDURE 10.2. Find $G^k := (G^- + G^+)/2$ and compute the gap function at this point. Like above, in case $|F(G^k)| \leq \varepsilon$ the process is finished. If $F(G^k) < -\varepsilon$, update $G^- := G^k$, $I(G^-) := I(G^k)$ and return to the beginning of Procedure 10.2. If $F(G^k) > \varepsilon$ but $I(G^k) = I(G^+)$, then put $G^+ := G^k$ and return again to the beginning of Procedure 10.2. Only if $F(G^k) > \varepsilon$ and $I(G^+) \subset I(G^k)$ then go to Procedure 10.1 after having updated $G^+ := G^k$, $I(G^+) := I(G^k)$.

Now we explain why, after having executed the *dichotomy step* resulting in the relationships $F(G^k) > \varepsilon$ and $I(G^k) = I(G^+)$, we use again the dichotomy but not the Newton refinement. The point is that the equality $I(G^k) = I(G^+)$ shows that the last iterate G^k belongs to the same smooth part of the curve $y = F(G)$ that has provided the last Newton refinement with the negative gap value. Since $G^k < G^+$ and the smooth part of the curve is the graph of the convex function (see Lemma 10.2), Newton step (10.14) would have driven us again into the negative gap area situated on the left of G^-. Having foreseen this outcome, we decline refinement (10.14) and make use of the dichotomy.

The latter analysis enables us to propose the following modification of the dichotomy procedure which can be useful when the gap function breakpoints G_i accumulate near the solution G^*. Then we can detect active producers with the help of the following procedure.

PROCEDURE 10.3. Consider the case when $F(G^k) < -\varepsilon$ after fulfilling Newton refinement (10.14). This outcome is possible only if the set of active producers at the point G^+ is thinner than that at the solution G^*, i.e. $I(G^+) \subset I(G^*)$. Indeed, if $I(G^*) \subseteq I(G^+)$, the Newton shift would not have driven the next iteration out of the limits of the interval (G^*, G^+). Furthermore, since $G^k < G^*$, we have $I(G^*) \subset I(G^k)$. According to Procedure 10.2, we then set $G^- := G^k$, $I(G^-) := I(G^k)$, and execute the dichotomy step. Suppose that it leads to the outcome $F(G^k) > \varepsilon$ and $I(G^k) = I(G^+)$. As $I(G^+) \subset I(G^*) \subseteq I(G^-)$, we call the producers with numbers $i \in I_d = I(G^-) \setminus I(G^+)$ as *doubtful* ones because we cannot be sure whether they belong or not belong to the set $I(G^+)$. In order to get closer to $I(G^*)$ we now choose a number $m \in I_d$ either at random or by the minimal unit cost factor $c_m = \min\{c_i \mid i \in I_d\}$. Then we compute the point G_m at which the m-th participant stops producing, as a (unique) solution of the equation

$$c_m - p(G) - \sigma_m p'(G)G = 0, \tag{10.15}$$

and calculate the gap $F(G_m)$. If $F(G_m) > \varepsilon$, then set $G^+ := G_m$, $I(G^+) := I(G_m)$ and go to the beginning of Procedure 10.1. If $|F(G_m)| \leq \varepsilon$, we stop the algorithm after having assigned $G^* := G_m$. At last, if $F(G_m) < -\varepsilon$, we assign $G^- := G_m$, $I(G^-) := I(G_m)$ and return to the beginning of Procedure 10.3. Now we have the thinner set of doubtful producers, as $m \notin I_d$ for sure.

10.4 CONVERGENCE RATE

We emphasize that there can occur only finite number of dichotomy steps (Procedure 10.2), as well as the steps involving the active producers (Procedure 10.3). As for the Newton refinement, it is executed at the points G^+ with the positive gap values. In order to formulate the following convergence result, assume that the procedures will not be interrupted after having achieved the prescribed precision, i.e. we consider $\varepsilon = 0$.

THEOREM 10.1. *Under conditions of Lemma* 10.2, *either the value* G^* *is found after a finite number of steps, or the infinite sequence* $\{G^k\}$ *converges to* G^* *from above at the quadratic rate.*

PROOF. It is readily verified that if $G^k \neq G^*\ \forall k$, then starting from some number k onwards, each step of the algorithm will be the Newton refinement (10.14) with the outcome $F(G^k) > 0$ and $I(G^k) = I(G^*)$. In the other words, since that moment, all the elements of the sequence $\{G^k\}$ belong to the interval $(G^*, \tilde{G})$ where $\tilde{G} = \min\{G_i \mid F(G_i) > 0\}$. As the gap function $F(G) = G - Q - \sum q_i$ is convex over that interval (see Lemma 10.2), and $F'(G^*) = 1 - \sum\limits_{i \in I(G^*)} q_i'(G^*) > 0$ due to (10.9), we can guarantee the convergence of the Newton method iterates G^k to G^* from above (*cf.* Berezin and Zhidkov,1962, pp.135–139 and/or Dennis and Schnabel,1983, Theorem 2.4.3). Moreover, since p'' is Lipschitz continuous, estimate (10.9) together with the Newton method convergence (*cf.* Bakhvalov,1973, p.412 and/or Dennis and Schnabel,1983, Theorem 2.4.3) imply the inequality

$$G^{k+1} - G^* \leq D(G^k - G^*)^2, \quad D > 0, \tag{10.16}$$

holding from some number k onward. The proof is complete. ■

Remark. It is well-known (Berezin and Zhidkov,1962, p.139) that the value $V_2/2v_1$ can serve as the constant D in (10.16); here

$$V_2 = \max_{[G^*, \tilde{G}]} |F''(G)|, \qquad v_1 = \min_{[G^*, \tilde{G}]} |F'(G)|. \tag{10.17}$$

Formulae (10.6),(10.10),(10.11) for functions $q_i(G)$ and their first and second derivatives clearly imply that the value $1/\alpha_i$ is the common factor for all those expressions. Hence, even small values α_i cannot spoil the constant D badly as presented in estimates (10.16)–(10.17) of the convergence rate. The latter proves the proposed algorithm to be robust enough when applied to the considered problem.

10.5 EXTENSION TO THE CASE OF NON-LINEAR COSTS

In this Section, we extend our algorithm to the case where the *marginal cost of production* may increase. In the other words, from here on, we assume that the cost functions $f_i(q_i)$, $i = 1,\dots,n$, need not being linear, but are twice continuously differentiable convex functions. In this case, the Newton refinement procedure can also be applied to the gap function $F(G)$, since the first derivatives of the producers' outputs are well-defined (see Chapter 4 of this book):

$$q_i'(G) = \frac{p'(1+\sigma_i) + \sigma_i p'' G + \alpha_i q_i p''}{f_i''(q_i) - \alpha_i p'}, \qquad i \in I(G), \tag{10.18}$$

even though the explicit formula (10.6) is not valid; here the argument G is omitted in the righthand side for the brevity purpose. However, in contrast to the previous sections, now we cannot warrant that the outputs $q_i = q_i(G)$ are piecewise concave. Therefore, the gap function $F(G)$ need not being piecewise convex which makes the above convergence proof invalid for this case.

Nevertheless, in what follows, we modify our algorithm and prove that it is convergent under slightly stricter assumptions than above. First of all, we note that if the cost functions are linear, i.e. when $f_i(q_i) = c_i q_i$, and if $\alpha_i + \sigma_i < 1$, $i = 1,\dots,n$, then the equilibrium total volume $G = G(c_1, c_2, \dots, c_n)$ is monotone decreasing with respect to each single variable c_k.

LEMMA 10.3. *Under assumptions* S1–S2, *together with the strict inequalities* $\alpha_i + \sigma_i < 1$, *the well-defined equilibrium total volume* $G = G(c_1, c_2, \dots, c_n)$ *is a monotone decreasing function with respect to every single variable* c_k.

PROOF. Since the cost functions are linear, we can use the explicit formula (10.6) and write down

$$q_i(G) = \frac{c_i - p(G) - \sigma_i G p'(G)}{\alpha_i p'(G)}, \qquad i \in I(G). \tag{10.19}$$

If $G = G(c_1, c_2, \dots, c_n)$ is the equilibrium total volume, the balance equality and (10.19) imply

$$G = Q + \sum_{i\in I(G)} q_i(G) =$$

$$= Q + \frac{1}{p'(G)} \sum_{i\in I(G)} \frac{c_i}{\alpha_i} - \frac{p(G)}{p'(G)} \sum_{i\in I(G)} \frac{1}{\alpha_i} - G \sum_{i\in I(G)} \frac{\sigma_i}{\alpha_i}. \tag{10.20}$$

Moving the last term in (10.20) to the lefthand side and multiplying the both sides by $p'(G) \neq 0$, we come to the expression

$$p'(G)G\left(1+\sum_{i\in I(G)}\frac{\sigma_i}{\alpha_i}\right)=p'(G)Q+\sum_{i\in I(G)}\frac{c_i}{\alpha_i}-p(G)\sum_{i\in I(G)}\frac{1}{\alpha_i}. \tag{10.21}$$

Now differentiating both sides of (10.21) by c_k, $k \in I$, we get

$$\frac{\partial G}{\partial c_k}\left[p''(G)G+p'(G)\right]\left(1+\sum_{i\in I(G)}\frac{\sigma_i}{\alpha_i}\right)=$$

$$=p''(G)Q+\frac{1}{\alpha_k}-\frac{\partial G}{\partial c_k}p'(G)\sum_{i\in I(G)}\frac{1}{\alpha_i}. \tag{10.22}$$

Rearrange the terms of (10.22) and come to the following equation with respect to $\partial G/\partial c_k$:

$$\frac{\partial G}{\partial c_k}\beta_k=\frac{1}{\alpha_k}, \tag{10.23}$$

where

$$\beta_k=\left[p''(G)G\left(1+\sum_{i\in I(G)}\frac{\sigma_i}{\alpha_i}\right)-p''(G)Q+\right.$$

$$\left.+p'(G)\left(1+\sum_{i\in I(G)}\frac{\sigma_i+1}{\alpha_i}\right)\right]. \tag{10.24}$$

Since $p''(G)>0$, we estimate

$$\beta_k\le\left[p''(G)G\left(1+\sum_{i\in I(G)}\frac{\sigma_i}{\alpha_i}\right)+p'(G)\left(1+\sum_{i\in I(G)}\frac{\sigma_i+1}{\alpha_i}\right)\right]. \tag{10.25}$$

As the function $p(G)G$ is concave, then $p''(G)G+2p'(G)\le 0$. Therefore, the quotient β_k is negative, if

$$\left(1+\sum_{i\in I(G)}\frac{\sigma_i+1}{\alpha_i}\right)>2\left(1+\sum_{i\in I(G)}\frac{\sigma_i}{\alpha_i}\right).$$

But the latter inequality is valid, since $\alpha_i+\sigma_i<1$, $i=1,\ldots,n$. Thus we have shown that $\beta_k<0$; then (10.23) implies that

$$\frac{\partial G}{\partial c_k}<0,$$

which completes the proof. ■

Now in order to extend our algorithm to the case of nonlinear production cost functions, we exploit a bit stronger assumption about the inverse demand function $p(G)$ which is widely used in the literature (*cf.* Sherali e.a., 1983).

S4. Namely, we assume that

$$p'(G) + p''(G)G \leq 0, \qquad \forall G > 0, \tag{10.26}$$

that clearly implies, by the way, the concavity of the product $p(G)G$.

Moreover, as it has been already proven in Chapter 4 of this book, assumption (10.26) also implies that the derivative $q_i'(G)$ determined by (10.18) takes only nonpositive values, i.e. $q_i'(G) \leq 0$, $\forall G > 0$. The latter means that the optimal i-th agent's output $q_i(G)$ does not increase as a function of the total output volume G.

We note that assumption (10.26) is clearly valid for a quite wide class of inverse demand power functions of the following type. Suppose that the maximal total market supply does not exceed some value $G_{\max} > 0$. If we choose a number $0 < \gamma < 1$ and positive constants A, B such that $G_{\max} \leq (A/B)^{1/\gamma}$, we define the inverse demand function as follows:

$$p(G) = \begin{cases} A - B \cdot G^{\gamma}, & 0 \leq G \leq (A/B)^{1/\gamma}, \\ 0, & G > (A/B)^{1/\gamma}. \end{cases}$$

It is easily checked that the function of this class satisfies both assumptions S2 and S3.

10.5.1 DESCRIPTION OF THE EXTENDED ALGORITHM

S5. From now onward, we allow the cost functions $f_i(q_i)$, $i = 1, \ldots, n$, to be not necesarily linear, but general twice continuously differentiable convex functions.

We consider the equilibrium problem: Find an $(n+1)$-tuple $(G, q_1, \ldots, q_n)$ with $G > 0$ satisfying the balance equality

$$G = Q + \sum q_i, \tag{10.27}$$

and such that for each $i = 1, \ldots, n$, the value $q_i \geq 0$ solves the complementarity problem

$$\varphi_i(G, q_i) = f_i'(q_i) - \alpha_i q_i p'(G) - \sigma_i G p'(G) - p(G) \geq 0, \tag{10.28}$$

and

$$q_i \varphi_i(G, q_i) = 0. \tag{10.29}$$

As obtained in Chapter 4, the solution to problem (10.27)–(10.29) exists uniquely under the assumptions S2 and S5. Denote the equilibrium total volume by G^*

and suppose that we have at hand an estimate G^0 for G^* from above, i.e. $G^* < G^0$. Then we may find the agents' optimal responses $q_i = q_i(G^0)$, $i = 1, \ldots, n$, as solutions to complementarity problems (10.28)–(10.29) with G^0 substituted for G. Due to the monotonicity of $q_i(G)$, we clearly have

$$q_i = q_i(G^0) \le q_i^*, \qquad i = 1, \ldots, n. \tag{10.30}$$

Therefore, we can estimate from below the equilibrium total volume G^* as follows:

$$\sum_{i \in I(G^0)} q_i(G^0) + Q \le G^*. \tag{10.31}$$

Now we suppose the assumptions S3 and S4 to be valid, too, and describe the steps of the extended algorithm.

STEP 1. Take $c_i = f_i'\left(q_i(G^0)\right)$, $i = 1, \ldots, n$, and consider problem (10.3)–(10.5) with the linear production costs $c_i q_i$. Apply the Newton-like algorithm to it, as described in Section 10.3, and find a solution $(\hat{G}, \hat{q}_1, \ldots, \hat{q}_n)$.

Since $c_i \le f_i'(q_i^*)$ because of the convexity of f_i, $i = 1, \ldots, n$, Lemma 10.3 implies $G^* \le \hat{G} < G^0$ (if $\hat{G} = G^0$, then it is evidently the equilibrium: $G^* = \hat{G} = G^0$).

STEP 2. Compute the agents' optimal responses $q_i = q_i(\hat{G})$, $i = 1, \ldots, n$, and find the nonnegative gap

$$F(\hat{G}) = \hat{G} - Q - \sum_i q_i(\hat{G}). \tag{10.32}$$

If $F(\hat{G}) \le \varepsilon$, then we stop and accept $\hat{G}$ as an approximation to G^*. Otherwise, update the quotients $c_i = f_i'\left(q_i(\hat{G})\right)$, $i = 1, \ldots, n$, and go to Step 1.

The convergence of the described algorithm can be verified by the standard technique for algorithms where the linearization is involved. Namely, the following theorem obtains.

THEOREM 10.2. *Under assumptions* S2–S5, *the extended algorithm converges linearly with respect to the gap function.*

PROOF. Denote by $q_i^0 = q_i^0(G)$ the optimal responses of the agents in the linearized problem, i.e. the solutions of problem (10.4)–(10.5) for a given G. If we define the subset of active agents in the linearized problem as $I_0(G) = \{1 \le i \le n \mid q_i^0(G) > 0\}$, then it is clear that $I_0(G^0) = I(G^0)$. Now define the gap function for the linearized problem as follows

$$F_0(G) = G - Q - \sum_{i \in I_0(G)} q_i^0(G), \tag{10.33}$$

and remark that

$$F(G^0) = F_0(G^0) > 0, \tag{10.34}$$

as implied by the definition of q_i^0, $i = 1, \dots, n$. It is easy to see, that Step 2 of the extended algorithm finds such a point $G^1 = \hat{G}$ that

$$F_0(G^1) = 0. \tag{10.35}$$

As it was noted above, we have $G^* \le G^1 < G^0$. Now we prove that $I(G) = I_0(G)$, $\forall G \in [G^1, G^0]$. Indeed, let $G \in [G^1, G^0]$, $i \in I(G)$, i.e. $q_i = q_i(G) > 0$. Then q_i satisfies the equation

$$f_i'(q_i) - (\alpha_i q_i + \sigma_i G)p'(G) - p(G) = 0. \tag{10.36}$$

As $G \le G^0$, then $q_i(G) \ge q_i(G^0)$, hence $c_i = f_i'\left(q_i(G^0)\right) \le f_i'(q_i)$ due to the convexity of f_i.

Now define a value q_i^0 as a solution of the equation

$$c_i - (\alpha_i q_i^0 + \sigma_i G)p'(G) - p(G) = 0. \tag{10.37}$$

Compare equations (10.36) and (10.37) and deduce the estimate $q_i^0 \ge q_i(G) > 0$, which means that q_i^0 is really the optimal response of the agent i to the price value $p(G)$ in the linearized problem. In the other words, $q_i^0 = q_i^0(G)$, and moreover, $i \in I_0(G)$.

Conversely, if $j \notin I(G)$, then

$$f_j'(+0) - \sigma_i G p'(G) - p(G) \ge 0. \tag{10.38}$$

Taking into account the estimate $c_j = f_j'\left(q_j(G^0)\right) \ge f_i'(+0)$, we deduce from (10.38)

$$c_j - \sigma_i G p'(G) - p(G) \ge 0,$$

which means that $j \notin I_0(G)$. Therefore, $I(G) = I_0(G)$ for all $G \in [G^1, G^0]$.

Each function $q_i^0(G)$, $i \in I_0(G)$, is differentiable over $[G^1, G^0]$, and its derivative takes the form

$$q_i'^0(G) = \frac{p'(1+\sigma_i) + \sigma_i p'' G + \alpha_i q_i^0 p''}{-\alpha_i p'}. \tag{10.39}$$

Making use of (10.18), we compare the values $q_i'(G)$ and $q_i'^0(G)$:

$$q_i'(G) = \frac{p'(1+\sigma_i) + \sigma_i p'' G + \alpha_i q_i p''}{f_i''(q_i) - \alpha_i p'} \le$$

$$\le \frac{p'(1+\sigma_i) + \sigma_i p'' G + \alpha_i q_i^0 p''}{f_i''(q_i) - \alpha_i p'} \cdot \frac{-\alpha_i p'}{-\alpha_i p'} =$$

$$= -\frac{\alpha_i p'}{f_i''\left(q_i(G)\right) - \alpha_i p'(G)} \cdot q_i'^0(G), \quad \forall G \in [G^1, G^0]. \tag{10.40}$$

The continuity of all the terms in the latter fraction and the strict negativity of $p'(G)$ imply

$$0 < \min_{G \in [G^1, G^0]} \left\{ \min_{i \in I(G)} \frac{-\alpha_i p'}{f_i''\left(q_i(G)\right) - \alpha_i p'(G)} \right\} = \delta \le 1. \tag{10.41}$$

From (10.40) and (10.41), we deduce the estimate

$$q_i'(G) \leq \delta \cdot q_i'^0(G), \quad \forall G \in [G^1, G^0],\ i \in I(G);$$

hence one easily obtains (here we recall that $I(G) = I_0(G)$ for all $G \in [G^1, G^0]$):

$$F'(G) = 1 - \sum_{i \in I(G)} q_i'(G) \geq 1 - \delta \cdot \sum_{i \in I_0(G)} q_i'^0(G) \geq$$

$$\geq \delta \left[1 - \sum_{i \in I_0(G)} q_i'^0(G) \right] = \delta \cdot F_0'(G) \geq 0. \tag{10.42}$$

Now making use of relationships (10.34), (10.35), and (10.42), we finally get

$$F(G^0) - F(G^1) = \int_{G^1}^{G^0} F'(G)dG \geq \delta \cdot \int_{G^1}^{G^0} F_0'(G)dG =$$

$$= \delta \cdot \left[F_0(G^0) - F_0(G_1) \right] = \delta \cdot F_0(G^0) = \delta \cdot F(G^0),$$

whence it follows that

$$F(G^1) \leq (1 - \delta) \cdot F(G^0). \tag{10.43}$$

Since $0 \leq (1 - \delta) < 1$, inequality (10.43) implies the linear convergence of the extended algorithm with respect to the gap function and thus completes the proof. ■

Remark. In order to accelerate the convergence of the extended algorithm, it is useful to abrupt the execution of Step 1 sooner than the exact solution $\hat{G}$ of the linearized problem (10.3)–(10.5) is achieved. Namely, we can select for the next iterate some value G^1 with $0 < F_0(G^1) \leq \eta$, where $\eta > 0$ is a reasonable accuracy parameter which can be updated after each step.

10.6 OPTIMAL CONTROL OF STEP PRECISION IN BILEVEL ITERATION PROCEDURES OF NEWTON TYPE

We consider a general iteration process

$$x_{k+1} = \Phi(x_k), \qquad k = 0, 1, \ldots, \tag{10.44}$$

with x_{k+1}, x_k from R^m. Assume $\Phi : R^m \to R^m$ to be a single-valued nonlinear mapping having a unique fixed point $x^* \in R^m$, i.e. such that $x^* = \Phi(x^*)$. Given x_k, suppose that in order to obtain x_{k+1}, the following iteration procedure

$$\begin{cases} z_s = & \Psi(z_{s-1}, x_k), \quad s = 1, 2, \ldots, \\ z_0 = & x_k \end{cases} \tag{10.45}$$

is used with a single-valued operator $\Psi : R^n \times R^m \to R^m$. We shall refer to (10.45) as the procedure of *lower level.*

Usually, the problem of finding x_{k+1} after having computed x_k is solved inexactly, i.e. procedure (10.44) is substituted with the following: take for x_{k+1} such a vector that

$$\|x_{k+1} - \Phi(x_k)\| \leq \gamma_k, \qquad k = 0, 1, ..., \tag{10.46}$$

for some *control parameter* $\gamma_k > 0$. Thus, the parameter γ_k determines the $(k+1)$-th step precision. Hence, the important problem of optimal or quasi-optimal (in a certain sense) choice of the control parameter γ_k at each step of the process (10.46) arises naturally. Our aim is to minimize the numerical expenditure of the whole procedure, accepting the latter to be proportional to the total number of steps.

We consider a typical case when the principal, or *upper level* procedure (10.44) converges linearly in the global sense, whereas its convergence is of the rate $(1 + \alpha)$ with some $0 < \alpha < 1$ (e.g., it is quadratic if $\alpha = 1$) in some neighborhood V of a fixed point x^* of the mapping Φ. In the other words, there exist positive parameters $\rho < 1$ and D, depending upon the initial point x^0, such that

$$\|\Phi(x_l) - x^*\| \leq \min\{\rho, D\|x_l - x^*\|^\alpha\}\, \|x_l - x^*\|, \quad l = 0, 1, 2... \tag{10.47}$$

Furthermore, suppose that procedure of lower level (10.45) converges linearly, i.e. for each $\ell \in N$ the following estimate is valid

$$\|z_s - \Phi(x_\ell)\| \leq \omega \|z_{s-1} - \Phi(x_\ell)\|, \quad s = 1, 2, ...,$$

with an appropriate parameter $0 < \omega < 1$. Then it can be shown that the problem of optimal control of the step precision γ_l aiming at the total calculation expenditure minimization, can be specified as follows:

$$\sum_{i=0}^{n} \left(\ln \frac{\eta_i}{\gamma_i} + B \right) \to \inf_{\gamma \in \Gamma}. \tag{10.48}$$

Here η_0 is a start error estimate of the upper level procedure (10.44), B is a constant, $B \geq \ln(1 + \rho)$, and Γ is a set of feasible sequences $\{\gamma_k\}_{k=0}^{\infty}$. We call a sequence *feasible*, if, first, it vanishes as $k \to \infty$, and second, it implies decreasing of the finite sequence of error estimates $\{\eta_k\}_{k=0}^{n-1}$ determined by the following recurrence relations

$$\varepsilon < \eta_k = \min\left\{\rho, D\eta_{k-1}^\alpha\right\} \eta_{k-1} + \gamma_{k-1}, \quad k = 1, 2, ..., n-1,$$

$$\eta_n = \min\left\{\rho, D\eta_{n-1}^\alpha\right\} \eta_{n-1} + \gamma_{n-1} \leq \varepsilon.$$

Here $\varepsilon > 0$ is a beforehand given final precision parameter that determines the stopping moment for procedure (10.44).

After having investigated some crucial properties of the feasible sequences $\{\gamma_k\}$, we solve problem (10.48) explicitly (Kalashnikov and Kalashnikova [3],

1988) and propose certain rules of choosing the parameter value γ_k at the $(k+1)$-th step of procedure (10.46). Namely, if the current error estimate η_k exceeds the threshold of $(\rho/D)^{1/\alpha}$, then one should take for γ_k the quantity

$$\gamma_k = \rho\xi_0\eta_k, \tag{10.49}$$

where ξ_0 is a unique positive root of the equation

$$\xi \ln \xi = \ln \rho + (1+\xi)\ln(1+\xi) + B\xi. \tag{10.50}$$

Otherwise, i.e. if

$$\eta_k \le \left(\frac{\rho}{D}\right)^{1/\alpha}, \tag{10.51}$$

then the value

$$\gamma_k = \xi_1 D\eta_k^{1+\alpha} \tag{10.52}$$

is recommended; here ξ_1 is a unique positive root of the equation

$$\ln\left(D^{1/\alpha}\eta_k\right) + \xi \ln r(\xi) + \frac{B\xi}{(1+\alpha)\ln(1+\alpha)} = 0. \tag{10.53}$$

Here the function $r(\xi)$ is determined by the formula

$$r(\xi) = \prod_{i=0}^{\infty}\left[1+(1+\alpha)^i\xi\right]^{\frac{1}{(1+\alpha)^{i+1}\xi}}. \tag{10.54}$$

This rule has an important feature: the control parameter value γ_k depends upon the error estimate η_k at the beginning of the $(k+1)$-th step of the iteration procedure (10.46) but does not depend explicitly on the desired final accuracy ε of the upper level process (10.44). The last property makes the above rules of the bilevel iteration procedure accuracy control rather convenient for practitioners.

In particular, these control techniques can be applied to the *inexact Newton method* by Pang (1986) solving the nonlinear complementarity problem (NCP): Find a vector $x \in R_+^m$ such that

$$f(x) \ge 0 \quad \text{and} \quad x^T f(x) = O, \tag{10.55}$$

where the mapping $f : R^m \to R^m$ is continuously differentiable.

First of all, we note that the exact Newton method applied to problem (10.55) consists in the following: given a $x^k \in R^n$, find $x^{k+1} \in R^n$ as a solution to the linear complementarity problem (LCP):

$$x \ge 0, \quad w = f(x^k) + f'(x^k)(x - x^k) \ge 0, \quad x^T w = 0. \tag{10.56}$$

Pang and Chan (1982) established a series of conditions which warrant the global convergence of the sequence $\{x^k\}$ generated by (10.56) to the solution x^* of problem (10.55). In partiular, they proved in Pang and Chan (1982) that if

$$f(x) = Gx + \varphi(x)$$

with a positive definite matrix G and a mapping $\varphi : R^n \to R^n$ the Jacobi matrix $\varphi'(x)$ of which is nonnegative definite for each $x \in F_+^n$, and if there exists a scalar β such that

$$\beta < \delta = \min_{\|y\|=1} y^T G y$$

and

$$\|\varphi'(x) - \varphi'(y)\| \le \beta \qquad \forall x, y \in R_+^n,$$

then $x^k \to x^*$ for any initial $x^0 \in R_+^n$.

We proved the following theorems.

THEOREM 10.3. *Let the mapping $f : R_+^n \to R_+^n$ be continuously differentiable and strongly monotone over R_+^n, that is, there exists a positive scalar μ such that*

$$(x-y)^T [f(x) - f(y)] \ge \mu \|x - y\|^2 \qquad \forall x, y \in R_+^n.$$

Moreover, let the components $f_i, i = 1, \dots, n$, be convex functions over R_+^n, and $f'(x)$ be a Z-matrix at any point $x \in R_+^n$, i.e. $f'_{ij}(x) \le 0$ if $i \ne j$.

Then for any initial $x^0 \in R_+^n$, the sequence $\{x^k\}$ generated by the Newton method converges to the unique solution x^* of problem (10.55). *Moreover, if the mapping f' is Lipschitz continuous over any compact subset of R_+^n, then the convergence is quadratic, i.e.*

$$\|x^{k+1} - x^*\| \le D\|x^k - x^*\|^2, k = 0, 1, ..., \tag{10.57}$$

for some scalar $D > 0$.

PROOF. See Kalashnikov and Kalashnikova [1] (1988). ■

THEOREM 10.4. *Let a mapping $f : R^n \to R^n$ have the form $f(x) = Ax + \varphi(x)$ where A is a real-valued positive definite $n \times n$-matrix, i.e. there exists a scalar $\mu > 0$ such that*

$$u^T A u \ge \mu \|u\|^2 \qquad \forall u \in R^n.$$

Furthermore, let the mapping φ be continuously differentiable with the bounded derivative:

$$\|\varphi'(x)\| \le q\mu \qquad \forall x \in R_+^n;$$

here $0 < q < 1/3$.

Then for any initial $x^0 \in R_+^n$, the sequence $\{x^k\}$ generated by (10.56) *converges to the unique solution x^* of problem* (10.55). *Moreover, if the gradient mapping φ' is Lipschitz continuous over each compact subset of R_+^n, then the convergence is quadratic, i.e. estimate* (10.57) *holds with some constant $D > 0$.*

PROOF. See Kalashnikov and Kalashnikova [1] (1988). ■

Now, a variant of the inexact Newton method can be described as follows Pang (1986). Let $x^k \in R^n$ be a current iterate, $k \geq 0$. Construct x^{k+1} by the rule

$$\|h_k(x^{k+1})\| \leq \gamma_k = \xi_k \|h(x^k)\|, \tag{10.58}$$

where

$$h_k(x) = \min\left(x, f(x^k) + f'(x^k)(x - x^k)\right), \qquad h(x) = \min\left(x, f(x)\right),$$

and $\{\xi_k\}$ is some positive sequence of reals.

For the purpose of the precision control, we proved some theorems concerning global convergence of the inexact Newton method. In particular, we obtained the following

THEOREM 10.5. *Let $x^* \in R^n_+$ be the unique solution of the nonlinear complementarity problem with the mapping $f(x) = Ax + \varphi(x)$ satisfying conditions of Theorem 10.4. Then*

(i) *for any initial $x^0 \in R^n_+$, the sequence $\{x^k\}$ generated by rule (10.58) converges to x^*, if*

$$\xi_k \left(1 + \|A + \varphi'(x^k)\|\right)\left(1 + \|A\| + q\mu\right) \leq \mu(1-q-\varepsilon_1) - (2-\varepsilon_1)\|\varphi'(x^k)\|, \quad k = 0, 1, \ldots,$$

for some

$$\varepsilon_1 \in \left(0, \frac{1-3q}{1-q}\right);$$

(ii) *moreover, if the gradient mapping φ' is Lipschitz continuous over any compact subset of R^n_+ and $\xi_k \to 0$ when $k \to \infty$, then the superlinear convergence takes place, i.e. the following limit relationship holds:*

$$\lim_{k\to\infty} \frac{\|x^{k+1} - x^*\|}{\|x^k - x^*\|} = 0;$$

(iii) *at last, if for k large enough, the inequality*

$$\xi_k \leq \lambda \|\min\left(x^k, f(x^k)\right)\|, \quad \lambda > 0,$$

is valid, then the convergence is of quadratic rate with some constant D.

PROOF. The proof is given in Kalashnikov and Kalashnikova [1] (1988). ∎

Like in Pang (1986), we use here $\min(a, b)$ to denote the componentwise minimum of two vectors.

Now we apply the above-described precision control scheme to the latter algorithm of the inexact Newton type as it was done in Kalashnikov and Kalashnikova [2] (1988). Namely, if the mapping $f(x) = Ax + \varphi(x)$ satisfies Theorem 10.4 conditions, then we set

$$D_1 = \frac{\mu(1-q)}{1 + \|A\| + q\mu}, \qquad D_2 = 1 + \|A\| + q\mu,$$

$$\rho = 1 - \varepsilon_1, \quad \alpha = 1, \quad B = \ln(2 - \varepsilon_1).$$

Now we can describe the precision control at the $(k+1)$-th step as follows.

Let x^k be the current iterate. First, we compute the value

$$\eta_k = D_1 D \| \min\left(x^k, f(x^k)\right) \|$$

and compare it to the quantity $\bar{\varepsilon} = D\varepsilon / D_2$, where $\varepsilon > 0$ is the final error bound specified beforehand. Now if $\eta_k \leq \bar{\varepsilon}$, then the process is finished, and x^k is taken as the approximate solution to problem (10.55). Otherwise, i.e. if $\eta_k > \bar{\varepsilon}$, we compare η_k to ρ and find the upper bound γ_k for the step error by the following formula that collects together the estimates and relationships (10.49)–(10.54):

$$\gamma_k = \begin{cases} \rho \xi_0 \eta_k / D, & \text{if } \eta_k > \rho; \\ \xi_1 \eta_k^2 / D, & \text{if } \eta_k \leq \rho; \end{cases}$$

here $\xi_0 > 0$ is the root of equation (10.50), and $\xi_1 > 0$ is the root of equation (10.51). Thus, here ξ_1 satisfies the equality

$$\ln \eta_k + \xi_1 \ln r(\xi_1) + \frac{B\xi_1}{2\ln 2} = 0,$$

where the function $r(\xi)$ is the infinite product

$$r(\xi) = \prod_{i=0}^{\infty} \left(1 + 2^i \xi\right)^{\frac{1}{2^{i+1}\xi}}.$$

Then we find an element $x^{k+1} \in R^n_+$ satisfying the inequality

$$\|h_k(x^{k+1})\| \leq \gamma_k / D_2$$

and determine the new error estimate η_{k+1}. With thus constructed x^{k+1} and η_{k+1}, the procedure described above is repeated.

At last, two algorithms finding the element x^{k+1} are examined. The first one, similar to that proposed by Pang and Chan (1982), consists in the following iteration procedure beginning with $z_0 = x^k$ as an initiate. Namely, if $z^t \in R^n_+$ is a current iterate, we put

$$z^{t+1} = \left[z^t - \beta(Az^t + b)\right]_+, \quad t = 0, 1, ..., \tag{10.59}$$

where $A = f'(x^k)$, $b = f(x^k) - f'(x^k)x^k$, and $\beta > 0$ is some scalar. We prove in Kalashnikov and Kalashnikova [2] (1988) that if $f'(x)$ is a positive definite matrix with the parameter $\mu > 0$ for each x in some compact subset of R^n, and

$$0 < \beta < \frac{2\mu}{\|A\|^2},$$

then the vectors z^{t+1} constructed by (10.59) converge linearly to the exact solution of LCP (10.56), i.e. to the element y^{k+1} with $\|h_k(y^{k+1})\| = 0$. Thus, we accept $x^{k+1} = z^{t+1}$ if $\|h_k(z^{t+1})\| \leq \gamma_k / D_2$.

As an alternate approach, we propose to find the vector x^{k+1} through the unconditioned minimization of the function

$$g(x) = \sum_{j=1}^{n} \left\{ (x_j)_-^2 + [(Ax+b)_j]_-^2 + (x_j)_+^2 \, [(Ax+b)_j]_+^2 \right\}; \tag{10.60}$$

here, as usual, $\nu_- = \min\{0, \nu\}$, $\nu_+ = \max\{0, \nu\}$. The two lemmas below were proved in Kalashnikov and Kalashnikova [2] (1988).

LEMMA 10.4. *Function g has a unique stationary point y^{k+1} solving the linear complementarity problem* (10.56). ■

LEMMA 10.5. *Function g level sets*

$$\Omega = \{x \in R^n \mid g(x) \le \tau\}$$

are bounded for each $\tau \ge 0$. ■

These statements allow one to apply a gradient type method to minimize function (10.60), as it is confirmed by the following theorem from Polyak (1963).

THEOREM 10.6. *If the set $\Omega = \{z \in R^n \mid g(z) \le \tau\}$ is compact for some τ and $z^0 \in \Omega$, then for the sequence $\{z^\ell)\}$ generated by the steepest descent method we have*

$$\lim_{\ell \to \infty} g'(z^\ell) = 0.$$ ■

Thus we apply the gradient method to minimize $g(z)$ and take z^ℓ for x^{k+1} when the following inequality holds

$$\|h_k(z^\ell)\| \le \frac{\gamma_k}{D_2}.$$

Finally, several test nonlinear complementarity problems were taken for numerical experiments which showed that the step precision control provided for considerable savings in number of operations needed to achieve the required final accuracy, in comparison with the process without such a control. Some of those test problems were original, and several of them were taken from the paper by Maugeri (1987). The detailed account of the computational tests can be found in Kalashnikov and Kalashnikova [2] (1988).

REFERENCES

Bakhvalov, N.S. *Numerical Methods* (in Russian), Part 1, Moscow: Nauka, 1973.

Berezin, I.S., and Zhidkov, N.G. *Computational Methods* (in Russian), Part 2, Moscow: Fizmatgiz, 1962.

Bulavsky VA and Kalashnikov VV.

1. Parameter driving method to study equilibrium. Ekonomika i Matematicheskie Metody *(in Russian)*. 1994; 30:129-138.

2. Equilibria in generalized Cournot and Stackelberg models. Ekonomika i Matematicheskie Metody *(in Russian)*. 1995; 31:164-176.

3. A Newton-like approach to solving an equilibrium problem. Annals Oper. Res. 1988; 81:115-128.

4. 'A Newton-like approach to solving an equilibrium problem'. -In: *High Performance Algorithms and Software in Nonlinear Optimization*, F. Giannessi e.a., eds. Dordrecht: Kluwer Academic Publishers, 1998.

Cohen G. Nash equilibria: Gradient and decomposition algorithms. Large Scale Systems. 1984; 12:173-184.

Dennis, J.E, Jr., and Schnabel, R.B. *Numerical Methods for Unconstrained Optimization and Nonlinear Equations*. Englewood Cliffs, NJ: Prentice-Hall, Inc., 1983.

Harker PT and Choi SC. A penalty function approach for mathematical programs with variational inequality constraints. Information and Decision Technologies. 1991; 17: 41-50.

Harker PT and Pang JS

1. Existence of optimal solutions to mathematical programs with equilibrium constraints. Oper. Res. Letters. 1988; 7: 61-64.

2. Finite-dimensional variational inequalities and nonlinear complementarity problems: a survey of theory, algorithms and applications. Math. Programming. 1990; 48:161-220.

3. 'A damped-Newton method for the linear complementarity problem.'- In: *Computational Solution of Nonlinear Systems of Equations*, E.L. Allgower and K. Georg, eds. AMS Lectures on Applied Mathematics. 1990; 26: 265-284.

Kalashnikov VV and Kalashnikova NI.

1. Global convergence of inexact Newton method solving complementarity problems.- In: *"Optimizatsia"*, Institute of Mathematics, Novosibirsk. 1988; 42(59): 66-85 (*in Russian*).

2. Step accuracy control for Newton method solving nonlinear complementarity problem.- In: *"Optimizatsia"*, Institute of Mathematics, Novosibirsk. 1988; 43(60): 27-40 (*in Russian*).

3. Kalashnikov VV and Kalashnikova NI. Optimal control of inner accuracy in bilevel iteration process. -In: *"Optimizatsia"*, Institute of Mathematics, Novosibirsk, 1988; 44(61): 27-55 (*in Russian*).

Kelley JE. The cutting plane method for solving convex programs. SIAM J. Optim. 1960; 8:703-712.

Maugeri A. Convex programming, variational inequalities and applications to the traffic equilibrium problem. Appl. Math. Optim. 1987; 16: 169-185.

Marcotte P.

1. Quelques notes et résultats nouveaux sur le problème d'équilibre d'un oligopole. R.A.I.R.O. 1984; 18:147-171.

2. A new algorithm for solving variational inequalities with application to the traffic assignment problem. Math. Programming. 1985; 33: 339-351.

3. Un algorithme général de calcul de l'état d'équilibre d'un oligopole. R.A.I.R.O. 1988; 22:225-241.

Pang JS. Inexact Newton methods for nonlinear complementarity problem. Math. Programming, 1986; 36: 54-71.

Pang JS and Chan D. Iterative methods for variational and complementarity problems. Math. Programming, 1982; 24: 284-313.

Polyak BT. Gradient methods to minimize functionals. Zhurnal Vychislit. Matemat. i Matem. Fiziki (Comput. Maths and Math. Phys.), 1963; 3: 643-653.

Sherali HD, Soyster AL and Murphy FH. Stackelberg -Nash- Cournot equilibria: characterizations and computations. Oper. Res. 1983; 31: 253-276.

Stackelberg, H. *Marktform und Gleichgewicht.* Vienna: Julius Springer, 1934.

Chapter 11

PARAMETRIZATION AND REDUCTION TO NONLINEAR EQUATIONS

This chapter is dedicated to two different types of methods solving nonlinear complementarity problems. The first one is a method of approximate solution of nonlinear complementarity problem with parameters: Given a continuous mapping $f : R^n \times R^m \to R^n$, and a fixed vector of parameters $u = (u_1, \ldots, u_m)^T$, find a $x \in R^n$ such that

$$x \geq 0, \quad f(x,u) \geq 0, \quad and \quad x^T f(x,u) = 0. \tag{11.1}$$

A perturbation method for solving problem (11.1) without parameters (i.e. with $m = 0$) was proposed by McLinden (1980) in the following form: for a positive vector $z \in R^n_{++}$ consider an auxiliary nonlinear equation

$$x_i f_i(x) = z_i, \qquad i = 1, \ldots, n; \tag{11.2}$$

If the perturbation vector z tends to zero, one can expect a sequence of solutions $x = x(z)$ of equation (11.2) to have the solution x^* of the initial complementarity problem (11.1) as its limit point.

In Section 11.1 of this chapter, we prove the existence of solutions to the perturbed complemenetary problem with parameters: For an arbitrary parameter vector $u = (u_1, \ldots, u_m)^T \in R^m$ and a positive perturbation vector $z > 0$, find a point $x \in R^n$ satisfying the following conditions:

$$x \geq 0, \quad x_i f_i(x,u) = z_i, \quad i = 1, \ldots, n. \tag{11.3}$$

In Section 11.2, we show the convergence of the solutions of the perturbed problem with parameters to the solution of the initial problem with parameters, and examine the convergence rate. Section 11.3 describes an algorithm solving the perturbed problem at each step and proves its convergence as well. All these results are taken from the paper by Kalashnikov (1991).

The second theme of the chapter concerns *regularization methods* which are standard ones for solving monotone complementarity problems. They usually have strong convergence properties. The papers by Kalashnikov and Kalashnikova (1988,1989) examined the inexact Newton-like algorithms applied to the nonlinear complementarity problems. The latter were based upon the merit functions introduced in Pang (1986), Pang and Chan (1982). In this chapter, we follow mainly the line developed in the recent papers by Facchinei and Kanzow (1997,1999), Kanzow and Kleinmichel (1998), Kanzow and Pieper (1999), B.Chen, X.Chen and C.Kanzow (2000), Kanzow and Zupke (1999), De Luca e.a. (1996) among others, in which the merit function by Fischer-Burmeister are utilized to reduce the complementarity problem to nonlinear equations. The Newton-type algorithms applied to thus obtained equation provide decent convergence rates. Section 11.4 is specifies the problem and restates some basic definitions, and describes the algorithm. Finally, in Section 11.5, some nice convergence results are extended to the class of P_0 nonlinear complementarity problems. In that part, we follow the paper by Facchenei and Kanzow (1999).

11.1 EXISTENCE OF SOLUTIONS TO PERTURBED PROBLEM

DEFINITION 11.1. A mapping $f : R^n \times R^m \to R^n$ is called *locally strongly monotone* with respect to the space variables (i.e. with respect to x) over a set $X \subseteq R^n$ if there exists a scalar function $\beta : R^1_+ \times R^m \to R^1_+$ such that $\beta(t,u) > 0$ for all $t > 0$ and $u \in R^m$, and satisfying the inequality

$$(x-y)^T \left[f(x,u) - f(y,u)\right] \geq \beta(t,u)\|x-y\|^2 \tag{11.4}$$

for any $x, y \in X \cap B_t(0)$, $u \in R^m$; here $B_t(0) = \{v \in R^n \mid \|v\| \leq t\}$ is a closed ball of radius t centered at the origin.

Note first that the function $\beta(t,u)$ is necessarily non-decreasing with respect to t, and second, any locally monotone mapping is strictly monotone.

THEOREM 11.1 *If the mapping $f : R^n \times R^m \to R^n$ is continuous and locally strongly monotone with respect to $x \in R^n_+$ for any $u \in R^m$ fixed, and the function $\beta(x,u)$ has the property*

$$\lim_{t \to +\infty} \beta(t,u)t = +\infty \tag{11.5}$$

for any $u \in R^m$, then problem (11.3) has a solution $x_z = x_z(u)$ that exists uniquely for each $u \in R^m$ and $z > 0$. Furthermore, if the mapping $f(x,u)$ is also continuous with respect to the parameters u then the solution $x_z(u)$ of problem (11.3) is continuous by u as well.

PROOF. Let us fix a positive vector $\hat{\varepsilon} = (\varepsilon, \ldots, \varepsilon)^T > 0$ and consider the mapping $F : R^n_+ \times R^m \to R^n$ defined by

$$F_i(x, u) = f_i(x + \hat{\varepsilon}, u) - \frac{z_i}{x_i + \varepsilon}, \qquad i = 1, \ldots, n. \tag{11.6}$$

It is easy to show that F is locally strongly monotone by x; indeed,

$$(x - y)^T [F(x, u) - F(y, u)] = (x - y)^T [f(x + \hat{\varepsilon}, u) - f(y + \hat{\varepsilon}, u)] -$$

$$-\sum_{i=1}^{n} (x_i - y_i) z_i \left(\frac{1}{x_i + \varepsilon} - \frac{1}{y_i + \varepsilon} \right) \geq \beta(t, u) \|x - y\|^2 +$$

$$+\sum_{i=1}^{n} \frac{z_i}{(x_i + \varepsilon)(y_i + \varepsilon)} (x_i - y_i)^2 \geq \beta(t, u) \|x - y\|^2$$

for any x, y in the set $R^n_+ \cap B_{t+\varepsilon\sqrt{n}}(0)$. As it was shown in Kalashnikov (1989), then for any fixed parameter vector $u \in R^m$, there exists a unique solution $x(\varepsilon)$ of the following complementarity problem: Find a $x \in R^n_+$ such that

$$F(x, u) \geq 0, \qquad x^T F(x, u) = 0. \tag{11.7}$$

Taking into account (11.6), rewrite (11.7) as follows:

$$x_i \geq 0, \; f_i(x + \hat{\varepsilon}, u) \geq \frac{z_i}{x_i + \varepsilon}, \; x_i f_i(x + \hat{\varepsilon}, u) = \frac{x_i z_i}{x_i + \varepsilon}, \; i = 1, \ldots, n. \tag{11.8}$$

The latter clearly implies that

$$x_i(\varepsilon) f_i\left(x(\varepsilon) + \hat{\varepsilon}, u\right) \leq z_i, \qquad i = 1, \ldots, n. \tag{11.9}$$

Now take a vanishing sequence of parameters $\{\varepsilon_k\}_{k \in \mathcal{N}}$ and show the sequence of points $\{x(\varepsilon_k)\}_{k=1}^{\infty}$ to be bounded. On the contrary, without loss of generality, suppose that

$$\|x(\varepsilon_k)\| \to +\infty \qquad \text{when} \quad k \to \infty. \tag{11.10}$$

Then, on one hand, (11.8) and (11.9) imply the inequalities

$$A_k \equiv [x(\varepsilon_k) - x(\varepsilon_1)]^T \left[f\left(x(\varepsilon_k) + \hat{\varepsilon}_k, u\right) - \right.$$

$$\left. -f\left(x(\varepsilon_1) + \hat{\varepsilon}_1, u\right)\right] \leq 2 \sum_{i=1}^{n} z_i, \quad k = 1, 2, \ldots; \tag{11.11}$$

on the other hand, due to the local strong monotonicity of f, we have the relationships

$$A_k \geq \beta(t_k, u) \|x(\varepsilon_k) - x(\varepsilon_1)\|^2, \quad k = 1, 2, \ldots, \tag{11.12}$$

where $t_k = \max\{\|x(\varepsilon_k)\|, \|x(\varepsilon_1)\|\}$. Next, (11.10) guarantees that $t_k \to +\infty$ as $k \to \infty$. Then taking into account (11.5), we obtain from (11.12) that $A_k \to +\infty$ as $k \to \infty$ which contradicts (11.11), and thus proves that the sequence

$\{x(\varepsilon_k)\}$ is bounded. Again without affecting generality, we can consider a limit point $x^* = \lim_{k \to \infty} x(\varepsilon_k)$.

Note that none of the components of the point x^* can equal zero: the latter follows from relationships (11.8) and the continuity of f with respect to x. Then making use of (11.8) we can obtain the limit relationships:

$$x_i^* > 0, \qquad x_i^* f_i(x^*, u) = z_i, \qquad i = 1, \ldots, n,$$

which prove the first assertion of the theorem in part of the existence of solutions to problem (11.3). Now we can deduce the uniqueness of x^* from the monotonicity of f with respect to x. Indeed, were there another solution $y^* > 0$ of problem (11.3) distinct from x^*, we could obtain the following series of relationships

$$0 \leq (x^* - y^*)^T \left[f(x^*, u) - f(y^*, u)\right] = \sum_{i=1}^{n} \left(2z_i - \frac{z_i}{x_i^*} y_i^* - \frac{z_i}{y_i^*} x_i^*\right) < 0$$

implying the impossible inequality $0 < 0$. This contradiction completes the proof of the first assertion of the theorem.

Now we prove the second assertion. Fix a perturbation vector $z > 0$ and consider the solution $x_z(u)$ of problem (11.3) for a parameter vector $u \in R^m$. Recall that now the mapping $f(x, u)$ is also continuous with respect to u. Assume that $u \to \hat{u}$ and show that then $x_z(u) \to x_z(\hat{u})$. In order to do it, write down the series of relationships in which the inequality is a consequence of the function $f(x, u)$ being monotone with respect to x:

$$\begin{aligned}
&[x_z(u) - x_z(\hat{u})]^T \left[f(x_z(\hat{u}), \hat{u}) - f(x_z(\hat{u}), u)\right] = \\
&= [x_z(u) - x_z(\hat{u})]^T \left[f(x_z(u), u) - f(x_z(\hat{u}), u)\right] - \\
&- [x_z(u) - x_z(\hat{u})]^T \left[f(x_z(u), u) - f(x_z(\hat{u}), \hat{u})\right] \geq \\
&\geq - \sum_{i=1}^{n} \left[2z_i - z_i \frac{x_z^i(u)}{x_z^i(\hat{u})} - z_i \frac{x_z^i(\hat{u})}{x_z^i(u)}\right].
\end{aligned}$$

Having transformed the right-hand side of the latter inequality, we come to the final estimate

$$\begin{aligned}
&[x_z(u) - x_z(\hat{u})]^T \left[f(x_z(\hat{u}), \hat{u}) - f(x_z(\hat{u}), u)\right] \geq \\
&\geq \sum_{i=1}^{n} z_i \frac{\left[x_z^i(u) - x_z^i(\hat{u})\right]^2}{x_z^i(u) x_z^i(\hat{u})}.
\end{aligned} \tag{11.13}$$

Now set

$$M(u) = \min_{1 \leq i \leq n} \frac{z_i}{x_z^i(u) x_z^i(\hat{u})} \tag{11.14}$$

and apply the Cauchy-Buniakowski inequality to (11.13) to get the following:

$$\|x_z(u) - x_z(\hat{u})\| \|f(x_z(\hat{u}), \hat{u}) - f(x_z(\hat{u}), u)\| \geq$$
$$\geq M(u)\|x_z(u) - x_z(\hat{u})\|^2,$$

or equivalently,

$$\|f(x_z(\hat{u}), \hat{u}) - f(x_z(\hat{u}), u)\| \geq M(u)\|x_z(u) - x_z(\hat{u})\|. \tag{11.15}$$

In order to complete the proof, it suffices to show that $\liminf_{u \to \hat{u}} M(u) > 0$. On the contrary, suppose that there is a sequence of parameters $\{u_k\} \subset R^m$ such that $u_k \to \hat{u}$ but $M(u_k) \to 0$ as $k \to \infty$. However, (11.14) implies that

$$M(u) \geq \frac{\delta}{\max_{1 \leq i \leq n} x_z^i(u)},$$

where $\delta = \min_{1 \leq i \leq n} \frac{z_i}{x_z^i(\hat{u})}$. Therefore,

$$M(u_k) \geq \frac{\delta}{\|x_z(u_k)\|}, \qquad k = 1, 2, \ldots,$$

hence $\|x_z(u_k)\| \to +\infty$ as $k \to \infty$. Then (11.15) allows us to conclude that

$$\|f(x_z(\hat{u}), \hat{u}) - f(x_z(\hat{u}), u_k)\| \geq \delta \frac{\|x_z(u_k) - x_z(\hat{u})\|}{\|x_z(u_k)\|},$$

which implies that

$$\liminf_{k \to \infty} \|f(x_z(\hat{u}), \hat{u}) - f(x_z(\hat{u}), u_k)\| \geq \delta.$$

The latter contradicts the continuity of $f(x, u)$ by u at the point $(x_z(\hat{u}), \hat{u})$ and thus shows that our assumption was wrong and actually, $\liminf_{u \to \hat{u}} M(u) > 0$. Therefore $x_z(u) \to x_z(\hat{u})$ as $u \to \hat{u}$ which completes the proof of the theorem. ∎

Remark. The continuity of the solution $x_z(u)$ of problem (11.3) with respect to the perturbation $z \in R^n_+$ for any parameter vector $u \in R^m$ is guaranteed by the continuity of the mapping $f(x, u)$ by the space variables x and the limit equation (11.5).

COROLLARY 10.1. *Under assumptions of Theorem* 11.1, *solution* $x_z(u)$ *of problem* (11.3) *is continuous with respect to the perturbation* $z \in R^n_+$ *for any parameter vector* $u \in R^m$.

PROOF. Indeed, let us fix a parameter vector $u \in R^m$ and consider a sequence of perturbations $\{z^k\}$ converging to a $z \in R^n_+$. First, we show that the sequence $\{x(z^k, u)\}_{k=0}^{\infty}$ of solutions to problem (11.3) is bounded. Indeed, we have the equalities

$$x_i(z^k, u) f_i(x(z^k, u), u) = z_i^k, \qquad i = 1, \ldots, n,$$

that allow us to write down the series of relationships below:

$$A_k = \left[x(z^k,u) - x(z^0,u)\right]^T \left[f\left(x(z^k,u),u\right) - f\left(x(z^0,u),u\right)\right] =$$

$$= \sum_{i=1}^{n} \left[z_i^k + z_i^0 - z_i^0 \frac{x_i(z^k,u)}{x_i(z^0,u)} - z_i^k \frac{x_i(z^0,u)}{x_i(z^k,u)}\right] \le$$

$$\le \sum_{i=1}^{n} \left(z_i^k + z_i^0\right) \le \sum_{i=1}^{n} \left(z_i + z_i^0\right) + \sigma, \tag{11.16}$$

for some $\sigma > 0$ and all $k = 0, 1, \ldots$. On the other hand,

$$A_k \ge \beta(t_k, u)\|x(z^k,u) - x(z^0,u)\|^2,$$

where $t_k = \max\{\|x(z^k,u)\|, \|x(z^0,u)\|\}$. Consequently, A_k could take arbitrarily large values if the sequence $\{x(z^k,u)\}$ were unbounded, as it follows from (11.5). Therefore, since the sequence $\{A_k\}$ is bounded by (11.16), the latter implies that the sequence $\{x(z^k,u)\}$ is bounded as well. Now the continuiuty of $f(x,u)$ with respect x permits us to conclude that any limit point of this sequence solves problem (11.3) with the perturbation vector z. As the solution of (11.3) is determined uniquely, the limit point of the sequence $\{x(z^k,u)\}$ is also unique which implies the continuity of the mapping $z \mapsto x(z,u)$ and thus completes the proof. ■

11.2 CONVERGENCE RATE

Now we return to problem (11.2) without parameters and consider a continuous mapping $f : R_+^n \to R^n$ assuming it to be also locally strongly monotone over R_+^n. As a consequence of Theorem 11.1 we have that the problem: Find a $x \in R_+^n$ such that

$$x_i f_i(x) = z_i, \qquad i = 1, \ldots, n, \tag{11.17}$$

is solvable uniquely for any $z > 0$ if the limit equality (11.5) holds: $\lim_{t \to +\infty} \beta(t)t = +\infty$. Let us fix two perturbation vectors $v > 0$ and $w > 0$, and denote by x and y the solutions of problem (11.17) in case of $z = v$ and $z = w$, respectively.

LEMMA 11.1. *Let the mapping $f : R_+^n \to R^n$ be continuous and locally strongly monotone, and limit relationship (11.5) be valid. Then for solutions x and y of problem (11.17) for perturbation vectors $v > 0$ and $w > 0$, respectively, the following estimates hold:*

$$\|x - y\| \le \frac{1}{\sqrt{\beta(t)}}\|\sqrt{v} - \sqrt{w}\|, \tag{11.18}$$

where $t = \max\{\|x\|, \|y\|\}$, and $\sqrt{z} = \left(\sqrt{z_1}, \ldots, \sqrt{z_n}\right)^T$ for $z > 0$.

PROOF. Due to (11.4) and (11.17), the following series of relationships holds:

$$\beta(t)\|x - y\|^2 \le (x - y)^T \left[f(x) - f(y)\right] =$$

$$= \sum_{i=1}^{n} [x_i f_i(x) + y_i f_i(y) - x_i f_i(y) - y_i f_i(x)] =$$

$$= \sum_{i=1}^{n} \left(v_i + w_i - w_i \frac{x_i}{y_i} - v_i \frac{y_i}{x_i} \right) \leq$$

$$\leq \sum_{i=1}^{n} (v_i + w_i - 2\sqrt{v_i}\sqrt{w_i}) = \sum_{i=1}^{n} (\sqrt{v_i} - \sqrt{w_i})^2 .$$

Extracting the square roots from first and the last terms of the series, we get the inequality

$$\sqrt{\beta(t)}\|x - y\| \leq \|\sqrt{v} - \sqrt{w}\|,$$

which implies (11.18) and completes the proof. ∎

Further, assume that the mapping f is Lipschitz continuous over every compact subset $S \subset R^n_+$, that is, for some positive scalar $L = L(S)$ and any $x, y \in S$ the following inequalities hold:

$$|f_i(x) - f_i(y)| \leq L\|x - y\|, \quad i = 1, \ldots, n. \tag{11.19}$$

Again fix perturbations $v > 0$ and $w > 0$ in problem (11.17) and denote by $x = x(v)$ and $y = y(w)$ its corresponding solutions.

LEMMA 11.2. *The following estimate is valid:*

$$\|v - w\|_1 \leq \sqrt{n}\,[L(S)\|x - y\| + \|f(y)\| + L(S)\|y\|]\,\|x - y\|, \tag{11.20}$$

where $\|p\|_1 = \sum_{i=1}^{n} |p_i|$ *and* $\|p\| = \|p\|_2 = \sqrt{\sum_{i=1}^{n} p_i^2}$ *for* $p \in R^n$.

PROOF. Consider an arbitrary compact subset $S \subset R^n_+$ that comprises both x and y. Write down the relationships

$$v_i - w_i = x_i f_i(x) - y_i f_i(y) = (x_i - y_i)\,[f_i(x) - f_i(y)] +$$

$$+ y_i f_i(x) + x_i f_i(y) - 2y_i f_i(y), \qquad i = 1, \ldots, n,$$

which together with (11.19) imply

$$|v_i - w_i| \leq |x_i - y_i||f_i(x) - f_i(y)| + |x_i - y_i||f_i(y)| +$$

$$+ |y_i||f_i(x) - f_i(y)| \leq L(S)|x_i - y_i|\|x - y\| +$$

$$+ |x_i - y_i||f_i(y)| + L(S)|y_i|\|x - y\| \leq$$

$$\leq \|x - y\|\,[L(S)|x_i - y_i| + |f_i(y)| + L(S)|y_i|] \;\; i = 1, \ldots, n.$$

Summing up the above inequalities for all $i = 1, \ldots, n$, we come to

$$\|v - w\|_1 \leq \|x - y\|\,[L(S)\|x - y\|_1 + \|f(y)\|_1 + L(S)\|y\|_1],$$

which gives us the needed estimate (11.20) as $\|p\|_1 \leq \sqrt{n}\|p\|_2$ for any $p \in R^n$. The proof is complete. ■

Exchanging positions of v and w, as well as $x(v)$ and $y(w)$ in both parts of (11.20), we come easily to the estimate

$$\|v - w\|_1 \leq \sqrt{n}\,[L(S)\|x - y\| + \|f(x)\| + L(S)\|x\|]\,\|x - y\|. \tag{11.21}$$

Estimates (11.20) and (11.21) taken together imply the inequality

$$\|v - w\|_1 \leq \sqrt{n}\,[L(S)\|x - y\| +$$

$$+ \min\{\|f(x)\| + L(S)\|x\|, \|f(y)\| + L(S)\|y\|\}]\,\|x - y\|.$$

Replacing y in (11.18) and (11.20) by the the solution x^* of the initial complementarity problem (11.1) (with $m = 0$), we get the respective estimates below:

$$\|x - x^*\| \leq \frac{1}{\sqrt{\beta(t)}}\|\sqrt{z}\| = \frac{1}{\sqrt{\beta(t)}}\sqrt{\|z\|_1}, \tag{11.22}$$

where $t = \|x^*\| + \|x - x^*\|$, and

$$\|z\|_1 \leq \sqrt{n}\,[L(S)\|x - x^*\| + \|f(x^*)\| + L(S)\|x^*\|]\,\|x - x^*\|, \tag{11.23}$$

where $S = B_t(0)$. From these two-sided estimates (11.22) and (11.23), and from the fact that $\|\hat{x} - x^*\| \leq \|x - x^*\|$ implies that $\hat{x}$ belongs to the ball $B_t(0)$ together with x and x^*, we obtain the following assertion: The sequence of solutions $x(z^k)$ of problem (11.17) with $z = z^k$ converges to the solution x^* of the initial problem (11.1) at the rate equal to the half of that of the convergence of perturbations z^k to zero. For instance, in case of quadratic convergence of z^k to zero, $x(z^k)$ will converge to x^* at least linearly (at the geometric rate).

11.3 SOLVING PERTURBED PROBLEMS

Let us fix a perturbation vector $z \in R^n_{++} = \{p \in R^n \mid p_i > 0,\ i = 1, \ldots, n\}$ and consider the perturbed problem (11.7):

$$x_i f_i(x) = z_i, \qquad i = 1, \ldots, n.$$

As $z > 0$, this problem can be rewritten as follows:

$$F(x) = 0, \tag{11.24}$$

where

$$F_i(x) = f_i(x) - \frac{z_i}{x_i}, \qquad i = 1, \ldots, n. \tag{11.25}$$

For convenience purpose, we represent F as $F(x) = f(x) + H(x)$, with

$$H_i(x) = -\frac{z_i}{x_i}, \qquad i = 1, \ldots, n.$$

It is easy to verify that the mapping $H : R^n_{++} \to R^n$ is monotone. Therefore, if the mapping f is continuous and locally strongly monotone over R^n, we can apply a somewhat modified Peaceman-Rachford method (*cf.* Ortega and Rheinboldt, 1970) to solve problem (11.24). Namely, having a current approximation x^k, we first find the vector $x^{k+1/2}$ as a positive solution of the equation

$$\lambda x^{k+1/2} + H(x^{k+1/2}) = \lambda x^k - f(x^k), \tag{11.26}$$

then having got $x^{k+1/2}$ we construct the vector x^{k+1} as a solution to the equation

$$\lambda x^{k+1} + f(x^{k+1}) = \lambda x^{k+1/2} - H(x^{k+1/2}). \tag{11.27}$$

Here $\lambda > 0$ is a fixed parameter.

One can prove the convergence of this algorithm by having adapted the proof of the global Peaceman-Rachford theorem in Ortega and Rheinboldt (1970).

THEOREM 11.2. *Let the mapping $f : R^n \to R^n$ be locally strongly monotone and Lipschitz continuos over each compact subset in R^n. Then for any initial $x^0 \in R^n$, any fixed $\lambda > 0$, the sequence $\{x^k\}$ is well-defined by formulas* (11.26)–(11.27) *and converges to the unique solution $x^* = x^*(z)$ of problem* (11.1).

PROOF. The existence and uniqueness of of the solution x^* is guaranteed by Theorem 11.1. The auxiliary point $x^{k+1/2}$ is well-defined due to the form of quadratic equations with respect to the components of $x_i^{k+1/2}$, induced by equation (11.26). Further, if $E : R^n \to R^n$ is the identity correspondence (i.e., $E(x) = x$ for all $x \in R^n$), then the mapping $(f + \lambda E)$ is continuous and strongly monotone on R^n and hence it is a homeomorphism of R^n onto itself: *cf.* Ortega and Rheinboldt (1970, Theorem 6.4.4). The latter implies that system of equations (11.27) has a unique solution x^{k+1} and justifies the iteration process (11.26)–(11.27). Let us rewrite it in the vector form:

$$x^{k+1} = Gx^k, \qquad k = 0, 1, \ldots, \tag{11.28}$$

where

$$Gx = (\lambda E + f)^{-1}(\lambda E - H)(\lambda E + H)_+^{-1}(\lambda E - f)x;$$

here $(\lambda E + H)_+^{-1}(\lambda E - f)$ is the operator that associates the vector $x^k \in R^n$ with the positive solution of equation (11.26). The relationships

$$F(x^*) = f(x^*) + H(x^*) = 0 \qquad \text{and} \qquad x^* > 0$$

imply the equalities below

$$(\lambda E + H)_+^{-1}(\lambda E - f)x^* = x^*,$$

and

$$(\lambda E + f)^{-1}(\lambda E - H)x^* = x^*.$$

Therefore, x^* is a fixed point of the mapping G.

Conversely, if x^* is a fixed point of the mapping G and $x^* > 0$, then x^* solves equation (11.24), that is, $f(x^*) + H(x^*) = 0$. Indeed, if $x^* = Gx^*$, denote by

$$y^* = (\lambda E + H)_+^{-1}(\lambda E - f)x^*,$$

and obtain the equality

$$(\lambda E + f)x^* = (\lambda E - H)y^*. \tag{11.29}$$

On the other hand, the definition of y^* implies that

$$(\lambda E + H)y^* = (\lambda E - f)x^*. \tag{11.30}$$

Let us rewrite equalities (11.29)–(11.30) in the form

$$\lambda(x^* - y^*) = -f(x^*) - H(y^*),$$

and

$$\lambda(x^* - y^*) = f(x^*) + H(y^*),$$

respectively. This immediately proves that $x^* = y^*$, and hence

$$f(x^*) + H(x^*) = 0.$$

Furthermore, let $S \subset R^n$ be an arbitrary compact subset. Then the subset $\hat{S} = (\lambda E + f)^{-1}S$ is also compact, and for any $u, v \in \hat{S}$ one has

$$(u - v)^T [f(u) - f(v)] \geq \beta(t)\|u - v\|^2, \tag{11.31}$$

where $t = \max\{\|w\| \mid w \in \hat{S}\}$. Next, denote by ν the Lipschitz coefficient of the mapping f over $\hat{S}$, and for arbitrary $x, y \in S$, $x \neq y$, set $u = (\lambda E + f)^{-1}x$, $v = (\lambda E + f)^{-1}y$. Then the following relationships are valid for the mapping $G_1 = (\lambda E - f)(\lambda E + f)^{-1}$ (here $\|\cdot\| = \|\cdot\|_2$ is the Euclidean norm in R^n):

$$\left[\frac{\|G_1x - G_1y\|}{\|x - y\|}\right]^2 =$$

$$= \frac{\lambda^2\|u - v\| - 2\lambda(u - v)^T [f(u) - f(v)] + \|f(u) - f(v)\|^2}{\lambda^2\|u - v\| + 2\lambda(u - v)^T [f(u) - f(v)] + \|f(u) - f(v)\|^2} \leq$$

$$\leq \frac{[\lambda^2 - 2\lambda\beta(t)]\,\|u - v\| + \|f(u) - f(v)\|^2}{[\lambda^2 + 2\lambda\beta(t)]\,\|u - v\| + \|f(u) - f(v)\|^2} \leq \frac{\lambda^2 - 2\lambda\beta(t) + \nu^2}{\lambda^2 + 2\lambda\beta(t) + \nu^2} < 1,$$

where the inequality before the last is based upon the monotonicity of the function $\frac{a+\tau}{b+\tau}$ over $\tau \geq 0$ if $b \geq a \geq 0$. Thus obtained estimate shows the mapping G_1 to be contractive one over S.

Now consider the mapping $G_2 = (\lambda E - H)(\lambda E + H)_+^{-1}$ and show it to satisfy the condition

$$\|G_2x - G_2y\| \leq \|x - y\| \tag{11.32}$$

for any $x, y \in R^n$. Indeed, we note first that the mapping H is monotone over R^n_{++} and Lipschitz continuous on each compact subset $S \subset R^n_{++}$. Moreover, for each pair $x, y \in R^n$ the points $u = (\lambda E + H)_+^{-1} x$ and $v = (\lambda E + H)_+^{-1} y$ belong to R^n_{++}, and the formula of quadratic equation roots implies that the set $\hat{S} = (\lambda E + H)_+^{-1} S$ is a compact subset of R^n_{++} whenever $S \subset R^n$ is such. Now it is clear that the validity of the relationship (11.32) for all $x, y \in R^n$ can be proved by an exact repetition of the arguments used to show the contractiveness of G_1.

Thus, the composition $G_2 G_1$ is a contractive mapping on every compact subset in R^n. Due to the identity

$$G = (\lambda E + f)^{-1} G_2 G_1 (\lambda E + f)$$

and the fact that G has the (unique) fixed point $x^* \in R^n_{++}$, the mapping $G_2 G_1$ has the fixed point $y^* = (\lambda E + f) x^*$. The latter allows one to conclude (*cf.* Ortega and Rheinboldt, 1970, Theorema 12.1.4) that the sequence $\{y^k\}$, constructed by the recurrence $y^{k+1} = G_2 G_1 y^k$, $y^0 = (\lambda E + f) x^0$, $k = 0, 1, \ldots$, will converge to y^*. However, the latter implies (*cf.* Ortega and Rheinboldt, 1970, Theorema 12.1.6) that the sequence $\{x^k\}$ will converge to x^*. The proof is complete. ∎

Remark. The point $x^{k+1/2}$ solving equation (11.26) is calculated easily by solving the quadratic equations. Namely,

$$x_i^{k+1/2} = \frac{x_i^k}{2} - \frac{f_i(x^k)}{2\lambda} + \frac{\sqrt{\left[\lambda x_i^k - f_i(x^k)\right]^2 + 4\lambda z_i}}{2\lambda}, \quad i = 1, \ldots, n.$$

Consequently, at each step of the described algorithm, one needs to solve only the nonlinear equation

$$\lambda x_i^{k+1} + f_i(x^{k+1}) = \lambda \frac{x_i^k}{2} - \frac{f_i(x^k)}{2} + \frac{\sqrt{\left[\lambda x_i^k - f_i(x^k)\right]^2 + 4\lambda z_i}}{2} - \frac{z_i}{x_i^{k+1/2}},$$

$i = 1, \ldots, n$. However, the mapping $(\lambda E + f)$ is strongly monotone, hence the latter system of equation can be treated efficiently by Newton-type algorithms.

11.4 REGULARIZATION METHODS

Now we consider the nonlinear complementarity problem (NCP) in the form: Find a vector $x \in R^n$ satisfying the conditions

$$x \geq 0, \quad , F(x) \geq 0, \quad x^T F(x) = 0,$$

where all inequalities are taken componentwise and $F : R^n \to R^n$ is any given function which is assumed to be continuosuly differentiable throughout this chapter. There are several methods to solve the NCP, see., e.g., the recent papers by De Luca, Facchinei and Kanzow (1996), Facchinei and Kanzow (1997), Kanzow and Kleinmichel (1998), B. Chen, X. Chen and C. Kanzow (2000), Kanzow and Pieper (1999), Kanzow and Zupke (1999), Ferris and Kanzow (to appear). Almost all of them are based upon the use of the Fischer-Burmeister merit function to reduce the NCP to nonlinear equations. The particular class of methods to be described in this and next sections of the chapter are the so-called *regularization methods* designed to handle *ill-posed* problems. These regularization-type algorithms have been recently applied successfully to improve the robustness of several complementarity solvers on difficult test problems; *cf.* Billups (1995), Billups and Ferris (1997). For a detailed discussion of ill-posedness in mathematical programming the reader is referred to Dontchev and Zolezzi (1993); roughly speaking, an ill-posed problem may be difficult to solve numerically as small errors in the computations can lead to a totally wrong solution.

Regularization methods substitute the solution of the original nonlinear complementarity problem with the solution of a family of well-posed problems whose solutions compose a trajectory converging to the solution of the original problem. More precisely, we apply the *Tikhonov regularization* solving a family of complementarity problems NCP(ε): Find a $x \in R^n$ such that

$$x \geq 0, \quad F_\varepsilon(x) \geq 0, \qquad x^T F_\varepsilon(x) = 0,$$

where $F_\varepsilon(x) = F(x) + \varepsilon x$ and ε is a positive parameter converging to zero.

The regularization methods of this type for complemntarity problems have already been examined in the literature; *cf.* Subramanyan (1988), Kalashnikov and Kalashnikova (1991, 1993), Cottle e.a. (1992, Theorem 5.6.2(b)). The basic results established in the monotone case, that are similar to those for convex programming problems, *cf.* Dontchev and Zolezzi (1993), Polyak (1987), are:

(a) The regularized problem NCP(ε) has a unique solution $x(\varepsilon)$ for every $\varepsilon > 0$.

(b) The trajectory $x(\varepsilon)$ is continuous for $\varepsilon > 0$.

(c) For $\varepsilon \to 0$, the trajectory $x(\varepsilon)$ converges to the least l_2-norm solution of NCP if the latter has a nonempty solutions set, otherwise it diverges.

In the last two sections of this chapter, we follow the paper by Facchinei and Kanzow (1999) to show that the above results can be extended to the larger class of P_0 nonlinear complementarity problems. Actually such a scheme has already been applied to P_0 linear CPs in Venkateswaran (1993), Cotlle e.a. (1992) and Ebiefung (1998). These results will be discussed below where the rather counterintuitive counterexample will be also provided that shows that if F is a nonlinear P_0-function, it does not necessarily imply that F_ε is a uniform P-function. The latter fact makes the extension of some well-known results for linear problems to nonlinear ones more difficult than one would expect. In what follows this task is accomplished by an approach based upon the Fischer-Burmeister's function and the Mountain Pass Theorem. The next section is dedicated to the extension of item (a) to the class of P_0-function complementarity problems and the partial generalization of items (b) and (c). Finally, an algorithm is examined which requires only an approximate solution of the perturbed problems. As far as we know this algorithm (due to Facchinei and Kanzow (1999)) is the first implementable algorithm which guarantees the computation of a solution to a P_0 complementarity problem under the mere assumption that the solutions set is nonempty and unbounded.

11.4.1 PRELIMINARIES

We first recall some definitions.

DEFINITION 11.2. A matrix $M \in R^{n\times n}$ is called a

(i) *P_0-matrix* if, for every $x \in R^n$ with $x \neq 0$, there is an index $i_0 = i_0(x)$ such that

$$x_{i_0} \neq 0 \qquad \text{and} \qquad x_{i_0}\left[Mx\right]_{i_0} \geq 0;$$

(ii) *P-matrix* if, for every $x \in R^n$ with $x \neq 0$, it holds that

$$\max_{1\leq i\leq n} x_i[Mx]_i > 0;$$

(iii) *R_0-matrix* if $x = 0$ is the only solution to NCP with $F(x) = Mx$.

We refer the reader to the book by Cottle e.a. (1992) for a discussion of several properties of these classes of matrices, some nonlinear generalizations of which are defined below.

DEFINITION 11.3. The function $F : R^n \to R^n$ is called a

(i) *P_0-function* if, for all $x, y \in R^n$ with $x \neq y$, there is an index $i_0 = i_0(x, y)$ such that

$$x_{i_0} \neq y_{i_0} \qquad \text{and} \qquad (x_{i_0} - y_{i_0})\left[F_{i_0}(x) - F_{i_0}(y)\right]_{i_0} \geq 0;$$

(ii) *P-function* if, for all $x, y \in R^n$ with $x \neq y$, it holds that

$$\max_{1\leq i\leq n}(x_i - y_i)\left[F_i(x) - F_i(y)\right] > 0;$$

(iii) *uniform P-function* if there is a constant $\mu > 0$ such that

$$\max_{1\leq i\leq n}(x_i - y_i)\left[F_i(x) - F_i(y)\right] \geq \mu\|x - y\|^2 \tag{11.33}$$

holds for all $x, y \in R^n$.

Obviously, every uniform P-function is a P-function and every P-function is a P_0-function. Moreover, an affine mapping $F(x) = Mx + q$ is a P_0- function [P-function] if and only if N is a P_0-matrix [P-matrix]. Furthermore, the class of P_0-functions includes the class of monotone functions. For further discussions the reader is referred to Moré and Rheinboldt (1973).

In the affine case, there are some known results for regularization methods which partially extend the above properties (a) and (c) of monotone complementarity problems to P_0 problems. We summarize these results in the following theorem.

THEOREM 11.3. *Assume that $F(x) = Mx + q$ with $M \in R^{n\times n}$ being a P_0-matrix and $q \in R^n$. Then*

(aa) The regularized problem NCP(ε) has a unique solution $x(\varepsilon)$ for every $\varepsilon > 0$.

(bb) If M is also an R_0-matrix then the family $x(\varepsilon)$ is bounded as $\varepsilon \to 0$, and every its limit point is a solution to NCP. ■

A proof if these results can be found in Cottle e.a. (1992, Theorem 5.6.2 (a)). Note also that in Venkateswaran (1993), item (bb) is proved under an assumption that implies that the original problem has a unique solution. A relaxation of the R_0 condition is also discussed in the recent paper by Ebiefung (1998). In the linear case the proof of statement (aa) is quite simple because if M is a P_0-matrix then $M + \varepsilon I$ is a P-matrix by Theorem 3.4.2 in Cottle e.a. (1992), so that NCP(ε) has a unique solution by Theorem 3.3.7 in Cottle e.a. (1992).

Therefore, in an attempt to extend the previous result from the linear to the nonlinear case, the following question naturally arises: Is F_ε a uniform P-function for every fixed $\varepsilon > 0$ if F itself is a P_0-function? If the answer were yes, then item (aa) above could readily be extended, as a complementarity problem with a uniform P-function has a unique solution; *cf.* Moré (1974, Corollary 3.2). Unfortunately, the following example by Facchinei and Kanzow (1999) shows that F_ε is not necessarily a uniform P-function over R^n_+ when F is nonlinear.

EXAMPLE 11.1. Consider the function $F : R^2 \to R^2$ defined by

$$F(x) = F(x_1, x_2) = \begin{pmatrix} 0 \\ -e^{x_1} \end{pmatrix}.$$

Since the Jacobian

$$F'(x) = \begin{pmatrix} 0 & 0 \\ -e^{x_1} & 0 \end{pmatrix}$$

is obviously a P_0-matrix for all $x \in R^2$, the function F itself is a P_0-function by Corollary 5.3 in Moré and Rheinboldt (1973). Now let $\varepsilon > 0$ and define

$$F_\varepsilon(x) = F(x) + \varepsilon x = \begin{pmatrix} \varepsilon x_1 \\ \varepsilon x_2 - e^{x_1} \end{pmatrix}.$$

Now it will be shown that F_ε is not a uniform P-function on R^n_+. This means to show that, given a fixed value $\varepsilon > 0$, one can find, for every fixed value $\mu > 0$, two distinct points in R^n_+ (possibly depending upon μ) for which inequality (11.33) from the definition of uniform P-function is not satisfied with that μ.

It will be actually shown that F_ε is not a uniform P-function for each positive ε. Indeed, suppose that $\varepsilon > 0$ is fixed and choose a positive μ. Pick up the following points

$$x_1 = 1, \qquad x_2 = \sqrt{\frac{\varepsilon}{\mu}}(c-1), \tag{11.34}$$

where c is a constant such that

$$c \geq 2, \tag{11.35}$$

$$\frac{\varepsilon^2}{\mu}(c-1)^2 - \sqrt{\frac{\varepsilon}{\mu}}\left(e^c - e^1\right) \leq \varepsilon. \tag{11.36}$$

Note that it is always possible to choose c large enough so the (11.36) be satisfied; indeed, the second term on the lefthand side of (11.36) is negative and its absolute value increases exponentially with c and hence dominates the first term. Multiplying (11.36) by $(c-1)^2$ we also obtain

$$\frac{\varepsilon^2}{\mu}(c-1)^4 - \sqrt{\frac{\varepsilon}{\mu}}(c-1)^2\left(e^c - e^1\right) \leq \varepsilon(c-1)^2. \tag{11.37}$$

By (11.34) one easily obtains

$$\varepsilon(c-1)^2 < \mu + \varepsilon(c-1)^2 = \mu\left[1 + \frac{\varepsilon}{\mu}(c-1)^2\right] = \mu\left(x_1^2 + x_2^2\right). \tag{11.38}$$

Now set $y = cx$. Then combining (11.34), (11.35), (11.37) and (11.38) leads to the following estimates:

$$\max_{1\leq i\leq 2}\left(x_i - y_i\right)\left[F_{\varepsilon,i}(x) - F_{\varepsilon,i}(y)\right] =$$

$$= \max\left\{\varepsilon(x_1-y_1)^2, \varepsilon(x_2-y_2)^2 + (x_2-y_2)\left(e^{y_1} - e^{x_1}\right)\right\} =$$

$$= \max\left\{\varepsilon(c-1)^2x_1^2, \varepsilon(c-1)^2x_2^2 + (c-1)x_2\left(e^{cx_1} - e^{x_1}\right)\right\} =$$

$$= \max\left\{\varepsilon(c-1)^2, \frac{\varepsilon^2}{\mu}(c-1)^4 - \sqrt{\frac{\varepsilon}{\mu}}(c-1)^2\left(e^c - e^1\right)\right\} =$$

$$= \varepsilon(c-1)^2 < \mu\left(x_1^2 + x_2^2\right) = \frac{\mu}{(c-1)^2}\|x-y\|_2^2 \leq \mu\|x-y\|_2^2,$$

which proves that F_ε is not a uniform P-function. ∎

Below it will be shown that, in spite of the fact that F_ε is not necessarily a uniform P-function, the regularized problem NCP(ε) has a unique solution $x(\varepsilon)$ for every $\varepsilon > 0$. However, as Example 11.1 demonstrates, the analysis turns out to be more complicated than one would expect.

11.4.2 EXISTENCE OF REGULARIZED SOLUTION

In this subsection, it will be shown that the regularized problem NCP(ε) has a unique solution $x(\varepsilon)$ for each $\varepsilon > 0$. The main tool to prove this result is the (nonsmooth) Fischer-Burmeister's function $\varphi : R^2 \to R$ defined by

$$\varphi(a,b) = \sqrt{a^2+b^2} - a - b.$$

This function was introduced by Fischer (1992), it has attracted much attention recently and has been extensively studied. Now it plays a central role in the design of several nonsmooth Newton-type methods solving NCP, *cf.* e.g.

Facchinei and Soares (1997), De Luca, Facchinei and Kanzow (1996). It is also interesting that this function can be used as a theoretical tool as we will see below. To this end, let us introduce the operator $\Phi : R^n \to R^n$ by

$$\Phi(x) = \begin{pmatrix} \varphi(x_1, F_1(x)) \\ \vdots \\ \varphi(x_n, F_n(x)) \end{pmatrix}$$

as well as the corresponding *merit function* $\Psi : R^n \to R$ by

$$\Psi(x) = \frac{1}{2}\Phi(x)^T\Phi(x) = \frac{1}{2}\|\Phi(x)\|^2.$$

Some of the elementary properties of these functions are summarized in the following result; see, e.g., Geiger and Kanzow (1996), De Luca, Facchinei and Kanzow (1996), Facchinei and Soares (1997).

PROPOSITION 11.1. *The following statements hold:*

(1) $x^* \in R^n$ solves NCP if and only if x^* solves the nonlinear equation $\Psi(x) = 0$.

(2) The merit function Ψ is continuously differentiable on the whole space R^n.

(3) If F is a P_0-function, then every stationary point of Ψ is a solution of NCP. ■

For the regularized problem, we define the corresponding operator and merit function similarly by

$$\Phi_\varepsilon(x) = \begin{pmatrix} \varphi(x_1, F_{\varepsilon,1}(x)) \\ \vdots \\ \varphi(x_n, F_{\varepsilon,n}(x)) \end{pmatrix}$$

and

$$\Psi_\varepsilon(x) = \frac{1}{2}\Phi_\varepsilon(x)^T\Phi_\varepsilon(x) = \frac{1}{2}\|\Phi_\varepsilon(x)\|^2,$$

where $F_{\varepsilon,i}$ denotes the i-th component function of F_ε. Now three preliminary results will be proven.

LEMMA 11.3. *Let $F : R^n \to R^n$ be a P_0-function, and $\varepsilon > 0$ be arbitrary. Then the Jacobian matrices $F_\varepsilon'(x)$ are P-matrices for all $x \in R^n$. Therefore, the function $F_\varepsilon : R^n \to R^n$ is a P-function.*

PROOF. Since F is a P_0-function, the Jacobian matrices $F'(x)$ are P_0-matrices for all $x \in R^n$ by Theorem 5.8 in Moré and Rheinboldt (1973). In view of Theorem 3.4.2 in Cottle e.a. (1992), the Jacobian matrices $F_\varepsilon'(x) = F'(x) + \varepsilon I$ are therefore P-matrices for all $x \in R^n$. Hence F_ε is a P-function by Theorem 5.2 in Moré and Rheinboldt (1973) which completes the proof. ■

LEMMA 11.4. *Let $\{a^k\}, \{b^k\} \subset R$ be any two sequences such that $a^k, b^k \to +\infty$ or $a^k \to -\infty$ or $b^k \to -\infty$ as $k \to \infty$. Then $|\varphi(a^k, b^k)| \to +\infty$ as $k \to \infty$.*

PROOF. A proof of this simple result can be found in Kanzow (1996). ■

PROPOSITION 11.2. *Suppose that F is a P_0-function and $\varepsilon > 0$. Then the merit function Ψ_ε is coercive, i.e.*

$$\lim_{\|x\| \to \infty} \Psi_\varepsilon(x) = +\infty.$$

PROOF. On the contrary, suppose that one can pick up an unbounded sequence $\{x^k\}$ such that $\{\Psi_\varepsilon(x^k)\}$ is bounded. Then the index set $J = \{1 \le i \le n \mid \{x_i^k\} \text{is unbounded}\}$ is nonempty. Without loss of generality one clearly can assume that $|x_j^k| \to +\infty$, for all $j \in J$, as $k \to \infty$. Let us construct the bounded sequence $\{y^k\}$ defined as follows:

$$y_i^k = \begin{cases} 0 & \text{if } i \in J; \\ x_i^k & \text{if } i \notin J. \end{cases}$$

From the definition of $\{y^k\}$ and the assumption that F is a P_0- function one obtains

$$\begin{aligned} 0 \le \max_{1 \le i \le n} \left(x_i^k - y_i^k\right) \left[F_i(x^k) - F_i(y^k)\right] &= \\ = \max_{i \in J} x_i^k \left[F_i(x^k) - F_i(y^k)\right] &= \\ = x_m^k \left[F_m(x^k) - F_m(y^k)\right], & \qquad (11.39) \end{aligned}$$

where m is one of the indices for which the maximum is attained, which one can, without affecting generality, assume to be the same for all k. As $m \in J$ one has that

$$|x_m^k| \to \infty \qquad \text{as} \qquad k \to \infty. \qquad (11.40)$$

Now two cases are to be considered.

Case 1: $x_m^k \to \infty$ as $k \to \infty$.

In this case, since $\{F_m(y^k)\}$ is bounded by the continuity of F_m, (11.39) implies that $F_m(x^k)$ cannot tend to $-\infty$ as $k \to \infty$. This in turn implies

$$\left\{\sqrt{(x_m^k)^2 + (F_m(x^k) + \varepsilon x_m^k)^2} - x_m^k - \left(F_m(x^k) + \varepsilon x_m^k\right)\right\} \to +\infty$$

by Lemma 11.4 since $F_m(x^k) + \varepsilon x_m^k$ tends to $+\infty$ as $k \to \infty$.

Case 2: $x_m^k \to -\infty$ as $k \to \infty$.

In this case it follows immediately from Lemma 11.4 that

$$\left\{\sqrt{(x_m^k)^2 + (F_m(x^k) + \varepsilon x_m^k)^2} - x_m^k - \left(F_m(x^k) + \varepsilon x_m^k\right)\right\} \to +\infty \text{as } k \to \infty,$$

either if $F_m(x^k) + \varepsilon x_m^k$ is unbounded or not.

In both cases it follows that $\Psi_\varepsilon(x^k) \to +\infty$ which contradicts the supposed boundedness of the sequence $\{\Psi_\varepsilon(x^k)\}$ and thus completes the proof. ∎

Note that Proposition 11.2 can also be stated in an alternative (equivalent) way by saying that the level sets $\mathcal{L}_\varepsilon(c) = \{x \in R^n \mid \Psi_\varepsilon(x) \leq c\}$ are compact for every $c \in R^n$. We are now in a position to prove the following existence and uniqueness result.

THEOREM 11.4. *Assume that F is a P_0-function. Then the regularized complementarity problem NCP(ε) has a unique solution $x(\varepsilon)$ for each $\varepsilon > 0$.*

PROOF. If $\varepsilon > 0$ then F_ε is a P-function by virtue of Lemma 11.3. Therefore NCP(ε) has at most one solution by Theorem 2.3 in Moré (1974).

In order to prove the existence of a solution, let $x^0 \in R^n$ be an arbitrary and define $c = \Psi_\varepsilon(x^0)$. Because of Proposition 11.2 the corresponding level set $\mathcal{L}_\varepsilon(c)$ is nonempty and compact. Hence the continuous function Ψ_ε attains the global minimum at a point x_ε on $\mathcal{L}_\varepsilon(c)$ which, in view of the definition of the level set, is also the global minimum of Ψ_ε on R^n. Therefore x_ε is a stationary point of Ψ_ε. But F_ε is a P-function, hence it is also a P_0- function, and Proposition 11.1 part (3) completes the proof. ∎

11.5 INEXACT REGULARIZATION METHODS

Before formulating the inexact regularization algorithm one needs examine first the properties of the solution path $\mathcal{P} = \{x(\varepsilon) \mid \varepsilon > 0\}$ and, in particular, conditions under which $x(\varepsilon)$ remains bounded when $\varepsilon \to 0$. This boundedness helps one to approximate solutions to the original problem NCP as it is demonstrated by the following easily verified result stated without proof.

THEOREM 11.5. *If a sequence $\{\varepsilon_k\} > 0$ converges to zero as $k \to \infty$ then any limit point $\bar{x}$ of the sequence $\{x(\varepsilon_k)\}$ solves NCP.* ∎

The first important property one can establish is the continuity of $x(\varepsilon)$.

LEMMA 11.5. *Suppose that F is a P_0-function. Then the mapping $\varepsilon \mapsto x(\varepsilon)$ is continuous at any $\varepsilon > 0$.*

PROOF. By Lemma 11.3, the Jacobian matrix $F_\varepsilon'(x)$ is a P-matrix for each $\varepsilon > 0$ and every $x \in R^n$; in particular, $M = F_\varepsilon'(x(\varepsilon))$ is a P-matrix. This immediately implies that every principal submatrix of M is again a P-matrix.

Moreover, using the same techniques of proof as for Lemma 2.3 in Chen and Harker (1993), it is easy to verify that any Schur-complement of a P-matrix is also a P-matrix. Hence the assertion follows from Theorem 3.1 in Kyparisis (1986). ∎

Note that Lemma 11.5 does not say anything about the continuity of the mapping $\varepsilon \mapsto x(\varepsilon)$ at $\varepsilon = 0$. Continuity at zero is equivalent to convergence of the solution path $x(\varepsilon)$ when ε goes to 0. This result holds if F is monotone and the complementarity problem admits a solution. In the more general setting considered here, a bit weaker result can be established.

THEOREM 11.6. *Let F be a P_0-function and assume that the solutions set $\mathcal{S}$ of NCP is nonempty and bounded. Then the path $\mathcal{P}_{\tilde{\varepsilon}} = \{x(\varepsilon) \mid \varepsilon \in (0, \tilde{\varepsilon}]\}$ is bounded for any positive $\tilde{\varepsilon}$ and*

$$\lim_{\varepsilon \downarrow 0} \operatorname{dist}(x(\varepsilon), \mathcal{S}) = 0. \qquad ∎$$

The proof of this result is postponed until the inexact regularization algorithm is constructed, when the former will follow from a more general result.

Next two immediate consequences of Theorem 11.6 are stated.

COROLLARY 11.2. *If F is a P_0-function and NCP has a unique solution $\bar{x}$, then $\lim_{\varepsilon \downarrow 0} x(\varepsilon) = \bar{x}$.* ∎

Due to a recent result in Facchinei (1998), the uniqueness of a solution to NCP is, for P_0 complementarity problem, equivalent to the existence of an isolated solution to NCP. Hence, alternatively, Corollary 11.2 could have been stated under the assumption that NCP has a locally isolated solution.

COROLLARY 11.3. *Let $F(x) = Mx + q$ be an affine mapping with $M \in R^{n \times n}$ being a P_0- and R_0-matrix. Then the path $\mathcal{P}_{\tilde{\varepsilon}}$ is bounded for any positive $\tilde{\varepsilon}$ and*

$$\lim_{\varepsilon \downarrow 0} \operatorname{dist}(x(\varepsilon), \mathcal{S}) = 0.$$

PROOF. As the solutions set of NCP is known to be nonempty and bounded under the stated assumptions (*cf.* Cottle e.a., 1992), the result follows immediately from Theorem 11.5. ∎

Note that the result of Corollary 11.3 is already known (see Theorem 5.6.2 (a) in Cottle e.a. (1992), restated in Theorem 11.3 of the previous section). However, the proof by Facchinei and Kanzow (1999) presented here is completely different from that given in Cottle e.a. (1992). It is also clear that Corollary 11.2 can be extended to nonlinear functions F if one assumes that F is a P_0-function and an R_0-function; the definition of the latter class of functions together with some of their properties the reader can find in Chen and Harker (1997) and Tseng (1997).

The counterexample below demonstrates that it is impossible to relax the boundedness assumption of the solutions set $\mathcal{S}$ in Theorem 11.5 without destroying the boundedness of the path $\mathcal{P}$. This constrasts with what happens in the case of monotone complementarity problem where one always has the boundedness of the regularized trajectory whenever the solutions set is nonempty.

EXAMPLE 11.2. Let $F : R^2 \to R^2$ be defined by $F(x) = Mx + q$ where

$$M = \begin{pmatrix} 0 & 1 \\ 0 & 0 \end{pmatrix} \quad \text{and} \quad q = \begin{pmatrix} -1 \\ 0 \end{pmatrix}.$$

Obviously F is a P_0-function. The solutions set is given by

$$\mathcal{S} = \{(x_1, x_2) \mid (x_1, 1), x_1 \geq 0\} \cup \{(x_1, x_2) \mid (0, x_2), x_2 \geq 1\},$$

i.e., the solution set is unbounded. It is easy to verify that $x(\varepsilon) = (1/\varepsilon, 0)$ is the unique solution of the corresponding regularized problem NCP(ε). Clearly $x(\varepsilon)$ is neither convergent nor bounded as $\varepsilon \to 0$. Even worse, the distance of $x(\varepsilon)$ to the solutions set $\mathcal{S}$ is constant and equal to 1. ■

All the above results suggest that the original problem NCP could be solved by calculating the exact solutions of a sequence of regularized problems NCP(ε) for a sequence of parameters ε_k converging to zero. From a practical point of view, however, it is usually impossible to solve the regularized problem NCP(ε) exactly for each $\varepsilon > 0$. Therefore, in what follows an algorithm is presented which only requires inexact solutions of these subproblems and which nevertheless preserves all the convergence properties of its exact counterpart. As before, we mainly follow here the paper by Facchinei and Kanzow (1999).

ALGORITHM 11.1. (*Inexact Regularization Method*).

Step 0. Choose $\varepsilon_0 > 0$, $\alpha_0 \geq 0$, and set $k := 0$.

Step 1. Compute an approximate solution $x^k \in R^n$ of NCP(ε) such that

$$\Psi_\varepsilon(x^k) \leq \alpha_k. \tag{11.41}$$

Step 2. Terminate the iteration if a suitable stopping criterion is satisfied.

Step 4. Choose $\varepsilon_{k+1} > 0$, $\alpha_{k+1} \geq 0$, set $k := k + 1$, and go to Step 1.

Obviously if one takes $\alpha_k = 0$ at each iteration, one has $x^k = x(\varepsilon_k)$. Note that a point x^k satisfying (11.41) can easily be obtained by applying any unconstrained minimization techniques to Ψ_{ε_k}. In fact, the level sets of Ψ_{ε_k} are compact and every stationary point $\bar{x}$ of Ψ_{ε_k} is such that $\Psi_{\varepsilon_k}(\bar{x}) = 0$. Therefore, every suitable minimization algorithm will produce a minimizing sequence and the point x^k can be surely found in a finite number of steps. This situation reflects the fact that the perturbed problems are well-posed and this, in turn, is one of the main advantages of using regularization methods.

To establish a result generalizing Theorem 11.6 to Algorithm 11.1, some further technical results are needed.

LEMMA 11.6. *Let $C \subset R^n$ be a compact set. Then, for every $\delta > 0$ there exists a $\tilde{\varepsilon} > 0$ such that*

$$|\Psi_\varepsilon(x) - \Psi(x)| \leq \delta$$

for all $x \in C$ and all $\varepsilon \in [0, \tilde{\varepsilon}]$.

PROOF. The function $\Psi_{\varepsilon_k}(x)$ viewed as a function of both x and ε is continuous on the compact set $C \times [0, \tilde{\varepsilon}]$. The lemma is then an immediate consequence of the fact that each continuous function on a compact set is uniformly continuous there. ■

We also recall a version of the famous Mountain Pass Theorem which is suitable for our purposes and which can easily be derived from the standard version of this theorem; e.g., *cf.* Theorem 9.2.7 in Palais and Terng (1988).

THEOREM 11.7. *Let $f : R^n \to R$ be a continuously differentiable and coercive function. Let $C \subset R^n$ be a nonempty and compact set and define ν be the least value of f on the (compact) boundary of C:*

$$\nu = \min_{x \in \partial C} f(x).$$

Assume further that there are two points $a \in C$ and $b \notin C$ such that $f(a) < \nu$ and $f(b) > \nu$. Then there exists a point $c \in R^n$ such that $\nabla f(c) = 0$ and $f(c) \geq \nu$. ■

In the convergence analysis of Algorithm 11.1 it will be implicitly supposed that Algorithm 11.1 generates an infinite sequence of points so that the termination criterion in Step 2 is never active. The following result is the main convergence theorem for Algorithm 11.1.

THEOREM 11.9. *Let F be a P_0-function and assume that the solutions set S of NCP is nonempty and bounded. Suppose that a sequence $\{x^k\}$ is generated by Algorithm 11.1. If $\varepsilon_k \to 0$ and $\alpha_k \to 0$ as $k \to \infty$, then $\{x^k\}$ remains bounded, and every accumulation point of $\{x^k\}$ solves NCP.*

PROOF. It follows from a simple continuity argument that each accumulation point of the sequence $\{x^k\}$ solves NCP. Hence it remains to be shown that $\{x^k\}$ is a bounded sequence. On the contrary, suppose that the sequence $\{x^k\}$ is unbounded. Subsequencing if necessary, one gets $\|x^k\| \to \infty$ as $k \to \infty$, whence there exists a compact set $C \subset R^n$ with $S \subset \text{int}C$ and

$$x^k \notin C \tag{11.42}$$

for all k sufficiently large. Let $a \in C$ be an arbitrary solution of NCP. Then one has

$$\Psi(a) = 0.$$

As

$$\bar{\nu} = \min_{x \in \partial C} \Psi(x) > 0,$$

one can apply Lemma 11.6 with $\delta = \bar{\nu}/4$ and conclude that

$$\Psi_{\varepsilon_k}(a) \leq \frac{1}{4}\bar{\nu} \tag{11.43}$$

and

$$\nu = \min_{x \in \partial C} \Psi_{\varepsilon_k}(x) \geq \frac{3}{4}\bar{\nu} \tag{11.44}$$

for all k sufficiently large. Since $\Psi_{\varepsilon_k}(x^k) \leq \alpha_k$ by Step 1 of Algorithm 11.1, one has

$$\Psi_{\varepsilon_k}(x^k) \leq \frac{1}{4}\bar{\nu} \tag{11.45}$$

for all k large enough because $\alpha_k \to 0$ as $k \to \infty$ by the theorem's assumptions. Now fix an index k such that (11.42)–(11.45) hold. Applying the Mountain Pass Theorem 11.7 with $b = x^k$, one obtains the existence of a vector $c \in R^n$ such that

$$\nabla \Psi_{\varepsilon_k}(c) = 0 \qquad \text{and} \qquad \Psi_{\varepsilon_k}(c) \geq \frac{3}{4}\bar{\nu} > 0.$$

By virtue of Proposition 11.1 (3), however, the stationary point c of Ψ_{ε_k} must be a global minimizer of Ψ_{ε_k} which gives the desired contradiction and thus completes the proof of the theorem. ■

Obviously Theorem 11.6 is deduced from Theorem 11.8 by setting $\alpha_k = 0$ for all k and then using Lemma 11.5. Also Corollaries 11.2 and 11.3 can easily be extended to the inexact framework.

COROLLARY 11.4. *Assume that F is a P_0-function and suppose that a sequence $\{x^k\}$ is generated by Algorithm 11.1. Suppose that $\varepsilon_k \to 0$ and $\alpha_k \to 0$ as $k \to \infty$. Then if NCP has a unique solution $\bar{x}$, one has*

$$\lim_{k \to \infty} x^k = \bar{x}.$$ ■

COROLLARY 11.5. *Let $F(x) = Mx + q$ be an affine mapping with $M \in R^{n \times n}$ being a P_0- and an R_0-matrix. Assume that $\{x^k\}$ is any sequence generated by Algorithm 11.1. such that $\varepsilon_k \to 0$ and $\alpha_k \to 0$ as $k \to \infty$. Then the sequence $\{x^k\}$ is bounded, and every accumulation point of the sequence $\{x^k\}$ is a solution to NCP.* ■

If F is a monotone function such that NCP is strictly feasible (i.e., there exists a vector $\hat{x} \in R^n$ such that $\hat{x} > 0$ and $F(\hat{x}) > 0$) then it is known (Harker and Pang, 1990, Theorem 3.4) that NCP has a nonempty and bounded solutions set. Hence one can also obtain the following corollary from the main result of this section.

COROLLARY 11.6. *Assume that F is a monotone function such that NCP is strictly feasible. Suppose that $\varepsilon_k \to 0$ and $\alpha_k \to 0$ as $k \to \infty$. Then any sequence $\{x^k\}$ generated by Algorithm* 11.1. *remains bounded, and every accumulation point of $\{x^k\}$ solves NCP.* ■

11.5.1 FINAL REMARKS

The second part of this chapter, following the recent paper by Facchinei and Kanzow (1999), shows that, under appropriate assumptions, regularization methods can be applied with success to P_0-complementarity problems. However, some properties which hold in the monotone case are lost. In particular, when the solutions set of the problem is unbounded one can no longer guarantee that the trajectory generated by the regularization method is bounded.

In the monotone case $x(\varepsilon)$ always converges to the least l_2-norm solution of NCP. In the P_0 case the least l_2-norm solution can even be not unique, since the solutions set is not necessarily convex.

The results presented in this part of the chapter have been recently developed by some authors. In particular, Sznajder and Gowda (1997) have shown that every limit point of $x(\varepsilon)$ is a weak Pareto minimum element of $\mathcal{S}$ and that $x(\varepsilon)$ actually converges if F is polynomial. It should be also mentioned that several results presented above have been generalized to a wider class of problems; *cf.* Gowda and Tawhid (1997), Qi (1997), Ravindran and Gowda (1997). Moreover, superlinearly convergent algorithms for the solution of P_0 complementarity problems with bounded solutions sets have been developed; see Qi (1997), Sun (1997) among others.

REFERENCES

Billups SC. "Algorithms for complementarity problems and generalized equations". PhD Thesis, Computer Science Department, University of Wisconsin, Madison, WI, 1995.

Billups SC and Ferris MC. QPCOMP: A quadratic programming based solver for mixed complementarity problems. Math. Programming. 1997; 76: 533-562.

Chen B and Harker PT.

1. A noninterior continuation method for quadratic and linear programming. SIAM J. Optim. 1993; 3: 503-515.

2. Smooth approximations to nonlinear complementarity problems. SIAM J. Optim. 1997; 7: 402-420.

Chen B, Chen X and Kanzow C. A penalized Fisher-Burmeister NCP-function: Theoretical investigation and numerical results. Math. Programming, 2000; 88: 211-216.

Cottle, R.W., Pang, J.-S. and Stone, R.E. *The Linear Complementarity Problem*. Boston: Academic Press, 1992.

De Luca T, Facchinei F, Kanzow C. A semismooth equation approach to the solution of nonlinear complementarity problems. Math. Programming, 1996; 75: 407-439.

Dontchev, A.L. and Zolezzi, T. *Well-Posed Optimization Problems*. Lecture Notes in Mathematics No. 1543, Springer-Verlag, 1993.

Ebiefung AA. New perturbation results for solving the linear complementarity problem with P_0-matrices. Appl. Math. Letters, 1998; 11: 37-39.

Facchinei F. Structural and stability properties of P_0 nonlinear complementarity problems. Math. Oper. Res. 1998; 23: 735-745.

Facchinei F and Kanzow C.

1. A nonsmooth inexact Newton method for the solution of large-scale nonlinear complementarity problems. Math. Programming, 1997; 76: 493-512.

2. Beyond monotonicity in regularization methods for nonlinear complementarity problems. SIAM J. Control Optim. 1999; 37: 1150-1161.

Facchinei F, Soares J. A new merit function for nonlinear complementarity problems and a related algorithm. SIAM J. Optim. 1997; 7: 225-247.

Ferris, M.C, Kanzow C. 'Complementarity and related problems.' - In: *Handbook on Applied Optimization*, P.M. Pardalos and M.G.C. Resende, eds. Oxford: Oxford University Press (to appear).

Fischer A. A special Newton-type optimization method. Optimization, 1992; 22: 269-284.

Geiger C, Kanzow C. On the resolution of monotone complementarity problems. Comput. Optim. Appl. 1996; 5: 155-173.

Gowda MS and Tawhid MA. "Existence and limiting behavior of trajectories associated with P_0-equations." Research Report 97-15, Department of Mathematics and Statistics, University of Maryland Baltimore County, Baltimore, Maryland 21250, USA, November 1997.

Harker PT and Pang JS. Finite-dimensional variational inequality and nonlinear complementarity problems: a survey of theory, algorithms and applications. Math. Programming, 1990; 48: 161-220.

Kalashnikov VV. Existence theorem for nonlinear complementarity problem. Optimization (Novosibirsk, Institute of Matehmatics). 1989; 45(62): 26-33. (*in Russian*).

Kalashnikov VV and Kalashnikova NI.

1. Convergence of Newton method for solving nonlinear complementarity problems. Optimization (Novosibirsk, Institute of Mathematics), 1991; 49(65): 69-79 (*in Russian*).

2. An iteration method to solve nonlinear complementarity problems. Optimization (Novosibirsk, Institute of Mathematics), 1993; 52(69): 42-54 (*in Russian*).

Kanzow C. Global convergence properties of some iterative methods for linear complementarity problems. SIAM J. Optim. 1996; 6: 326-341.

Kanzow C and Kleinmichel H. A new class of semismooth Newton-type methods for nonlinear complementarity problems. Comput. Optim. Appl. 1998; 11: 227-251.

Kanzow C and Pieper H. Jacobian smoothing methods for nonlinear complementarity problems. SIAM J. Optim. 1999; 9: 342-372.

Kanzow C and Zupke M. 'Inexact trust region methods for nonlinear complementarity problems. - In: *Reformulation: Nonsmooth, Piecewise Smooth, Semismooth and Smoothing Methods*, M. Fukushima and L. Qi, eds. Dordrecht: Kluwer Academic Press, 1999. - P. 211-233.

Kyparisis J. Uniqueness and differentiability of parametric nonlinear complementarity problems. Math. Programming, 1986; 36: 105-113.

McLinden, L. 'The Complementarity Problem for Maximal Monotone Multifunctions '. - In: R.W. Cottle, F. Giannessi, and J.L. Lions, eds. *Variational Inequalities and Complementarity Problems*. New York: Academic Press, 1980.- P. 251-270.

Moré JJ.

1. Coercivity conditions in nonlinear complementarity problems. SIAM Review, 1974; 16: 1-16.

2. Classes of functions and feasibility conditions in nonlinear complementarity problems. Math. Programming, 1974; 6: 327-338.

Moré JJ and Rheinboldt WC. On P- and S-functions and related classes of n-dimensional nonllnear mappings. Lin. Alg. Appl. 1973; 6: 45-68.

Ortega, J., Rheinboldt, W.C. *Iterative Solution of Nonlinear Equations in Several Variables.* New York: Academic Press, 1970.

Palais, R.S. and Terng, C.-L. *Critical Point Theory and Submanifold Geometry.* Lecture Notes in Mathematics, No. 1353. Berlin: Springer-Verlag, 1988.

Polyak, B.T. *Introduction to Optimization.* New York: Optimization Software Inc., 1987.

Qi HD. "A regularized smoothing Newton method for box constrained variational inequality problems with P_0 functions." Research Report, Chinese Academy of Sciences, Institute of Computational Mathematics and Scientific/Engineering Computing, Beijing, China, July 1997.

Ravindran G and Gowda MS. "Regularization of P_0-functions in box variational inequality problems." Research Report 97-07, Department of Mathematics and Statistics, University of Maryland Baltimore County, Baltimore, Maryland 21250, USA, August (revised October), 1997.

Subramanyan PK. A note on least two norm solutions of monotone complementarity problems. Appl. Math. Letters, 1988; 1: 395-397.

Sun D. "A regularization Newton method for solving nonlinear complementarity problems". Research Report, School of Mathematics, The University of New South Wales, Sydney 2052, Australia, July 1997.

Sznajder R and Gowda MS. "On the limiting behavior of the trajectory of regularized solutions of a P_0-complementarity problem". Research Report 97-08, Department of Mathematics and Statistics, University of Maryland Baltimore County, Baltimore, Maryland 21250, USA, October, 1997.

Tseng P. An infeasible path-following method for monotone complementarity problems. SIAM J. Optim. 1997; 7: 386-402.

Venkateswaran V. An algorithm for the linear complementarity problem with a P_0-matrix. SIAM J. Matrix Anal. Appl. 1993; 14: 967-977.

PART IV

EFFICIENCY IN ABSTRACT SPACES

Chapter 12

EFFICIENCY

12.1 INTRODUCTION

A very popular domain of applied mathematics is *optimization*, because the diversity of its applications to economics, engineering and sciences. Certainly the applications to practical problems stimulated the impressive development of this domain. Between the chapters of optimization, one is *the optimization of vector-valued functions*, known also under the name of *Pareto optimization*. In 1906 V. Pareto wrote: *"Principeremo con definire un termine di cui è comodo fare uso per scansare lungaggini. Diremo che i componenti di una colletività godono, in una certa postione, del massimo di ofelimita, quando è impossibile allontanarsi pochissimo da quella positione giovando, o nuocendo, a tutti i componenti la collectività; ogni picolissimo spostamento da quella positione avendo necessariamente per effetto di giovare a parte dei componenti la collectività e di nuocere ad altri."* (Pareto, 1919).

Vector optimization (i.e. vector maximization or minimization) problems arise when $m \geq 2$ non-comparable criterion functions, are to be simultaneously maximized or minimized over a given feasibility region. The concept of *"efficiency"* has played a useful role in the analysis of such problems. The necessity of the notion of efficiency, is the fact that in the case of optimization of vector-valued functions, we must use an ordering given in a vector space. It is well known that this ordering is *not total*, as it is the ordering of the field of real numbers, used in optimization of real-valued functions.

The development of optimization of vector-valued functions is based on the study of several kinds of efficiency.

In this part of the book we will present the most important notions of efficiency and we will study the relations between these notions. We will study several existence theorems related to efficiency and some properties of solution set, will be also considered. We will present also some Arrow-Barankin -Blackwell type theorems. The nuclear cones will be applied to the study of efficiency. We will consider also several topological properties of efficient point sets. A general constructive test for Pareto efficiency will be also given.

12.2 PRELIMINARIES: A MINIMAL BACKGROUND ON CONVEX CONES

In this section we will introduce the notion of convex cone and we will present several remarkable classes of convex cones, necessary to study the concepts of efficiency. We will study their properties. For this chapter we suppose that the reader has a minimal background on the general topology, on linear algebra and on topological vector spaces. We will use the Euclidean space, Banach spaces, locally convex spaces and topological vector spaces. The convex cones will be considered in a such vector space defined over the real field. The n-dimensional Euclidean space will be denoted by $(R^n, <\cdot,\cdot>)$, a Banach space by $(E, ||\cdot||)$, a locally convex space by $(E(\tau), \{p_\alpha\}_{\alpha\in A})$, where the topology τ will be defined by a *sufficient family of seminorms* and a general topological vector space by $E(\tau)$. The topologies will be Hausdorff topologies.

If $E(\tau)$ is a topological vector space and D a non-empty subset of E we will denote by $\bar{D}$ or by cl$D$ *the closure* of D, by int(D) *the interior* of D (if it is non-empty) and by ∂D *the boundary* of D, all with respect to topology τ.

Let E be a real vector space and τ a topology on E. The pair (E,τ) (denoted by $(E(\tau))$) is called a *topological vector space over* R if the following axioms are satisfied:

$(TVS)_1 : (x,y) \longrightarrow x+y$ is continuous as a mapping from $E\times E$ into E,
$(TVS)_2 : (\lambda, x) \longrightarrow \lambda x$ is continuous as a mapping from $R\times E$ into E.

Let U, A be subsets of E. We say that U *absorbs* A if there exists $\rho\in R$ such that $A\subset\lambda U$ whenever $|\lambda|\geq|\rho|$. The subset U is called *radial* (or *absorbing*) if U absorbs every finite subset of E.

We say that U is *circled* if $\lambda U\subset U$ whenever $|\lambda|\leq 1$.

A subset D of E is said to be *convex* if $x\in D$, $y\in D$ imply that $\lambda x+(1-\lambda y\in D$ for all real numbers $\lambda\in]0,1[$.

If $D\subset E$ is any radial subset the non-negative real function on E:

$$x \longrightarrow p_D(x) = inf\{\lambda>0 | x\in\lambda D\}$$

is called *the Minkowski functional* of D.

We say that $p: E\longrightarrow R$ is a *seminorm* if the following conditions are satisfied:

$(S_1) : p(x)\geq 0$, for all $x\in E$,
$(S_2) : p(\lambda x) = |\lambda| p(x)$, for all $x\in E$ and $\lambda\in R$,
$(S_3) : p(x+y)\leq p(x)+p(y)$, for all $x,y\in E$.

We can show that if $D\subset E$ is a radial, circled and convex subset, then the Minkowski functional p_D is a seminorm on E.

The following result is well known:
if $p: E\longrightarrow R$ is a seminorm and $E(\tau)$ is a topological vector space, then the following properties are equivalent:
p is continuous at the origin $0\in E$
$D_0 = \{x\in E | p(x)<1\}$ is open in E
p is uniformly continuous on E. A *norm* on a vector space E is a seminorm

$p : E \longrightarrow R$ such that $p(x) = 0$ if and only if $x = 0$. If p is a norm on E we denote $p(x)$ by $||x||$.

Thus a norm $||\cdot||$ on E is characterized by the following analytical properties:
$(h_1) : ||x|| \geq 0$ for all $x \in E$ and $||x|| = 0$ if and only if $x = 0$.
$(h_2) : ||\lambda x|| = |\lambda| ||x||$, for all $x \in E$ and $\lambda \in R$
$(h_3) : ||x + y|| \leq ||x|| + ||y||$, for all $x, y \in E$.

A *normed vector space* will be denoted by $(E, ||\cdot||)$. A *Banach space* is a *complete normed space*, i.e. a normed vector space such that any Cauchy sequence, with respect to the norm is convergent.

A topological vector space $E(\tau)$ over R is called *locally convex*, if it is a Hausdorff space such that every neighborhood of any point $x \in E$ contains a convex neighborhood of x. Equivalently $E(\tau)$ is a locally convex topological vector space if the convex neighborhood of 0 form a base at 0 with intersection $\{0\}$.

Analytically a locally convex topology on E is determined by an arbitrary family $\{p_\alpha\}_{\alpha \in A}$ of seminorms as follows: for each $\alpha \in A$, let $U_\alpha = \{x \in E | p_\alpha(x) \leq 1\}$ and consider the family $\{n^{-1}U\}_{n \in N}$, where U runs through all finite intersections of sets U_α with $\alpha \in A$. This family $\mathcal{B}$ is a base at 0 for a locally convex topology τ on E, called the topology generated by the family of seminorms $\{p_\alpha\}_{\alpha \in A}$. Conversely, every locally convex topology on E is generated by a suitable family of seminorms.

Indeed it suffices to take the Minkowski functionals of a family of convex, circled 0-neighborhoods whose positive multiples form a base at 0. When the topology τ on E is defined by a family $\{p_\alpha\}_{\alpha \in A}$ of seminorms we denote $(E(\tau), \{p_\alpha\}_{\alpha \in A})$.

For many problems, in locally convex spaces, it is good to have the topology defined by a *"sufficient"* family of seminorms (Marinescu, 1963).

We say that the family of seminorms $\{p_\alpha\}_{\alpha \in A}$ is *sufficient* if the following properties are satisfied:
(1): for any $x \in E \backslash \{0\}$ there exists $\alpha \in A$ such that $p_\alpha(x) \neq 0$,
(2): for any $\alpha, \beta \in A$ there exists $\gamma \in A$ such that $\sup\ (p_\alpha(x), p_\beta(x)) \leq p_\gamma(x)$, for all $x \in E$.

Let E, E^* be a pair of vector spaces over R and let $< \cdot, \cdot >$ be a bilinear form on $E \times E^*$. The triplet $(E, E^*, < \cdot, \cdot >)$ is called a *dual system* or *duality* over R if the following axioms are satisfied:
$(d_1) :< x_0, y >= 0$ for all $y \in E^*$ implies $x_0 = 0$,
$(d_2) :< x, y_0 >= 0$ for all $x \in E$ implies $y_0 = 0$.

The triplet $(E, E^*, < \cdot, \cdot >)$ is more conveniently denoted by $< E, E^* >$.

The duality is an important chapter in the theory of topological vector spaces. If $< E, E^* >$ is a dual system the *weak topology* $\sigma(E, E^*)$ is the coarsest topology on E for which the linear forms $x \longrightarrow < x, y >$, $y \in E^*$ are continuous.

The space E is a topological vector space with respect to topology $\sigma(E, E^*)$.

By a similar method we can define a weak topology on E^*. Generally this topology is denoted by $\sigma(E^*, E)$ and it is called the *weak-(*)- topology* on E^*.

For more information and details about topological vector spaces the reader is referred to Schaefer (1966).

CONVEX CONES

The concept of *convex cone* in a real vector space E is related to an *ordering* in E, usually denoted by $\leq$.

An ordering in E is a binary relation in E which is *reflexive, transitive* and *antisymmetric.* If the vector space E is endowed with an ordering, we say that $(E, \leq)$ is an *ordered vector space over* R if the following axioms are satisfied:
$(OVS)_1 : x \leq y$ *implies* $x + z \leq y + z$ *for all* $x, y, z \in E$,
$(OVS)_2 : x \leq y$ *implies* $\lambda x \leq \lambda y$ *for all* $x, y \in E$ *and* $\lambda \in R_+$.

The axioms $(OVS)_1$ and $(OVS)_2$ imply that the set $E_+ = \{x \in E \mid x \geq 0\}$ satisfies the following conditions:
$E_+ + E_+ \subseteq E_+$
$\lambda E_+ \subseteq E_+$, for all $\lambda \in R_+$
$E_+ \cap (-E_+) = \{0\}$. We say that E_+ is a *pointed convex cone.*

Now we can introduce the following general concept.

DEFINITION 12.1. We say that a non-empty subset $K \subset E$ is a *pointed convex cone* if the following conditions are satisfied:
$(c_1) K + K \subseteq K$,
$(c_2) \lambda K \subseteq K$, for all $\lambda \in R_+$,
$(c_3) K \cap (-K) = \{0\}$.

If for a non-empty subset $K \subset E$ only (c_1) and (c_2) are satisfied, we say that K is a *convex cone.* If $K \subset E$ is a convex cone and $\bar{K} \cap (-\bar{K}) = \{0\}$ we say that K is an *acute convex cone.*

Remark. If $K \subset E$ is a pointed convex cone then we can define on E an ordering by:"$x \leq y$ if and only if $y - x \in K$". This ordering satisfies the axioms $(OVS)_1$ and $(OVS)_2$. In this case we have $E_+ = K$.

EXAMPLES OF POINTED CONVEX CONES

1. $E = R^n, K = R^n_+ = \{x = (x_1, ..., x_n) \in R^n | x_1 \geq 0, ..., x_n \geq 0\}$.

2. E=$\mathcal{F}(\Omega, R) = \{f : \Omega \longrightarrow R\}$, where Ω is an arbitrary non-empty set. K=$\{f \in \mathcal{F}, K) | f(t) \geq 0$ for all $t \in \Omega\}$.

3. E=$\mathcal{C}(\Omega, R) = \{f : \Omega \longrightarrow R \| f$ continuous $\}$, where Ω is a topological space. K=$\{f \in \mathcal{C}, K) | f(t) \geq 0$ for all $t \in \Omega\}$.

4. E=$\omega = \{x = (x_n)_{n \in N} | x_n \in R$ for all $n \in N\}$, K= $\{x = (x_n)_{n \in N} \in \omega | x_n \geq 0$ for all $n \in N\}$.

5. E=$l^p = \{x = (x_n)_{n \in N} \in \omega | \sum_{n=1}^{\infty} |x|^p < +\infty\}$ where $1 \leq p < +\infty$, K=$\{x = (x_n)_{n \in N} \in l^p | x_n \geq 0$ for all $n \in N\}$.

6. E=$c = \{x = (x_n)_{n \in N} \in \omega | \lim_{n \to \infty} x_n$ exists $\}$, K=$\{x = (x_n)_{n \in N} \in c | x_n \geq 0$ for all $n \in N\}$.

7. E=$c_0 = \{x = (x_n)_{n \in N} \in \omega | \lim_{n \to \infty} x_n = 0\}$, K=$\{x = (x_n)_{n \in N} \in c_0 | x_n \geq 0$ for all $n \in N\}$.

8. Let (Ω, μ) be an arbitrary measure and $E = L^p(\Omega, \mu)$, where $1 \leq p < +\infty$. K=$\{f \in L^p(\Omega, \mu) | f \geq 0$ μ-almost everywhere$\}$ is a pointed convex cone.

9. Let $(E, ||\cdot||)$ be a Banach space and $D \subset E$ a closed bounded convex set such that $0 \notin D$. The set K=$\{\lambda x | \lambda \in R_+$ and $x \in D\}$ is a pointed convex cone.

10. Let $< E, E^* >$ be a dual system of locally convex vector spaces and $\{x_n, f_n\}_{n \in N}$ a biorthogonal system. Suppose that $\{x_n, f_n\}_{n \in N}$ is a Schauder base for E. The set K=$\{x \in E | x = \sum_{n=1}^{\infty} f_n(x) x_n$ with $f_n(x) \geq 0$ for all $n \geq 1\}$ is a pointed convex cone.

Let $(E(\tau), K)$ be a topological vector space and $K \subset E$ a pointed convex cone. If K is τ-closed we say that $(E(\tau), K)$ is a *topological ordered vector space.* The cone $K \subset E$ is said to be *generating* if $E = K - K$.

Let (E, K) be an ordered vector space. An *order interval* is a set of the form $[x, y]_0 = \{z \in E | x \leq z \leq y\}$. Every order interval is a convex set.

A subset $D \in E$ is *majorized* [*minorized*] if there is an element $x_0 \in E$ such that $x \leq x_0$ for all $x \in D$ [$x_0 \leq x$ for all $x \in D$].

A subset $D \in E$ is *order bounded* if D is contained in an order interval. Let D be a non-empty subset of E and $x \in E$. We say that x is the *supremum* of D and we write x=sup(D) if the following properties are satisfied:
(1).$y \leq x$ for all $y \in D$,
(2).$x \leq z$ whenever $y \leq z$ for all $y \in D$.

The *infimum* of D, written inf(D), is defined by duality, that is, by replacing $\leq$ by $\geq$ in (1) and (2).

Let $< E, F >$ be a dual system consisting of a pair of real vector spaces. In particular we may consider the dual system $< E, E^* >$ where E is locally convex and the bilinear functional $< \cdot, \cdot >$ is $(x, u) \longrightarrow < x, u >= u(x)$ for all $x \in E$ and $u \in E^*$.

If $K \subset E$ is a convex cone the *dual* of K is:

$$K^* = \{y \in F | \langle x, y \rangle \geq 0 \text{ for all } x \in K\}.$$

It is easy to show that K^* is a convex cone in F and K^* is a pointed cone if and only if $K - K$ is dense in E. If K^0 is the polar cone of K, i.e.

$$K^0 = \{y \in F | < x, y > \leq 0, \text{ for all } x \in K\},$$

it is true that $K^* = -K^0$.

The convex cone K^* is always a $\sigma(F,E)$-closed subset of F. Moreover the *Bipolar Theorem* (see Schaefer, 1966) implies that the bidual cone $K^{**} = (K^*)^*$ coincides with the $\sigma(E,F)$-closure of the given cone K. We will use also the following notation:

$$K^{*i} = \{y \in K^* | \langle x, y \rangle > 0 \text{ for all } x \in K \backslash \{0\}\}.$$

The following results can be easily established:

1. *the convex cone $K \subset E$ is closed for a topology τ compatible with the duality $< E, F >$ if and only if $x \in K$ precisely when $\langle x, y \rangle \geq 0$ for all $y \in K^*$,*

2. *if $x \in E$ there is an element $y \in K^*$ such that $\langle x, y \rangle < 0$ if and only if x is not in the $\sigma(E,F)$-closure of K,*

3. *if K is a pointed convex cone in E then the closure $\bar{K}$ of K for a topology τ compatible with the duality $< E, F >$ is a pointed convex cone in E if and only if $K^* - K^*$ is dense in F for $\sigma(E,F)$.*

For more results on convex cones the reader is referred to Schaefer (1966), Peressini (1967) and Jameson (1970).

NORMAL CONES

The most important class of cones, used currently in the theory of ordered topological vector spaces, is the class of *normal* cones.

We will present this notion and several classical results related to this notion. The proofs are in Schaefer (1966), Peressini (1967) and Jameson (1970). Let $E(\tau)$ be a topological vector space, let $K \in E$ be a pointed convex cone and let $A \subset E$ be a non-empty subset. The $K-$*saturated* hull of A is denoted by $[A]$ and it is defined by

$$[A] = (A + K) \cap (A - K) = \bigcup \{[x, y]_0 | x \in A, y \in A\}.$$

If $A = [A]$ we say that A is K-saturated. We have $A \subseteq [A]$ and $[[A]] = [A]$ for any subset $A \in E$. The K-saturated hull of a convex set [circled set] is convex [circled].

DEFINITION 12.2. If $E(\tau)$ is a topological vector space and $K \subset E$ a pointed convex cone, we say that K is *normal* (for the topology τ) if there is a neighborhood basis of zero for τ consisting of K-saturated sets.

In a general topological vector space we have the following result.

THEOREM 12.1. *Let $E(\tau)$ be a topological vector space and $K \subset E$ a pointed convex cone. The following assertions are equivalent:*
(n_1). K is normal (for τ),
(n_2). there exists a neighborhood basis $\mathcal{U}$ of zero for τ consisting of sets U for which $0 \le x \le y \in U$ implies $x \in U$,
(n_3). for any nets $\{x_i\}_{i\in I}$ and $\{y_i\}_{i\in I}$ in E, if $0 \le x_i \le y_i$ for all $i \in I$ and $\{y_i\}_{i\in I}$ converges to zero for τ then $\{x_i\}_{i\in I}$ converges to zero for τ too,
(n_4). given a τ-neighborhood U of zero, there exists a τ- neighborhood V of zero such that $0 \le x \le y \in V$ implies $x \in U$. ■

In locally convex vector spaces we have the following characterization of normality.

THEOREM 12.2. *Let $E(\tau)$ be a locally convex vector space and $K \subset E$ a pointed convex cone. The following assertions are equivalent:*
(n_5). K is normal for τ,
(n_6). there exists a family $\{p_\alpha\}_{\alpha\in A}$ of seminorms generating the topology τ such that $0 \le x \le y$ implies $p_\alpha(x) \le p_\alpha(y)$ for all $\alpha \in A$. ■

Remark Assertion (n_6) of Theorem 12.2 is equivalent to:
(n_6'): there exists a family $\{p_\alpha\}_{\alpha\in A}$ of seminorms generating the topology τ, such that for all $x, y \in K$ and $\alpha \in A$ we have $p_\alpha(x) \le p_\alpha(x+y)$.

If the space E is a normed vector space, then the normality is characterized by the following results.

THEOREM 12.3. *Let $(E, ||\cdot||)$ be a normed vector space and $K \subset E$ a pointed convex cone. The following assertions are equivalent:*

1. *K is normal (for the topology generated by the norm $||\cdot||$)*
2. *there exists an equivalent norm $||\cdot||_*$ on E such that $0 \le x \le y$ implies $||x||_* \le ||y||_*$,*
3. *there exists a constant $\delta > 0$ such that $0 \le x \le y$ implies $\delta||x|| \le ||y||$,*
4. *there exists a constant $\delta > 0$ such that $\delta \max\{||x||, ||y||\} \le ||x+y||$ for all $x, y \in K$,*
5. *the set $\{||x|| \mid 0 \le x \le y; ||y|| \le 1\}$ is bounded above.* ■

The theorems presented above have several important consequences. As examples we cite the following results

PROPOSITION 12.1. *If $E(\tau)$ is a topological vector space and $K \subset E$ is a normal cone, then every order interval in E is τ-bounded.*

PROOF. Indeed, if U is any K-saturated, circled neighborhood of zero for τ and if $[x, y]_0$ is a given order interval in E, choose $\lambda > 0$ such that $\lambda x \in U$ and

$\lambda y \in U$. Then $\lambda[x, y]_0 \subset U$. ■

PROPOSITION 12.2. *If $E(\tau)$ is a topological vector space, then every normal cone $K \subseteq E$ is pointed.*

PROOF. If $x \in K \cap (-K)$, then $x \in [\{0\}] \subset [U]$ for each neighborhood U of zero and because E is supposed to be Hausdorff it follows that $x = 0$. ■

PROPOSITION 12.3. *Let $E(\tau)$ be a locally convex space and $K \subset E$ a pointed convex cone. If K is normal, then the closure $\bar{K}$ of K is a normal cone.*

PROOF. It is evident that $\bar{K}$ is a convex cone. Let $\{p_\alpha\}_{\alpha \in A}$ be a generating family of seminorms for τ satisfying assertion in Theorem 12.2. If $x \in \bar{K} \cap (-\bar{K})$, then $0 = p_\alpha(x + (-x)) \geq p_\alpha(x)$ for all $\alpha \in A$ which implies that $x = 0$ (since τ is a Hausdorff topology). Hence $\bar{K}$ is pointed. Applying again assertion (n_6) of Theorem 12.2 we can show that $\bar{K}$ is normal for τ. ■

One of the most important consequences of normality for locally convex spaces is the following result.

THEOREM 12.4. *Let $E(\tau)$ be a locally convex space and $K \subset E$ a closed pointed convex cone. If K is a normal cone (for τ) then every continuous linear functional on E is the difference of two positive continuous linear functionals on E.* ■

Remark. In a non-normable locally convex space it is impossible to have a normal cone with non-empty interior. Indeed, let $E(\tau)$ be a non-normable locally convex space and $K \subset E$ a normal cone such that $\text{int}(K) \neq \emptyset$. Let $x_0 \in \text{int}(K)$ be an arbitrary element. The order interval $[-x_0, x_0]_0 = (-x_0 + K) \cap (x_0 - K)$ is a τ-neighborhood of zero which is bounded since K is normal. By *Kolmogorov's theorem* $E(\tau)$ must be normable, which is impossible.
Now we give several examples of normal cones.

EXAMPLES:

1. Let $(R^n, \langle \cdot, \cdot \rangle)$ be the Euclidean space. The convex cone $K = \{x = (x_i) \in R^n \mid x_i \geq 0, i = 1, 2, .., n\}$ is a normal cone. In this topological vector space every closed pointed convex cone is normal.

2. Let Ω be a compact topological space and $E = \mathcal{C}(\Omega, R)$ the vector space of continuous functions from Ω into R. The pointed convex cone $K = \{f \in \mathcal{C}(\Omega, R) | f(x) \geq 0 \text{ for all } x \in \Omega\}$ is a normal cone with respect to the topology defined by the norm $||f|| = \sup\{|f(x)|,\ x \in \Omega\}$.

3. Let (Ω, μ) be an arbitrary measure space and $E = L_p(\Omega, \mu)$ where $p \geq 1$. The convex cone $K = \{f \in L_p(\Omega, \mu) | f \geq 0$, μ-almost everywhere$\}$ is a normal cone with respect to the topology defined by the norm $||f|| = [\int_\Omega |f(t)|^p d\mu]^{1/p}$.

4. If E is one of the following topological vector spaces l_p, l_∞, c, c_0, each with the usual topology, then in every space, the pointed convex cone formed by the sequences $x = \{x_n\}_{n \in N}$ such that $x_n \geq 0$ for all $n \in N$, is a normal cone.

5. Let $(H, < \cdot, \cdot >)$ be a Hilbert space and $\mathcal{L}(H) = \{T : H \rightarrow H \mid$T is continuous and linear$\}$. The convex cone $K = \{T \in \mathcal{L} \mid T$ is self-adjoint and $< T(x), x > \geq 0$ for all $x \in H\}$ is a normal cone with respect to the uniform convergence topology on $\mathcal{L}(H)$.

6. Let $E = \mathcal{D}$ the vector space of infinite differentiable real valued functions with compact support in R^n. The convex cone $K = \{f \in \mathcal{D} | f \geq 0\}$ is not normal with respect to the Schwartz topology on $\mathcal{D}$.

REGULAR AND COMPLETELY REGULAR CONES

The origin of regularity is the fact that in an ordered topological vector space, the monotonicity and boundedness of a sequence do not imply its strong or weak convergence.

Let $E(\tau)$ be a locally convex space and $K \subseteq E$ a closed pointed convex cone.

DEFINITION 12.3. We say that the cone K is *regular* [*sequentially regular*] if every net [every sequence] of elements of K, monotone increasing and order bounded is convergent.

DEFINITION 12.4. The cone K is called *completely regular* [*sequentially completely regular*] if every net [every sequence] of elements of K, monotone increasing and topological bounded is convergent.

In a normed vector space there is no distinction between regular [completely regular] and sequentially regular [sequentially completely regular]. In this case we say that K is *regular* or *completely regular.*

From definitions we obtain immediately the following result.

PROPOSITION 12.4. *In a locally convex space, every pointed normal, completely regular convex cone is regular.*

THEOREM 12.5. *In a locally convex space $E(\tau)$ any weakly complete [weakly sequentially complete] normal, pointed convex cone is completely regular [sequentially completely regular].*

PROOF. Let $K \subset E$ be a weakly complete, normal pointed convex cone.

The cone K is τ-closed since it is weakly closed.
The normality of K implies that $E^* = K^* - K^*$. Let $\{x_i\}_{i\in I}$ be a net in K. Suppose that $\{x_i\}_{i\in I}$ is monotone increasing and topologically bounded. Therefore, for every $f \in K^*$, $\{f(x_i)\}_{i\in I}$ is a Cauchy net (since it is monotone increasing and bounded). It follows that $\{x_i\}_{i\in I}$ is a weakly Cauchy net and because K is weakly complete it is weakly convergent to an element $x_* \in K$. Applying Peressini (1967, Proposition 3.4), we deduce that $\{x_i\}_{i\in I}$ is τ-convergent to the same limit x_* and the proof is complete. ■

COROLLARY 12.1. *In a semireflexive locally convex space, every pointed normal closed convex cone is sequentially regular.*

PROOF. It is known that every semireflexive locally convex space is $\sigma(E, E^*)$-quasicomplete (Schaefer, 1966). Therefore, every semireflexive locally convex space is weakly sequentially complete. The corollary is a consequence of Theorem 12.5. ■

THEOREM 12.6. *In a Banach space $(E, ||\cdot||)$ a pointed convex cone $K \subset E$ is a completely regular if and only if the relations $x_n \in K$ and $||x_n|| \geq \delta > 0$ for all $n \in N$ imply that $\lim_{n\to\infty} ||x_1 + x_2 + ... + x_n|| = +\infty$.*

PROOF. A proof of this classical result is in Krasnoselskii e.a. (1989). ■

A consequence of Theorem 12.6 is the completely regularity of the cone K of non-negative functions in L^p $(1 \leq p < +\infty)$. Indeed, since $(a+b)^p \geq a^p + b^p$ for $a, b \geq 0$ and $1 \leq p < +\infty$, then we have $||x_1 + x_2 + \cdots + x_n||^p \geq ||x_1||^p + ||x_2||^p + \cdots + ||x_n||^p$ for all $x_i \in K \subset L^p$, $i = 1, 2, ..., n$ and Theorem 12.6 is applicable.

About completely regular cones it is well to cite also the following result.

THEOREM 12.7 [KARLIN]. *Let $(E, ||\cdot||)$ be a Banach space with E^* separable with respect to the strong topology $\beta(E^*, E)$. In this case every normal closed convex cone $K \subset E^*$ is completely regular and regular (with respect to the topology $\beta(E^*, E)$).*

PROOF. The reader can find a proof of this result in Karlin (1959). ■

EXAMPLES.

1. Every closed pointed convex cone in R^n is regular and completely regular.

2. If $p \in [1, +\infty[$, then the convex cone $K = \{x = (x_n)_{n\in N} \in l^p | x_n \geq 0$ for all $n \in N\}$ is regular and completely regular in l^p.

3. Let Ω be a compact topological space and $C(\Omega, R)$ the Banach space of continuous functions from Ω into R, where the norm is $||x|| = sup\{|x(t)| | t \in \Omega\}$ for any $x \in C(\Omega, R)$. The convex cone $K = \{x \in C(\Omega, R) | x \geq 0\}$ is neither regular nor completely regular.

4. If I is an arbitrary non-empty set, we consider on the vector space R^I the product topology. The convex cone $K = R^I_+$ is a regular cone.

5. If E=l^∞, then the convex cone $K = \{x = (x_n)_{n\in N} \in l^\infty | x_n \geq 0$ for all $n \in N\}$ is not regular.

DANIELL CONE

Some authors working in Pareto optimization named the regular cone, *Daniell cone*, as in measure theory.

We recall that a Daniell cone is a pointed closed convex cone such that every decreasing net which has a lower bound converges to its infimum.

In a topological vector space the following result is known.

PROPOSITION 12.5. *If $\{x_i\}_{i\in I}$ is a monotone increasing (respectively, decreasing) net in a topological vector space $E(\tau)$ ordered by a closed convex cone K and if $\{x_i\}_{i\in I}$ converges to an element x_0 for τ, then $x_0 = \sup\{x_i\}_{i\in I}$ (respectively, $x_0 = \inf\{x_i\}_{i\in I}$).*

PROOF. This result is Corollary 3.2 in Peressini (1967). ∎

From Proposition 12.5 we deduce that Daniell cone is another name used for regular cone.

WELL BASED CONES

Let $E(\tau)$ be a locally convex space and $K \subset E$ a pointed convex cone. Suppose that $K \neq \{0\}$. If B is a non-empty convex subset of K we say that K is generated by B if $K = \cup\lambda B = \{x = \lambda b | \lambda \in R_+$ and $b \in B\}$.

A particular case of generating set for a convex cone is given by the following definition.

DEFINITION 12.5. We say that a non-empty convex subset B of K is a *base* for K if each element $x \in K\backslash\{0\}$ has a unique representation of the form $x = \lambda b$, with $\lambda > 0$ and $b \in B$.

If B is a base for a convex cone, we can show that $0 \notin B$. A convex cone $K \subset E$ which has a base is necessarily pointed, i.e. $K \cap (-K) = \{0\}$.

We say that $f \in E^*$ is *strictly positive* on E (with respect to K) if, for any $x \in K\backslash\{0\}$ we have $f(x) > 0$.

THEOREM 12.8 [KREIN-RUTMAN]. *In a separable Banach space, every closed pointed convex cone has a base.*

PROOF. The reader can find a proof of this classical result in Krein and Rutman (1948). ∎

THEOREM 12.9. *Let $E(\tau)$ be a locally convex space and $K \subset E$ a convex cone. A subset $B \in K$ is a base for K if and only if there is a strictly positive linear functional f on E such that $f^{-1}(1) \cap K = B$.*

PROOF. A proof of this theorem is in Peressini (1967). ■

If a cone K in a locally convex space $E(\tau)$ has a base, it is not necessary to have that this base is a bounded set.

A convex cone which has a bounded base has remarkable properties.

DEFINITION 12.6. We say that a convex cone K in a locally convex space $E(\tau)$ is *well based*, if it has a bounded base B such that $0 \notin \bar{B}$.

Remark. By using a classical separation theorem we can show that $K \subset E(\tau)$ is a well based convex cone if and only if there exists a convex bounded set A such that $0 \notin \bar{A}$ and $K = \cup_{\lambda \geq 0} \lambda A$.

The well based cones have the following properties:

1. Any pointed convex cone which has a closed bounded base is closed.
2. Let $E(\tau)$ be a locally convex space and $K \subset E$ a pointed convex cone. The cone K is well based if and only if its dual K^* has an interior point with respect to the strong topology $\beta(E^*, E)$.
3. Let $E(\tau)$ be a locally convex space and $K \subset E$ a pointed convex cone. If K has a non-empty interior, then K^* has a $\sigma(E^*, E)$- compact base.
4. Let $E(\tau)$ be a locally convex space and $K \subset E$ a pointed convex cone. If K is well based then it is normal (with respect to τ).
5. Let $E(\tau)$ be a locally convex space. Any complete, well based convex cone $K \subset E$ is completely regular.

A remarkable class of well based cones is formed by locally compact cones.

PROPOSITION 12.6. *Let $E(\tau)$ be a locally convex space and $K \subset E$ a pointed convex cone. The cone K is locally compact if and only if there exists a τ- neighborhood U of zero such that $U \cap K$ is compact.*

The following theorem is a classical result.

THEOREM 12.10[KLEE]. *Let $E(\tau)$ be a locally convex space and $K \subset E$ a pointed convex cone. The cone K is locally compact if and only if it has a compact base.* ■

COROLLARY 12.2. *In R^n any pointed closed convex cone is well based.* ■

We finish the results about well based cones with the following result due to Isac (1980) and which is useful in many applications.

THEOREM 12.11. *Let $(E(\tau), \{p_i\}_{i \in I})$ be a locally convex space, with the topology τ defined by a sufficient family of seminorms and let $K \subset E$ be a closed pointed convex cone.*
The cone K is well based if and only if there exists $f \in K^*$ such that for every $i \in I$ there exists a real number $\delta_i > 0$ with the property that $\delta_i p_i(x) \leq f(x)$ for all $x \in K$. ■

For the proofs of properties (1)-(5), Proposition 12.6 and Theorems 12.10 and 12.11 the reader is reffered to Peressini (1967), Hyers, Isac and Rassias (1997) and Isac (1980).

CONES OF CLASS $\mathcal{C}$

The following class of cone was introduced in Sterna-Karwat [2] (1986). The cones of class $\mathcal{C}$ have interesting applications to the study of efficiency.

Let $E(\tau)$ be a topological vector space.

DEFINITION 12.7. We say that a convex cone $K \subset E$ belongs to the class $\mathcal{C}$ if the following condition is satisfied: for every vector subspace F contained in E, $K \cap F$ is a vector subspace whenever $\overline{K \cap F}$ is a vector subspace.

In Sterna-Karwat [2] (1986), the following results are proved.

Let K be a convex cone in a topological vector space $E(\tau)$. Denote by $l(K)$ the largest vector space contained in K, i.e. $l(K) = K \cap (-K)$.

We have that $K \in \mathcal{C}$, whenever one of the following conditions holds:
(i) E is finite dimensional
(ii) the vector subspace $l(\bar{K})$ is finite dimensional; in particular if $\bar{K}$ is pointed
(iii) K is closed
(iv) $K \setminus \{0\}$ is open
(v) K admits a continuous K-strictly positive functional on E, i.e. there exists $f \in E^*$ such that $f(x) > 0$ for all $x \in K \setminus \{0\}$.

Remark. From statement (v) we deduce that in a locally convex space any well based cone is in $\mathcal{C}$.

PROPOSITION 12.7. *If K is a convex cone in a topological vector space $E(\tau)$ such that $l(K) = K \cap l(\bar{K})$, then $K \in \mathcal{C}$.*

PROOF. Suppose that $l(K) = K \cap l(\bar{K})$ and $K \notin \mathcal{C}$. Then there exist a vector subspace F of E and $x \in K \cap F$ such that $\overline{K \cap F}$ is a vector subspace and $-x \notin K$.
Because $-x \in \overline{K \cap F} \subseteq \bar{K}$ we must have that $x \in K \cap l(\bar{K}) = l(K)$, which contradicts $-x \notin K$. ■

Now we introduce some special cones necessary in the study of efficiency.

BISHOP-PHELPS CONE

Let $(E, ||\cdot||)$ be a Banach space. Given $\phi \in E^*$ and a real number α such that $0 < \alpha \leq 1$, the set $K = \{x \in E \mid \alpha||x|| \leq \phi(x)\}$ is a convex cone called by some authors the *Bishop-Phelps cone*. Obviously, the Bishop-Phelps cone is a well based cone.

HENIG'S DILATING CONE

Let $(E, ||\cdot||)$ be a Banach space and K a convex cone. Suppose that there exists a closed convex set B such that $0 \notin B$ and $K = cone(B)$. Since $0 \notin B$ we have that

$$\delta := \inf\{||b|| \mid b \in B\} > 0.$$

For each integer n satisfying $n \geq \frac{2}{\delta}$ denote by $B_n = cl(B + S_n)$ and $K_n = cone(B_n)$, where $S_n = \{e \in E \mid ||e|| \leq \frac{1}{n}\}, n = 1, 2, ..$

If $c = b + e$, with $b \in B$ and $e \in S_n$, then

$$||c|| \geq ||b|| - ||e|| \geq \delta - \frac{1}{n} \geq \delta - \frac{\delta}{2} = \frac{\delta}{2}.$$

Hence $\inf\{||c|| \mid c \in B_n\} \geq \frac{\delta}{2} > 0$ and B_n is such that $0 \notin B_n$ and K_n is generated by B_n. The cone K_n is called *Henig's dilating cone* and it is important in the study of proper efficiency.

D-CONE

The notion of D-cone was defined in Gallagher and Saleh (1993). Let $< E, E^* >$ be a dual system of locally convex spaces. Let E^* be equipped with a topology induced by E.

DEFINITION 12.8. A pointed convex cone $K \subset E$ is said to be a D-cone if there exists a non-empty subset $D \subset K^{*i}$ satisfying the following conditions:

1. the set D is contained in a compact convex subset of E^*,
2. for every $f, g \in D$ there exists $h \in D$ such that $h \leq g$,
3. if $< x, f > \geq 0$ for all $f \in D$, then $x \in K$

THEOREM 12.12. *Let $E(\tau)$ be a locally bounded Hausdorff topological vector space such that the topological dual E^* separates points of E, (i.e. if $< x_0, f >= 0$ for all $f \in E^*$ then $x_0 = 0$ and if $< x, f_0 >= 0$ for all $x \in E$ then $f_0 = 0$). If $K \subset E$ is a weakly closed convex cone such that K^{*i} is non-empty, then K is a D-cone.*

PROOF. Let U be a bounded neighborhood of the origin of E. By the Banach-Alaoglu Theorem (Rudin, 1973, Theorem 3.15), the polar

$$U^0 = \{f \in E^* \mid f(x) \leq 1, \text{ for all } x \in U\}$$

of U is a weak-*-compact set.
Also, since U is bounded, it follows that $\sup\{|f(x)|\ ,\ x \in U\}$ is finite for all $f \in E^*$.
Thus $K^{*i} \cap U^0$ is non-empty. Now let $p \in K^{*i} \cap U^0$ and define

$$D = \bigcup\left(\left\{\frac{1}{n}p\right\} + K^*\right)\bigcap U^0.$$

Then D is a non-empty subset of K^{*i} contained in the weak-*-compact set U^0. It must be shown that conditions (2), (3) of Definition 12.8 are satisfied. Indeed consider two arbitrary elements $f, g \in D$. Then $f = \left(\frac{1}{r}\right)p + k_1^*$ and $g = \left(\frac{1}{s}\right)p + k_2^*$ for some positive integers r and s and some k_1^*, $k_2^* \in K^*$.
Choose t=max$\{r, s\}$ and define $h = \left(\frac{1}{t}\right)p$. Then $h \in D$, $h \leq f$ and $h \leq g$. Hence, condition (2) is satisfied. Suppose $x \notin K$. Since E_σ (E with the weak topology) is a locally convex space and since K is weakly closed and convex there exists $q \in (E_\sigma)^*$ such that $q(x) < 0 \leq q(k)$ for all $k \in K$.
The fact that E^* separates points of E implies that $E^* = (E_\sigma)^*$ (Rudin, 1973, p. 62-63) and hence $q \in K^*$. Now since U is bounded, we may assume that $\sup\{|q(x)|\ ,\ x \in U\} < \frac{1}{2}$. Then for n sufficiently large, we have $\left(\left(\frac{1}{n}\right)p + q\right)(x) \in D$ and $\left(\left(\frac{1}{n}\right)p + q\right)(x) < 0$. Therefore, condition (3) of Definition 12.8 is also satisfied. ■

Remark. From Theorem 12.12 we obtain that in a normed space any closed convex cone that admits strictly positive continuous linear functionals is a D-cone. Thus, in particular the non-negative orthant in R^n and the usual cones in $C([a, b])$, R, l^p, L^p (for $1 \leq p \leq +\infty$) are D-cones. Also, the nonnegative usual cone in the non-locally convex spaces l^p $(0 < p < 1)$ are D-cones.

TANGENT CONE

Let $E(\tau)$ be a locally convex topological vector space and $C \subset E$ a nonempty set. The *tangent cone* to C at $\bar{x} \in C$ is defined to be the set of limits of the form $h = \lim_{n\to+\infty} t_n(x_n - \bar{x})$ with $\{t_n\}_{n\in N}$ a sequence of nonnegative real numbers and $\{x_n\}_{n\in N} \subset C$ a sequence with limit $\bar{x}$. It is denoted by $T(C, \bar{x})$.

RECESSION CONES

The concept of recession cones has been first developed to study convex sets and convex functions and it is an indispensable concept in convex analysis.

Let $E(\tau)$ be a topological vector space and $D \subset E$ a non-empty subset. It is known that, if D is convex and closed, then the set

$$\{v \in E \mid x + tv \in D \text{ for all } x \in D \text{ and } t \in R_+\}$$

is a closed convex cone. This cone reflects the behavior of the set at infinity and it is called the *recession cone* of D.

This concept has been extended for nonconvex sets. Now we will give the definition of recession cone for a general non-empty set. We need to introduce some notations. If $D \subset E$ is a non-empty subset, then D^c is the complement of D in E, clD its closure, $conv(D)$ is the convex hull of D and $cone(D)$ is the cone generated by D, i.e.

$$cone(D) = \{\lambda x \mid x \in D \text{ and } \lambda \in R_+\}.$$

Let U be a non-empty set in E. To the space E we add the point ∞. *We say that U is an open [closed] neighborhood of ∞ if it is open [closed] and its complement U^c in E is bounded.*

Let $\mathcal{U}$ be the filter of neighborhoods of ∞.

DEFINITION 12.9. The *recession* cone of the set $D \subset E$ is the cone

$$Rec(D) = \bigcap_{U \in \mathcal{U}} clcone(D \cap U).$$

(If the set $D \cap U$ is empty we set $clcone(D \cap U) = \{0\}$).

Remark. If the space E is normable then there exists a bounded balanced absorbing neighborhood V of zero. In this case we have

$$Rec(D) = \bigcap_{n=1}^{\infty} clcone\left(D \cap (nV)^c\right).$$

Given a set $D \subset E$ we define also two other cones:

$$D_\infty = \bigcap_{\alpha > 0} cl\{]0, \alpha]D\},$$

where $]0, \alpha]D = \{tx \mid x \in D,\ 0 < t \leq \alpha\}$ and
$A_S(D) = \{u \in E \mid$ there are a net $\{x_\alpha\} \subset D$ and a net of positive numbers $\{t_\alpha\}$ converging to 0 such that $u = \lim t_\alpha x_\alpha\}$.

We can show that if E is a normable space, then $D_\infty = A_S(D) = Rec(D)$. In the general case we have only the relation $D_\infty = A_S(D) \subseteq Rec(D)$.

PROPOSITION 12.8. *A non-zero vector $v \in E$ belongs to $Rec(D)$ if and only if for every $U \in \mathcal{U}$ and every neighborhood U_0 of zero in E we have that $cone(v + U_0) \cap D \cap U \neq \emptyset$.*

PROOF. Considering the definition we have that $v \in Rec(D)$ if and only if $v \in clcone(D \cap U)$ for each $U \in \mathcal{U}$. In this case since v is non-zero, the intersection $D \cap U$ must be non-empty. Therefore, $v \in Rec(D)$ if and only if for every neighborhood U_0 of zero in E and every element $U \in \mathcal{U}$ we have that $(v + U_0) \cap cone(D \cap U) \neq \emptyset$, or equivalently, $cone(v + U_0) \cap D \cap U \neq \emptyset$. ∎

PROPOSITION 12.9. *The following assertions are true:*

1. $Rec(D) = \{0\}$, *if D is bounded,*
2. $Rec(D_1) \subseteq Rec(D_2)$, *if* $D_1 \subseteq D_2 \subseteq E$,
3. $Rec(D_1 \cup D_2) = Rec(D_1) \cup Rec(D_2)$, *for any* $D_1, D_2 \subseteq E$,
4. $Rec(D_1 \cap D_2) \subseteq Rec(D_1) \cap Rec(D_2)$, *for any* $D_1, D_2 \subseteq E$,
5. $conv(Rec(D)) \subseteq Rec(conv(D))$,
6. $Rec(cl(D) = Rec(D)$.

PROOF. The proposition is a consequence of Definition 12.9 and Proposition 12.8. ∎

Remark. If D_1 and D_2 are convex and closed with $D_1 \cap D_2 \neq \emptyset$, then $Rec(D_1 \cap D_2) = Rec(D_1) \cap Rec(D_2)$. This fact is not true when D_1 and D_2 are arbitrary sets.

Also, the inverse inclusion of the assertion (5), in general, does not hold even in finite dimensions.

PROPOSITION 12.10. *If a set* $B \subset E$ *is bounded, then* $Rec(D + B) = Rec(D)$.

PROOF. The proposition is a consequence of Definition 12.9 and Proposition 12.8. ∎

We recall that a nontrivial cone $K \subset E$ is generated by a non-empty set $B \subset E$ if $0 \notin B$ and $K = cone(B)$.

DEFINITION 12.10. We say that a non-empty subset $D \subset E$ satisfies condition (CB) if there exists a neighborhood U_* of ∞ such that the cone $clcone(D \cap U_*)$ is generated by a compact set.

DEFINITION 12.11. We say that a non-empty subset $D \subset E$ satisfies condition (CD) if for each $v \in Rec(D)$ there is a bounded set A such that $(tv + A) \cap D \neq \emptyset$ for all $t \geq 0$.

Remark. The condition (CB) holds for every set in finite dimensional spaces. In infinite dimensional spaces if a set is convex and locally compact, then both conditions (CB) and (CD) are satisfied.

The reader can find the proofs of the following two results in Luc (1991).

PROPOSITION 12.11. *If the set $D \subset E$ satisfies condition (CB), then D is bounded if and only if $Rec(D) = \{0\}$.* ∎

THEOREM 12.13. *Given two non-empty subsets D_1 and D_2 of the space E, we have the following relations:*

1. $Rec(D_1), Rec(D_2) \subseteq Rec(D_1 + D_2)$,
2. *if the condition (CD) holds for at least one of the two sets then* $Rec(D_1) + Rec(D_2) \subseteq Rec(D_1 + D_2)$,
3. *if the condition (CB) is satisfied for at least one of the two sets and in addition* $Rec(D_1) \cap (-Rec(D_2)) = \{0\}$ *then we have that* $Rec(D_1 + D_2) \subseteq Rec(D_1) + Rec(D_2)$ ∎

12.3 SOME NOTIONS OF EFFICIENCY

It is known that vector optimization problems (i.e. optimization of vector valued functions), are originated from decision-making problems appearing in economics, management science and social science. In such domains it is often required that decision making be based on optimizing several criteria. Obviously, one of the most important problems in vector optimization is to identify the *"efficient points"*, with respect to some vector partial order.

Related to this problem we must study under what conditions the efficient points set is non-empty and we must study also some topological properties of this set. Because some efficient points exhibit certain abnormal properties, which cause arbitrarily large marginal trade off, or may not be expressed as a solution of an appropriate linear scalar optimization, various concepts of proper efficiency have been introduced to eliminate such anomalous efficient points.

In this sense, we will introduce in this section several kinds of efficiency. Let $F(\tau_1)$ and $E(\tau)$ be topological vector spaces and let $K \subset E$ be a closed pointed convex cone. Suppose given a mapping $f : A \to E$, where A is a non-empty subset of F, and consider the *vector maximization problem*:

$$\begin{cases} VMax \quad f(x) \\ x \in A \end{cases} \tag{12.1}$$

DEFINITION 12.12. A point $x_0 \in A$ is said to be an *efficient* (*Pareto maximum*) *solution* of problem (12.1) if and only if

$$[f(A) - f(x_0)] \cap K = \{0\}, \tag{12.2}$$

or equivalently

$$f(A) \cap [K + f(x_0)] = \{f(x_0)\}. \tag{12.3}$$

Similarly, if we consider the *vector minimization problem*:

$$\begin{cases} VMin & f(x) \\ x \in A \end{cases} \tag{12.4}$$

then we have the following definition.

DEFINITION 12.13. A point $x_0 \in A$ is said to be an *efficient* (*Pareto minimum*) *solution* of problem (12.4) if and only if

$$[f(A) - f(x_0)] \cap (-K) = \{0\}, \tag{12.5}$$

or equivalently

$$f(A) \cap [f(x_0) - K] = \{f(x_0)\}.$$

Definitions 12.12 and 12.13 inspired the following general notions. Let $X \subset E$ be an arbitrary non-empty subset.

DEFINITION 12.14. We say that $x_* \in X$ is an *efficient* (*maximal*) *point* of X with respect to K if and only if $X \cap [x_* + K] = \{x_*\}$.

Remark. If in Definition 12.14 we replace the cone K by $-K$ we obtain that x_* is an *efficient* (*minimal*) *point.*

We observe that x_* is an efficient (maximal) point if and only if $(X - x_*) \cap K = \{0\}$ or if and ony if $X \cap (x_* + K \backslash \{0\}) = \emptyset$.

The notion of efficiency, in the sense of Definition 12.14 has been studied by many authors. (See the references of this chapter). The efficiency defined in Definition 12.14 is the most general notion of efficiency. The set of all efficient points of X will be denoted by $E[X, K]$.

Let $(R^n, < \cdot, \cdot >)$ be the Euclidean space and let $K \subset R^n$ be a pointed closed convex cone. Consider the vector optimization of a function $f : A \rightarrow R^n$, where $A \subset R^n$ is a non-empty subset and $f(x) = (f_1(x), ..., f_n(x))$. In this case $x_* \in A$ is a *Pareto efficient* point (in the sense used in Economics) if $f(x_*)$ is *efficient* (maximal or minimal) in the sense of Definition 12.14 for $X = f(A)$.

Suppose that the cone K is such that $K^{*i} \neq \emptyset$, i.e. K is strictly supported.

DEFINITION 12.15. We say that $x_* \in E[X, K]$, where $X = f(A)$, is *strictly efficient* if it solves the maximization problem

$$\max\{< y, x > \mid x \in X\}$$

for some $y \in K^{*i}$.

Any other efficient point is said to be *nonstrict efficient point.* The set of strictly efficient points will be denoted by $E_s[X, K]$.

For more results on strictly efficient points the reader is referred to Bitran and Magnanti (1979).

While the concept of efficiency has played a useful role in the analysis of several practical vector-valued optimization problems, a slightly restricted definition of efficiency, namely, that of *proper* efficiency, has been proposed in order to eliminate efficient solutions of certain anomalous type. The first variant of proper efficiency was proposed, as it seems, in Kuhn and Tucker (1950). We will introduce this notion. For this, consider a vector-valued criterion function

$$f(x) = [f_1(x), f_2(x), ..., f_p(x)]$$

defined over a set A given by

$$A = \{x \in R^n \mid g_i(x) \geq 0, \text{ for all } i \in I\},$$

where $p \geq 2$, $f_j : A \to R$ for all $j \in J = \{1, 2, .., p\}$, $I = \{1, 2, .., m\}$ and $g_i : R^n \to R$ for all $i \in I$. Consider the ordering on R^n defined by R^n_+. For any $k \in J$ denote by $J_k = J \setminus \{k\}$, and consider the following vector maximization problem

$$(P) : \begin{cases} VMax & f(x) \\ x \in A \end{cases}$$

Problem (P) is the problem of finding all solutions there are efficient in the sense of Definition 12.14, where $X = f(A)$ or equivalently in the sense of the following definition.

DEFINITION 12.16. We say that x^0 is an *efficient solution* of problem (P) if $x^0 \in A$ and $f_i(x) > f_i(x^0)$ for some $x \in A$ and some $i \in J$ implies that there exists at least one $j \in J_i$ such that $f_j(x) < f_j(x^0)$.

The next definition, proposed in Geoffrion (1968) is a refined version of the definition proposed initially by Kuhn and Tucker.

DEFINITION 12.17. An element $x^0 \in A$ is said to be a *properly efficient solution* of problem (P), when it is efficient for (P) and there exists a scalar $M > 0$ such that for each $i \in J$ and each $x \in A$ satisfying $f_i(x) > f_i(x^0)$, there exists at least one $j \in J_i$ with $f_j(x) < f_j(x^0)$ and

$$\frac{f_i(x) - f_i(x^0)}{f_j(x^0) - f_j(x)} \leq M$$

An efficient solution that is not properly efficient is said to be *improperly efficient.* The following two definitions are useful for the next main result.

DEFINITION 12.18. An element $x^0 \in A$ is said to be a *k-th-entry efficient solution* of problem (P), where $k \in J$, if $f_k(x) > f_k(x^0)$ for some $x \in A$ implies that there exists at least one $j \in J_k$ such that $f_j(x) < f_j(x^0)$.

DEFINITION 12.19. We say that $x^0 \in A$ is a *properly k-th-entry efficient solution* of problem (P), where $k \in J$, if it is k-th-entry efficient for problem (P) and there exists a scalar $M_k > 0$ such that for each $x \in A$ satisfying $f_k(x) > f_k(x^0)$ there exists at least one $j \in J_k$ with $f_j(x) < f_j(x^0)$ and

$$\frac{f_k(x) - f_k(x^0)}{f_j(x^0) - f_j(x)} \leq M_k.$$

From Definitions 12.18 and 12.19 it is easy to deduce the following results.

PROPOSITION 12.12. *An element $x^0 \in A$ is an efficient solution of problem (P) if and only if, it is a k-th-entry efficient solution of problem (P) for each $k \in J$.* ■

PROPOSITION 12.13. *An element $x^0 \in A$ is a properly efficient solution of problem (P) if and only if, it is a properly k-th-entry efficient solution of problem (P) for each $k \in J$.* ■

Now we will give some necessary and sufficient conditions for proper efficiency. In this sense, we associate to problem (P) the following two problems:

$$(P_u): \begin{cases} \max \left[f_k(x) + \sum_{j \in J_k} u_j f_j(x) \right] \\ \text{subject to } x \in A, \end{cases} \qquad \text{where } k \in J$$

and for $k \in J$,

$$(P_{b,k}): \begin{cases} \max f_k(x) \\ \text{subject to } f_j(x) - b_j \geq 0, \forall j \in J_k \text{ and } x \in A, \\ \text{where } \{b_j \mid j \in J_k\} \text{is any set of } p-1 \text{ real numbers.} \end{cases}$$

Denote the set of optimal solutions for (P_u) by A_u^*. The following lemma characterizes properly k-th-entry efficient solutions for (P) in terms of solutions to (P_u).

LEMMA 12.1.

1. *If $x^0 \in A_u^*$ for some $u = u_0 \in R_+^{p-1}$, then x^0 is a properly k-th-entry efficient solutions for problem (P).*

2. *Suppose that $f_1(x), f_2(x), ..., f_p(x)$ are concave functions on the convex set A. Then x^0 is a properly k-th-entry efficient solution for problem (P), if and only if, there exists an element $u = u_0 \in R_+^{p-1}$ such that $x^0 \in A_u^*$.*

PROOF. A proof of this lemma is in Benson and Morin (1977). ■

We assume throughout the next results that A is non-empty.
Consider the mapping

$$v(y) = \sup_{x \in A}\{f_k(x) | f_j(x) - b_j \geq y_j \text{ for all } j \in J_k\}$$

defined for all $y \in R^{p+1}$. The mapping v is a perturbation of a problem $(P_{b,k})$. Notice that when the optimal value of $(P_{b,k})$ exists, it is equal to $v(0)$.

DEFINITION 12.20. We say that $(P_{b,k})$ is *stable* when $v(0)$ is finite and there exists a scalar $M > 0$ such that for all $y \neq 0$

$$\frac{v(y) - v(0)}{||y||} \leq M.$$

Remark. If the problem $(P_{b,k})$ is not stable the ratio of the improvement in its optimal value to the amount of perturbation can be made as large as desired. The choice of the particular norm $||\cdot||$ used to measure the amount of perturbation is arbitrary.

The next lemma shows that if concavity holds, then the stability of a problem of the form $(P_{b,k})$ constitutes a necessary and sufficient condition for problem (P) to be properly k-th-efficient for (P).

LEMMA 12.2. *If x^0 is a k-th-entry efficient solution for problem (P) and $f_1, f_2, ..., f_p$ are concave functions on the non-empty convex set A, then x^0 is a properly k-th-entry efficient solution for (P), if and only if the problem $(P_{b^0,k})$ (where $b_j^0 = f_j(x^0)$, for all $j \in J_k$) is stable.*

PROOF. The reader can find a proof of this lemma in Benson and Morin (1968). ■

Now we can give the following main result due to Benson and Morin (1968).

THEOREM 12.14. *If x^0 is an efficient solution for problem (P) and $f_1, f_2, ...,$ f_p are concave functions on the non-empty convex set A, then x^0 is a properly efficient solution for (P) if and only if $(P_{b^0,k})$, (where $b_j^0 = f_j(x^0)$, for all $j \in J_k$) is stable for each $k \in J$.*

PROOF. By Proposition 12.12 x^0 is k-th-entry efficient for each $k \in J$. By Lemma 12.2 then x^0 is properly k-th-entry efficient for each $k \in J$ if and only if the problem $(P_{b^0,k})$ is stable for each $k \in J$.

By Proposition 12.13, x^0 is properly k-th-entry efficient for each $k \in J$ if and only if x^0 is properly efficient. Combining the latter two statements yields the desired conclusion. ■

COROLLARY 12.3. *If f_j and g_i are linear for all $j \in J$ and for all $i \in I$, respectively, then any efficient solution x^0 of problem (P) is a properly efficient solution for (P).* ∎

The notion of proper efficiency was generalized by Borwein (1977). By this generalization the notion of proper efficiency was extended to an arbitrary cone in a general locally convex vector space.

Let $F(\tau_1)$ and $E(\tau)$ be locally convex vector spaces and $K \subset E$ a non-trivial closed convex cone which induces a partial ordering $\leq$ on E. Suppose given a non-empty subset $A \subset F$ and a mapping $f : F \to E$.

Consider again problem (12.1) in which we are interested in finding the efficient points. The efficiency is in the sense of Definition 12.12.

The following definition was initially introduced in Borwein (1977) . We recall that if $C \subset E$ is a non-empty subset and $x_* \in \bar{C}$, $T(C, x_*)$ denotes the tangent cone to C at x_*. When E is metrizable then $T(C, x_*)$ is closed.

DEFINITION 12.21. A point $x_0 \in A$ will be said a *proper efficient point* for problem (12.1) if it is efficient and

$$clT(f(A) - K, f(x_0)) \cap K = \{0\}.$$

In many papers, it is not x_0 but $f(x_0)$ that is called efficient point.

The proper efficiency in the sense of Definition 12.21 is more general as the proper efficiency in the sense of Definition 12.17. Indeed, suppose that $f(x) = (f_1(x), ..., f_p(x))$ maps R^n into R^p and $A \subset R^n$. We have the following result.

PROPOSITION 12.14. *If x_* is proper efficient for f over A in the sense of definition 12.17, then x_* satisfies*

$$T(f(A) - R^p_+, f(x_*)) \cap R^p_+ = \{0\}.$$

PROOF. Suppose $0 \neq u \in R^p_+ \cap T(f(A) - R^p_+, f(x_*))$. Without loss of generality we suppose that $u_1 > 1, u_i \geq 0, i = 2, .., p$. Let

$$t_n(f(x_n) - r_n - f(x_*)) \to u$$

where $r_n \in R^p_+$, $t_n > 0$ and $f(x_n) - r_n \to f(x_*)$, with $x_n \in A$. Taking a subsequence we can assume that

$$I^* = \{i | f_i(x_n) < f_i(x^*)\}$$

is constant for all n (and non-empty since x^* is Pareto efficient). Set $M > 0$. Then for all $n > n_0$,

$$f_1(x_n) - f_1(x^*) \geq \frac{t_n^{-1}}{2}$$

$$f_i(x_n) - f_i(x^*) \geq \frac{-t_n^{-1}}{2M}.$$

Then for all $i \in I^*$, one has

$$0 < f_i(x^*) - f_i(x_n) \leq \frac{t_n^{-1}}{2M},$$

and for $n \geq n_0$,

$$\frac{f_1(x_n) - f_1(x^*)}{f_i(x^*) - f_i(x_n)} \geq \frac{t_n^{-1}/2}{t_n^{-1}/2M} = M$$

which contradicts Definition 12.17. ∎

For more results about Borwein's proper efficiency, the reader is reffered to Borwein (1977). It is known that when $K = R_+^p$ and for all $j \in J$, f_j is a concave function on the convex set A, Definition 12.21 coincides with Definition 12.17 (Borwein, 1977).

However, when $K = R_+^p$, and at least one objective function is non-concave on A, solutions may exist which are proper according to Borwein's definition (Definition 12.21) but improper by Geoffrion's definition (Definition 12.17). This situation can be eliminated by modifying Definition 12.21 as it was proposed in Benson (1984).

If $D \subset R^p$, we denote by $cone(D)$ the cone generated by D, i.e.,

$$cone(D) = \{\lambda d \mid \lambda \geq 0, d \in D\}.$$

DEFINITION 12.22. Let $K \subset R^p$ be a nontrivial, closed convex cone. A point x_* is said to be a *proper efficient solution* of problem (12.6), when x_* is an efficient solution of (12.6) and $clcone[f(A) - K - \{f(x_*)\}] \cap K = \{0\}$.

We remark that if $D \subset R^p$ and $d \in D$, then $T(D,d) \subseteq clcone[D - \{w\}]$, but the converse is not true.

PROPOSITION 12.15. *Let $K \subset R^p$ be a non-trivial, closed convex cone. Suppose that f is a concave function with respect to K on the convex set A. then Definition 12.21 and Definition 12.22 are equivalent.*

PROOF. Let $x_* \in A$. Since K is a non-trivial, closed convex cone and f is concave with respect to K on the set A, we have that $D = f(A) - K$ is a convex set. In this case it is known (*cf.* Varaiya, 1967) that $clcone[D - \{f(x_*)\}] = T[D, f(x_*)]$. ∎

We note that, when $K = R_+^p$, Definitions 12.17 and 12.22 are equivalent [see Benson (1984)].

Now we present a survey of notions of efficiency introduced by several authors.

Let $F(\tau_1), E(\tau)$ be topological vector spaces, $A \subseteq F$ non-empty subset and $f : A \to E$ a mapping. Suppose that E is ordered by a closed pointed convex cone $K \subset E$.

Let us consider again the vector maximization problem (12.1). Denote by $E[f(A), K]$ the set of efficient points for problem (12.1) in the sense of Definition 12.12. The notion of proper efficiency introduced by Hurwicz in 1958 has been strictly related to the scalarization of a vector optimization problem (Hurwicz, 1958).

DEFINITION 12.23. A point $y_0 \in f(A)$ is said to be a *positive proper efficient* (or "*scalar maximum*" in Hurwicz's terminology) of $f(A)$, with respect to K, if and only if there exists $\phi \in K^{*i}$ such that $\phi(y_0) \geq \phi(y)$ for all $y \in f(A)$.

The set of positive proper efficient points will be denoted by $PPE[F(A), K]$. The following result is due to J. P. Dauer and R. J. Gallagher.

THEOREM 12.15. *A point $y_0 \in f(A)$ is a positive proper efficient of $f(A)$ if and only if there exists an open convex set U in E such that*

1. $K \subseteq cone(U)$;
2. $convcone(f(A) - \{y_0\}) \cap U = \emptyset$.

PROOF. Suppose that y_0 is a positive proper efficient point of $f(A)$. Hence, there exists $\phi \in K^{*i}$ such that $\phi(y_0) \geq \phi(y)$ for all $y \in f(A)$. Consider the set $U = \{y \in E | 1 < \phi(y) < 2\}$.

Then U is an open convex set such that $K \subseteq cone(U)$. Now if $y \in f(A)$, then $\phi(y - y_0) \leq 0$. Thus

$$f(A) - \{y_0\} \subseteq \{z \in E | \phi(z) \leq 0\}$$

which implies that $convcone(f(A) - \{y_0\}) \cap U = \emptyset$.

Conversely, suppose that U is an open convex set satisfying conditions 1 and 2. By applying Theorem 3.4, proved in Rudin (1973), we obtain a linear mapping $\phi \in E^*$ and a number $\gamma \in R$ such that

$$\phi(x) \leq \gamma < \phi(u),$$

for all $x \in convcone(f(A) - \{y_0\})$ and all $u \in U$. Since $0 \in convcone(f(A) - \{y_0\})$ we have $\gamma \geq 0$. We have $\phi(u) > 0$ for all $u \in U$. From (1) it follows that $\phi \in K^{*i}$. A contradiction argument shows that $\phi(x) \leq 0$ for all $x \in convcone(f(A) - \{y_0\})$. Thus $\phi(y) \leq \phi(y_0)$ for all $y \in f(A)$, which means that y_0 is a positive proper efficient point of $f(A)$. ∎

The following definition is also due to Hurwicz.

DEFINITION 12.24. A point $y_0 \in f(A)$ is said to be *properly efficient (in Hurwicz's sense)* if and only if

$$clconvcone[(f(A) - \{y_0\}) \cup -K] \cap K = \{0\}$$

The set of points properly efficient in the sense of Definition 12.24 will be denoted by $HUPE[f(A), K]$. It is known (Hurwicz, 1958) that

$PPE[f(A), K] \subseteq HUPE[f(A), K]$. The converse result can be proved if some suitable hypothesis are satisfied. For example, if E is a separable normed vector space then $HUPE[f(A), K] \subseteq PPE[f(A), K]$. (Hurwicz, 1958).

THEOREM 12.16. *If E is a locally convex vector space and $K \subseteq E$ has a compact base, then*

$$HUPE[f(A), K] \subseteq PPE[f(A), K].$$

PROOF. If $y_* \in HUPE[f(A), K]$ then we have

$$clconvcone[(f(A) - \{y_*\}) \cup (-K)] \cap K = \{0\}.$$

Then from Borwein (1977, Proposition 3) there exists $\phi \in K^{*i}$ such that

$$\phi\left(clconvcone\left[(f(A) - \{y_*\}) \cup (-K)\right]\right) \leq 0 < \phi(K \backslash \{0\})$$

Hence, it follows that $\phi(f(A) - y_*) \leq 0$, and therefore $y_* \in PPE[f(A), K]$. ∎

The following definitions are due respectively to R. Hartley, H. P. Benson, J. M. Borwein and M. I. Henig.

DEFINITION 12.25. Let $(E, ||\cdot||)$ be a normed vector space. A point $y_0 \in f(A)$ is *properly efficient in Hartley's sense*, if and only if it is efficient and there is a constant $\delta > 0$ such that , whenever there is $\phi \in K^*$ with $\phi(y - y_0) > 0$ for some $y \in f(A)$, one can find $\psi \in K^*$ with

$$\frac{\phi(y - y_0)}{||\phi||} \leq -\delta \frac{\psi(y - y_0)}{||\psi||}.$$

The set of properly efficient points in Hartley's sense will be denoted by $HaPE[f(A), K]$. The next definition is a slight variant of Definition 12.22.

DEFINITION 12.26. We say that $y_0 \in f(A)$ is *properly efficient point in Benson's sense* if it is efficient and $clcone[f(A) - K - y_0] \cap K = \{0\}$.

The set of properly efficient points in Benson's sense will be denoted by $BePE[f(A), K]$.

DEFINITION 12.27. A point $y_0 \in f(A)$ is said to be *properly efficient in Geoffrion-Borwein's sense* if and only if $clcone[f(A) - y_0] \cap K = \{0\}$.

The set of points satisfying Definition 12.27 will be denoted by $GBOPE[f(A), K]$.

DEFINITION 12.28. A point $y_0 \in f(A)$ is *globally properly efficient in Henig's sense* if and only if it is efficient with respect to some cone K_0 with $K \backslash \{0\} \subseteq \text{int}(K_0)$.

The set of efficient points in the sense of Definition 12.28 will be denoted by $GHePE[f(A), K]$.

Finally the next definition is due to Borwein and Zhuang (1993).

DEFINITION 12.29. Suppose that $(E, ||\cdot||)$ is a normed vector space. We say that a point $y_0 \in f(A)$ is a *super-efficient point* (and we denote $y_0 \in SE[f(A), K]$, if and only if there exists $\delta > 0$ such that

$$clcone[f(A) - y_0] \cap (K + U) \subseteq \delta U$$

where U is the unit ball.

About the notions defined above we give now several inclusions.

THEOREM 12.17. *If K is well based, then we have*

$$SE[f(A), K] \subseteq GHePE[f(A), K].$$

PROOF. Because K is well based, there exists a bounded convex set $B \subset E$, such that $0 \notin cl(B)$ and $K = \cap_{\lambda \geq 0} \lambda B$.
Let $(f(A) - y_0) \cap cone(B + \epsilon U) \backslash \{0\} \neq \emptyset$ for all $\epsilon > 0$, (where U is the unit ball).
This relationship is equivalent to

$$cone(f(A) - y_0) \cap (B + \epsilon U) \neq \emptyset$$

for all $\epsilon > 0$.
Therefore there exists some

$$u_n \in cone(f(A) - y_0) \cap (B + \frac{1}{n}U).$$

We have $||u_n|| \geq inf||b|| - \dfrac{1}{n} > 0$, $b \in B$. Hence, we have $||nu_n|| \to +\infty$ and $nu_n \in cone(f(A) - y_0) \cap (K + U)$. This contradicts the assumption $y_0 \in SE[f(A), K]$.
Therefore there exists $\varepsilon > 0$ such that $(f(A) - y_0) \cap cone(B + \varepsilon U) = \{0\}$ and this implies that $y_0 \in GHePE[f(A), K]$. ■

THEOREM 12.18. *The following inclusion is satisfied*

$$SE[f(A), K] \subseteq HaPE[f(A), K].$$

PROOF. This inclusion is Proposition 3.5 in Borwein and Zhuang (1993). ■

THEOREM 12.19. *The inclusion* $HaPE[f(A), K] \subseteq BePE[f(A), K]$ *is true.*

PROOF. This inclusion is Proposition 5, section 3 in Borwein (1980). ■

THEOREM 12.20. *We have* $BePE[f(A),K] \subseteq GBOPE[f(A),K]$.

PROOF. This inclusion follows from definitions. ■

THEOREM 12.21. *If* $(E, \|\cdot\|)$ *is a normed vector space and* $K \subset E$ *a closed pointed convex cone, weakly locally compact then we have* $GBOPE[f(A),K] \subseteq SE[f(A),K]$.

PROOF. This inclusion follows from Borwein and Zhuang (1993, Proposition 3.7). ■

From Proposition 3.6 proved in Borwein and Zhuang (1993), we deduce the following results.

THEOREM 12.22. *If* $(E, \|\cdot\|)$ *is a normed vector space and* $K \subset E$ *a normal, closed pointed convex cone, then* $HaPE[f(A),K] \subseteq SE[f(A),K]$. ■

THEOREM 12.23. *The following inclusion is true*

$$HUPE[f(A),K] \subseteq GBOPE[f(A),K].$$

PROOF. This inclusion follows from the definitions. ■

THEOREM 12.24. *The following inclusion is valid*

$$PPE[f(A),K] \subseteq GHePE[f(A),K].$$

PROOF. Let $y_0 \in PPE[f(A),K]$. By Theorem 12.15, there exists an open convex set $U \subset E$ such that $(f(A)-y_0) \cap cone(U) = \emptyset$. Obviously, $cone(U)$ is a convex cone such that $K\backslash\{0\} \subseteq \text{int}cone(U)$ and the inclusion follows. ■

For the next theorem we need to recall the following fact.

PROPOSITION 12.16. *If* $0 \in H \subset K$ *then* $E[f(A),K] = E[f(A)-H,K]$.

PROOF. A proof of this result is in Jahn (1986,[2]). ■

THEOREM 12.25. *The inclusion* $GHePE[f(A),K] \subseteq BePE[f(A),K]$ *is true.*

PROOF. Let $y_0 \in GHePE[f(A),K]$. We have $y_0 \in E[f(A),K_0]$, where K_0 is a convex cone such that $K\backslash\{0\} \subseteq int(K_0)$. By Proposition 12.16 we have that

$$[f(A)-K-y_0] \cap K_0 = \{0\}$$

and consequently

$$cone[f(A)-K-y_0] \cap K_0 = \{0\}.$$

Then we obtain

$$clcone[f(A) - K - y_0] \cap K = \{0\}.$$

Indeed, from

$$u \in clcone[f(A) - K - y_0] \cap K \backslash \{0\}$$

it follows that

$$u \in clcone[f(A) - K - y_0] \cap \operatorname{int}(K_0).$$

Therefore, a net $\{u_i\}_{i \in I}$ exists such that

$$u_i \in cone[f(A) - K - y_0] \cap K_0$$

with $\lim_{i \in I} u_i = u$. But such a net cannot exist which completes the proof. ∎

By the results presented above we established the following inclusions:

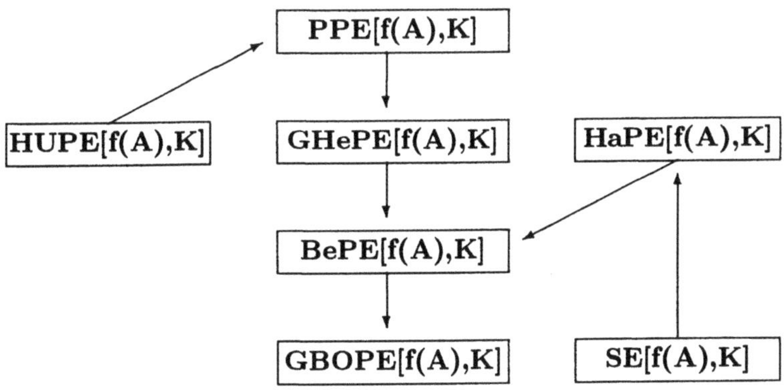

Because the proper efficiency in the sense of Definition 12.21 is due to J.M. Borwein, we will denote the set of such efficient points by $BOPE[f(A), K]$. The following notion of efficiency is a generalization to linear topological vector spaces of a similar notion of efficiency introduced in Henig [1](1982).

DEFINITION 12.30. We say that a point $y_0 \in f(A)$ is *locally properly efficient* if and only if for every neighborhood U of the origin, there exists a convex cone K_u with $K \backslash \{0\} \subseteq K_u$ such that

$$y_0 \in E[(f(A) - K) \cap (y_0 + U), K_u].$$

We denote the set of such efficient points by $LHePE[f(A), K]$.

Remark. Henig calls this efficiency *"local"*, because of the presence of a neighborhood of the point y_0 in the definition. We note that the properties of the efficient point y_0 also depend on the shape of the feasible region $f(A)$ outside the neighborhood of y_0.
We have the following result.

THEOREM 12.26. *If $(E, ||\cdot||)$ is a normed space, then $LHePE[f(A), K] \subseteq BOPE[f(A), K]$. Moreover, if K has a compact base, then $BOPE[f(A), K] \subseteq LHePE[f(A), K]$.*

PROOF. For a proof of this result the reader is referred to Guerraggio e.a. (1994). ■

For other results about the notions of efficiency in abstract spaces, introduced above, the reader is referred to Guerraggio e.a. (1994) and Zheng (1997,[1]).

The next local notions of proper efficiency are affected only by what happens in a neighborhood of the point y_0.

Suppose that $(F, ||\cdot||)$ and $(E, ||\cdot||)$ are normed vector spaces and $f, g : F \to E$ two functions.

The space E is supposed to be ordered by a closed pointed convex cone $K \subset E$.

The admissible region is expressed by

$$A = \{x \in F \mid g(x) \in K_1\}$$

where K_1 is another closed pointed convex cone.

The following notion of proper efficiency is a generalization to normed spaces of a similar notion introduced in Kuhn and Tucker (1950).

DEFINITION 12.31. If f and g are Frechet differentiable functions on the normed space $(F, ||\cdot||)$ then we say that $x_0 \in A \subset F$ is a *properly efficient point in the sense of Kuhn and Tucker*, if and only if there is no vector $u \in F$ such that:

1. $f'(x_0)[u] \in K \backslash \{0\}$ and
2. $g'(x_0)[u] \in clT(K_1, g(x_0))$

The set of efficient points in the sense of Definition 12.31 will be denoted by $KT_\infty[f(A), K]$.

The next notion is a variant of a similar notion considered in Borwein (1980).

DEFINITION 12.32. A point $y_0 \in f(A)$ is said to be *locally properly efficient in Borwein's sense* and we write $y_0 \in LBO[f(A), K]$ if and only if it is efficient and $clT(f(A), y_0) \cap K = \{0\}$.

We have the following result.

THEOREM 12.27. *Let $(F, ||\cdot||)$, $(E, ||\cdot||)$ be Banach spaces, $f, g : F \to E$ Frechet differentiable mappings, K, $K_1 \subset F$, closed pointed convex cones and $A = \{x \in F | g(x) \in K_1\}$. Consider the constraint qualification $G(A, x_0) = \{u \in F | g'(x_0)[u] \in clT(K_1, g(x_0))\}$ and suppose that $G(A, x_0) \subseteq T(A, x_0)$. If $y_0 = f(x_0) \in LBO[f(A), K]$ then $x_0 \in KT_\infty[f(A), K]$.*

PROOF. If $f(x_0) \in LBO[f(A), K]$ then we have

$$T(f(A), f(x_0)) \cap K = \{0\}.$$

We recall the following property of tangent cone:

$$f'(x_0)[T(A, x_0)] \subseteq T(f(A), f(x_0)).$$

Therefore we have

$$f'(x_0)[T(A, x_0)] \cap K = \{0\}$$

and by virtue of the constraint qualification, we obtain

$$f'(x_0)[G(A, x_0)] \cap K = \{0\},$$

which is an equivalent formulation of Definition 12.31. ■

For the next notion we need to recall the notion of radial tangent cone. Let $D \subset F$ be a non-empty subset and $x_0 \in cl(D)$. The *radial tangent cone* to D at x_0 is the set $\mathcal{F}(D, x_0) = \{y \in F |$ for each $\lambda > 0$ there exists $t \in]0, \lambda[$ such that $x_0 + ty \in D\}$.

The following definition is a generalization to general normed vector spaces of a definition introduced in Klinger (1967).

DEFINITION 12.33. If $(F, ||\cdot||)$ is a normed vector space and $f : F \to E$ is Frechet differentiable, then a point $x_0 \in A \subset F$ is said to be *properly efficient in Klinger's sense*, and we write $x_0 \in K_\infty[f(A), K]$ if and only if it is efficient and

$$f'(x_0)[\mathcal{F}(A, x_0)] \cap K = \{0\}.$$

The following two inclusions are known (Guerraggio e.a., 1994).

THEOREM 12.28. *If f and g are Frechet differentiable on the normed vector space $(F, ||\cdot||)$ then $KT_\infty[f(A), K] \subseteq K_\infty[f(A), K]$.*

PROOF. Suppose that $x_0 \in KT_\infty[f(A), K]$. We have

$$f'(x_0)[G(A, x_0)] \cap K = \{0\}.$$

The theorem will be proved if we show that $\mathcal{F}(A, , x_0) \subseteq G(A, x_0)$.
Indeed, let $u \in \mathcal{F}(A, x_0)$; then there exists a sequence $\{\alpha_n\} \to 0^+$ such that $a_n = x_0 + \alpha_n u \in A$. Setting $\lambda_n = \dfrac{1}{\alpha_n}$, we have
$g'(x_0)[u] = g'(x_0)[\lambda_n(a_n - x_0)] = \lim_{n\to\infty} \lambda_n(g(a_n) - g(x_0)) \in clT(K_1, g(x_0))$, and the theorem follows. ■

THEOREM 12.29. *If f is Frechet differentiable on the normed vector space $(F, ||\cdot||)$ and $y_0 = f(x_0) \in LBO[f(A), K]$, then $x_0 \in K_\infty[f(A), K]$.*

PROOF. If $f(x_0) \in LBO[f(A), K]$ then from the inclusion

$$f'(x_0)[T(A, x_0)] \subseteq T(f(A), f(x_0))$$

we have

$$f'(x_0)[T(A, x_0)] \cap K = \{0\}.$$

Since $\mathcal{F}(A, x_0) \subseteq T(A, x_0)$ we obtain

$$f'(x_0)[\mathcal{F}(A, x_0)] \cap K = \{0\},$$

which is equivalent to Definition 12.33. ■

Finally, it is interesting to consider also the following inclusion.

THEOREM 12.30. *If F is a locally convex vector space, K has a compact base and $f(A)$ is convex, then $LBO[f(A), K] \subseteq PPE[f(A), K]$.*

PROOF. Suppose $y_0 \in LBO[f(A), K]$. Since $f(A)$ is convex we have $clT(f(A), y_0) = clcone(f(A) - y_0) = clconvcone(f(A) - y_0)$. Therefore, we deduce

$$clconvcone(f(A) - y_0) \cap K = \{0\}.$$

Applying Borwein (1977, Proposition 3) we obtain that $y_0 \in PPE[f(A), K]$. ■

We will finish this section giving other generalizations of efficiency. Let $E(\tau)$ be a Hausdorff locally convex space. We assume that E is ordered by a closed pointed convex cone $K \subset E$. Let $D \subset E$ be a non-empty subset. We denote by τ-cl(D) the closure of D with respect to the topology τ and by int(D) the τ-interior of D.

We recall that $x \in E$ is an *efficient* (minimal) element of D if and only if $x \in D$ and $[x - (K \setminus \{0\})] \cap D = \emptyset$. The set of all efficient elements of D will be denoted by $E(D, K)$.

DEFINITION 12.34. We say that x is a *τ-infimal element* of D if and only if $x \in \tau$-cl(D) and $[x - (K \setminus \{0\})] \cap D = \emptyset$.
The set of all τ-infimal elements of D will be denoted by τ-$INF(D)$.

DEFINITION 12.35. Suppose that an element $e \in K \setminus \{0\}$ is given. We say that x is an *e-minimal element* of D if and only if $x \in D$ and $[x - e - (K \setminus \{0\})] \cap S = \emptyset$.

We denote by e-$MIN(D)$ the set of all e-minimal elements of D.

Remark. The notion defined in Definition 12.34 is the infimality concept introduced in Bacopoulos e.a. (1978) or the $(\tau - \lambda)$-extremum defined in Cesari and Suryanarayana (1978).

The concept defined in Definition 12.35 was initially introduced in Loridan (1984) and investigated in Valyi (1985).

About these notions we have the following situations:

(1). $E(D,K) \subset e\text{-}MIN(D)$, for all $e \in K$ with equality for $e = 0$,

(2). $E(D,K) = \bigcap_{e \in (K \setminus \{0\})} e\text{-}MIN(D)$,

(3). $E(D,K) \subset \tau\text{-}INF(D)$ with equality if D is closed,

(4). $E(\tau\text{-cl}(D),K) \subset \tau\text{-}INF(D)$,

(5). $\tau\text{-}INF(D) \subset \tau\text{-cl}(D) \setminus \tau\text{-int}(D)$.

THEOREM 12.31. *Let $D \subset E$ be a non-empty subset. Suppose that there exist some $f \in K^* = \{g \in E \mid g(x) \geq 0$ for all $x \in K\}$ and some $x \in \tau$- $cl(D)$ satisfying*

$$f(x) = \inf f(D). \tag{12.6}$$

If either $f \in K^{*i} = \{g \in K^* \mid g(x) > 0$ for all $x \in K \setminus \{0\}\}$ or if the set $D+K$ is open it follows that $x \in \tau\text{-}INF(D)$.

PROOF. Assume that equality (above) holds and that $x \notin \tau\text{-}INF(D)$. Then there is some $u \in K\setminus\{0\}$ with $x-u \in D$. If $f \in K^{*i}$ we get $f(x-u) < \inf f(D)$, which is a contradiction. If the set $D+K$ is open the inequality $f(x-u) \leq \inf f(D) = \inf f(D+K)$, which follows from (above), is a contradiction to

$$x - u \in D \subset D + K = \text{int}(D+K). \qquad \blacksquare$$

THEOREM 12.32. *Let $D \subset E$ be a non-empty subset. If D is bounded or has a lower bound, then for any $e \in K \setminus \{0\}$ the set e-$MIN(D)$ is non-empty.*

PROOF. Separating the compact set $\{-e\}$ and the closed convex set K, by a classical separation theorem, we obtain a functional $f \in E^*$ such that $f(-e) < \inf f(K) = 0$. From the last inequality it follows that $f(e) > 0$ and $f \in K^*$ and from the boundedness of D we obtain that $\inf f(D) > -\infty$ (if D has a lower bound u_0, we have $f(D) \geq f(u_0)$, because $f \in K^*$. Denote by $m := \inf(D) > -\infty$. The theorem will be proved if we show that

$$D(f,e) := \{v \in D \mid f(v) < m + f(e)\} \subset e\text{-}MIN(D).$$

Since $f \in K^*, f(e) > 0$ and $m > -\infty$, we have that $D(f,e)$ is non-empty. To show the inclusion we suppose that $x \in D(f,e)$ and $x \notin e\text{-}MIN(D)$. In this case there exists $u \in K \setminus \{0\}$ with $x - e - u \in D$ and from $x \in D(f,e)$ and $f \in K^*$ we obtain

$$f(x-e-u) < f(x-e) < m = \inf(D),$$

which is a contradiction. $\blacksquare$

DEFINITION 12.36. Let $D \subset E$ be a non-empty subset such that $D+K$ has a non-empty interior. We say that an element $x \in E$ is a *weak minimal* *element* of D if and only if $x \in \tau\text{-cl}(D)$ and $(x - K) \cap \text{int}(D + K) = \emptyset$.

We will denote by $WINF(D)$ the set of all weak minimal elements of D.

DEFINITION 12.37. Let $D \subset E$ be a non-empty subset and $e \in K$ an arbitrary element. We say that $x \in E$ is an *e-weak minimal element* of D if and only if $x \in D$ and $(x - e - K) \cap \text{int}(D + K) = \emptyset$.

The set of all e-weak minimal elements of D will be denoted by e-$WMIN(D)$. It is easy to show that

$$e\text{-}MIN(D) \subset e\text{-}WMIN(D)$$

and

$$INF(D) \subset WINF(D)$$

if the set $D + K$ has a non-empty interior. Moreover if the ordering cone K has non-empty interior we have

$$e\text{-}WMIN(D) = \{x \in D \mid [x - e - \text{int}(K)] \cap D = \emptyset\},$$

which means that for $e = 0$, e-$WMIN(D)$ equals the set of weakly minimal elements of D.

We mention the following result without proof.

THEOREM 12.33. *Let $E(\tau)$ be a locally convex space ordered by a closed convex cone $K \subset E$. Let $D \subset E$ be a non-empty subset such that $D + K$ is convex and has a non-empty interior. Then $x \in WINF(D)$ if and only if there exists $f \in K^*$, $f \neq 0$ with $f(x) = \inf f(D)$ and for a given element e of K, $x \in$ e-$WMIN(D)$ if and only if there exists $f \in K^*, f \neq 0$ satisfying*

$$f(x - e) \leq \inf f(D).$$

PROOF. A proof of this result is in Staib (1988). ■

The notion of e-minimal point can be generalized in the following sense.

Let $E(\tau)$ be a topological vector space ordered by a closed pointed convex cone $K \subset E$ and $H \subset K \setminus \{0\}$ a non-empty subset.

DEFINITION 12.38. Given a set $D \subset E$, we say that a point $x \in D$ is an *H-efficient point* of D if there is no $x \in D$ such that $x \leq x_0 - h$ for some $h \in H$, i.e. $x_0 \in D$ and $(x_0 - H - K) \cap D = \emptyset$.

The notion of H-efficient point was introduced by A.B. Németh and the reader can find a very interesting study of this notion in Németh (1989).

A new development in the theory of efficiency is the study of ε -efficiency as introduced and studied in several papers by Tammer (1993), Tammer [2],[3]

(1996), Göpfert and Tammer (1995,1997,1998), Göpfert, Tammer and Zalinescu [1],[2] (1998), and by Göpfert, Sekatzek and Tammer (1999).

The notion of ε-efficiency is strongly related to some generalizations of Ekeland's variational principle Isac (1996,1997). We will develop this subject in Chapter 13 of this book.

12.4 SOME EXISTENCE THEOREMS FOR EFFICIENCY

In this section we will consider a topological vector space $E(\tau)$. In some situations the topology τ will be a locally convex topology. In this particular case we will have a locally convex topological vector space.

For our results is appropriate to consider a locally convex space as a couple $(E, Spec(E))$, where E is a real vector space and $Spec(E)$ is a set of seminorms on E satisfying the folowing properties:
(1) $\lambda p \in Spec(E)$, whenever $\lambda \in R_+$ and $p \in Spec(E)$,
(2) if $p \in Spec(E)$ and q is a seminorm on E such that $q \leq p$, then $q \in Spec(E)$, and
(3) for every $p_1, p_2 \in Spec(E)$, $\sup(p_1, p_2) \in Spec(E)$ where $\sup(p_1, p_2)(x) = \sup(p_1(x), p_2(x))$ for any $x \in E$.

If $Spec(E)$ is given, then there exists a locally convex topology τ on E such that $E(\tau)$ is a topological vector space such that a seminorm p on E is τ-continuous if and only if $p \in Spec(E)$.

We say that a subset $\mathcal{B} \subset Spec(E)$ is a *base* of $Spec(E)$ if and only if for every $p \in Spec(E)$, there exist $q \in \mathcal{B}$ and a real number $\lambda > 0$ such that $p \leq \lambda q$.
We will suppose that $Spec(E)$ has a *Hausdorff base* $\mathcal{B}$, i.e.
$\{x \in E \mid p(x) = 0, \forall p \in \mathcal{B}\} = \{0\}$.

If $K \subset E$ is a closed pointed convex cone we denote by $\leq$ the ordering defined by K i.e. $x \leq y$ if and only if $y - x \in K$, for any $x, y \in E$. Denote by E^* the topological dual of E and consider a duality $\langle E, E^* \rangle$ between E and E^*. We recall that the topological dual of K is $K^* = \{y \in E^* | \langle y, x \rangle, \forall x \in K\}$. If K is given we can show that K is normal (with respect to the topology τ) if and only if the following assertion is satisfied:
there exists a base $\mathcal{B}$ of $Spec(E)$ such that for every $p \in \mathcal{B}$ and $x, y \in K$ with $x \leq y$ we have $p(x) \leq p(y)$.

Let $A \subset E$ be a non-empty subset and $K \subset E$ a closed pointed convex cone. We recall that a point $a \in A$ is an efficient (maximal) point of A if $A \cap (a + K) = \{a\}$.

We denote by $E[A, K]$ the set of efficient points of A with respect to K.

After 1975, several authors, as J. M. Borwein, J. M. P. Penot, D. H. Wagner, H. W. Corley, A. Sterna-Karwat, R. Hartley, among others have studied the existence of efficient points supposing that K is a Daniell (i.e. regular cones), or exploiting the compactness, or some generalizations of compactness as for example the semicompactness or the K-compactness. In 1983, G. Isac obtained a general existence result, considering an efficient point of a set as a stationary (critical) point for a general dynamical system and using the completness instead of the compactness Isac (1983).

Let $(E(\tau), Spec(E))$ be a locally convex space. We say that a set-valued mapping $\Gamma : E \to E$ is a *dynamical system* (in generalized sense), if for every $x \in E$, $\Gamma(x)$ is a non-empty set. A point $x_* \in E$ is called a *stationary* point for the dynamical system Γ if and only if $\Gamma(x_*) = \{x_*\}$.

Let $D \subset E$ be a non-empty set. Consider a dynamical system $\Gamma : D \to 2^D$. We say that Γ is:

1. *reflexive* if and only if $x \in \Gamma(x)$ for each $x \in D$,

2. *antisymmetric* if and only if $x_2 \in \Gamma(x_1)$ and $x_1 \in \Gamma(x_2)$ imply $x_1 = x_2$, and

3. *transitive* if and only if $x_2 \in \Gamma(x_1)$ and $x_3 \in \Gamma(x_2)$ imply $x_3 \in \Gamma(x_1)$.

If $x \in D$ we denote by $\Gamma^C(x)$ the complement of $\Gamma(x)$ with respect to E. We say that Γ is *closed-valued* if $\Gamma(x) \subset D$ is closed in E for every $x \in D$.

DEFINITION 12.39. We say that D is Γ-*semicompact* if Γ is closed-valued and every open cover of D of the form $\{\Gamma^C(x_\alpha) | x_\alpha \in D, \alpha \in A\}$ has a finite subcover.

We note that the notion of Γ-semicompactness was introduced and considered in Corley (1986). Every compact subset of E is Γ-semicompact but the converse is not true. The following result is due to H.W. Corley.

THEOREM 12.34. *Let $\Gamma : D \to 2^D$ be a closed-valued, reflexive, antisymmetric and transitive dynamical system. If D is Γ -semicompact, then Γ has a stationary (critical) point.*

PROOF. The reader can find a proof of this result in Corley (1986) or in Hyers, Isac and Rassias (1997). ■

If $A \subset E$ is a non-empty set and $K \subset E$ a closed pointed convex cone, it is easy to verify that the set-valued mapping defined by $\Gamma_A(x) = A \cap (x + K)$, for every $x \in A$ is a dynamical system. The reader can verify the following result.

PROPOSITION 12.17. *An element $x_* \in A$ is an efficient point of A with respect to the ordering defined by K if and only if x_* is a stationary point for Γ_A.* ■

Let $(E(\tau), Spec(E))$ be a locally convex space and $K \subset E$ a closed pointed convex cone and $M \subset E$ a non-empty subset. We denote by M^C the complement of M with respect to E. If $A \subset E$ is a non-empty subset and Γ_A is the dynamical system defined as above we have that Definition 12.39 has the following form.

DEFINITION 12.40. We say that A is *K-semicompact* if and only if every open cover of A of the form $\{(x_i + K)^C | x_i \in A\}_{i \in I}$ has a finite subcover.

The semicompactness implies the existence of efficient points. The next result proved initially in Corley (1986) is in this sense.

THEOREM 12.35. *Let $(E(\tau), Spec(E))$ be a locally convex space and $K \subset E$ a closed pointed convex cone and $A \subset E$ a non-empty subset. If A is K-semicompact, then the set of efficient points of A is non-empty (i.e. $E[A, K] \neq \emptyset$).*

PROOF. We consider the dynamical system $\Gamma_A(x) = A \cap (x + K)$, for all $x \in A$. Since K is a closed pointed convex cone we have that all the assumptions of Theorem 12.34 are satisfied and we obtain that Γ_A has a stationary point. By applying Proposition 12.17 the theorem is proved. ∎

COROLLARY 12.4. *If $A \subset E$ is a compact set and $K \subset E$ is a closed pointed convex cone then the set of efficient points of A is non-empty.* ∎

We defined an efficient point for A as a maximal element. If we consider the cone $(-K)$ we obtain an *efficient minimal* point. Because of this fact some authors studied the existence of efficient maximal points and others the existence of efficient minimal points for a given set. In Corollary 12.4 we supposed that K is *pointed* but in several problems studied in economics the cone K is not pointed.

If K is not pointed, we write $x_1 \mathcal{R}_K x_2$ if $x_1 \leq_K x_2$ and $x_2 \leq_K x_1$ (where $\leq_K$ is the preordering defined by K).

In this case *we say that $a \in A$ is an efficient (minimal) point for A with respect to K (not necessarily pointed), if $x \in A$ and $x \leq_K a$ imply $x \mathcal{R}_K a$.*

We denote by $E_m[A, K]$ the set of all efficient (minimal) points for A with respect to K and by $A(x; \leq_K)$ the *section* of A determined by x, i.e.

$$A(x; \leq_K) = \{u \in A \mid u \leq_K x\}.$$

The following result was independently proved in Penot (1978) and Borwein (1983).

THEOREM 12.36. *Let $(E(\tau), Spec(E))$ be a locally convex space and $K \subset E$ a closed pointed convex cone and $A \subset E$ a non-empty subset. If one of the following assumptions are satisfied:*

1. *A has a minorized closed section and K is regular (Daniell),*

2. *A is closed and bounded, K is regular (Daniell) and E is boundedly order complete (i.e. every bounded decreasing net has an infimum),*

3. *A has a compact section,*

then $E_m[A, K]$ is non-empty.

PROOF. Let $A(x; \leq_K)$ be a section. We consider on $A(x; \leq_K)$ the preordering defined by K. It is sufficient to show that for an appropriate section $A(x; \leq_K)$ every chain $\{x_i\}_{i \in I}$ in this section admits a lower bound in $A(x; \leq_K)$. We obtain the conclusion of the theorem applying Zorn's Lemma. Let $\mathcal{F}$ denote the finite subsets of I ordered by inclusion and define, for each S in $\mathcal{F}$, $x_S = \inf\{x_i | i \in S\}$. For every $S \in \mathcal{F}$, x_S is well defined since $\{x_i\}_{i \in I}$ is a chain and lies in $A(x; \leq_K)$.

Then $\{x_S | S \in \mathcal{F}\}$ is a decreasing net and in the case of assumption 1, it is minorized and hence, in case 1 or 2, has an infimum x_* to which it converges. Since $A(x; \leq_K)$ is closed, $x_* \in A$ and is an appropriate minorant.
We remark that any point in $E_m[A(x; \leq_K), K]$ is actually in $E_m[A, K]$. We have that if 3 is satisfied, the conclusion also follows since $\{A(x; \leq_K)\}_{i \in I}$ is a family of compact subsets of $A(x; \leq_K)$ with the finite intersection property. Any intersection point is then a minorant for the chain and the proof is complete. ∎

COROLLARY 12.5. *If $E(\tau)$ is a topological vector space and $K \subset E$ a closed convex cone, not necessarily pointed, then $E_m[A, K]$ is non-empty whenever a subset A of E is non-empty and compact.* ∎

Remark. A similar result as Corollary 12.5 is available for $E_M[A, K]$ where $E_M[A, K]$ is the set of all efficient maximal points of A.

In the following results we will abbreviate $E_M[A, K]$ by $E[A, K]$. A natural question is the following:
Is it true that for every convex not pointed cone $K \subset E$ the set $E[A, K]$ is non-empty, whenever A is non-empty and compact?

A complete answer to this question was given by Sterna-Karwat [1],[2] (1986). We give now her result without proof.

Let $E(\tau)$ be a topological vector space. By Definition 12.7 we introduced the class $\mathcal{C}$ of convex cones defined and studied by Sterna-Karwat [1],[2] (1986). We indicated also several situations when a convex cone $K \subset E$ belongs to the class $\mathcal{C}$. We have the following existence result.

THEOREM 12.37. *Let $E(\tau)$ be a Hausdorff topological vector space and $K \subset E$ a convex cone such that K is of class $\mathcal{C}$. Then for every non-empty compact set $A \subset E$, the set $E[A, K]$ is non-empty.*

PROOF. The proof of this result is long and it is based on several technical results. The reader can find the proof of this theorem in Sterna-Karwat [1],[2] (1986). ∎

By Definitions 12.23, 12.28 and 12.29 we introduced respectively the notions of positive proper effieciency, global Henig's proper efficiency and super-efficiency for a vector-valued optimization problems. Now we will give some existence theorems obtained recently by Truong (1999), related to these kinds of efficiency. Before giving these results, we need to introduce the cited notions for an abstract set, but in the sense of Pareto minimality.

Let $(E(\tau), Spec(E))$ be a locally convex space and $K \subset E$ a closed pointed convex cone and $A \subset E$ a non-empty subset. In Section 12.2 we defined the *strong dual* K^{*i} of K. Denote by $E[A, K]$ the set of efficient (Pareto minimal) points of A with respect to K.

DEFINITION 12.41. A point $a \in A$ is said to be:

1. a *positive proper efficient point of A with respect to K* ($a \in PPE[A, K]$) if there exists $\phi \in K^{*i}$ such that $\phi(x) \geq \phi(a)$ for all $x \in A$;
2. a *global Henig's proper efficient point of A with respect to K* ($a \in GHePE[A, K]$) if there exists a cone K_0 not equal to the whole space such that $K \setminus \{0\} \subseteq \text{int}(K_0)$ and $a \in E[A, K_0]$;
3. a *super-efficient point of A with respect to K* ($a \in SE[A.K]$) if for each neighborhood V of 0 in E, there exists a neighborhood U of 0 such that $cone(A - a) \cap (U - K) \subseteq V$.

It is known that $PPE[A, K] \subseteq GHePE[A, K] \subseteq E[A, K]$. Because we supposed that E is a locally convex space we have also $SE[A, K] \subseteq E[A, K]$. Moreover, if E is a Banach space and K is well based, we have $SE[A, K] \subseteq GHePE[A, K]$ (Borwein and Zhuang, 1991).

We recall that in Definition 12.40 we introduced the semicompactness for maximality. Now we consider this notion for minimality.

DEFINITION 12.42. A set $A \subset E$ is said to be:

1. K-compact (in Luc's sense) if any cover of A of the form $\{U_i + K\}_{i \in I}$ (where U_i is open in E) admits a finite subcover;
2. K-semicompact if any cover of A of the form $\{(x_i - K)^C\}_{i \in I}$ (where $x_i \in A$) admits a finite subcover;
3. K-bounded if there is a bounded set $M \subset E$ such that $A \subseteq M + K$.

It is known from Luc [4] (1989) that if A is K-compact, then it is K- semicompact and K-bounded.

THEOREM 12.38. *Let $A \subset E$ a non-empty subset. If A is K-compact and K^{*i} is non-empty, then PPE[A,K] is non-empty.*

PROOF. Since $(K^*)i$ is non-empty, we consider an arbitrary element $\phi \in (K^*)i$ and we set $D = \phi(A)$. The K-compactness of A implies the existence of a non-empty bounded set $M \subset E$ such that $A \subseteq M + K$. The strict positivity of ϕ implies that $D \subseteq \phi(M) + R_+$.

We have that $\phi(M)$ is bounded, since ϕ is continuous and linear. Therefore

we have $-\infty < m = \inf D < +\infty$. We claim that $m \in D$. Indeed if m is not in D, we have $\phi(a) > m$ for every $a \in A$, and consequently, there is a scalar $\beta_a > 0$ depending on a such that $\phi(a) - \beta_a > m$.
Because ϕ is continuous there exists an open neighborhood U_a of a such that $\phi(U_a) \subseteq \{x \in E || \phi(a) - x| < \beta\}$. Observe that $A \subseteq \cup \{U_a + K | a \in A\}$.
The K-compactness of A implies that there is a natural number $n_* > 0$ such that $\{U_{aj} + K | a_j \in A\}_{j=1,2,..,n_*}$ is a cover of A. We can show that

$$D \subseteq \phi\left(\cup_{j=1}^{n_*}\left(U_{aj} + K\right)\right) \subseteq [t_*, +\infty[,$$

where $t_* = \min\{\phi(a_j) - \beta_a \mid j = 1, 2, .., n_*\}$. By the definition of β_a we have that $t_* > m$, which contradicts the definition of m. Hence, $m = \min D$ and the point $a \in A$ satisfying $\phi(a) = m$ belongs to $PPE[A, K]$. ■

THEOREM 12.39. *Let $(E(\tau), Spec(E))$ be a locally convex space and $K \subset E$ a closed pointed convex cone and $A \subset E$ a K-compact subset. If K is well based, then SE[A,K] is non-empty.*

PROOF. Since K is well based K^{*i} is non-empty. Let $\phi \in K^{*i}$ be an arbitrary element. Let $B = \{x \in K \mid \phi(x) = 1\}$, $V = \left\{x \in E \mid |\phi(x)| \leq \frac{1}{4}\right\}$ and $K_V = cone(B + V)$. We can show that B is a base for K and $\phi \in K_V^{*i}$. It is known that A is also K_V-compact Luc [4] (1989). Theorem 12.38 implies that PPE[A,K] is non-empty and therefore $E[A, K_V]$ is non-empty.

Considering Propositions 3.2 and 3.5 in Borwein and Zhuang (1993) and the remark at the page 472 in the same work we deduce that $E[A, K_V] \subseteq SE[A, K]$ and thus complete the proof. ■

THEOREM 12.40. *If E is a weakly complete Banach space, $K \subset E$ is a completely regular cone with a closed base and $A \subset E$ is a closed K-bounded set, then $GHePE[A, K]$ is non-empty.*

PROOF. Denote by K_n the Henig's dilating cone associated with K, where $n \geq \frac{2}{\delta}$, $\delta = \inf\{||b||\ ,\ b \in B\} > 0$ and B is a closed base of K (see the definition of Henig's dilating cone given in Section 12.2). Because K_n is well based it is normal and since E is weakly complete then the complete regularity of K implies the same property for K_n. If $a \in A$ is an arbitrary element, we denote by A_n the set $A \cap (a - K_n)$. The K-boundedness of A implies the existence of a bounded set M of E such that $A \subseteq M + K$. Since $A_n \subseteq (M + K) \cap (a - K_n)$, we can show that A_n is bounded. Obviously A_n is also closed. Applying Isac (1983, Proposition 5 and Theorem 2) we have that $E[A, K_n]$ is non-empty and because $E[A, K_n \subseteq GHePE[A; K]$, the proof is complete. ■

In the next results we will present some existence theorems for efficient points (in the sense of Pareto maximal points) for more general sets such as compact sets. The following results are due to G. Isac.

THEOREM 12.41. *Let $(E, Spec(E))$ be a locally convex space, $\{p_i\}_{i\in I}$ a base of Spec(E), $A \subset E$ a non-empty complete subset and $\Gamma : A \to 2^A$ a dynamical system. Assume that for every $i \in I$, there exists a lower semicontinuous mapping $\Phi_i : A \to R_+$ and a constant $c_i > 0$ such that for every $x \in A$ and every $u \in \Gamma(x)$, $c_i p_i(x-u) \le \Phi_i(x) - \Phi_i(u)$. Then Γ has a stationary point in A.*

PROOF. For a proof of this theorem the reader is referred to Isac (1983). ∎

Let $(E(\tau), Spec(E))$ be a locally convex space and $K \subset E$ a closed pointed convex cone and $A \subset E$ a non-empty subset. Let $A_0 \subset A$ be a non-empty subset. We define the dynamical system $\Gamma_0 : A_0 \to 2^A$ by $\Gamma_0(x) = A \cap (K + x)$ for all $x \in A_0$ and the set $K(A_0)$ by $K(A_0) = \{v \in K | v = v_1 - v_2$, with $v_2 \in A_0$ and $v_1 \in \Gamma(v_2) \cap A_0\}$. We note that for some A_0, $K(A_0)$ can be empty, but if for every $x \in A_0$, $\Gamma_0(x) \cap A$ is non-empty, we have that $K(A_0)$ is non-empty.

THEOREM 12.42. *Let $(E(\tau), Spec(E))$ be a locally convex space $\{p_i\}_{i\in I}$ a base of Spec(E), $K \subset E$ a closed pointed convex cone and $A \subset E$ a non-empty subset. If there exists a non-empty subset $A_0 \subset A$ such that*
(1).A_0 is complete,
(2).$\Gamma_0(A_0) \subseteq A_0$,
(3).for every $i \in I$, there exists $f_i \in E^*$ such that:
(i). $p_i(v) \le f_i(v)$, for every $v \in K(A_0)$,
(ii). $\sup\{f_i(x) | x \in A_0\} < +\infty$,
then E[A,K] is non-empty.

PROOF. By assumption (2) we have that $\Gamma_0 : A_0 \to 2^{A_0}$ is a generalized dynamical system. The theorem will be proved if we show that Γ_0 has a stationary point in A_0 (by the definition of Γ_0, a stationary point of Γ_0 in A_0 is an efficient point of A with respect to K). We obtain the existence of a stationary point for Γ_0 by applying Theorem 12.41.
Indeed, for an arbitrary $i \in I$, let $f_i \in E^*$ such that $\sup\{f_i(x) | x \in A_0\} < +\infty$ and $p_i(v) \le f_i(v)$ for every $v \in K(A_0)$.
We denote by $m_i = \sup\{f_i(x) | x \in A_0\}$, and for every $i \in I$ we set $\Phi_i(x) = m_i - f_i(x)$, for every $x \in A_0$.

If $x \in A_0$ and $u \in \Gamma_0(x) = A \cap (K + x) \subseteq A_0$, there exists $v \in K$ such that $u = v + x \in A_0$, and hence $u \in \Gamma_0(x) \cap A_0$, which implies that $v = u - x \in K(A_0)$. By assumption (3), we have that for every $i \in I$,

$$p_i(x-u) = p_i(u-x) = p_i(v) \le f_i(v) = f_i(v + x - x)$$

$$= f_i(v+x) - f_i(x) = [m_i - f_i(x)] - [m_i - f_i(v+x)]$$

$$= \Phi_i(x) - \Phi_i(u).$$

Since for every $i \in I$, $\Phi_i : A_0 \to R_+$ is continuous and A_0 is a complete set, we have that all the assumptions of Theorem 12.41 are satisfied and we obtain that Γ_0 has a stationary point and consequently $E[A, K]$ is non-empty. ∎

The converse of Theorem 12.42 is also true but in a more general case.

THEOREM 12.43. *Let $(E(\tau), Spec(E))$ be a locally convex space, $\{p_i\}_{i\in I}$ a base of Spec(E), $K \subset E$ a closed pointed convex cone and $A \subset E$ a non-empty subset. If E[A,K] is non-empty, then there exists a subset $A_0 \subset A$ such that:*
(1).$\Gamma_0(A_0) \subseteq A_0$,
(2).for every $i \in I$, there exists $f_i \in E^*$ such that $p_i(v) \leq f_i(v)$, for every $v \in K(A_0)$.

PROOF. We take the set $E[A, K]$ for the set A_0. By assumption we have that A_0 is non-empty and $\Gamma_0(x) = A \cap (K + x) = \{x\}$, for every $x \in A_0$ since every $x \in A_0$ is an efficient point for A. Thus we have $\Gamma(A_0) \subseteq A_0$, that is conclusion (1) is satisfied. If $v_2 \in A_0$ and for every $v_1 \in \Gamma(v_2) \cap A_0$ we have $v_1 = v_2$, that is $K(A_0) = \{0\}$. This fact implies that conclusion (2) is satisfied if, for every $i \in I$ we choose an arbitrary $f_i \in E^* \setminus \{0\}$. ∎

Remark. If, in the proof of Theorem 12.43 we take a finite subset of $E[A, K]$ as the set A_0, we obtain that the converse of Theorem 12.42 is also true.

PROPOSITION 12.18. *Let $(E(\tau), Spec(E))$ be a locally convex space, $\{p_i\}_{i\in I}$ a base of Spec(E), $K \subset E$ a closed pointed convex cone and $A \subset E$ a non-empty subset. If there exists a non-empty subset $A_0 \subset A$ such that:*
(1).A_0 is complete,
(2).$\Gamma_0(A_0) \subseteq A_0$,
(3).for every $i \in I$, there exists $f_i \in E^*$ such that: $p_i(v) \leq f_i(v)$, for every $v \in K(A_0)$,
then E[A,K] is non-empty.

PROOF. The proposition is an immediate consequence of Theorem 12.42. ∎

12.5 NUCLEAR CONES AND EFFICIENCY

The notion of *nuclear cone* is due to G. Isac and its origine is the following fact. Considering Proposition 12.18 we have the following result
"every bounded and complete subset of a locally convex space $(E, Spec(E))$ has an efficient point with respect to a closed pointed convex cone $K \subset E$, if the cone K satisfies the following property: there exists a base $\mathcal{B} = \{p_i\}_{i\in I}$ of Spec(E) such that, for every $p \in \mathcal{B}$ there exists $f_p \in E^*$ with the property that $p(x) \leq f_p(x)$ for every $x \in K$."

DEFINITION 12.43. We say that the convex cone $K \subset E$ is a *nuclear cone* if and only if there exists a base $\mathcal{B} = \{p_i\}_{i\in I}$ of $Spec(E)$ such that for every $p \in \mathcal{B}$ there exists $f_p \in E^*$ satisfying the inequality $p(x) \leq f_p(x)$ for every $x \in K$.

The notion of nuclear cone was introduced in Isac (1983).

When the cone K is nuclear, Proposition 12.18 implies the following result.

THEOREM 12.44. *Let $(E, Spec(E))$ be a locally convex space, $A \subset E$ a non-empty subset, and $K \subset E$ a nuclear cone. If there exists a bounded and complete subset $A_0 \subset A$, such that $A \cap (K + x) \subseteq A_0$, for every $x \in A_0$, then $E[A, K]$ is non-empty.* ∎

In the study of some practical problems, we need to know under what conditions the set $E[A, -K]$ is non-empty. The notion of nuclear cone equally solves this problem.

We note that the notion of nuclear cone, discovered in optimization, has also applications to the fixed point theory , to the best approximation theory, to the study of conically bounded sets and to the study of nuclearity of topological vector spaces [*cf.* Isac (1983,[3]-[4],1985,1987,1994,1997), Bahya (1989,1991,1995,1997), Postolica (1986-1989,1992,1994,1995,1997), Truong [1],[2] (1994), Truong (1997),(1999), Pontini (1990)].

Before giving other applications of nuclear cones to the study of efficiency, we present some important properties and examples of nuclear cones.

PROPOSITION 12.19. *In a locally convex space $(E, Spec(E))$ every nuclear cone $K \subset E$ is normal.*

PROOF. Indeed, let $\{x_i\}_{i\in I}$ and $\{y_i\}_{i\in I}$ be two nets in E such that for every $i \in I$, $0 \le x_i \le y_i$ and $\lim_{i\in I} y_i = 0$. Since K is nuclear there exists a base $\mathcal{B}$ of $Spec(E)$ such that for every $p \in \mathcal{B}$, there exists $f_p \in E^*$ with $p(x) \le f_p(x)$ for every $x \in K$. We have $0 \le p(x_i) \le f_p(y_i)$ for every $p \in \mathcal{B}$ and $i \in I$. Since $\lim_{i\in I} f_p(y_i) = 0$ we deduce $\lim_{i\in I} p(x_i) = 0$ and, as $\mathcal{B}$ is a base for $Spec(E)$, we have that $\{x_i\}_{i\in I} \to 0$. The normality of K is thus proved. ∎

PROPOSITION 12.20. *Let $(E(\tau), Spec(E))$ be a locally convex space and $K \subset E$ a convex cone. If K is τ-normal, then it is $\sigma(E, E^*)$-nuclear.*

PROOF. Let p be a $\sigma(E, E^*)$-continuous seminorm. Then there exists a constant $C > 0$ and n functionals $f_1, f_2, ..., f_n \in E^*$, such that we have

$$p(x) \le C \sup_{1\le i\le n} \{|f_i|(x)\}, \text{ for every } x \in E.$$

Because K is τ-normal we have that for every $i = 1, 2, ..., n$ there exist $h_i, g_i \in K^*$ such that h_i and g_i are positive on K and $f_i = h_i - g_i, i = 1, 2, ..., n$. Hence for every $x \in K$ we have

$$p(x) \leq C \sup_{1\leq i\leq n} \{|f_i \mid (x)\} \leq C \sup_{1\leq i\leq n} \{(|h_i| + |g_i|)(x)\} \leq$$

$$\leq C \sup_{1\leq i\leq n} (h_i(x) + g_i(x)) \leq C \sum_{i=1}^{n} (h_i + g_i)(x),$$

which implies that K is $\sigma(E, E^*)$-nuclear. ■

COROLLARY 12.6. *A convex cone $K \subset E$ is $\sigma(E, E^*)$-nuclear if and only if, it is $\sigma(E, E^*)$-normal.*

PROOF. The corollary is a consequence of Propositions 12.19 and 12.20. ■

We recall that a locally convex space $(E, Spec(E))$ is *quasi-complete* if every closed and bounded subset of E is complete.

PROPOSITION 12.21. *In a quasi-complete locally convex space $(E, Spec(E))$ every τ-nuclear closed convex cone $K \subset E$ is completely regular.*

PROOF. Let $\{x_i\}_{i\in I} \subset K$ be a τ-bounded and monotone increasing net. We can suppose that $\{x_i\}_{i\in I} \subset B_0 \subset K$, where B_0 is a τ-bounded and τ-closed set. Let $\mathcal{B}$ be a base of $Spec(E)$ such that, for every $p \in \mathcal{B}$ there exists $f_p \in E^*$ satisfying the inequality $p(x) \leq f_p(x)$ for every $x \in K$. Since f_p is continuous and positive on K we have that $\{f_p(x_i)\}_{i\in I}$ is a monotone increasing and bounded net of real numbers. Hence it is convergent. For every $\epsilon > 0$, there exists $i_0 \in I$ such that for all $i, j \in I$ satisfying $i_0 \leq i, j$ we have

$$p(x_i - x_j) \leq p(x_i - x_{i_0}) + p(x_j - x_{i_0}) \leq$$

$$\leq f_p(x_i - x_{i_0}) + f_p(x_j - x_{i_0}) \leq \frac{\varepsilon}{2} + \frac{\varepsilon}{2} = \varepsilon.$$

We deduce that $\{x_i\}_{i\in I}$ is a τ-Cauchy net in the subset B_0 and, because E is a quasi-complete space, we obtain that $\{x_i\}_{i\in I}$ is τ-convergent and the proof is complete. ■

PROPOSITION 12.22. *Let (E, Spec(E)) be a Hausdorff locally convex space. Every well based convex cone $K \subset E$ is nuclear.*

PROOF. The proposition is a consequence of Theorem 12.11. ■

Remark. It is not true that in a locally convex space every nuclear cone is well based. The class of nuclear cone is strictly larger than the class of well based cones.

The following corollaries are consequences of Proposition 12.22.

COROLLARY 12.7. *In a normed vector space* $(E, \|\cdot\|)$, *a pointed convex cone* $K \subset E$ *is nuclear if and only if* K *is well based.* ∎

COROLLARY 12.8. *In a locally convex space any locally compact or weakly locally compact convex cone is nuclear.* ∎

COROLLARY 12.9. *Every closed pointed convex cone in* R^n *is nuclear.* ∎

Another class of cones which are nuclear is the class of *semi-complete cones.* This concept was introduced in 1967 by G. Mokobodski (1967/1968).

Let $(E, Spec(E))$ be a Hausdorff locally convex space, $K \subset E$ a convex cone and $\{f_m\}_{m\in N}$ a sequence in K^*.

DEFINITION 12.44. The convex cone K is *semi-complete* with respect to the sequence $\{f_m\}_{m\in N}$ if and only if for every sequence $\{x_n\}_{n\in N} \subset K$ such that $\sum_{n=1}^{\infty} f_m(x_n) < +\infty$, for every $m \in N$ we have that $\sum_{n=1}^{\infty} x_n$ is convergent to an element $x \in K$.

The semi-complete cones are important in the axiomatic theory of potential. For example, the cone $\mathcal{H}^+$ of positive harmonic functions in a locally compact space Ω, with respect to Bauer's, Brelot's or Constantinescu-Cornea's axioms system is a semi-complete cone with respect to a sequence $\{\mu_n\}_{n\in N}$ of positive Radon measures with compact support on Ω.

We cite without proof the following two results.

THEOREM 12.45. *Let (E, Spec(E)) be a Hausdorff locally convex space and* $K \subset E$ *a convex cone. If* K *is semi-complete with respect to a sequence* $\{f_m\}_{m\in N} \subset K^*$, *then it is nuclear.*

PROOF. For a proof of this theorem the reader is referred to Isac (1983) or Bahya (1995). ∎

THEOREM 12.46. *Let (E, Spec(E)) be a Frechet space and* $K \subset E$ *a convex cone. If* K *is nuclear then it is semi-complete with respect to a sequence* $\{f_m\}_{m\in N} \subset K^*$.

PROOF. A proof of this result is in Bahya (1997). ∎

The following characterizations of nuclear cones may be considered as nuclearity tests.

THEOREM 12.47. *Let* $(E(\tau), Spec(E))$ *be a locally convex space and* $K \subset E$ *a convex cone. Then the following assertions are equivalent:*

1. *K is nuclear,*

2. *for every $V \in \mathcal{V}_\tau(0)$, there exists $W \in \mathcal{V}_\sigma(0)$ such that $0 \leq x \leq y \in W$ implies $x \in V$,*

3. *K is normal and every net in K is τ-convergent to zero whenever it is σ-convergent to zero,*

4. *there exists a base $\mathcal{B}$ of Spec(E) such that for every $p \in \mathcal{B}$, there exists $\mu_p > 0$ such that $\mu_p p(x) + p(y) \leq p(x+y)$ for all $x, y \in K$.*

PROOF. These results are due to A.O. Bahya and the reader is referred to Bahya (1997). ■

For other examples of nuclear cones the reader is referred to Isac [3],[4] (1983), Isac (1985,1987,1994,1997), Bahya (1989,1991,1995,1997), Postolica (1986-1989, 1992,1994,1995,1997), Truong [1],[2] (1994), Truong (1997,1999).

We note that the origin of the name "*nuclear cone*" is the fact that *in a nuclear space (in Grothendieck's sense) any normal convex cone is nuclear in our sense.* A proof of this result is in Isac (1983). It is also known that under some special conditions if K is a nuclear cone and $E = K - K$, then the space E is a nuclear space in Grothendieck's sense.

Using the concept of nuclear cone, we give now some existence theorems for efficient points.

THEOREM 12.48. *Let (E, Spec(E)) be a locally convex space and $K \subset E$ a nuclear cone. If, for a non-empty set $A \subset E$, there exists a point $x_0 \in A$ such that $A \cap (K + x_0)$ is bounded and complete, then E[A,K] is non-empty.*

PROOF. The proposition is a consequence of Theorem 12.44. ■

The next result was initially proved by F.E. Browder [see the reference in Isac (1983)].

THEOREM 12.49. *Let $(E, \|\cdot\|)$ be a Banach space and $A \subset E$ a closed bounded set. Then, with respect to every well based closed convex cone $K \subset E$, the set E[A,K] is non-empty.*

PROOF. If $K \subset E$ is an arbitrary well based closed convex cone then by Proposition 12.22, K is nuclear and we can apply Theorem 12.44. ■

Remark. The result presented in Theorem 12.49 was used by F.E. Browder in the study of the normal solvability of nonlinear operators.

The next result is based on the notion of *recession cone* introduced in Section 12.2. We recall only the well-known fact that if $\{X_i\}_{i\in I}$ is a family of closed convex subsets in R^n such that $\bigcap_{i\in I} X_i \neq \emptyset$, then $Rec(\bigcap_{i\in I} X_i) = \bigcap_{i\in I} Rec(X_i)$.

The next result is due to Bitran and Magnanti (1979).

THEOREM 12.50. *Let $K \subset R^n$ be a pointed closed convex cone. If $A \subset R^n$ is a non-empty closed convex set, then $E[K,A]$ is non-empty if and only if $K \cap Rec(A) = \{0\}$.*

PROOF. If $K \cap Rec(A) \neq \{0\}$ then there exists $0 \neq y \in K \cap Rec(A)$ and, since $x + y \in (x + K) \cap A$ for every $x \in A$, we deduce that it is impossible to have an efficient point in A with respect to K. To prove the converse, we suppose that $K \cap Rec(A) = \{0\}$. Then, if $x_0 \in A$ we have

$$Rec[(x_0 + K) \cap A] = K \cap Rec(A) = \{0\}$$

which implies that $(x_0 + K) \cap A$ is bounded. Because K is nuclear, we apply Theorem 12.48 and the proof is complete. ∎

The next concept is due to Hartley (1978). Let $K \subset R^n$ be a convex cone and $A \subset R^n$ a non-empty subset.

DEFINITION 12.45. We say that A is *K-compact (in Hartley's sense)* if and only if for every $x \in A$, the set $(x - \overline{K}) \cap A$ is compact.

The following result is due to Hartley (1978).

THEOREM 12.51. *Let $K \subset R^n$ be a pointed closed convex cone. If $A \subset R^n$ is a non-empty K-compact (in Hartley's sense) set, then $E[A, -K]$ is non-empty.*

PROOF. The cone $-K$ is nuclear and we can apply Theorem 12.48 to the set A and the cone $-K$. ∎

The next notions are due to T. Tanino and Y. Sawaragi [see the references in Isac (1983)]. Let $(E, Spec(E))$ be a locally convex space and $K \subset E$ a pointed closed convex cone.

DEFINITION 12.46.

1. We say that a set $A \subset E$ is K-closed if and only if A+K is closed,
2. We say that a set $A \subset E$ is K-bounded if and only if there exists $a_0 \in E$ such tht $A \subseteq a_0 + K$,
3. We say that a set $A \subset E$ is K-(TS)-semicompact if and only if it is K-bounded and K-closed.

We have the following result.

THEOREM 12.52. *If $(E, Spec(E))$ is quasi-complete locally convex space and $K \subset E$ is a closed nuclear cone, then, for every non-empty K-(TS)-semicompact subset $A \subset E$, we have that $E[A, -K]$ is non-empty. Moreover, $A \subseteq E[A, -K] + K$.*

PROOF. We consider the set $B = (B + K) \cap (a_* - K)$, where a_* is an arbitrary element. Since A is K-closed and the cone K is closed we have that B is closed. The set B is also bounded. Indeed, since A is K-bounded, there exists $a_0 \in E$ such that

$$A + K \subseteq a_0 + K + K \subseteq a_0 + K,$$

that is $B \subseteq (a_0 + K) \cap (a_* - K)$, which implies $B \subseteq [a_0, a_*]_0 := \{x \in E | a_0 \leq x \leq a_*\}$. Because K is nuclear, it is normal and consequently $[a_0, a_*]_0$ is topologically bounded. On the other hand, the nuclearity of K implies the existence of a base $\{p_\alpha\}_{\alpha\in\Lambda}$ of $Spec(E)$ such that, for every $\alpha \in \Lambda$, there exists $f_\alpha \in E^*$ satisfying $p_\alpha(x) \leq f_\alpha(x)$, for every $x \in K$.
Since $p_\alpha(-x) = p_\alpha(x) \leq f_\alpha(x) = -f_\alpha(-x)$, for every $x \in K$, we deduce that $-K$ is a nuclear cone.
We have $B \cap (-K + x) \subseteq B$ for every $x \in B$ and because E is quasicomplete, we apply Theorem 12.44 and we obtain that $E[B, -K]$ is non-empty. To obtain the first conclusion of the theorem we must show that $E[B, -K] \subseteq E[A, -K]$. In order to show that this inclusion holds, we must prove two facts about any $u \in E[B, -K]$:
(i). $u \in A$ and
(ii). $u \in E[A, -K]$.
Indeed,
(i). if $u \in E[B, -K]$ and $u \notin A$, then there exists $r \in K$, $r \neq 0$ and $a \in A$ such that $u = a + r$, which implies $a < u$. Since $a \in A$ we have

$$a \in A + K. \tag{12.7}$$

On the other hand, we have $u = a_* - r_1$, where $r_1 \in K$, which implies $a = u - r = a_* - (r_1 + r)$ and hence

$$a \in a_* - K \tag{12.8}$$

Considering (12.7) and (12.8) we have that there exists $a \in B$ such that $a < u$ which is impossible since $u \in E[B, -K]$. The conclusion is that for every $u \in E[B, -K]$ we have $u \in A$.
(ii). Suppose now that $u \in E[B, -K]$ but $u \notin E[A, -K]$. In this case there exists $u_0 \in A$ such that $u_0 < u$. We have $u_0 \in B$ since $u_0 \in u - K \subseteq a_* - K$ and $u_0 \in A \subseteq A + K$. Hence there exists $u_0 \in B$ such that $u_0 < u$, which is impossible since $u \in E[B, -K]$. The final conclusion is $E[A, -K] \neq \emptyset$.

The proof will be complete if we show that $A \subseteq E[A, -K]$. Indeed, let $a_* \in A$ be an arbitrary element. Consider again the set $B = (B + K) \cap (a_* - K)$. We know that there exists $a_0 \in E$ such that $B \subseteq [a_0, a_*]_0$. Since we have shown that $E[B, -K] \neq \emptyset$, we can pick $u \in E[B, -K]$ and we get $u \leq a_*$, which implies $a_* \in E[B, -K] + K \subseteq E[A, -K] + K$. Because a_* was an arbitrary element of A we have $A \subseteq E[A, -K] + K$. ∎

Let $(E(\tau), Spec(E))$ be a locally convex space and (F, K) an ordered vector space with the ordering defined by a pointed convex cone K. We say that a mapping $f : E \to F$ is ***τo-bounded*** if, for every τ-bounded set, $B \subset E$, we have that $f(B)$ is order bounded, i.e. there exists $u_1, u_2 \in F$ such that

$$f(B) \subset [u_1, u_2]_0 = \{y \in F | u_1 \leq y \leq u_2\}.$$

COROLLARY 12.10. *Let $(E(\tau), Spec(E))$ and $(F, Spec(F))$ be locally convex spaces. Suppose that F is quasicomplete and ordered by a closed convex nuclear cone $K \subset F$. Let $B \subset E$ be a bounded set and $f : B \to F$ a mapping such that f(B) is K-closed. If there exists a τo-bounded mapping $g : E \to F$ such that $g(x) \leq f(x)$ for all $x \in B$, then $E[f(B), -K]$ is non-empty.*

PROOF. We observe that $f(B)$ is a K-semicompact set and we apply Theorem 12.52. ■

Another aspect of the importance of nuclear cones in the study of existence theorems for efficiency, is the *possibility to use the weakly complete cones in Pareto optimization.*

The weakly complete cone is an essential mathematical tool in Choquet's integral representation theory, *cf.* Choquet (1969), and in the axiomatic theory of potential. The reader can find examples of weakly complete cones in, among others, Choquet (1969), Becker (1989) and Constantinescu (1970). There exist weakly complete cones without weakly compact base Becker (1989). For efficient points we have the following result.

THEOREM 12.53. *Let $(E, Spec(E))$ be a locally convex space and $K \subset E$ a closed normal cone. If the cone K is weakly complete then for every non-empty weakly closed set $A \subset E$ such that there exists a point $x_0 \in A$ with the property that $A \cap (K + x_0)$ is bounded, then we have that E[A,K] is non-empty.*

PROOF. Since K is normal, it is $\sigma(E^*, E)$-nuclear and we observe that Theorem 12.48 is applicable. ■

Let $(E, Spec(E))$ be a locally convex space and $K \subset E$ a closed pointed convex cone. The following notion is due to Isac (1997).

DEFINITION 12.47. We say that the convex cone K *has property (EBC)* if and only if, every bounded complete set of E has an efficient point with respect to K.

Since every nuclear cone has property (EBC), the following open problem was defined in Isac (1997).

OPEN PROBLEM *Under what conditions a convex cone $K \subset E$ having property (EBC) is a nuclear cone?*

This problem was considered by Truong [1] (1994), Troung (1999).

Now we inform the reader about the state concerning the solvability of this problem by giving, without proof, some results obtained recently by Truong [1] (1994).

If K is a convex cone in a general topological vector space $E(\tau)$, we say that K *satisfies property (**)* if any bounded increasing net which is contained in K and in a complete subset of E has a limit.

THEOREM 12.54. *Let $E(\tau)$ be a topological vector space and $K \subset E$ a convex cone. If the following assumptions are satisfied:*
(1). the cone K is acute
(2). K satisfies property (EBC) then, K satisfies property (**). ■

Motivated by property (EBC), Truong (1999) introduced the following variant of this property.

DEFINITION 12.48. We say that a cone K in a locally convex space E *satisfies property (SEBC)*, if and only if every bounded complete set of E has a super-efficient point with respect to K.

THEOREM 12.55. *If $(E, \|\cdot\|)$ is a Banach space and $K \subset E$ a closed pointed convex cone such that $K^{*i} \neq \emptyset$, then K is nuclear if and only if K satisfies property (SEBC).* ■

OPEN PROBLEM *Is Theorem* 12.55 *valid in a locally convex space?*

12.6 A GENERAL CONSTRUCTIVE EXISTENCE PRINCIPLE FOR EFFICIENCY

In this section, we will present a general constructive existence test for K-efficient points, for non-empty closed subsets. This test is due to Isac (1998).

THEOREM 12.56. *Let $(E(\tau), \{p_\lambda\}_{\lambda\in\Lambda})$ be a sequentially complete locally convex space, $K \subset E$ a pointed closed convex cone and $D \subset E$ a non-empty closed subset. Consider the generalized dynamical system $\Gamma : D \to 2^D$ defined by $\Gamma(x) = (x + K) \cap D$. Given $x_0 \in D$, if there exists a sequence $\{x_n\}_{n=0}^{\infty} \subset D$ such that $x_{n+1} \in \Gamma(x_n)$ for all $n = 0, 1, 2, ..$ and $\lim_{n\to\infty} \delta_\lambda(\Gamma(x_n)) = 0$ for all $\lambda \in \Lambda$, where $\delta_\lambda(A) = \sup\{p_\lambda(x - y) | x, y \in A\}$, then D has a K-efficient point x_* such that $x_0 \le x_*$. Moreover, $\bigcap_{n=0}^{\infty} \Gamma(x_n) = \{x_*\}$ and $\lim_{n\to\infty} x_n = x_*$.*

PROOF. First we observe that

$$\Gamma(x_{n+1}) \subseteq \Gamma(x_n), \text{ for all } n = 0, 1, 2.3...$$

Indeed, if $x \in \Gamma(x_{n+1}) = (x_{n+1} + K) \cap D$, then $x = x_{n+1} + k_1$, with $k_1 \in K$ and $x \in D$.
Since $x_{n+1} \in \Gamma(x_n)$, there exists $k_2 \in K$ such that $x_{n+1} = x_n + k_2$. Thus we

have $x \in D$ and $x = x_n + k_1 + k_2 \in x_n + K$, that is $x \in \Gamma(x_n)$.
Therefore $\{\Gamma(x_n)\}_{n=0}^{\infty}$ is decreasing. The sequence $\{x_n\}_{n\in N}$ is a Cauchy sequence. Indeed, since $\lim_{n\to\infty} \delta_\lambda(\Gamma(x_n)) = 0$ for all $\lambda \in \Lambda$, we have that for each $\varepsilon > 0$ and $\lambda \in \Lambda$, there exists $n_\varepsilon \in N$ such that

$$\sup_{x,y\in\Gamma(x_n)} p_\lambda(x-y) < 0, \text{ for all } n > n_\varepsilon.$$

Because $x_m \in \Gamma(x_{m-1}) \subseteq \Gamma(x_n)$, for all $m, n \in N$ such that $m > n$ we have that $p_\lambda(x_m - x_n) < \varepsilon$, whenever $m > n > n_\varepsilon$. This fact implies that $\{x_n\}_{n=0}^{\infty}$ is a Cauchy sequence in $E(\tau)$. By the sequential completness of E there exists a point $x_* \in E$ such that $\lim_{n\to\infty} x_n = x_*$. We have also $x_m \in \Gamma(x_n)$ for all $m > n, n = 0, 1, 2...$ Hence we obtain $x_* \in \Gamma(x_n)$ for all $n = 0, 1, 2..$, i.e. $x_* \in \bigcap_{n=0}^{\infty} \Gamma(x_n)$. (We used also the fact that $\Gamma(x_n)$ is closed for each $n = 0, 1, 2...$).

The element x_* is unique. Indeed, assume that there exists $u_* \in \bigcap_{n=0}^{\infty} \Gamma(x_n)$, with $u_* \neq x_*$. Since the family of seminorms $\{p_\lambda\}_{\lambda\in\Lambda}$ is sufficient we have that $p_\lambda(x) = 0$ for all $\lambda \in \Lambda$ implies $x = 0$. Hence because $u_* - x_* \neq 0$, there exists some $\lambda \in \Lambda$ such that $p_\lambda(u_* - x_*) = \alpha > 0$. Since $\lim_{n\to\infty} \delta_\lambda(\Gamma(x_n)) = 0$ for all $\lambda \in \Lambda$, there exists $n_0 \in N$ such that $\delta_\lambda(\Gamma(x_n)) < \alpha$. Because $x_*, u_* \in \Gamma(x_{n_0})$ we have

$$p_\lambda(u_* - x_*) \leq \delta_\lambda(\Gamma(x_{n_0})) < p_\lambda(u_* - x_*),$$

which is a contradiction. Therefore we have that $\bigcap_{n=0}^{\infty} \Gamma(x_n) = \{x_*\}$. We must show that $\Gamma(x_*) = \{x_*\}$. Indeed, suppose that $\hat{x} \in \Gamma(x_*)$. In this case we have $\hat{x} = x_* + k$, with $k \in K$, and $\hat{x} \in D$.
Since $x_* \in \Gamma(x_n)$ for all $n = 0, 1, 2, ...$ we deduce that $x_* - x_n \in K$ for all $n = 0, 1, 2, ...$, which implies that $\hat{x} = x_* + k = x_n + k + k_1 \in \Gamma(x_n)$, for all $n = 0, 1, 2, ...$(where $k_1 \in K$).
It follows $\hat{x} \in \bigcap_{n=0}^{\infty} \Gamma(x_n) = \{x_*\}$. Obviously, we have $\hat{x} = x_*$. Since in particular $x_* \in x_0 + K$, we have $x_0 \leq x_*$ and the proof is complete. ∎

Applications of Theorem 12.56 to particular problems are dependent of the ability to construct a sequence $\{x_n\}_{n=0}^{\infty}$ with the properties indicated in Theorem 12.56.

Now we give some realizations of Theorem 12.56.
(I). Let $(E, ||\cdot||)$ be a Banach space and $K \subset E$ a solid closed pointed convex cone. For any $k_0 \in \text{int}(K)$ and $\lambda \in]0, 1[$, we can define the set $K(\lambda, k_0) = \{x \in E \mid x - \lambda||x||k_0 \in K\}$. We can show that $K(\lambda, k_0)$ is a closed pointed convex cone. Because, for every $x \in K(\lambda, k_0)$ there exists $k \in K$ such that $x - \lambda||x||k_0 = k$, it follows that $x = \lambda||x||k_0 + k \in K$, i. e., $K(\lambda, k_0) \subseteq K$. The cone $K(\lambda, k_0)$ was recently defined and used in Göpfert and Tammer (1997).

The next result, proved by Göpfert and Tammer (1997), can be considered in particular as a realisation of Theorem 12.56.
We denote by $\leq_{K(\lambda,h_0)}$ the ordering defined by $K(\lambda, k_0)$ and by $E[D, K(\lambda, k_0)]$ the set of $K(\lambda, k_0)$- efficient points of D.

THEOREM 12.57. *Let $(E, ||\cdot||)$ be a Banach space and $K \subset E$ a solid, closed pointed convex cone. Let $k_0 \in int(K)$ and $\lambda \in]0,1[$. If $D \subset E$ is an arbitrary closed K-bounded (i. e. $D \subset y - K$ for some $y \in E$) set, then for any $x_0 \in D$ there exists $x_* \in E[D, K(\lambda, k_0)]$ such that $x_0 \leq_{K(\lambda,h_0)} x_*$. Moreover, x_* is obtained as the limit of a sequence $\{x_n\}_{n\in N} \subset D$.*

PROOF. We consider on D the generalized dynamical system $\Gamma(x) = [K(\lambda, k_0)+ x] \cap D$, for all $x \in D$. Let $\{x_n\}_{n\in N}$ be the sequence obtained by the construction proposed by Göpfert and Tammer (1997), i.e.,
(i). $x_1 = x_0$,
(ii). if $x_1, x_2, ..., x_n$ are defined, we choose $x_{n+1} \in \Gamma(x_n)$ such that does not exists $x \in \Gamma(x_n)$ with $x \in x_{n+1} + \dfrac{1}{n+1}k_0 + (K \setminus \{0\})$.

In Göpfert and Tammer (1997) it is proved that this sequence is well defined and for every $n \in N, \Gamma(x_{n+1}) \subseteq \Gamma(x_n)$. If $x \in \Gamma(x_n)$ is an arbitrary element, we have that $x \in x_{n+1}+K(\lambda, k_0)$, which implies that $(x-x_{n+1})-\lambda||x-x_{n+1}||k_0 \in K$. We have

$$x \in x_{n+1} + \lambda||x - x_{n+1}||k_0 + K. \tag{12.9}$$

Since $\Gamma(x_{n+1}) \in \Gamma(x_n)$, we have that $x \in \Gamma(x_n)$ and considering (ii) we deduce

$$x \notin x_{n+1} + \frac{1}{n+1}k_0 + (K \setminus \{0\}). \tag{12.10}$$

Hence, taking into account (12.9) and (12.10) we have $\lambda||x - x_{n+1}|| \leq \dfrac{1}{n+1}$ which implies that $\delta(\Gamma(x_{n+1})) \leq \dfrac{2}{\delta(n+1)}$, where $\delta(\Gamma(x_{n+1})) = diam(\Gamma(x_{n+1}))$.
Applying Theorem 12.56 the proof is complete. ∎

We will show now that a result similar to Theorem 12.57 follows from a more general construction, certainly, modulo another definition of the sequence $\{x_n\}_{n\in N}$.

THEOREM 12.58. *Let $(E, ||\cdot||)$ be a Banach space, ε a strictly positive real number and $K \subset E$ a closed pointed convex cone such that*

$$K \subseteq \{x \in E | \Psi(x) + \varepsilon||x|| \leq 0\},$$

where $\Psi : E \longrightarrow R$ is a subadditive continuous mapping. Let D be a non-empty closed subset of E. Then, for every $x_\alpha \in D$ such that Ψ is bounded from below on $(x_\alpha + K) \cap D$, there exists $x_* \in E[D, K]$ such that $x_\alpha \leq_K x_*$. Moreover, x_* is obtained as the limit of a sequence $\{x_n\}_{n\in N} \subset D$.

PROOF. We consider on D the generalized dynamical system $\Gamma(x) = (x + K) \cap D$, for all $x \in D$, and we define a sequence $\{x_n\}_{n \in N} \subset D$ inductively as follows.
We start the process taking $x_0 = x_\alpha$ and suppose that $x_n \in D$ is defined and $\Gamma(x_{k+1}) \subset \Gamma(x_k)$, for all $k = 0, 1, 2, ..., n-1$.
We have two posibilities:
(i). either $\Gamma(x_n) = (x_n + K) \cap D = \{x_n\}$ or,
(ii). there exists $x \neq x_n$ with $x \in \Gamma(x_n)$.
If (i) is the case, we have that $x_n = x_*$. Suppose that (ii) is true. Since $K \subseteq \{x \in E | \Psi(x) + \varepsilon ||x|| \leq 0\}$ we deduce from (ii) and using the fact that Ψ is subadditive,

$$\Psi(x) - \Psi(x_n) \leq \Psi(x - x_n),$$

$$\Psi(x) - \Psi(x_n) + \varepsilon ||x - x_n|| \leq \Psi(x - x_n) + \varepsilon ||x - x_n|| \leq 0,$$

and

$$\Psi(x) \leq \Psi(x_n) - \varepsilon ||x - x_n|| < \Psi(x_n). \tag{12.11}$$

Thus,

$$\Psi(x_n) - \inf_{v \in \Gamma(x_n)} \Psi(v) \geq \Psi(x_n) - \Psi(x) > 0.$$

We pick a point $x_{n+1} \in \Gamma(x_n)$ such that

$$\Psi(x_{n+1}) \leq \inf_{v \in \Gamma(x_n)} \Psi(v) + \frac{1}{2} \left[\Psi(x_n) - \inf_{v \in \Gamma(x_n)} \Psi(v) \right], \tag{12.12}$$

and we have $\Gamma(x_{n+1}) = (x_{n+1} + K) \cap D$, and so on.
Since K is a convex cone we can show that $\Gamma(x_{n+1}) \subseteq \Gamma(x_n)$, for all $n = 0, 1, 2....$
The sequence $\{\Gamma(x_n)\}_{n=0}^{\infty}$ (if it is not stopped after a finite number of steps) is decreasing. From (12.12) we obtain

$$\Psi(x_{n+1}) - \inf_{v \in \Gamma(x_{n+1})} \Psi(v) \leq \Psi(x_{n+1}) - \inf_{v \in \Gamma(x_n)} \Psi(v) \frac{1}{2} \left[\Psi(x_n) - \inf_{v \in \Gamma(x_n)} \Psi(v) \right]. \tag{12.13}$$

Let $z \in \Gamma(x_n)$, then applying (12.13) n times we obtain

$$\Psi(x_n) - \Psi(z) \leq \Psi(x_n) - \inf_{v \in \Gamma(x_n)} \Psi(v) \leq$$

$$\leq \frac{1}{2} [\Psi(x_{n-1}) - \inf_{v \in \Gamma(x_{n-1})} \Psi(v)] \leq ... \leq \frac{1}{2^n} \left[\Psi(x_\alpha) - \inf_{v \in \Gamma(x_\alpha)} \Psi(v) \right].$$

Since $K \subseteq \{x \in E | \Psi(x) + \varepsilon ||x|| \leq 0\}$ and $\Gamma(x_n) = (x_n + K) \cap D$, we have for any $z \in \Gamma(x_n)$,
$\Psi(z) - \Psi(x_n) \leq \Psi(z - x_n)$, which implies $-\Psi(z - x_n) \leq \Psi(x_n) - \Psi(z)$ and finally,

$$||x_n - z|| \leq \frac{1}{\varepsilon} [\Psi(x_n) - \Psi(z)] \leq \frac{1}{\varepsilon 2^n} \left[\Psi(x_\alpha) - \inf_{v \in \Gamma(x_\alpha)} \Psi(v) \right].$$

Therefore, $\delta(\Gamma(x_n)) \leq \frac{1}{\varepsilon 2^{n-1}}\left[\Psi(x_\alpha) - \inf_{v\in\Gamma(x_\alpha)} \Psi(v)\right]$, which implies that $\delta(\Gamma(x_n)) \to 0$ as $n \to +\infty$. The theorem is now a consequence of Theorem 12.56. ∎

In Dauer and Saleh (1993), the following result is proved.

THEOREM 12.59. *Let $E(\tau)$ be a topological vector space and K a solid pointed convex cone in E. If $y \in int(K)$ then the mapping $\Psi_y(x) = \inf\{\alpha \in R | x \in (\{-\alpha y\} + K)\}$ for all $x \in E$, is a continuous sublinear functional on E which satisfies the following properties:*
(1). $int(K) = \{x \in E \mid \Psi_y(x) < 0\}$,
(2). $\overline{K} = \{x \in E \mid \Psi_y(x) \leq 0\}$,
(3). Ψ_y is K-decreasing on E. ∎

Applying Theorem 12.59 we obtain the following variant of Theorem 12.58.

THEOREM 12.60. *Let $(E, \|\cdot\|)$ be a Banach space , $K \subset E$ a solid closed, pointed convex cone and Ψ_y the continuous sublinear mapping defined in Theorem 12.59. Let ε be a strictly positive (eventually sufficiently small) number. Consider the closed convex cone $K(\Psi_y) = \{x \in E \mid \Psi_y(x) + \varepsilon\|x\| \leq 0\}$.*
If $D \subset E$ is an arbitrary K-bounded set, then for any $x_\alpha \in D$ there exists $x_* \in E[D, K(\Psi_y)]$ such that $x_\alpha \leq_{K(\Psi_y)} x_*$. Moreover, x_* is obtained as the limit of a sequence $\{x_n\}_{n=0} \subset D$ defined by the method given in Theorem 12.58.

PROOF. We observe that $K(\Psi_y) \subseteq K$. Since D is K-bounded (i.e., $D \subseteq u - K$ for some $u \in K$), we have that Ψ_y is bounded from below on $(x_\alpha + K(\Psi_y)) \cap D$. (This is a consequence of property (3) of mapping Ψ_y.) The theorem is now a consequence of Theorem 12.58. ∎

Remark. Because for every small $\varepsilon > 0$, the cone $K(\Psi_y)$ is, in some sense, very close to K, the $K(\Psi_y)$-efficiency obtained by Theorem 12.60 is a kind of ε-efficiency. For ε-efficiency the reader is referred to Németh (1989), Isac (1996) and Tammer (1993).

Now we consider the case of locally convex spaces.
Let $(E(\tau), \{p_\lambda\}_{\lambda\in\Lambda})$ be a locally convex space as defined in this chapter. If $K \subset E$ is a well based pointed closed convex cone, then there exists $f \in K^*$ with the property that for every $\lambda \in \Lambda$ there exists $\varepsilon_\lambda > 0$ such that $K \subseteq \{x \in E \mid f(x) + \varepsilon_\lambda p_\lambda(x) \leq 0\}$.

We have the following result.

THEOREM 12.61. *Let $(E(\tau), \{p_\lambda\}_{\lambda\in\Lambda})$ be a sequentially complete locally convex space and $K \subset E$ a closed pointed convex cone. Suppose that there exists a subadditive continuous mapping $\Psi : E \to R$ with the property that for every $\lambda \in \Lambda$ there exists $\varepsilon_\lambda > 0$ such that $K \subseteq \{x \in E | \Psi(x) + \varepsilon_\lambda p_\lambda(x) \leq 0\}$.*

Let D be a non-empty closed subset of E. Then for every $x_\alpha \in D$ such that Ψ is bounded from below on $(x_\alpha + K) \cap D$, there exists $x_* \in E[D, K]$ such that $x_\alpha \leq_K x_*$. Moreover, x_* is obtained as the limit of a sequence $\{x_n\}_{n \in N} \subset D$.

PROOF. We consider on D the generalized dynamical system $\Gamma(x) = (x + K) \cap D$, $x \in D$ and we define a sequence $\{x_n\}_{n=0} \subset D$ inductively. We take $x_0 = x_\alpha$. Suppose that $x_n \in D$ is defined and $\Gamma(x_{k+1}) \subseteq \Gamma(x_k)$ for all $k = 0, 1, 2, .., n-1$. We have two possibilities:
(i). either $\Gamma(x_n) = (x_n + K) \cap D = \{x_n\}$, or
(ii). there exists $x \neq x_n$ with $x \in \Gamma(x_n)$.
If (i) is the case, we have that $x_n = x_*$. Suppose that (ii) is true. Because the family of seminorms $\{p_\lambda\}_{\lambda \in \Lambda}$ is suffiecient, there exists $\lambda_0 \in \Lambda$ such that $p_{\lambda_0}(x - x_n) \neq 0$ (i.e., $p_{\lambda_0}(x - x_n) > 0$). Because Ψ is subadditive we have

$$\Psi(x) - \Psi(x_n) + \varepsilon_{\lambda_0} p_{\lambda_0}(x - x_n) \leq \Psi(x - x_n) + \varepsilon_{\lambda_0} p_{\lambda_0}(x - x_n) \leq 0,$$

which implies

$$\Psi(x) \leq \Psi(x_n) - \varepsilon_{\lambda_0} p_{\lambda_0}(x - x_n) < \Psi(x_n).$$

Thus,

$$\Psi(x_n) - \inf_{v \in \Gamma(x_n)} \Psi(v) \geq \Psi(x_n) - \Psi(x) > 0.$$

We pick a point $x_{n+1} \in \Gamma(x_n)$ such that

$$\Psi(x_{n+1}) \leq \inf_{v \in \Gamma(x_n)} \Psi(v) + \frac{1}{2}\left[\Psi(x_n) - \inf_{v \in \Gamma(x_n)} \Psi(v)\right].$$

We have $\Gamma(x_{n+1}) = (x_{n+1} + K) \cap D$ and $\Gamma(x_{n+1}) \subset \Gamma(x_n)$, for all $n = 0, 1, 2, ...$ It is evident that

$$\Psi(x_{n+1}) - \inf_{v \in \Gamma(x_{n+1})} \Psi(v) \leq \frac{1}{2}\left[\Psi(x_n) - \inf_{v \in \Gamma(x_n)} \Psi(v)\right]. \tag{12.14}$$

Let $x \in \Gamma(x_n)$. Applying (12.14) n times we obtain

$$\Psi(x_n) - \Psi(x) \leq \frac{1}{2^n}\left[\Psi(x_\alpha) - \inf_{v \in \Gamma(x_n)} \Psi(v)\right].$$

Since for every $\lambda \in \Lambda$ we have

$$K \subseteq \{x \in E \mid \Psi(x) + \varepsilon_\lambda p_\lambda(x) \leq 0\},$$

we can show as in the proof of Theorem 12.58 that

$$p_\lambda(x_n - x) \leq \frac{1}{\varepsilon_\lambda 2^n}\left[\Psi(x_\alpha) - \inf_{v \in \Gamma(x_n)} \Psi(v)\right],$$

which implies that

$$\delta_\lambda(\Gamma(x_n)) \leq \frac{1}{\varepsilon_\lambda 2^n}\left[\Psi(x_\alpha) - \inf_{v \in \Gamma(x_n)} \Psi(v)\right].$$

We have that $\delta_\lambda(\Gamma(x_n)) \to 0$ as $n \to \infty$ for all $\lambda \in \Lambda$. The theorem is now a consequence of Theorem 12.56. ■

COROLLARY 12.11. *Let $(E(\tau), \{p_\lambda\}_{\lambda\in\Lambda})$ be a sequentially complete locally convex space and $K \subset E$ a well based closed pointed convex cone. Let D be a non-empty closed subset of E. Then for every $x_\alpha \in D$ such that $(x_\alpha + K) \cap D$ is bounded, there exists $x_* \in E[D, K]$ such that $x_\alpha \leq_K x_*$. Moreover, x_* is obtained as the limit of a sequence $\{x_n\}_{n\in N} \subset D$.* ■

If $(E(\tau), \{p_\lambda\}_{\lambda\in\Lambda})$ is a locally convex space and $K \subset E$ is a nuclear cone, we know that in this case for every $\lambda \in \Lambda$ there exists $f_\lambda \in K^*$ such that $p_\lambda(x) \leq f_\lambda(x)$, for all $x \in K$. Therefore, in this case, K has the following property: *for all $\lambda \in \Lambda$ there exists $h_\lambda \in K^*$ such that $K \subseteq \{x \in E | h_\lambda(x) + p_\lambda(x) \leq 0\}$.*

Thus it is interesting to consider the following problem.

OPEN PROBLEM *Under what conditions it is possible to extend Theorem 12.61 to nuclear cones?*

12.7 SOME TOPOLOGICAL PROPERTIES OF EFFICIENT POINT SETS

Because, generally, the efficient points are not unique for many practical, vector-valued optimization problems, it is important to know some topological properties of efficient point sets.

This section of the chapter is dedicated to this subject.

(I). DENSITY THEOREMS FOR EFFICIENCY

The starting point, in the study of density theorems for efficiency point sets, is the following classical result proved in 1953 by K.J. Arrow, E.W. Barankin and D. Blackwell.

We recall that, if $A \subset R^n$ is a non-empty subset, then we say that $x_0 \in A$ is supported by a linear functional $f : R^n \to R$, if $f(x_0) \geq f(x)$, for all $x \in A$.

THEOREM 12.62 [ARROW-BARANKIN-BLACKWELL]. *If R^n is equipped with its natural ordering and if A is a closed convex subset in R^n, then the set of points in A that can be supported by strictly positive linear functionals is dense in the set of all efficient points of A.*

PROOF. The reader can find a proof of this theorem in Arrow, Barankin and Blackwell (1953). ■

This theorem has important implications in vector-valued optimization and in mathematical economics. In vector-valued optimization, we are interested in identifying the set of all efficient points of a given set A; the theorem implies that *"nearly every"* efficient point of A can be found as the solution of a suitable single-objective real-valued optimization problem. In mathematical economics, a strictly positive linear functional $f : R^n \to R$ can be regarded as a pricing system, and a point $x_0 \in A$ supported by f represents an efficient allocation of resources that can be sustained by the pricing system f. Hence, the theorem states that under appropriate conditions, "nearly every" efficient allocation of resources can be sustained by a suitable pricing system.

Over the past 30 years, several authors have generalized Arrow-Barankin-Blackwell theorem in various ways.

For example, Hartley (1978), as well as Bitran and Magnanti (1979), showed that Arrow-Barankin-Blackwell theorem remains valid if R^n is partially ordered by an arbitrary closed pointed convex cone.

We note that three mathematical economists, Radner (1967), Majundar (1972) and Peleg (1972), obtained density results in the infinite dimensional normed vector spaces, namely, ℓ_∞, L_∞, and ℓ_p, $1 \le p \le \infty$, respectively. Chichilnisky and Kalman (1980) proved a density theorem in a Hilbert space. Finally, we note that Salz (1976), Borwein (1980,1983), Jahn (1988), Dauer and Gallagher (1990), and Petsche (1990) proved density results in general normed spaces.

Generally, the density results obtained by many authors require strong assumptions concerning the ordering cone and relatively weak assumptions on the set A, or vice versa, relatively weak assumptions on the ordering cone and quite strong assumptions about the set A.

However, very recently some authors proved certain density results in more general situations. In this section, we will present the most significant density results as well as their relation to other density results.

We start with a density result which can be considered as a generalization of the Arrow-Barankin-Blackwell theorem and which is due to Jahn (1988).

THEOREM 12.63. *Let $(E, \|\cdot\|)$ be a Banach space ordered by a Bishop-Phelps cone $K \subset E$. Let S be a nonempty convex subset of E, and let x_* be an efficient (minimum) element of S [i.e., $(x_* - K) \cap S = \{x_*\}$], for which the set $\hat{S} := \{x \in S \mid \|x - x_*\| \le 1\}$ is weakly compact.*

Then, for every $\varepsilon \in]0,1[$ there exists some $\varphi_\varepsilon \in K^*$ and some $x_\varepsilon \in S$ with $\varphi_\varepsilon(x_\varepsilon) \le \varphi_\varepsilon(x)$ for all $x \in S$ such that $\|x_\varepsilon - x_*\| \le \varepsilon$.

PROOF. Because the cone K is a Bishop-Phelps cone, we have that

$$K = \{x \in E \mid \alpha\|x\| \le \varphi(x)\},$$

for some $\varphi \in K^*$ with $\|\varphi\| = 1$ and some $\alpha \in]0,1[$. For an arbitrary $\beta \in]0,\alpha[$ we define the mapping $\varphi_\beta : E \to R$, by

$$\varphi_\beta(x) = \beta\|x - x_*\| + \varphi(x), \qquad \text{for all} \quad x \in E.$$

The mapping φ_β is weakly lower semicontinuous and therefore, it attains its minimum on $\hat{S}$, i.e., there exists some $x_\beta \in \hat{S}$ with

$$\varphi_\beta(x_\beta) \le \varphi_\beta(x), \quad \text{for all} \quad x \in \hat{S}.$$

If in the last inequality we take $x = x_*$, we obtain

$$\beta\|x_\beta - x_*\| + \varphi(x_\beta) \leq \varphi(x_*)$$

which implies

$$\|x_\beta - x_*\| \leq \frac{1}{\beta}\varphi(x_* - x_\beta). \tag{12.15}$$

As a necessary optimality condition there exists a subgradient $\psi_\beta \in E^* \setminus \{0\}$ with

$$\psi_\beta(x - x_\beta) \geq 0, \qquad \text{for all} \quad x \in \hat{S},$$

and

$$\begin{aligned} \psi_\beta(x - x_\beta) &\leq \varphi_\beta(x) - \varphi_\beta(x_\beta) = \\ &= \beta\|x - x_*\| + \varphi(x) - \beta\|x_\beta - x_*\| - \varphi(x_\beta) \leq \\ &\leq \beta\|x - x_\beta\| + \varphi(x - x_\beta), \quad \text{for all} \quad x \in E. \end{aligned}$$

For every $x \in K \setminus \{0\}$ we get

$$\psi_\beta(-x) \leq \beta\| - x\| + \varphi(-x) <$$

$$< \alpha\|x\| - \varphi(x) \leq \varphi(x) - \varphi(x) = 0.$$

Therefore, we have $\psi_\beta \in K^*$ and

$$\psi_\beta(x_\beta) \leq \psi_\beta(x), \qquad \text{for all} \quad x \in \hat{S}. \tag{12.16}$$

Because x_* is assumed to be a minimal element of S, we notice that

$$x_* - x \notin K, \qquad \text{for all} \quad x \in S \setminus \{x_*\}.$$

Using the fact that K is a Bishop-Phelps cone, we deduce

$$\alpha_i\|x_* - x\| > \varphi(x_* - x), \quad \text{for all} \quad x \in S \setminus \{x_*\},$$

or

$$\alpha_i\|x_* - x\| - \varphi(x_* - x) > 0, \quad \text{for all} \quad x \in \hat{S} \setminus \{x_*\}. \tag{12.17}$$

Taking an arbitrary $\varepsilon \in]0, 1[$, we consider the case that the set

$$\hat{\hat{S}} := \left\{x \in \hat{S} \mid \varphi(x_* - x) \geq \frac{\varepsilon}{2}\right\}$$

is non-empty. Because $\hat{\hat{S}}$ is weakly compact and the functional on the left-hand side of inequality (12.17) is weakly lower semicontinuous, there exists some $\delta > 0$ such that

$$\alpha_i\|x_* - x\| - \varphi(x_* - x) > \delta, \quad \text{for all} \quad x \in \hat{\hat{S}}.$$

Then, for all $\beta \geq \alpha_i - \delta$ we have

$$\beta\|x_* - x\| \geq (\alpha_i - \delta)\|x_* - x\| >$$

$$> \varphi(x_* - x) + \delta - \delta\|x_* - x\| \geq$$
$$\geq \varphi(x_* - x), \qquad \text{for all} \qquad x \in \hat{\hat{S}}.$$

From the last inequality and considering (12.15), we have $x_\beta \notin \hat{\hat{S}}$, i.e., $\varphi(x_* - x_\beta) < \frac{\varepsilon}{2}$. In the case that the set $\hat{\hat{S}}$ is empty, it is evident that the previous inequality is satisfied. Hence, from (12.15) we have

$$\|x_\beta - x_*\| \leq \frac{1}{\beta} \cdot \frac{\varepsilon}{2} \leq \varepsilon$$

for $\beta \geq \alpha_i - \delta$ and $\beta \geq \frac{1}{2}$. Since $\varepsilon < 1$ and S is convex, we conclude, considering (12.16), that x_β is also a global minimizer, i.e.,

$$\psi_\beta(x_\beta) \leq \psi_\beta(x), \qquad \text{for all} \qquad x \in S,$$

and the proof is complete. ■

Let $(E, \|\cdot\|)$ be a normed vector space.

DEFINITION 12.49. We say that a non-empty subset D of E is *representable as a Bishop-Phelps cone*, if there is some $\varphi \in E^*$ and a norm $p : E \to R_+$, which is equivalent to $\|\cdot\|$ and $D = \{x \in E \mid p(x) \leq \varphi(x)\}$.

The following result is due to M. Petschke.

THEOREM 12.64. *Let $(E, \|\cdot\|)$ be a real normed vector space. Let D be a non-empty subset of E with $D \neq \{0\}$.*

Then the following assertions are equivalent:

(1). D is representable as a Bishop-Phelps cone;

(2). D is a convex cone with a closed bounded base.

PROOF. For a proof of this result, the reader is referred to Petschke (1990). ■

The following generalization of the Arrow-Barankin-Blackwell theorem is also due to Petschke (1990).

THEOREM 12.65. *Let $(E, \|\cdot\|)$ be a closed pointed convex cone $K \subset E$ with a closed bounded base. Let A be a non-empty convex subset of E and let $x_* \in A$ be an efficient (minimal) element of A such that the set $A_1 = \{x \in A \mid \|x - x_*\| \leq 1\}$ is weakly compact.*

Then, for every $\varepsilon > 0$ there is some $\varphi_\varepsilon \in K^*$ and some $x_* \in A$ such that $\varphi_\varepsilon(x_\varepsilon) = \min_{x \in A} \varphi_\varepsilon(x)$ and $\|x_\varepsilon - x_*\| \leq \varepsilon$.

PROOF. We suppose that x_* is an efficient (minimal) element of A, i.e., $A \cap (x_* - K) = \{x_*\}$. Let $\varepsilon > 0$ be an arbitrary positive real number. By Theorem 12.64, there exists $\varphi \in E^*$ and a norm $p : E \to R_+$ that is equivalent to $\|\cdot\|$ such that

$$K = \{x \in E \mid p(x) \leq \varphi(x)\}.$$

Since p is equivalent to $\|\cdot\|$, there are numbers $m > 0$, $M > 0$ with

$$m\|x\| \leq p(x) \leq M\|x\|, \qquad \text{for every} \quad x \in E.$$

If we denote by $q = \frac{p}{m}$, we have that q is also a norm equivalent to $\|\cdot\|$. Let $D := \{x \in A \mid q(x - x_*) \leq 1\}$. We can show that D is a subset of A_1. The set D is closed since q is equivalent to $\|\cdot\|$. We have that D is a weakly compact set as a closed convex subset of the weakly compact set A_1. Then D is also weakly compact in the space (E, q), since the weak topologies of (E, q) and $(E, \|\cdot\|)$ coincide. Certainly, K^* remains the same for both spaces $(E, \|\cdot\|)$ and (E, q). We define

$$\delta := \sup\left\{\frac{\varphi(x)}{m} \mid x \in E, q(x) \leq 1\right\},$$

$$\alpha_i := \frac{1}{\delta} \in]0, 1[\qquad \text{and} \quad \varphi_0 = \frac{\alpha_i}{m}\varphi.$$

We observe that φ_0 has norm one in the dual space of (E, q) and $K = \{x \in E \mid \alpha_i q(x) \leq \varphi_0(x)\}$. Therefore K is a Bishop-Phelps cone in (E, q), and D is a weakly compact set in (E, q). Applying Theorem 12.63 to the set A as a subset of the space (E, q) we obtain that for $\varepsilon > 0$ there is some x_ε and some $\varphi_\varepsilon \in K^*$ such that

$$\varphi_\varepsilon(x_\varepsilon) \leq \varphi(x) \qquad \text{for every} \qquad x \in A,$$

and $q(x_* - x_\varepsilon) \leq \varepsilon$. Since we have

$$\|x_\varepsilon - x_*\| \leq \frac{1}{m}p(x_\varepsilon - x_*) = q(x_\varepsilon - x_*) \leq \varepsilon,$$

the proof is complete. ■

From Theorem 12.65, we obtain the following two corollaries. The first corollary covers two similar results obtained in Borwein (1980) and in Jahn (1988), respectively. The second corollary covers a similar result obtained in Bitran and Magnanti (1979).

COROLLARY 12.12. *Let $(E, \|\cdot\|)$ be a normed vector space, ordered by a closed pointed well based convex cone, and let $A \subset E$ be a weakly compact convex subset. Then, the set of properly efficient (minimal) elements of A is dense (with respect to the norm topology) in the set of efficient (minimal) elements.* ■

COROLLARY 12.13. *Let R^n be ordered by a closed convex pointed cone. Let $A \subset R^n$ be a closed convex subset of R^n. Then the set of properly efficient (minimal) elements of A is dense in the set of efficient (minimal) elements of A.* ■

Remark. We note that another proof of Theorem 12.65, in the case of locally convex space, was proposed by R.J. Gallagher and O.A. Saleh (see Theorem 3.1 in Gallagher and Saleh (1993).

The following density result is also based on weakly compactness. Before giving this result, we need to recall some definitions.

Let $(E, \|\cdot\|)$ be a normed vector space. We recall that a non-empty subset $A \subset E$ is an *affine* set, if for all $x, y \in A$ and all $\lambda \in R$ we have that $(1-\lambda)x + \lambda y \in A$. If $D \subset E$ is a non-empty subset, we denote by aff(D) the affine hull of D (i.e., the intersection of all affine sets of E containing the subset D). Let $D \subset E$ be a convex set. The *relative interior* (or the *algebraic interior*) of D, denoted by ri(D), is the set of all points of D which are interior with respect to aff(D). It is well-known that if $K \subset E$ is a pointed convex cone, then we have

$$\mathrm{int}(K^*) \subseteq \mathrm{ri}(K^*) \subseteq K^{*i}.$$

Given $A \subset E$ and $x^* \in K^{*i}$, we define

$$\mathcal{P}_A(x^*) = \{x_0 \in A \mid x^*(x_0) \leq x^*(x), \text{ for all } x \in A\},$$

and

$$\mathcal{P}_A = \bigcup_{x^* \in K^{*i}} \mathcal{P}_A(x^*).$$

Obviously, $\mathcal{P}$ is empty whenever K^{*i} is empty.

Denote by $E_m[A, K]$ the set of efficient elements in the sense of minimality, i.e.,

$$E_m[A, K] = \{u \in A \mid (u - K) \cap A = \{u\}\}.$$

We have $\mathcal{P}_A \subseteq E_m[A, K]$. Indeed, if $\mathcal{P}_A$ is empty, then the inclusion is trivial. Otherwise, let $u_0 \in \mathcal{P}_A$. If $u_0 \notin E_m[A, K]$, then there exists $x \in A$ such that $u_0 - x \in K \setminus \{0\}$. Therefore, for every $x^* \in K^{*i}$, we have $x^*(u_0) > x^*(x)$, which implies $u_0 \notin \mathcal{P}_A$. This contradiction proves the inclusion.

Denote by w-cl(A) the closure of A with respect to the weak topology. The following result obtained in 1991 is due to F. Ferrot. We will give this result without proof, since we will also present another, more general theorem which contains the mentioned one as a particular case.

THEOREM 12.66. *Let $(E, \|\cdot\|)$ be a normed vector space and $K \subset E$ a closed pointed convex cone. If $A \subset E$ is a weakly compact convex set and* ri(K^*) *is non-empty, then we have*

$$\mathcal{P}_A \subseteq E_m[A, K] \subseteq \text{w-cl}(\mathcal{P}_A).$$

PROOF. For the proof of this result, the reader is referred to Ferrot (1991). ∎

For the next theorem, we need to recall the following classical minimax theorem due to Ky Fan.

THEOREM 12.67. *Let U, V be compact convex sets, each in a topological vector space, and let $f : U \times V \to R$ be a function. Suppose that for each $v \in V$,*

$f(u,v)$ is a continuous convex function on U and for each $u \in U$, $f(u,v)$ is a continuous concave function on V.

Then, there exists a point $(u_0, v_0) \in U \times V$ satisfying

$$f(u_0, v) \leq f(u_0, v_0) \leq f(u, v_0)$$

for all $u \in U$ and $v \in V$.

PROOF. A proof of this classical result is in Ky Fan (1953). ■

The following theorem is also due to Ferrot.

THEOREM 12.68. *Let $(E, \|\cdot\|)$ be a normed vector space, and let $K \subset E$ be a closed pointed convex cone, such that K^{*i} is non-empty. If A is a weakly compact convex set in E, then $\mathcal{P}_A \subseteq E_m[A, K] \subseteq$ w-cl$(\mathcal{P})$.*

PROOF. Let $u_0 \in E_m[A, K]$ be an arbitrary element. Denote by $A_0 = A - u_0$. We have that A_0 is convex, weakly compact and $0 \in E_m[A, K]$. Let $x_0^* \in K^{*i}$ be such that $\|x_0^*\| < 1$ and define

$$x_n^* = \frac{x_0^*}{n}, \qquad \text{for every} \qquad n \in N \setminus \{0\}.$$

The set

$$K^*(x_n^*) = \{y^* \in x_n^* + K^* \mid \|y^*\| \leq 1\}$$

is non-empty, since $x_m^* \in K^*(x_m^*)$, and we can show that it is convex and contained in K^{*i}. Moreover, it is w^*-compact in E^*, because it is the intersection of the w^*-closed set $x_m^* + K^*$ and the w^*-compact unit ball in E^*. We observe that if $n_1 < n_2$, then

$$K^*(x_{n_1}^*) \subseteq K^*(x_{n_2}^*). \tag{12.18}$$

Let $x^* \in K^*$ be such that $\|x_{n_1}^* + x^*\| \leq 1$. We have

$$x_{n_1}^* + x^* = x_{n_2}^* + \frac{n_2 - n_1}{n_1 n_2} x_0^* + x^* \in K^*(x_{n_2}^*).$$

Consider the function $f : K^*(x_n^*) \times A_0 \to R$ defined by

$$f(y^*, r) = y^*(r).$$

We can show that the function f satisfies all the assumptions of Theorem 12.67 (with f replaced by $-f$). According to this theorem, there exist $y_n^* \in K^*(x_n^*)$ and $r_n \in A_0$ such that

$$y^*(r_n) \leq y_n^*(r_n) \leq y_n^*(r), \tag{12.19}$$

for every $y^* \in K^*(x_n^*)$ and $r \in A_0$. Since $0 \in A_0$, applying (12.19) we deduce that

$$y^*(r_n) \leq 0, \qquad \text{for every} \qquad y_n^* \in K^*(x_n^*).$$

Applying the classical Smulian Theorem [see, for example, Edwards (1965, Theorem 8.12.1)], we have that there exist $r_0 \in A_0$ and a subsequence $\{r_{n_i}\}_{i \in N}$ such that $\{r_{n_i}\} \to r_0$ weakly. Let $x^* \in K^*$ be such that $\|x^*\| < \frac{1}{2}$. Then there exists i_0 such that

$$\|x^*_{n_i} + x^*\| \leq 1, \qquad \text{for} \quad i \geq i_0,$$

and so

$$x^*_{n_i} + x^* \in K^*(x^*_{n_i}), \qquad \text{for} \quad i \geq i_0.$$

Therefore,

$$\left(x^*_{n_i} + x^*\right)(r_{n_k}) \leq 0 \qquad \text{for} \quad k \geq i \geq i_0,$$

and for $k \to +\infty$ we have

$$\left(x^*_{n_i} + x^*\right)(r_0) \leq 0.$$

If $i \to +\infty$, we obtain $x^*(r_0) \leq 0$ for every $x^* \in K^*$ such that $\|x^*\| < \frac{1}{2}$ and so $x^*(r_0) \leq 0$, for every $x^* \in K^*$. This implies that $r_0 \in -K$, and since $0 \in E_m[A, K]$, we obtain that $r_0 = 0$.

Now, we observe that, by the definition of A_0, for every n there exists $u_n \in A$ such that $r_n = u_n - u_0$ and $u_{n_i} \to u_0$ weakly. By (12.19) we have $y^*_{n_i}(u_{n_i}) \leq y^*_{n_i}(u)$, for every $u \in A$, and the proof is complete. ∎

Remark. The conclusion of Theorem 12.68 means that, if $K^{*i} \neq \emptyset$, then almost all efficient (minimal) points of a weakly compact convex set $A \subset E$ can be obtained through a scalarization.

We note that in a recent paper (a preprint) F. Ferrot improved Theorem 12.68 in the sense that if A is weakly compact in a Banach space $(E, \|\cdot\|)$ and K has a bounded base, then in this case we have

$$\mathcal{P}_A \subseteq E_m[A, K] \subseteq \text{cl}(\mathcal{P}_A).$$

The following density result due to R.J. Gallagher and O.A. Saleh is more interesting, because the class of cones used in this theorem is strictly larger than the class of cones used in Theorem 12.68.

THEOREM 12.69. *Let $\langle E, F\rangle$ be a dual system of real vector spaces, such that F is equipped with the topology induced by E, and E is equipped with a topology such that $\langle \cdot, f\rangle$ is continuous for all $f \in F$. Suppose that E is ordered by a D-cone K. If $A \subset E$ is a non-empty subset such that there exists a compact convex subset $C \subset E$ satisfying*

$$A \subset C \subset A - K,$$

then $E[A, K] \subseteq \text{cl}\,(PPE[A, K])$, where $E[A, K]$ is the set of efficient (maximal) points of A with respect to K.

PROOF. We follow the ideas of the proof proposed in Gallagher and Saleh (1993).

We observe that it is sufficient to show that if $0 \in E[A,K]$, then $0 \in \mathrm{cl}(PPE[A,K])$. Indeed, since K is a D-cone, then there exists a non-empty subset D of K^{*i} satisfying equations (1), (2) and (3) in Definition 12.8. Let B be the compact convex subset containing D, and for each $p \in D$ define

$$B(p) = \{f \in B \mid p \leq f\}.$$

Then $B(p)$ is a compact convex subset of F. Since C is also compact and convex, applying Theorem 12.67 we obtain the existence of an element $c_p \in C$ and $f_p \in B(p)$ satisfying

$$\langle c, f_p \rangle \leq \langle c_p, f_p \rangle \leq \langle c_p, f \rangle, \tag{12.20}$$

for all $c \in C$ and all $f \in B(p)$. Since $f_p \in K^{*i}$, it follows that $c_p \in PPE[C,K]$. Also, since $A \subset C \subset A - K$, we have that $c_p \in PPE[A,K]$. [See Gallagher and Saleh (1993,Lemma 2.3)]. Since the set D satisfies condition (2) in Definition 12.8, the pair $(D, \leq)$ is a directed set. Hence, the set $\{c_p\}_{p \in D}$ is a net in $PPE[A,K]$. Because C is compact, the net has a cluster point, say $c_* \in C$. It follows that $c_* \in \mathrm{cl}(PPE[A,K])$. The proof will be finished if we show that $c_* = 0$. Since $0 \in E[A,K] = E[C,K]$, it is sufficient to show that $c_* \in K$. Thus, by (3) in Definition 12.8, we need only show that $\langle c_*, g \rangle \geq 0$ for all $g \in D$. For this, let $g \in D$ and $\varepsilon > 0$ be given. Since c_* is a cluster point of $\{c_p\}_{p \in D}$ and since $\langle \cdot, g \rangle$ is continuous, there exists $r \in D$, $r \leq g$ such that $\langle c_*, g \rangle > \langle c_r, g \rangle - \varepsilon$. Since $r \leq g$ we have $g \in B(r)$. Hence, since $0 \in C$, inequality (12.20) implies that $\langle c_r, g \rangle \geq 0$. Thus, $\langle c_*, g \rangle > -\varepsilon$. Since ε and g are arbitrary, it follows that $\langle c_*, g \rangle \geq 0$ for all $g \in D$. Hence, $c_* = 0$ and the proof is complete. ∎

From Theorem 12.69, we have the following result reproved in 1996 in Wantao (1996,Theorem 3).

COROLLARY 12.14. *Let E be a locally convex space and $K \subset E$ a closed convex cone. Suppose that K has a base. Then for any compact convex set $A \subset E$ we have $E[A,K] \subseteq \mathrm{cl}(PPE[A,K])$.*

We finish this section by citing the following result due also to F. Wantao.

THEOREM 12.70. *Let E be a locally convex space and $K \subset E$ a closed convex cone. Suppose that K has a base. Then for any compact or weakly compact set $A \subset E$, we have*

$$E[A,K] \subseteq \mathrm{cl}\,(GHePE[A,K]).$$

PROOF. For a proof of this result the reader is referred to Wantao (1996). ∎

For other density theorems the reader is referred to the following papers: Zhuang (1994), Song (1997), Gong (1994), Makarov and Rachkovski (1996), and Chen (in press).

(II). DOMINATION PROPERTY

Another problem which is important in the theory of decision making concerns the existence of an efficient alternative which is smaller (with respect to the ordering cone) than a given alternative. This is the *domination property* or something called *external stability* [see (Sawaragi, Nakayama and Tanino (1985)], introduced in Vogel (1977), and investigated by several authors as Benson (in 1983), Luc (in 1984, 1988, 1990), Henig (in 1986), Zheng (in 1997), and others. Clearly, an alternative is either efficient, or else it is dominated by another alternative. However, the preferred alternative may not be efficient. We say that a set of alternatives has the *domination property* if every alternative either is efficient, or else there exists a preferred alternative which is efficient.

First, we consider the case of finite dimensional vector spaces. We consider the vector spaces R^n and R^m and we suppose that $K \subset R^m$ is a nontrivial convex cone. Given a non-empty subset $D \subset R^n$ and a mapping $f : D \to R^m$, we consider the vector maximization problem

$$\begin{cases} VMax \quad f(x) \\ x \in D \end{cases} \tag{12.21}$$

About problem (12.21), we are interested in finding the set D_E of all efficient solutions, i.e.,

$$D_E = \{x_* \in D \mid f(x) - f(x_*) \in K, \text{ for some } x \in D,$$
$$\text{implies} \quad f(x) = f(x_*)\}.$$

DEFINITION 12.50. We say that the vector maximization problem (12.21) satisfies the *domination property* with respect to K, if for each $x \in D$ there exists a point $x_* \in D_E$ such that $f(x_*) - f(x) \in K$.

We recall that the cone K is *acute*, if $\mathrm{cl}K \cap (-\mathrm{cl}K) \subseteq \{0\}$. Obviously, if K is closed and pointed then it is acute.

The following result was remarked in Luc (1984).

THEOREM 12.71. *If the convex cone $K \subset R^m$ is closed and acute and $f(D) - K$ is closed convex, then the following statements are equivalent:*

(1). D_E is non-empty;

(2). The vector maximization problem (12.21) has the domination property.

PROOF. This result is a consequence of Corollary 2.2 proved in Henig [2] (1982). Another direct proof was proposed in Luc (1984). ■

The domination property can be defined also in a general topological vector space. Let E be a topological vector space and $K \subset E$ a closed pointed convex cone. If A is a non-empty subset of E, denote by $E_m[A, K]$ the efficient (minimal) points of A with respect to K. Denote by $\leq$ the ordering defined by K. If $x \geq y$ and $x \neq y$, we denote $x > y$.

DEFINITION 12.51. We say that a set $A \subset E$ has the *domination property,* if for every point $y \in A$ there is some $x \in E_m[A, K]$ such that $x \leq y$.

PROPOSITION 12.23. *A set A has the domination property if and only if $A \subseteq E_m[A, K] + K$.*

PROOF. The proposition is a consequence of the definition. ∎

If $x \in A$, denote by A_x the section

$$A(x; \leq) = \{u \in A \mid u \leq x\} = A \cap (x - K).$$

A net $\{x_\alpha\}_{\alpha \in I}$ in E is said to be *decreasing,* if $x_\alpha > x_\beta$ whenever $\alpha, \beta \in I$ and $\beta > \alpha$.

We say that a set $A \subset E$ is *K-strongly complete* (or *K-complete* since K is supposed to be closed) if it has no cover of the form $\{(x_\alpha - K)^c\}_{\alpha \in I}$, with $\{x_\alpha\}_{\alpha \in I}$ being a decreasing net in A.

THEOREM 12.72. *The domination property is satisfied for $A \subset E$, if and only if for each $y \in A$ there is some $x \in A_y$ such that A_x is K-complete.*

PROOF. The theorem is a particular case of Theorem 4.3 proved in Luc [4] (1989, Chapter 2). ∎

Remark. The assumptions of Theorem 12.72 are satisfied if A is convex, locally compact, $\text{Rec}(A) \cap (-K) = \{0\}$, K is regular and E is boundedly order-complete.

The following result is due to X.Y. Zheng.

THEOREM 12.73. *Let E be a locally convex space, $K \subset E$ a well based closed convex cone and $A \subset E$ a sequentially compact set. Suppose that B is a bounded base of K such that $0 \notin cl(B)$. If there exists $f \in K^*$ lower bounded on A and such that $\inf\{f(b) \mid b \in B\} > 0$, then A satisfies the domination property with respect to K. Moreover, $E_m[A, K]$ is non-empty.*

PROOF. For each $x_0 \in A$ we have

$$\inf\{f(x) \mid x \in A \cap (x_0 - K)\} \geq \inf\{f(x) \mid x \in A\} > -\infty.$$

Hence, for each $\varepsilon > 0$, there is $x_\varepsilon \in A \cap (x_0 - K)$ such that

$$f(x_\varepsilon) < \inf\{f(x) \mid x \in A \cap (x_0 - K)\} + \varepsilon. \tag{12.22}$$

Obviously we have

$$x_\varepsilon \leq x_0. \tag{12.23}$$

Using (12.22) and (12.23) we can construct inductively a sequence $\{x_n\}_{n \in N}$ in A such that

(a): $x_0 \geq x_1 \geq \ldots \geq x_n \geq \ldots$;
(b): $f(x_n) < \inf\{f(x) \mid x \in A \cap (x_{n-1} - K)\} + \frac{1}{n}$, for each $n \in N$.

The sequence $\{x_n\}_{n \in N}$ is a Cauchy sequence in A. Indeed, $\{f(x_n)\}_{n \in N}$ is lower bounded. Let $\alpha_i = \inf\{f(b) \mid b \in B\}$. Then $\alpha_i > 0$. For each neighborhood U of 0, there is $\delta > 0$ such that $0 \leq t \leq \delta$ implies $tB \subset U$ (because B is bounded). By (a), the scalar sequence $\{f(x_n)\}_{n \in N}$ is convergent. Hence, there is an index n_0 such that whenever $n \geq m \geq n_0$, then $f(x_m - x_n) < \alpha_i \delta$. Since, whenever $n \geq m$, then $x_m - x_n \geq 0$, there are $\lambda_{m,n} \geq 0$ and $b_{m,n} \in B$ such that $x_m - x_n = \lambda_{m,n} b_{m,n}$. Therefore, whenever $n \geq m \geq n_0$, then

$$\alpha_i \delta > f(x_m - x_n) = \lambda_{m,n} f(b_{m,n}) \geq \lambda_{m,n} \alpha_i,$$

and so, $0 \leq \lambda_{m,n} < \delta$. Hence, whenever $n \geq m \geq n_0$, then $x_m - x_n = \lambda_{m,n} b_{m,n} \in U$. It follows that $\{x_n\}_{n \in N}$ is a Cauchy sequence in A.

Since A is sequentially complete, there is $x_* \in A$ such that $\lim_{n \to \infty} x_n = x_*$. By (a) and using the fact that K is a closed cone, we have that $x_n \geq x_*$, for all $n \in N$ (which implies also that $x_* \leq x_0$). We claim that $x_* \in E_m[A, K]$. Indeed, if $x_* \notin E_m[A, K]$, there would be $y_* \in A$ such that $y_* \leq x_*$ and $x_* \neq y_*$. This implies that $f(x_*) > f(y_*)$ and $x_n \geq y_*$ for all $n \in N$. Hence, we have $y_* \in A \cap (x_n - K)$. By (b), we have also

$$f(y_*) \geq \inf\{f(x) \mid x \in A \cap (x_n - K)\} > f(x_{n+1}) - \frac{1}{n}.$$

Let $n \to \infty$. One has $f(y_*) \geq f(x_*)$, which is a contradiction and the theorem is proved. ■

The study of domination property is an interesting and important subject, it is a nice subject for new investigations.

(III). CONNECTEDNESS OF EFFICIENT POINT SETS

One of the most important problem in vector optimization problem is to investigate topological properties of efficient point sets. There exist many papers dedicated to this topic. Among others, we cite the following Bergstresser e.a. (1976), Bitran and Magnanti (1979), Choo and Atkins (1983), Luc (1980), Luc [4] (1989), Luc [6],[8] (1987), Luc (1985), Naccache (1978), Morozov (1977), Peleg (1972), Sawaragi e.a. (1985), Warburton (1983), Molho and Zaffaroni (1998), Gong (1994).

An interesting topological property is connectedness. In this section we will give some significant results on this subject.

Let R^n be the Euclidean space and $K \subset R^n$ a closed pointed convex cone. If $A \subset R^n$ is a non-empty subset, the set of efficient (minimal) points of A, with respect to K will be denoted, as in the previous section by $E_m[A, K]$. We recall that $y_* \in E_m[A, K]$, if and only if, $(A - y_*) \cap (-K) = \{0\}$, or $(y_* - K) \cap A = \{0\}$. In this section, we will study the connectedness of the set $E_m[A, K]$.

A set $D \subset R^n$ is said to be *separated* if it can be represented as $D = D_1 \cup D_2$, where $\text{cl}D_1 \cap D_2 = D_1 \cap \text{cl}D_2 = \emptyset$, with two non-empty sets D_1 and

D_2. Equivalently, D is separated if there exist two open sets U_1 and U_2 such that $D \subset U_1 \cup U_2$, $D \cap U_1 \neq \emptyset$, $D \cap U_2 \neq \emptyset$, and $D \cap U_1 \cap U_2 = \emptyset$.

If the set D is not separated, then D is said to be *connected.* We recall also the classical definition of a connected set. We say that a subset D of R^n is connected if and only if, it is impossible to find a representation of D of the form $D = U_1 \cup U_2$, where U_1 and U_2 are open and $U_1 \cap U_2 = \emptyset$.

The following two results are well-known.

LEMMA 12.3. *If $D \subset R^n$ is connected and $D \subseteq A \subseteq \bar{D}$, then the set A is also connected.* ■

LEMMA 12.4. *If $\{D_i\}_{i\in I}$ is a family of connected subsets of R^n such that $\cap_{i\in I} D_i \neq \emptyset$, then $\cup_{i\in I} D_i$ is connected.* ■

Lemmas 12.3 and 12.4 are proved, for example, in (Bourbaki N. *General Topology.* Paris: Hermann, 1965).

Let $D \subset R^n$ be a closed compact subset. We denote by

$$S(\varphi, D) = \left\{ x_* \in D \mid \langle \varphi, x_* \rangle = \inf_{x\in D} \langle \varphi, x \rangle \right\}$$

and by

$$\mathcal{D}(D, K) = \bigcup_{\varphi \in \mathrm{int} K^*} S(\varphi, D).$$

We note that since K is a closed pointed convex cone, it is well based and consequently, K^* has a non-empty interior. We have that

$$\Delta(D, K) \subseteq E_m[A, K] \subseteq \mathrm{cl}\Delta(D, K). \tag{12.24}$$

For a proof of these inclusions the reader is referred to Sawaragi e.a. (1985) or Naccache (1978).

PROPOSITION 12.24. *If D is a compact convex set, then $\Delta(D, K)$ is connected.*

PROOF. Indeed, suppose that $\Delta(D, K)$ is not connected. Then there exist two open sets D_1 and D_2 such that

$$D_i \cap \Delta(D, K) \neq \emptyset, \ i = 1, 2; \quad D_1 \cap D_2 \cap \Delta(D, K) = \emptyset$$

and $\Delta(D, K) \subset D_1 \cap D_2$.

Let

$$F_i = \{\varphi \in \mathrm{int} K^* \mid S(\varphi, D) \cap D_i \neq \emptyset, \} \ i = 1, 2.$$

Since $S(\varphi, D)$ is connected by convexity, we have

$$F_i = \{\varphi \in \mathrm{int} K^* \mid S(\varphi, D) \subset D_i, \} \ i = 1, 2.$$

It is easy to see that $F_1 \cap F_2 = \emptyset$.

Because $D_i \cap \Delta(D,K) \neq \emptyset$, we have

$$F_i \cap \text{int}K^* \neq \emptyset \qquad \text{for} \quad i = 1,2;$$

and because $\Delta(D,K) \subset D_1 \cup D_2$ we observe that $\text{int}K^* \subset F_1 \cup F_2$ (in fact, we have $\text{int}K^* = F_1 \cup F_2$). Now, we can show that F_1 and F_2 are open. To show this fact, it is sufficient to establish that F_1 is open. Suppose the contrary, i.e. that F_1 is not open. In this case, there exist $\varphi_* \in F_1$ and $\varphi_k \in (\text{int}K^*)\backslash F_1 = F_2$ $(k = 1,2,\ldots)$, such that $\{\varphi_k\}_{k\in N} \to \varphi_*$. Let $x_k \in S(\varphi_k, D)$ be arbitrary and $k = 1,2,\ldots$. Since D is compact we can suppose, without loss of generality that the sequence $\{x_k\}_{k\in N}$ converges to an element $x_* \in D$, which is contained in $S(\varphi_*, D)$. Indeed, if $x_* \notin S(\varphi_*, D)$, there exists an element $x' \in D$ such that $\langle \varphi_*, x'\rangle < \langle \varphi_*, x_*\rangle$. Considering the continuity of the inner product we obtain that for sufficiently large k, $\langle \varphi_k, x'\rangle < \langle \varphi_k, x_k\rangle$, which contradicts the facts that $x_k \in S(\varphi_k, D)$. Since $x_k \in S(\varphi_k, D) \subset D_2 \cap \Delta(D,K)$ and $D_1 \cap D_2 \cap \Delta(D,K) = \emptyset$, then $x_k \in \complement D_1$ (the complement of D_1) for all $k = 1,2,\ldots$. Since $\complement D_1$ is a closed set, the limit x_* of $\{x_k\}_{k\in N}$ is also in $\complement D_1$. Therefore, we have $x_* \notin D_1$, which is a contradiction to the fact that $\varphi_* \in F_1$. Consequently, F_1 and F_2 are open and hence $\text{int}K^*$ is not connected, which is a contradiction. Finally, we conclude that $\Delta(D,K)$ should be connected and the proof is complete. ∎

DEFINITION 12.52. We say that a subset D of R^n is K-*compact* if and only if, $(x - K) \cap D$ is compact for all $x \in D$.

THEOREM 12.74. *If K is a closed pointed convex cone with non-empty interior and A is a closed convex K-compact subset of R^n, then the set $E_m[A,K]$ is connected.*

PROOF. Let $u \in \text{int}K$ be an arbitrary element. We define $x(\alpha_i) = \alpha_i u$, $\alpha_i \in R$. We claim that, for any $x \in R^n$, there exists $\alpha_i > 0$ such that

$$x \in x(\alpha) - K.$$

Indeed, if this is not true, then we can apply the classical separation theorem to the ray $\{x - \alpha u \mid \alpha > 0\}$ and the convex cone $-K$ to deduce the existence of $v \in R^n \setminus \{0\}$ such that

$$\langle v, x - \alpha u\rangle \geq 0 \qquad \text{for all} \quad \alpha > 0$$

and

$$\langle v, u'\rangle \geq 0 \qquad \text{for all} \quad u' \in K.$$

Because $u \in \text{int}K$, the latter inequality implies $\langle v, u\rangle > 0$, and we have $\langle v, x - \alpha_i u\rangle < 0$ for sufficiently large α_i which is a contradiction.

Therefore we can choose $\alpha_* > 0$ such that

$$(x(\alpha_*) - K) \cap E_m[A,K] \neq \emptyset.$$

Let

$$A(\alpha) = E_m\left[(x(\alpha) - K) \cap A, K\right].$$

The claim proven shows that

$$E_m[A, K] = \bigcup_{\alpha \geq \alpha_*} A(\alpha).$$

Since $(x(\alpha) - K) \cap A$ is compact convex from the assumptions, $A(\alpha)$ is connected, from Lemma 12.3, Proposition 12.24 and formula (12.24). Moreover, if $\alpha \geq \alpha_*$, then $A(\alpha_*) \subseteq A(\alpha)$, which implies that

$$\bigcap_{\alpha \geq \alpha_*} A(\alpha) = A(\alpha_*) \neq \emptyset.$$

Because $E_m[A, K]$ is represented as the union of a family of connected sets having a non-empty intersection, we obtain that $E_m[A, K]$ is connected and the proof is complete. ∎

We note that Theorem 12.74 is due to P.H. Naccache.

The following result is also based on compactness.

Consider the Euclidean space R^n ordered by the closed pointed convex cone R^n_+. Given a non-empty closed convex set $S \subset R^n$ and n real-valued continuous mappings $f_i : S \to R$, $(i = 1, 2, \ldots, n)$, we consider the vector maximization problem

$$\begin{cases} VMax & (f_1(x), \ldots, f_n(x)) \\ x \in S & \end{cases} \tag{12.25}$$

Denote by $E_M\left[f(S), R^n_+\right]$ the Pareto efficient (maximal) points of $f(S)$ with respect to R^n_+, and we define

$$E[S, f] = \left\{x \in S \mid f(x) \in E_M\left[f(S), R^n_+\right]\right\}.$$

Suppose that each mapping $f_i : S \to R$, $(i = 1, 2, \ldots, n)$ is quasiconcave, i.e., for each real number λ, the set $\{x \in S \mid f_i(x) \geq \lambda\}$ is convex. For each $i = 1, 2, \ldots, n$, we denote by $A_i = f_i(S)$ and we put

$$A = A_1 \times A_2 \times \cdots \times A_n, \qquad \text{and} \quad S(a) = \{x \in S \mid f(x) \geq a\},$$

for each $a \in A$.

THEOREM 12.75. *If $f_1, f_2, \ldots, f_n$ are continuous and quasiconcave on the closed convex set $S \subset R^n$ and for each $a \in A$, the set $S(a)$ is compact, then $E[S, f]$ is non-empty and connected.*

PROOF. For a proof of this result, the reader is referred to Warburton (1983). We note that the proof is based on several technical results. ∎

The following result is related to the connectedness of Pareto efficient points set for a vector optimization problem defined by a set-valued mapping.

Let $X(\tau)$ be a Hausdorff topological vector space and $(Y, \|\cdot\|)$ a Banach space. We suppose that Y is ordered by a closed convex pointed cone $K \subset Y$. If Y^* is the topological dual of Y then the *dual* of K is

$$K^* = \{\varphi \in Y^* \mid \varphi(y) \geq 0, \quad \text{for all} \quad y \in K\}.$$

Let A denote a non-empty subset of X, and let F denote a set-valued mapping from A to Y with $F(x) \neq \emptyset$ for all $x \in A$.

We consider the following vector optimization problem for the set-valued mapping F:

$$V - \min\{F(x) \mid x \in A\}. \tag{12.26}$$

We will denote the image of A under F by $F(A) = \bigcup_{x \in A} F(x)$, and the set of all the efficient solutions of (12.25) will be denoted by

$$\mathrm{Min}(A, K, F) = \bigcup \left\{F^{-1}(y) \mid y \in \mathrm{Min}(F(A), K)\right\},$$

where

$$\mathrm{Min}(F(A), K) = \{y \in F(A) \mid (y - K) \cap F(A) = \{y\}\}.$$

We suppose that K has a base (not necessarily bounded), i.e., there exists a convex set B such that $0 \notin \bar{B}$ and for each $y \in K \setminus \{0\}$, there exist a unique $\lambda > 0$ and an element $b \in B$ such that $y = \lambda b$.

We recall that F is upper semicontinuous at $x_0 \in A$ if for any neighborhood V of $F(x_0)$ in Y there exists a neighborhood U of x_0 in X such that $F(x) \subseteq V$ for all $x \in U \cap A$. If F is upper semicontinuous at every point $x \in A$, then we say that F is upper semicontinuous on A. Also, we say that F is K-*convex* on A if for any $x_1, x_2 \in A$ and $\lambda \in [0, 1]$ we have

$$\lambda F(x_1) + (1 - \lambda)F(x_2) \subseteq F(\lambda x_1 + (1 - \lambda)x_2) + K.$$

The following result is due to A.R. Warburton. Initially this result was proved in finite dimensional vector space.

PROPOSITION 12.25. *Let $(X, \|\cdot\|)$ and $(Y, \|\cdot\|)$ be Banach spaces, $W \subset X$ and $V \subset Y$ non-empty subsets such that W is connected. Let T be an upper semicontinuous set-valued mapping from W to V such that $T(x)$ is connected for each $x \in W$. Then $T(W) = \bigcup_{x \in W} T(x)$ is connected.*

PROOF. Suppose that $T(W)$ is separated, that is, there exist open sets A and B such that

$$T(W) \subset A \cup B; \qquad A_1 = A \cap T(W) \neq \emptyset;$$

$$B_1 = B \cap T(W) \neq \emptyset, \qquad \text{and} \quad A_1 \cap A_2 = \emptyset.$$

Let x be an arbitrary element in W. As $T(x)$ is connected by assumptions and $\emptyset \neq T(x) \subset T(W)$, it follows that either $T(x) \subset A_1$ or $T(x) \subset B_1$, but not both. Let

$$W_A = \{x \in W \mid T(x) \subset A_1\}$$

and

$$W_B = \{x \in W \mid T(x) \subset B_1\}.$$

Obviously, $W = W_A \cup W_B$. Let x_* belong to the closure of W_A (i.e., $x_* \in \text{cl}W_A$). If $x_* \in W_B$, then we have $T(x_*) \subset B_1 \subset B$. Thus, by the upper semicontinuity of T and the fact that $W = W_A \cup W_B$, there exists a neighborhood U of x_* such that

$$T(x) \subset B_1, \qquad \text{for all} \quad x \in U \cap W.$$

But U must contain elements of W_A, as $x_* \in \text{cl}W_A$. Obviously, for such elements, $T(x) \subset A_1$, which contradicts to $T(x) \subset B_1$. Hence, $T(x_*) \subset A_1$, that is, $x_* \in W_A$. Similarly, no point of the closure of W_B belongs to W_A. We have that $W = W_A \cup W_B$ is separated, which is a contradiction, and the proof is complete. ■

The following result is due to X.H. Gong.

THEOREM 12.76. *Let $(X, \|\cdot\|)$ and $(Y, \|\cdot\|)$ be Banach spaces and A a non-empty compact convex set in X. Let K be a closed pointed convex cone in Y with a base B and let F be an upper semicontinuous, K-convex set-valued mapping from A to Y with compact values. Assume that for each $h \in K^*$ and $x \in A$, the set*

$$P(F(x);h) := \{y \in F(x) \mid h(x) = \min\{h(z) \mid z \in F(x)\}\}$$

is a connected set. Then $\text{Min}(A, K, F)$ *is a connected set.*

PROOF. First, we note that by definition we have

$$\text{Min}(A, K, F) = \bigcup \{F^{-1}(y) \mid y \in \text{Min}(F(A), K)\},$$

where $F^{-1}(y) = \{x \in A \mid y \in F(x)\}$.

We define a set-valued mapping $T : \text{Min}(F(A), K) \to A$ by $T(y) = \{x \in A \mid y \in F(x)\}$. Hence,

$$\text{Min}(A, K, F) = \bigcup \{T(y) \mid y \in \text{Min}(F(A), K)\}. \tag{12.27}$$

We can show that for each $y \in \text{Min}(F(A), K)$, $T(y)$ is a convex set. Indeed, let $x_1, x_1 \in T(y)$ and $\lambda \in [0,1]$. Then we have $x_1, x_2 \in A$, $y \in F(x_1)$, $y \in F(x_2)$. Since F is K-convex, $y \in F(\lambda x_1 + (1-\lambda)x_2) + K$, and hence there exists some $k \in K$ such that

$$y - k \in F(\lambda x_1 + (1-\lambda)x_2). \tag{12.28}$$

Then, we have

$$y - k \in (y - K) \cap F(A) = \{y\},$$

and thus, $k = 0$. By (12.28), we have

$$\lambda x_1 + (1-\lambda)x_2 \in T(y).$$

Therefore, $T(y)$ is a convex set, and hence, $T(y)$ is a connected set.

Next, we will show that T is upper semicontinuous on $\mathrm{Min}(F(A),K)$. If not, then there exist some $y_0 \in \mathrm{Min}(F(A),K)$, an open neighborhood U of $T(y_0)$ in X, and a sequence $\{y_n\}_{n\in N}$ in $\mathrm{Min}(F(A),K)$ converging to y_0 such that $T(y_n) \not\subseteq U$, for all $n \in N$. Therefore, there exists some $x_n \in T(y_n)$ such that

$$x_n \notin U, \qquad \text{for each} \quad n \geq 1. \tag{12.29}$$

Since $x_n \in T(y_n)$, we have $y_n \in F(x_n)$, $x_n \in A$. As A is a compact set, we may assume that $\{x_n\}_{n\in N} \to x_0 \in A$. Because by assumptions we can deduce that F has a closed graph, we have that $y_0 \in F(x_0)$. Hence, $x_0 \in T(y_0) \subset U$, which is a contradiction. In view of Gong (1994, Theorem 4.1), we know that $\mathrm{Min}(F(A),K)$ is a connected set. By Proposition 12.25 and (12.27), it follows that $\mathrm{Min}(A,K,F)$ is a connected set. ■

(IV). CONTRACTIBILITY

Let $(E,\|\cdot\|)$ be a normed vector space ordered by a closed pointed convex cone $K \subset E$. The order defined by by K will be denoted by $\leq$. We have $x \leq y$ if and only if $y - x \in K$. With this order one can define the set of efficient (minimal) points of a set

$$X \subset E, \quad E_m[A,K] = \{x \in X \mid (x-K) \cap X = \{x\}\}.$$

We recall that a set $A \subset E$ is called *contractible* if there is a continuous mapping $H : A \times [0,1] \to A$ such that $H(x,1) = x$, $H(x,0) = x_0$, for all $x \in A$ and some $x_0 \in A$.

Let $\varphi : E \to R$ be a mapping. We say that the mapping φ is *level-strictly convex* if for each $\lambda \in R$, the level set $\varphi_\lambda = \{x \in E \mid \varphi(x) \leq \lambda\}$ is strictly convex, i.e., it is convex and for any $x, y \in \varphi_\lambda$, $x \neq y$, we have that $\varphi\left(\frac{x+y}{2}\right) < \lambda$.

In particular, the norm $\|\cdot\|$ of E can be level-strictly convex. In this case we say that $(E,\|\cdot\|)$ is a normed vector space with a norm level-strictly convex. In Functional Analysis, a level-strictly convex norm is called a *strict norm*. Obviously, not every norm on E is strict norm, even in the finite dimensional case. The norm of a Hilbert space is a strict norm. It is well-known that every separable Banach space possesses a strict norm.

The following result is useful in this section and it is due to D.T. Luc. Prior to presenting it, we recall that a mapping $f : E \to R$ is *increasing* if and only if, $x < y$ implies $f(x) < f(y)$ (here $x < y$ means $x \leq y$ and $x \neq y$).

THEOREM 12.77. *Let $(E,\|\cdot\|)$ be a normed vector space with the norm $\|\cdot\|$ being level-strictly convex and let $K \subset E$ be a well based closed convex cone. Then one can construct a continuous mapping φ from E into R which is increasing and level-strictly convex on E.*

PROOF. A proof of this result is in Luc (1980). The proof is long and it is based on several technical results. ■

PROPOSITION 12.26. *Let $(E, \|\cdot\|)$ be a normed vector space with the norm $\|\cdot\|$ being level-strictly convex and let $K \subset E$ be a well based closed convex cone. Let X be a weakly compact set in E with $X+K$ being convex. Then the mapping φ obtained by Theorem 12.77 attains its minimum on X at a unique point.*

PROOF. Considering the properties of φ, we have that every level set of φ is closed convex, hence weakly closed. This means that φ is lower semicontinuous in the weak topology. Applying the Weierstrass Theorem we have that φ attains its minimum on the weakly compact set X. Let λ_0 be the minimal value of φ on X and let $x, y \in X$ with $\varphi(x) = \varphi(y) = \lambda_0$. If $x \neq y$, by the level-strict convexity, we have $\varphi\left(\frac{x+y}{2}\right) < \lambda_0$. Since $X+K$ is convex, the point $\frac{x+y}{2}$ belongs to $X+K$, which means that there are some $u \in X$, $k \in K$, such that $\frac{x+y}{2} = u + k$. Since φ is increasing we have the contradiction $\varphi(z) \leq \varphi\left(\frac{x+y}{2}\right) < \lambda_0$ and the proof is complete. ■

PROPOSITION 12.27. *If X is a closed convex set in E with $X+K$ having a non-empty relative interior* $\mathrm{ri}(X+K)$*, then the set-valued mapping G from $X+K$ to E defined by*

$$G(x) = (x-K) \cap (x+K), \qquad x \in X$$

is continuous at every point of the set $\mathrm{ri}(X+K) \cup E_m[X,K]$.

PROOF. By passing to a subspace if necessary, we may assume that $\mathrm{int}(X+K)$ is non-empty. Obviously G is upper semicontinuous. Now, we show the lower semicontinuity. If $E_m[X,K]$ is non-empty let $x_0 \in E_m[X,K]$ be an arbitrary element. In this case $G(x_0) = \{x_0\}$. Any open set V meeting $G(x_0)$ contains it. Hence, by the upper semicontinuity, there is a neighborhood U of x_0 such that $G(x) \subseteq V$, in particular, $G(x) \cap V \neq \emptyset$, for all $x \in U \cap X$, which means that G is lower semicontinuous at x_0. Now, we consider $x_0 \in \mathrm{int}(X+K)$ and suppose to the contrary that the map is not lower semicontinuous at that point, i.e., there are a neighborhood V of some point $y_0 \in G(x_0)$, and a sequence $\{x_n\}_{n \in N}$ in $X+K$ converging to x_0 such that

$$G(x_n) \cap V = \emptyset, \qquad \text{for all} \quad n \in N. \tag{12.30}$$

Since $x_0 \in \mathrm{int}(X+K)$, there is a convex neighborhood, say W of x_0 with $W \subseteq \mathrm{int}(X+K)$. Let W_0 be the convex hull of y_0 and W. This set belongs to $X+K$, because $y_0 \in X+K$ and $X+K$ is convex. Moreover,

$$y_\lambda = \lambda y_0 + (1-\lambda)x_0 \in (\mathrm{int} W_0) \cap V \tag{12.31}$$

whenever λ is smaller than and close to 1. Let U be a neighborhood of y_λ in the set of the right-hand side of (12.31). Then $x_0 \in U + K$ and there is some n_0 such that

$$x_n \in U + K, \qquad \text{for all} \quad n \geq n_0. \tag{12.32}$$

Let further $u_n \in U$, $k_n \in K$ with $x_n = u_n + k_n$. Since $U \subseteq W_0 \subseteq X + K$, we have that

$$u_n \in X + K, \qquad \text{for all} \quad n \geq n_0. \tag{12.33}$$

It follows from (12.32) that $u_n = x_n - k_n \in x_n - K$, which together with (12.33) gives the inclusion $u_n \in G(x_n)$ for all $n \geq n_0$. This contradicts (12.30) and the proof is complete. ■

PROPOSITION 12.28. *Assume that X is a convex set and $G(x_0)$ is non-empty for some $x_0 \in E$. If φ attains its minimum on $G(x_0)$ at some point y_0, then $y_0 \in E_m[X, K]$. (Here G is the same as in Proposition 12.27.*

PROOF. The proposition is a consequence of the fact that φ is increasing. ■

The following theorem is the main result in this section.

THEOREM 12.78. *Let $(E, \|\cdot\|)$ be a normed vector space with the norm $\|\cdot\|$ being level-strictly convex and $K \subset E$ a well based closed convex cone. Assume that $X \subset E$ is a convex set with the property that $X + K$ has a non-empty relative interior and $(x - K) \cap X$ is weakly compact for each $x \in E$. Then the set $E_m[X, K]$ is contractible.*

PROOF. We consider the mapping $h : X + K \to E_m[X, K]$ defined by

$$\begin{cases} h(x) = \{y \in G(x) \mid \varphi(y) = \min\{\varphi(z) \mid z \in G(x)\}\} \\ \text{for all} \quad x \in X + K, \end{cases} \tag{12.34}$$

where $G(x) = (X - K) \cap (X + K)$, for all $x \in X$, and φ is the mapping defined by Theorem 12.77. Since φ is increasing, we have the equality

$$\min\{\varphi(z) \mid z \in G(x)\} = \min\{\varphi(z) \mid z \in (x - K) \cap X\}.$$

The value in the right-hand side of the above equality is finite because φ is continuous level-strictly convex, hence lower semicontinuous in the weak topology and because the set $(x - K) \cap X$ is weakly compact. By Proposition 12.26, the set in the right-hand side of (12.34) is a single point, which in view of Proposition 12.28 belongs to $E_m[X, K]$. Hence, h is well defined. Moreover, by Proposition 12.27 and by continuiuty of φ, the map h is continuous on $\mathrm{ri}(X + K) \cup E_m[X, K]$. We have also,

$$h(x) = x, \quad \text{for every} \quad x \in E_m[X, K]. \tag{12.35}$$

We consider now the mapping

$$H : E_m[X, K] \times [0, 1] \to E_m[X, K]$$

defined by:

$$H(x, t) = h\left(tx + (1 - t)a\right),$$

where a is an arbitrary but fixed element in $\mathrm{ri}(X+K)$. The mapping H is continuous in both variables. Moreover, considering (12.35), for every $x \in E_m[X,K]$ we have

$$H(x,1) = h(x) = x,$$

$$H(x,0) = h(a) \in E_m[X,K].$$

Therefore $E_m[X,K]$ is a contractible set and the proof is complete. ∎

It is interesting to extend the study of contractibility of the set $E_m[X,K]$ to more general classes of sets X.

12.8 EFFICIENT POINTS AND THE CHOQUET BOUNDARY

In this section, we will study an interesting relation between the efficient (Pareto) points set of a compact set in a locally convex space and the Choquet boundary of the same set with respect to a convenient cone of continuous functions on this set.

The main result of this section is due to V. Postolica and I. Bucur.

Let $E = E(\tau)$ be a Hausdorff locally convex space and let $K \subset E$ be a closed pointed convex cone. We consider on E the order relation $\leq_K$ defined by K, i.e., $x \leq_K y$ if and only if $y - x \in K$. Obviously, this order relation is closed [see Nachbin (1965)], that is, the set

$$G_K = \{(x,y) \in E \times E \mid x \leq_K y\}$$

is a closed subset in $E \times E$. Let X be a non-empty compact subset of E. We recall again that an element $x_* \in X$ is called *efficient (Pareto minimum)* point with respect to K if $X \cap (x_* - K) = \{x_*\}$, or, equivalently, $K \cap (x_* - X) = \{0\}$. We denote by $E_m[X,K]$ the set of all efficient points of X with respect to K.

We recall also some basic considerations in Axiomatic Potential Theory, that is, the definition and some properties of the Choquet boundary for X with respect to a convex cone of continuous functions on X. If S is a convex cone of real continuous functions on X, we say that S is *min-stable* if for every $f_1, f_2 \in S$, it follows that $\inf(f_1, f_2) \in S$. If S is a convex cone of real continuous functions on X such that the constant function on X belongs to S, it is min-stable and it separates the points of X, then on the set $\mathcal{M}_+(X)$ of all positive Radon measures on X we define the following order relation: if $\mu_1, \mu_2 \in \mathcal{M}_+(X)$, then $\mu_1 \leq_S \mu_2$, if and only if, $\mu_1(s) \leq \mu_2(s)$, for all $s \in S$.

A measure $\mu \in \mathcal{M}_+(X)$ is *minimal* with respect to the above order relation, if for any continuous function $f : X \to R$, we have $\mu(Q_S f) = \mu(f)$, where by definition $Q_S f = \inf\{s \in S \mid f \leq s\}$; *cf.* Mokobodski [1] (1967/1968).

In particular, if $x \in X$, then the Dirac measure ε_x is minimal if and only if $\varepsilon_x(Q_S f) = \varepsilon_x(f)$, that is, $Q_S f(x) = f(x)$ for every continuous function $f : X \to R$. The set of all points of $x \in X$ such that ε_x is a minimal measure with respect to $\leq_S$ is named the *Choquet boundary* of X with respect to S, and it is denoted by $\partial_S X$.

Hence, if $S = C(X)$ is the set of all real continuous functions on X, then

$$\partial_S X = \{x \in X \mid Q_S f(x) = f(x), \quad \text{for all} \quad f \in C(X)\}.$$

We recall also that a set $A \subset E$ is a G_δ-set if there exists a denumerable family of open sets $\{U_n\}_{n\in N}$ such that $A = \cap_{n\in N} U_n$.

THEOREM 12.79. *Let $E = E(\tau)$ be a Hausdorff locally convex space, $K \subset E$ a closed pointed convex cone and $A \subset E$ a compact subset. Then $E_m[A, K]$ coincides with the Choquet boundary of A with respect to the convex cone of all real continuous functions on A, which are increasing with respect to the order relation defined by K. Consequently, the set $E_m[A, K]$ endowed with the trace topology τ_A generated by the locally convex topology τ of E is a Baire space. Moreover, if (A, τ_A) is metrizable, then $E_m[A, K]$ is a G_δ-set of (A, τ_A).*

PROOF. Let $S_i = \{f \in C(A) \mid f$ is increasing with respect to $\leq_K\}$. For this special cone we have $\partial_{S_i} A = E_m[A, K]$. Indeed, if $x \in \partial_{S_i} A$ and $x' \in A$ is such that $x' \leq_K x$, then $s(x') \leq s(x)$ for all $s \in S_i$, therefore $\varepsilon_{x'} \leq_{S_i} \varepsilon_x$. Because ε_x is minimal with respect to $\leq_{S_i}$, we deduce that $\varepsilon_{x'} = \varepsilon_x$, that is $x' = x$. Hence $x \in E_m[A, K]$ and $\partial_{S_i} A \subseteq E_m[A, K]$.

Conversely, let $f \in C(A)$ be arbitrary and $f_* : A \to R$ defined by

$$f_*(x) = \sup\{f(x') \mid x' \in A \quad \text{and} \quad x' \leq_K x\}.$$

Obviously, f_* is increasing with respect to $\leq_K$ and since for any $x \in A$ the set $\{x' \in A \mid x' \leq_K x\}$ is compact, it follows that there exists $x^0 \in A$, $x^0 \leq_K x$ such that $f_*(x) = f(x^0)$. Moreover, f_* is upper semicontinuous. Indeed, let $\{x_i\}_{i\in I}$ be a net in A converging to $x \in A$ and for any $i \in I$, let x_i^0 be an element of A such that $x_i^0 \leq_K x_i$ and $f_*(x_i) = f(x_i^0)$. Without loss of generality we may suppose that the net $\{x_i^0\}_{i\in I}$ is also convergent to $y \in A$. Since f is continuous and the order relation $\leq_K$ is closed, we obtain

$$y = \lim_{i\in I} x_i^0 \leq_K \lim_{i\in I} x_i = x,$$

and

$$\lim_{i\in I} f_*(x_i) = \lim_{i\in I} f(x_i^0) = f(y) \leq f_*(x),$$

and therefore f_* is upper semicontinuous. [See also Nachbin (1965, Appendix, Lemma 3)].

We have $f \leq f_*$ and moreover for any continuous function $g : A \to R$ which is increasing with respect to $\leq_K$ with $f \leq g$, it follows that $f_* \leq g$. Indeed, we have

$$f_*(x) = \sup\{f(x') \mid x' \in A \quad \text{and} \quad x' \leq_K x\} \leq$$
$$\leq \sup\{g(x') \mid x' \in A \quad \text{and} \quad x' \leq_K x\} = g(x).$$

In particular, for any $s \in S_i$ with $f \leq s$ we have $s \geq f_*$ and $Q_{S_i} f \geq f_* \geq f$.

On the other hand, applying Nachbin (1965, Appendix, Theorem 3), we deduce that $Q_{S_i} f_* = f_*$. Since $f \leq f_*$ we have

$$f_* \leq Q_{S_i} f \leq Q_{S_i} f_* = f_*.$$

Hence, $Q_{S_i}f = f_*$ for all $f \in C(A)$, which implies

$$\partial_{S_i}A = \{x \in A \mid f(x) = f_*(x), \quad \text{for all} \quad f \in C(A)\}.$$

Let $x_* \in E_m[A, K]$ and $f \in C(A)$. We have

$$Q_{S_i}f(x_*) = f_*(x_*) =$$

$$= \sup\{f(x) \mid x \in A \quad \text{and} \quad x \leq_K x_*\} = f(x^0).$$

Therefore $x_* \in \partial_{S_i}A$ and $E_m[A, K] \subseteq \partial_{S_i}A$. Therefore we have that

$$E_m[A, K] = \{x \in A \mid f(x) = f_*(x), \quad \text{for all } f \in C(A)\} = \partial_{S_i}A.$$

Since it is well-known (see the potential theory) that $\partial_{S_i}A$ endowed with the trace topology of A is a Baire space and if, in addition, A is metrizable, then $\partial_{S_i}A$ is a G_δ-subset of A, the proof is complete. ∎

It is interesting to extend Theorem 12.79 to more general situations.

12.9 NUCLEAR CONES, PARETO EFFICIENCY AND A GEOMETRICAL ASPECT OF EKELAND'S PRINCIPLE

Let $(E(\tau), \text{Spec}(E))$ be a locally convex space, $K \subset E$ a closed pointed convex cone, and $A \subset E$ a non-empty subset. An *efficient* point of A will be a point $x_* \in A$ such that $A \cap (x_* + K) = \{x_*\}$, i.e., x_* is an *efficient maximal* point. The set of all efficient points of A will be denoted by $E_M[A, K]$.

The results of this section are due to G.Isac. The following general existence test for efficient points is a generalization of Theorem 12.67 proved in this chapter.

THEOREM 12.80. *Let $(E(\tau), \text{Spec}(E))$ be a complete locally convex space, $\{p_i\}_{i\in I}$ a base of* Spec(E)*and $K \subset E$ a closed pointed convex cone. If for every seminorm p_i with $i \in I$, there exist a continuous subadditive mapping $\Psi_i : E \to R$ and a real number $\varepsilon_i > 0$ such that*

$$K \subset \{x \in E \mid \Psi_i(x) + \varepsilon_i p_i(x) \leq 0\},$$

then, for every non-empty closed subset $D \subset E$ such that every Ψ_i, with $i \in I$, is bounded from below on D and for every $x_0 \in D$, there exists an element $x_* \in E_M[D, K]$, such that $x_* \in x_0 + K$ (i.e., $x_0 \in x_*$).

PROOF. The proof follows from the proof of Theorem 12.43, by taking $A_0 := (x_0 + K) \cap D$, $A := D$ and replacing f_i by $-\Psi_i$ for every $i \in I$. We note that the subadditivity of Ψ_i $(i \in I)$ is sufficient. ∎

Applying the concept of *nuclear cone*, we obtain from Theorem 12.80 the following result.

PROPOSITION 12.29. *Let $(E(\tau), \mathrm{Spec}(E))$ be a complete locally convex space and $K \subset E$ a closed pointed nuclear cone. Let $\mathcal{B} = \{p_i\}_{i \in I}$ be a base of* Spec(E) *and $\{f_i\}_{i \in I}$ the functionals, both defined by the nuclearity of K. Every non-empty closed subset D of E, such that each functional $f_i (i \in I)$ is bounded from above on D, has the property that for every $x_0 \in D$ there exists an element $x_* \in E_M[D, K]$ such that $x_0 \leq x_*$.*

PROOF. The proposition is a consequence of Theorem 12.80 if we take for every $i \in I$, $\Psi_i = -f_i$. ∎

Now, we recall the Ekeland's ε-variational principle.

THEOREM 12.81. *Let $(E, \|\cdot\|)$ be a Banach space and $f : E \to R \cup \{+\infty\}$ a lower semicontinuous and proper function which is bounded from below. If $x_0 \in E$ is an element such that for some $\varepsilon > 0$,*

$$f(x_0) \leq \inf f + \varepsilon,$$

then, for any $\lambda > 0$, there exists some point $x_* \in E$ such that
(i). $f(x_*) \leq f(x_0)$;
(ii). $\|x_* - x_0\| \leq \lambda$;
(iii). $f(x) > f(x_*) - (\varepsilon/\lambda)\|x - x_*\|$, for all $x \neq x_*$.

We will give a geometrical proof of Ekeland's principle for real-valued functions defined on complete locally convex spaces. This geometrical proof is based on the concept of nuclear cone, and in this way we establish an interesting relation between *Ekeland's principle* and the Pareto efficiency.

Let $(E(\tau), \mathrm{Spec}(E))$ be a complete locally convex space and $f : E \to R \cup \{+\infty\}$ a proper, lower semicontinuous function which is bounded from below. Let $\varepsilon > 0$ be arbitrary and $\{\lambda_i\}_{i \in I}$ a sequence of real numbers such that $\lambda_i > 0$ for all $i \in I$. We recall that the epigraph of f is $\mathrm{epi}(f) = \{(x, t) \in E \times R \mid f(x) \leq t\}$.

DEFINITION 12.53. *We say that a point $x_* \in E$ is an $\{\varepsilon/\lambda_i\}$-extremal point of f, if for every $x \in E$, $x \neq x_*$, there exists $i_0 \in I$ such that*

$$f(x) > f(x_*) - (\varepsilon/\lambda_{i_0})\, p_{i_0}(x - x_*).$$

We denote by $\{\varepsilon/\lambda_{i_0}\}$-Ext$(f)$ the set of $\{\varepsilon/\lambda_{i_0}\}$-extremal points of f.

THEOREM 12.82. *Let $(E(\tau), \mathrm{Spec}(E))$ be a complete locally convex space, $\{p_i\}_{i \in I}$ a base of the* Spec(E), *and $f : E \to R \cup \{+\infty\}$ a proper, lower semicontinuous function which is bounded from below. Then, for every $\varepsilon > 0$ and every sequence $\{\lambda_i\}_{i \in I}$ of strictly positive real numbers, there exists a nuclear closed pointed convex cone K in $E \times R$ such that $E_M[\mathrm{epi}\,(f), K]$ is non-empty and $(x_*, f(x_*)) \in E_M[\mathrm{epi}(f), K]$ if and only if $x_* \in \{\varepsilon/\lambda_i\}$*–Ext$(f)$. *Moreover, for every $x_0 \in E$ satisfying*

$$f(x_0) \leq \inf f + \varepsilon, \tag{12.36}$$

there exists $(x_*, f(x_*)) \in E_M[\text{epi}(f), K]$ *such that*

(i). $f(x_*) \leq f(x_0)$ *and*

(ii). $p_i(x_* - x_0) \leq \lambda_i, \quad (i \in I)$.

PROOF. For every $i \in I$, we define the set

$$K_i = \{(x,t) \in E \times R \mid \lambda_i t + \varepsilon p_i(x) \leq 0\}$$

and $K = \cap_{i \in I} K_i$. Since every K_i is a closed convex cone, we have that K has the same properties. Because $\{p_i\}_{i \in I}$ is a Hausdorff base of Spec(E) we can show that K is pointed. We consider on $E \times R$ the locally convex topology defined by the family of seminorms

$$q_i[(x,t)] = p_i(x) + |t|, \qquad (i \in I),$$

and the family of linear mappings on $E \times R$

$$\Psi_i[(x,t)] = (\varepsilon + \lambda_i)t, \qquad (i \in I).$$

The space $E \times R$ is complete with respect to the topology defined by the family of seminorms $\{q_i\}_{i \in I}$, and with respect to this topology the mappings $\{\Psi_i\}$ are continuous. We have that $K \neq \{(0,0)\}$, since $(0,t) \in K$ for all $t < 0$. Because we can show that

$$K = \bigcap_{i \in I} \{(x,t) \in E \times R \mid \Psi_i[(x,t)] + \varepsilon q_i[(x,t)] \leq 0\},$$

we obtain that K is a nuclear cone if we take for every q_i the linear mapping $f_i = -\Psi_i$. We consider the set D =epi(f) as a subset of $E \times R$. Since f is lower semicontinuous and considering the definition of seminorms q_i, we can show that D is closed in $E \times R$ endowed with the topology defined by the family of seminorms $\{q_i\}_{i \in I}$. Moreover, we have

$$\inf\{\Psi_i[(x,t)] \mid (x,t) \in D\} =$$

$$= \inf\{(\varepsilon + \lambda_i)t \mid f(x) \leq t\} = (\varepsilon + \lambda_i)\inf f,$$

which is a finite number by our assumptions. We can verify that all the assumptions of Proposition 12.29 are satisfied. Hence, by the cited proposition, $E_M[\text{epi}(f), K]$ is non-empty. Let $(x_*, t_*) \in E_M[\text{epi}(f), K]$ be an arbitrary efficient point.

We have

$$y[(x_*, t_*) + K] \cap \text{epi}(f) = \{(x_*, t_*)\}. \tag{12.37}$$

For any $x \in E$, $x \neq x_*$, we have

$$(x - x_*, f(x) - f(x_*)) \notin K,$$

that is, for every $x \neq x_*$, there is $i_0 \in I$ such that

$$\lambda_{i_0}[f(x) - f(x_*)] + \varepsilon p_{i_0}(x - x_*) > 0. \tag{12.38}$$

From (12.38), we deduce that

$$x_* \in \left\{\frac{\varepsilon}{\lambda_i}\right\} - \text{Ext}(f).$$

Conversely, we suppose given $x_* \in \{\varepsilon/\lambda_i\}-\text{Ext}(f)$. If $(x_*, f(x_*)) \notin E_M[\text{epi}(f), K]$, then there exists $(x,t) \in K$ such that $(x,t) \neq (0,0)$ and $(x_*, f(x_*)) + (x,t) \in\text{epi}(f)$. We have $f(x) \leq t$. We observe that $x \neq 0$. Indeed, if we suppose that $x = 0$ and $t \neq 0$, by considering the fact that $(x_* + 0, f(x_*) + t) \in\text{epi}(f)$, we deduce that $t \geq 0$, which is impossible since, from the definition of K we have that $t < 0$. (We consider that $t \neq 0$). Thus, the couple (x,t) must be such that $x \neq 0$. In this case, $x + x_* \neq x_*$ and because we supposed $x_* \in \{\varepsilon/\lambda_i\}-\text{Ext}(f)$, there exists $i_0 \in I$ such that

$$f(x + x_*) > f(x_*) - \frac{\varepsilon}{\lambda_{i_0}} p_{i_0}(x),$$

which implies

$$t \geq f(x) - \frac{\varepsilon}{\lambda_{i_0}} p_{i_0}(x),$$

that is,

$$\lambda_{i_0} t + \varepsilon p_{i_0}(x) > 0,$$

which contradicts the definition of K_{i_0}. Hence, we proved that

$$x_* \in \{\varepsilon/\lambda_i\} - \text{Ext}(f) \quad \text{implies} \quad (x_*, f(x_*)) \in E_M\left[\text{epi}(f), K\right].$$

Let $x_0 \in E$ be an arbitrary element such that inequality (12.36) is satisfied. By Proposition 12.29, there exists $(x_*, t_*) \in\text{epi}(f)$ such that

$$(x_* - x_0, t_* - f(x_0)) \in K, \tag{12.39}$$

and

$$\left[(x_*, t_*) + K\right] \cap \text{epi}(f) = \left\{(x_*, t_*)\right\}. \tag{12.40}$$

From equality (12.40) we deduce that $t_* = f(x_*)$. Indeed, since $(x_*, t_*) \in\text{epi}(f)$, we have that $f(x_*) \leq t_*$. If $f(x_*) < t_*$, then we obtain $f(x_*) = t - \alpha_i$ for some $\alpha_i > 0$, which implies that

$$(x_*, f(x_*)) \neq (x_*, t_*),$$

and

$$(x_*, f(x_*)) = (x_*, t_*) + (0, -\alpha_i) \in (x_*, t_*) + K,$$

which is impossible since $(x_*, f(x_*)) \in\text{epi}(f)$. Hence $t_* = f(x_*)$. From this and (12.39), we have

$$\lambda_i \left[f(x_*) - f(x_0)\right] + \varepsilon p_i(x_* - x_0) \leq 0, \text{ for all } i \in I, \tag{12.41}$$

that is, $f(x_*) \leq f(x_0)$, i.e., conclusion (i) is proved. Finally, applying inequality (12.36), we obtain from (12.41)

$$\varepsilon p_i(x_* - x_0) \leq \lambda_i \left[f(x_0) - f(x_*)\right] \leq \lambda_i \varepsilon,$$

for all $i \in I$, that is, conclusion (ii) is also satisfied and the proof is complete. ■

REFERENCES

Arrow, K.J., Barankin, E.W. and Blackwell, D. 'Admissible Points of Convex Sets'.- In: *Contributions to the Theory of Games, 2.* K.W. Kuhn and A.W. Tucker, eds., Princeton University Press: Princeton, New Jersey, 1953.- P. 87-92

Bacopoulos A, Godini G. and Singer I. Infima of sets in the plane and applications to vectorial optimization. Rev. Roumaine de Math. Pures et Appl. 1978; 23: 343-360

Bahya AO.

1. Étude des cônes nucléaires. Ann. Sci. Math. Quebec. 1991; 15: 123-133

2. Ensembles côniquement bornés et cônes nucléaires dans les éspaces localement convexes séparés'. Thèse (3-eme cycle), Ècole Normale Supérieure, Rabat, Morocco, 1989.

3. Résultats sur des cônes séquentiellement complètement réguliers'.Preprint, 1995.

4. Études des Cônes Nucléaires dans les Éspaces Localement Convexes Séparés. Thèse (Docteur d'Etat Es. Sciences Mathématiques), Université Mohammed V, Faculté des Sciences, Rabat, Morocco, 1997.

Becker R. Sur les cônes (faiblement complets) contenus dans le dual d'un espace de Banach non reflexif. Bull. Austral. Math. Soc. 1989; 39: 321-328.

Benson HP.

1. An improved definition of proper efficiency for vector minimization with respect to cones. J. Math. Anal. Appl. 1979; 71: 232-241.

2. On a domination property for vector maximization with respect to cones. J. Optim. Theory Appl. 1983; 39: 125-132; Errata Corrige, J. Optim. Theory Appl. 1984; 43: 477-479.

3. Efficiency and proper efficiency for vector minimization with respect to cones. J. Math. Anal. Appl. 1983; 93: 273-289.

Benson HP and Morin TL. The vector maximization problem: proper efficiency and stability. SIAM J. Appl. Math. 1977; 32: 64-72.

Bergstresser K, Charnes A and Yu PL. Generalization of domination structures and nondominated solutions in multicriteria decision making. J. Optim. Theory Appl. 1976; 18: 3-13.

Bitran GR and Magnanti TL. The structure of admissible points with respect to cone dominance. J. Optim. Theory Appl. 1979; 29: 573-614.

Borwein JM.

1. Proper efficient points for maximization with respect to cones. SIAM J. Control Opt. 1977; 15: 57-63.

2. The geometry of Pareto efficiency over cones. Math. Operationsforsch. Scr. Optim. 1980; 11: 235-248.

3. On the existence of Pareto efficient points. Math. Oper. Res. 1983; 8: 68-75.

Borwein JM and Zhuang DM.

1. Super-efficiency in vector optimization. ZOR - Methods and Models of Operations Research. 1991; 35: 175-184.

2. Super-efficiency in vector optimization. Trans. Amer. Soc. 1993; 338: 105-122.

Bucur I and Postolica V. A coincidence result between sets of efficient points and Choquet boundaries in separated locally convex spaces. Optimization. 1996; 36: 231-234.

Cesary L and Suryanarayana MB. Existence theorems for Pareto optimization, multivalued and Banach space valued functional. Trans. Amer. Math. Soc. 1978; 244: 37-65.

Chen Guang-Ya. On generalized Arrow-Barankin-Blackwell theorems in locally convex spaces. J. Optim. Theory Appl. (in press)

Chen GY and Huang XX.

1. Stability results for Ekeland's ϵ-varational principle for vector valued functions. Math. Meth. Oper. Res. 1998; 48: 97-103.

2. A unified approach to the existing three types of variational principles for vector valued functions. Math. Meth. Oper. Res. 1998; 48: 349-357.

Chew KL. Maximal points with respect to cone dominance in Banach spaces and their existence. J. Optim. Theory Appl. 1984; 44: 1-53.

Chichilnisky G and Kalman PJ. Applications of functional analysis to models of efficient allocation of economic resourses. J. Optim. Theory Appl. 1980; 30: 19-32.

Choo EV and Atkins DR. Connectedness in multiple linear fractional programming. Management Science. 1983; 29: 250-255.

Choquet, G. *Lectures on Analysis*, Vol. 1-3, Mathematics Lecture Notes Series, New York-Amsterdam: Benjamin, 1969.

Constantinescu G. L'ensemble des fonctions surharmoniques positives sur un espace harmonique est faiblement complet. C.R.Acad. Sci. Paris. 1970; 271: A549-A551.

Corley HW.

1. An existence result for maximization with respect to cones. J. Optim. Theory Appl. 1980; 31: 277-281.

2. Some hybrid fixed point theorems related to optimization. J. Math. Anal. Appl. 1986; 120: 528-532.

Dauer JP and Gallagher RJ. Positive proper efficient points and related cone results in vector optimization theory. SIAM J. Control Opt. 1990; 28: 158-172.

Dauer JP and Saleh OA. A characterization of proper minimal points as solutions of sublinear optimization problems. J. Math. Anal. Appl. 1993; 178: 227-246.

Dauer JP and Stadler W. A survey of vector optimization in infinite dimensional spaces. Part II. J. Optim. Theory Appl. 1986; 51: 205-241.

Durier R and Michelot C. Sets of efficient points in a normed space. J. Math. Anal. Appl. 1986; 117: 506-528.

Edwards, R.E. *Functional Analysis*. New York: Holt Reinhart and Winston, 1965.

Ekeland I.

1. Sur les problèmes variationnels. C.R. Acad. Sci. Paris. 1972; 275: A1057-A1059.

2. On some variational principle. J. Math. Anal. Appl. 1974; 47: 324-354.

3. Nonconvex minimization problems. Bull. Amer. Math. Soc. 1979; 1(3): 443-474.

4. Some lemmas about dynamical systems. Math. Scan. 1983; 52: 262-268.

5. 'The ϵ-variational principle revised' (notes by S. Terracini). - In: *Methods on Nonconvex Analysis*, A. Cellina, ed., Lecture Notes in Math. Berlin-Heidelberg: Springer-Verlag, 1990, Nr.1446.- P. 1-15.

Fan K. Minimax theorems. Proc. Nat. Acad. Sci. U.S.A. 1953; 39: 42-47.

Ferrot F.

1. A generalization of the Arrow-Barankin-Blackwell theorem in normed spaces. J. Math. Anal. Appl. 1991; 158: 47-54.

2. General form of the Arrow-Barankin-Blackwell theorem in normed spaces and in the l^∞-case. J. Math. Anal. Appl. 1993; 79: 127-138.

3. A new ABB theorem in Banach spaces. Preprint, Universita di Genova, Italy, 1999.

Fu WT. On the density of proper efficient points. Proceed. Amer. Math. Soc. 1996; 124: 1213-1217.

Gallagher RJ and Saleh OA. Two generalizations of a theorem of Arrow-Barankin-Blackwell. SIAM J. Control Opt. 1993; 31: 247-256.

Georffrion AM. Proper efficiency and the theory of vector maximization. J. Math. Anal. Appl. 1968; 22: 618-630.

Göpfert A and Tammer Chr.

1. A new maximal point theorem. ZAA. 1995; 14: 379-390.

2. "ϵ-Approximate solutions and conicall support points. A new maximal point theorem". Preprint, University of Halle, Germany, 1997.

3. "Maximal point theorems in product spaces and applications for multicriteria approximation problems". Preprint Nr.26, University of Halle, Germany, 1998.

Göpfert A, Tammer Chr. and Zalinescu C.

1. "On the vectorial Ekeland's variational principle and minimal points in product spaces." Preprint, University of Halle, Germany, 1998.

2. "Maximal point theorems in product spaces." Preprint, University of Halle, Germany, 1998.

Göpfert A, Sekatzek M and Tammer Chr. "Novelties around Ekeland's variational principle." Preprint, University of Halle, Germany, 1999.

Gong XH. Connectedness of efficient solution set for set-valued maps in normed spaces. J. Optim. Theory Appl. 1994; 83: 83-96.

Gotz A and Jahn J. The Lagrange multiplier rule in set-valued optimization. (To appear in SIAM J. Opt.)

Guerraggio A, Molho E and Zaffaroni A. On the notion of proper efficiency in vector optimization. J. Optim. Theory Appl. 1994; 82: 1-21.

Hartley R. On cone-efficiency, cone-convexity and cone-compactness. SIAM J. Appl. Math. 1978; 34: 211-222.

Henig MI.

1. Proper efficiency with respect to cones. J. Optim. Theory Appl. 1982; 36: 387-407.

2. Existence and characterization of efficient decisions with respect to cones. Math. Programming. 1982; 23: 111-116.

Hurwicz L. 'Programming in Linear Spaces'.- In: *Studies in Linear and Nonlinear Programming*, K.J. Arrow, L. Hurwicz and H. Uzawa, eds., Stanford: Standford University Press, 1958.

Hyers, D.H., Isac, G. and Rassias, Th.M. *Topics in Nonlinear Analysis and Applications.* New Jersey-London-Hong Kong: World Scientific, Singapore, 1997.

Isac, G.

1. *Cônes Localment Bornés et Cônes Complètement Réguliers. Application a l'Analyse Nonlinéare.* Seminaire d'Analyse Moderne Nr.17, Université de Sherbrooke, 1980.

2. The (M)-(L) type duality for locally convex lattices. Rev. Roumaine Math. Pures Appl. 1971; 16: 217-223.

3. Sur l'éxistence de l'optimum de Pareto. Rèv. Mat. Univ. Parma. 1983; (4) 9: 303-325.

4. Un critère de sommabilité absolue dans les espaces localement convexes ordonnés: cônes nucléaires. Mathematica. 1983; 25(48): 159-169.

5. Supernormal cones and absoulte summability. Libertas Math. 1985; 5: 17-31.

6. Supernormal cones and fixed point theory. Rocky Mountain J. Math. 1987; 17: 219-226.

7. Pareto optimization in infinite dimensional spaces: the importance of nuclear cones. J. Math. Anal. Appl. 1994; 182: 393-404.

8. 'The Ekeland's principle and Pareto ϵ-efficiency'. - In: *Multi-Objective Programming and Goal Programming*, M. Tamiz, ed., Lecture Notes in Econom. Math. Systems Nr.432, Springer-Verlag, 1996. - P. 148-162.

9. Ekeland's principle and nuclear cones: a geometrical aspect. Math. Comput. Modelling. 1997; 26: 111-116.

10. "On Pareto efficiency. A general constructive existence principle." Preprint, 1998.

Isac, G. and Postolica, V. *The Best Approximation and Optimization in Locally Convex Spaces*, Verlag Peter Lang: Approximation and Optimization, Vol.2, Frankfurt am Main, 1993.

Isermann H. Proper efficiency and the linear vector maximum problem. Operations Res. 1974; 22: 189-191.

Jahn J.

1. Existence theorems in vector optimization. J. Optim. Theory Appl. 1986; 50: 397-406.

2. *Mathematical Vector Optimization in Partially Ordered Linear Spaces.* Frankfurt: Peter Lang, 1986.

3. A generalization of a theorem of Arrow, Barankin and Blackwell. SIAM J. Control and Opt. 1988; 26: 999-1005.

Jameson, G.O.*Ordered Linear Spaces*, Springer-Verlag, Lecture Notes in Mathematics Nr.141, 1970.

Karlin S. Positive operators. J. Math. Mech. 1959; 8: 907-937.

Klinger A. Improper solutions of the vector maximum problem. Operations Research. 1967; 15: 570-572.

Krasnoselskii, M.A., Lifshits, Je.A. and Sobolev, A.V. *Positive Linear Systems. The Method of Positive Operators.* Berlin: Heldermann Verlag, 1989.

Krein MG and Rutman MA. Linear operators which leave a cone in a Banach space invariant. Uspeki Mat. Nauk. 1948; 3: 3-95.

Kuhn HW and Tucker AW. 'Nonlinear Programming'.- In: *Proceedings of the 2-nd Berkeley Symposium on Mathematical Statistics and Probability*, J. Neyman, ed., University of California Press, Berkeley, Ca, 1950. P. 481-492.

Loridan P. ϵ-solution in vector minimization problems. J. Opt. Theory Appl. 1984; 42: 265-276.

Luc DT.

1. An existence theorem in vector optimization. Math. Oper. Res. 1989; 14: 693-699.

2. Contractibility of efficient point sets in normed spaces. Nonlinear Anal. Theory Math. Appl. 1990; 15: 527-535.

3. Recession cones and the domination property in vector optimization. Math. Programming. 1991; 49: 113-122.

4. *Theory of Vector Optimization*, Lecture Notes in Economics and Math. Systems, Vol.319. New York - Berlin: Springer-Verlag, 1989.

5. On the domination property in vector optimization. J. Opt. Theory Appl. 1984; 43: 327-330.

6. Connectedness of the efficient point sets in quasiconcave vector maximization. J. Math. Anal. Appl. 1987; 12: 346-354.

7. Structure of the efficient point set. Proc. Amer. Math. Soc. 1985; 19: 433-440.

8. Scalarization of vector optimization problems. J. Optim. Theory Appl. 1987; 55: 85-102.

Majundar M. Some general theorems on efficiency prices with an infinite dimensional commodity space. J. Econom. Theory. 1972; 5: 1-13.

Makarov EK and Rachkovski NN. Density theorems for generalized Henig proper efficiency. J. Optim. Theory Appl. 1996; 91: 419-437.

Marinescu, Gh. *Espaces Pseudotopologiques et Théorie des Distributions*. Berlin: D.V.W., 1963.

Molho, E. and Zaffaroni, A. 'Quasiconvexity of sets and connectedness of the efficient frontier in ordered vector spaces'.- In: *Generalized Convexity, Generalized Monotonicity: Recent Results*, J.P. Crouzeix et al., eds., Dordrecht: Kluwer Academic Publishers, 1998.- P. 408-424.

Mokobodski, G.

1. 'Cônes normaux et éspaces nucléaires: Cônes semi-complets'. - In: *Seminaire Choquet, Iniation a l'Analyse 7-ème Année*, 1967/1968, B-6.

2. 'Principe de Balayage, principe de domination'. In: *Seminaire de Choquet 1.*, 1967/1968, B-6.

Morozov VV. On properties of the set of nondominated vectors. Vestnik Moskov. Univ. Comput. Sci. Cyber. 1977; 4: 51-61.

Naccache PH. Connectedness of the set of nondominated outcomes in multicriteria optimization. J. Optim. Theory Appl. 1978; 25: 459-467.

Nachbin, L. *Topology and Order*. New York: Van Nostrand, 1965.

Németh AB. Between Pareto efficiency and Pareto ϵ-efficiency. Optimization. 1989; 20: 615-637.

Pareto, V.

1. *Manuale di Economia Politica*. Società Editrice Libraria, Milano, Italy (1906); Piccola Biblioteca Scientifica Nr.13, Societa Editrice Libraria Milano, Italy, 1919.

2. *Sociological Writings*. Selected and introduced by S. E. Finer, Translated by D. Mirfin, Frederick A. Praeger, New York, New York, 1966.

Peleg B. Topological properties of the efficient points set. Proc. Amer. Math. Soc. 1972; 35: 531-536.

Penot JP. L'optimization à la Pareto: Deux ou trois chose que je sais d'elle. Publ. Math. Univ. Pan, 1978.

Peressini, A.L. *Ordered Topological Vector Spaces*. New York: Harper and Row Publisher, 1967.

Petschke M. On a theorem of Arrow, Barankin and Blackwell. SIAM J. Control Opt. 1990; 28: 395-401.

Pontini C. Inclusion theorems for non-explosive and strongly exposed cones in normed spaces. J. Math. Anal. Appl. 1990; 148: 275-286.

Popovici N. 'Contribution à l'Optimization Vectorielle', Thèse, Université de Limoges, France, 1995.

Postolica V.

1. Vectorial optimization programs with multifunctions and duality. Ann. Sci. Math. Quebec 1986; 10: 85-102.

2. 'Some existence results concerning the efficient points in locally convex spaces'. - In: *Babes-Bolyai Univ. Faculty of Math., Seminar on Math. Analysis*, 1987. -P. 75-80.

3. 'Existence results for the efficient points in locally convex spaces ordered by supernormal cones and conically bounded sets'. - In: *Babes-Bolyai Univ. Faculty of Math., Seminar on Math. Anal.*, 1988. - P. 187-192.

4. Existence conditions of efficient points for multifunctions with values in locally convex spaces. Stud. Cerc. Mat. 1989; 41: 325-331.

5. New existence results for efficient points in locally convex spaces ordered by supernormal cones. J. Global Optim. 1993; 3: 233-242.

6. Properties of Pareto sets in locally convex spaces. Optimization 1995; 34: 223-229.

7. An extension to sets of supernormal cones and generalized subdifferential. Optimization. 1994; 29: 131-139.

8. "Recent conditions for Pareto efficiency in locally convex spaces". Preprint, 1997.

Preocupanu Th. Scalar minimax properties in vectorial optimization. International Series of Numerical Mathematics. 1992; 107: 299-306.

Radner, R. 'A note on maximal points of convex sets in l^{∞}'. - In: *Proc. Fifth Berkeley Symposium on Mathematical Statistics and Probability*, University of California Press, Berkeley, CA, 1967. - P. 351-354.

Rudin, W. *Functional Analysis*. New York: McGraw-Hill, 1973.

Salz W. Eine topologische Eigenschaft der effizienten Punkte konvexer Mengen. Operat. Res. Verfahren. 1976; 23: 197-203.

Sawaragi, Y., Nakayama, H. and Tanino, T. *Theory of Multiobjective Optimization*. Academic Press, 1985.

Schäffer, H.H. *Topological Vector Spaces*. New York: MacMillan Company, 1966.

Song W. Generalization of the Arrow-Barankin-Blackwell theorem in a dual space setting. J. Optim. Theory Appl. 1997; 95: 225-230.

Stadler W. A survey of multicriteria optimization or the vector maximum problem. Part (I): 1776-1960. J. Optim. Theory Appl. 1979; 29: 1-52.

Staib T. On two generalization of Pareto minimality. J. Optim. Theory Appl. 1988; 59: 289-306.

Sterna-Karwat A.

1. On existence of cone-maximal points in a real topological linear space. Israel J. Math. 1986; 54: 33-41.

2. 'On the existence of cone-efficient points'. - In: *Recent Advances and Historical Development of Vector Optimization*, J. Jahn and W. Krabs, eds., Lecture Notes in Economics and Mathematical Systems, Nr. 294, Springer-Verlag, 1986. - P. 233-240.

Tammer, Chr.

1. 'Existence results and necessary conditions for ϵ-efficient elements'. -In: *Multicriteria Decision-Proceedings of the 14-th Meeting of the German Working Group "Mehrkriterielle Entsch"*, Brosowski e.a., eds., Frankfurt/Main-Bern: Lang-Verlag, 1993. - P. 97-110.

2. "A variational principle and applications for vectorial control approximation problems". Report Nr.9, 1996, Martin-Luher-Universitat, Halle-Wittenberg, Germany.

3. "Approximate solutions of vector-valued control-approximation problems". Report Nr.27, 1996, Martin-Luher-Universitat, Halle-Wittenberg, Germany.

Tang Y. Conditions for constrained efficient solutions of multiobjective problems in Banach spaces. J. Math. Anal. Appl. 1983; 96: 505-519.

Trevis, F. *Locally Convex Spaces and Linear Partial Differential Equations*. New York: Springer-Verlag, 1967.

Truong XDH.

1. On the existence of efficient points in locally convex spaces. J. Global Optim. 1994; 4: 265-278.

2. A note on a class of cones ensuring the existence of efficient points in bounded complete sets. Optimization. 1994; 31: 141-152.

3. Cones admitting strictly positive functionals and scalarization of some vector optimization problems. J. Optim. Theory Appl. 1997; 93: 355-372.

4. "Existence and density results for proper efficiency in cone compact sets". Preprint, Hanoi Institute of Mathematics, 1999.

Valyi, I. 'Approximate solutions of vector optimization problems'. - In: *Systems Analysis and Simulation*, A. Sydow, M. Thoma and R. Vychnevetski, eds., Berlin DDR: Akademic-Verlag, Vol. 1, 1985. - P. 246-250.

Varaiya PP. Nonlinear programming in Banach spaces. SIAM J. Appl. Math. 1967; 15: 284-293.

Vogel, W. 'Vektoroptimierung in Produkträumen', *Math. Systems in Economics* 35, Verlag Anton Hain, Meisenheim am Glan, 1977.

Wagner DH. Semi-compactness with respect to an Euclidean cone. Canad. J. Math. 1997; 29: 29-36.

Wantao F. On the density of proper efficient points. Proc. Amer. Math. Soc. 1996; 124: 1213-1217.

Warburton AR. Quasiconcave vector maximization: connectedness of the sets of Pareto optimal and weak Pareto optimal alternatives. J. Optim. Theory Appl. 1983; 40: 537-557.

Wong, Y.Ch. *Schwartz Spaces, Nuclear Spaces and Tensor Product*, Lecture Notes in Math. Vol. 726. New York - Berlin: Springer-Verlag, 1979.

Yu PL. Cone convexity, cone extreme points and nondominated solutions in decision problems with multiobjective. J. Optim. Theory Appl. 1974; 14: 319-377.

Zheng XY.

1. Proper efficiency in locally convex topological vector spaces. J. Optim. Theory Appl. 1997; 94: 469-486.

2. The domination property for efficiency in locally convex topological vector spaces. J. Math. Anal. Appl. 1997; 213: 455-467.

Zhuang DM.

1. Bases of convex cones and Borwein's proper efficiency. J. Optim. Theory Appl. 1991; 71: 613-620.

2. Density results for proper efficiency. SIAM J. Control Opt. 1994; 32: 51-58.

Chapter 13

APPROXIMATIVE EFFICIENCY

A new direction in the study of efficiency, which is now in developing, is the study of ε-efficiency or of other forms of approximative efficiency. In particular, the ε-efficiency is related to the study of ε-solutions in vector optimization problems. Begining with the paper of Loridan (1984) several concepts for approximately efficient solutions of a vector optimization problem were published in the last years. We mention the works by Németh (1986), Staib (1988), Valyi (1985), Gerth (Tammer) (1978), Tammer (1992), Tammer [1]-[3] (1993), Helbig e.a. (1992), Isac (1984, 1986), Göpfert and Tammer [1],[3] (1995), Göpfert and Tammer (1998) among others. This chapter is dedicated to the study of approximative efficiency.

13.1 ε-EFFICIENT POINTS

In this section we will introduce the notion of *ε-efficient point* for a set in R^n. For this, we need to recall the following notations.

Let $(R^n, +, \cdot, R^n_+)$ be the Euclidean space ordered by the closed pointed convex cone R^n_+.

Let $x = (x_1, x_2, ..., x_n), y = (y_1, y_2, ..., y_n)$ be vectors in R^n. Then,

1. $x < y \Leftrightarrow x_i < y_i$ for all $i = 1, 2, ..., n$

2. $x \leq y \Leftrightarrow x_i \leqq y_i$ for all $i = 1, 2, ..., n$ and $x \neq y$

3. $x \leqq y \Leftrightarrow x_i \leqq y_i$ for all $i = 1, 2, ..., n$

The zero vector in R^n will be denoted by 0. We recall also that

$$R^n_+ = \{x \in R^n | x \geqq 0\} \text{ and } R^n_- = \{x \in R^n | x \leqq 0\}.$$

The interior of R^n_+ is $\text{int} R^n_+ = \{x \in R^n | x > 0\}$.

Let ε be an element of R^n_+ and a subset A of R^n.

DEFINITION 13.1. We say that an element $x_* \in R^n$ is a *lower ε-efficient point* of A if $x_* \in A$ and there is no $x \in A$ such that $x \leq x_* - \varepsilon$.
We denote by ε-$E_m[A, R^n_+]$ the set of all lower ε-efficient points of A.

Remark. If $\varepsilon = 0$, Definition 13.1 is the definition of an *efficient* (minimal) point of A.

DEFINITION 13.2. We say that an element $x_* \in R^n$ is an *upper ε-efficient point* of A if $x_* \in A$ and there is no $x \in A$ such that $x \geq x_* + \varepsilon$.
The set of all upper ε-efficient points in A will be denoted by ε-$E_M[A, R^n_+]$.
We can show easily the following properties:

(a). $\varepsilon\text{-}E_m[A, R^n_+] = -[\varepsilon\text{-}E_M[A, R^n_+]]$;

(b). $\varepsilon\text{-}E_m[A, R^n_+] = \varepsilon\text{-}E_m[A + R^n_+], R^n_+]$;

(c). $\varepsilon\text{-}E_M[A, R^n_+] = \varepsilon\text{-}E_m[A - R^n_+, R^n_+]$;

(d). $\varepsilon\text{-}E_m[x_0 + A, R^n_+] = x^0 + [\varepsilon\text{-}E_m[A, R^n_+]]$;

(e). $\varepsilon\text{-}E_M[x_0 + A, R^n_+] = x^0 + [\varepsilon\text{-}E_m[A, R^n_+]]$.

PROPOSITION 13.1. *Let ε be an element of R^n_+ such that $\varepsilon \neq 0$. Let A be an arbitrary non-empty subset of R^n. If there exists $\lambda \in intR^n_+$ such that* $\inf\{\langle\lambda, x\rangle | x \in A\} > -\infty$, *then ε-$E_m[A, R^n_+]$ is non-empty.*

PROOF. We consider the set

$$D(\lambda, \varepsilon) = \{x_* \in A \mid \langle\lambda, x_*\rangle \leqq \inf\langle\lambda, x\rangle + \langle\lambda, \varepsilon\rangle\}.$$

From assumptions we have that $D(\lambda, \varepsilon) \neq \emptyset$. If we show that $D(\lambda, \varepsilon) \subset \varepsilon$-$E_m[A, R^n_+]$ then the proposition will be proved. Indeed, assume that there exists $x_* \in D(\lambda, \varepsilon)$ which is not in ε-$E_m[A, R^n_+]$. Then there is $x_0 \in A$ such that $x_0 \leq x_* - \varepsilon$. In this case we have $\langle\lambda, x_0\rangle < \langle\lambda, x_*\rangle - \langle\lambda, \varepsilon\rangle \leqq \inf\langle\lambda, x\rangle$. We have a contradiction and the proof is complete. ∎

We recall that a set $A \subset R^n$ is called R^n_+-bounded if there exists a point $b \in R^n$ such that $A \subset b + R^n_+$.

PROPOSITION 13.2. *If $\varepsilon \in R^n_+ \setminus \{0\}$, then for any non-empty R^n_+-bounded set $A \subset R^n$, we have ε-$E_m[A, R^n_+]$ is non-empty.*

PROOF. Because A is R^n_+-bounded we have that there exists $x^0 \in R^n$ such that

$$\inf\{\langle\lambda, x\rangle \mid x \in A\} \geqq \langle\lambda, x^0\rangle > -\infty$$

for any $\lambda \in \text{int}R^n_+$. [See also Tanino and Sawaragi (1980)]. Now the proposition is a consequence of Proposition 13.1. ■

Let E be an arbitrary vector space and X a non-empty subset of E. Consider as given a real n-dimensional vector-valued function $f = (f_1, f_2, ..., f_n)$ defined on X, i.e. $f : X \to R^n$. We are concerned with the following vector minimization problem

$$(P) : \underline{Eff}\{f(x)|x \in X\}.$$

Denote by $\underline{Eff}(P)$ the set of lower efficient values for the problem (P). Every $x \in X$ such that $f(x) \in \underline{Eff}(P)$ is called a *Pareto solution of (P)*. For $\varepsilon \in \text{int}R^n_+$ we will consider the set

$$\underline{\varepsilon\text{-}Eff}(P) = \varepsilon\text{-}E_m[f(X), R^n_+].$$

DEFINITION 13.3. We say that an element x_ε is an *ε-Pareto solution* of problem (P) if and only if the following conditions are satisfied:

1. $x_\varepsilon \in X$,
2. $f(x_\varepsilon) \in \underline{\varepsilon\text{-}Eff}(P)$.

Remark. In 1979, S.S. Kutateladze gave an analogous definition to Definition 13.3. His definition is the following: *x_ε is an ε-Pareto solution of the problem (P) if for any $x \in X$ such that $f(x) - f(x_\varepsilon) \leqq \varepsilon$, we have $f(x) = f(x_\varepsilon) - \varepsilon$.* For more details about this definition and its applications the reader is referred to Kutateladze (1979).

Obviously, if $n = 1$, we obtain the definition of an ε-solution in classical mathematical programming problem; *cf.* Loridan (1982), Strodiot e.a. (1980).

THEOREM 13.1. *Let $\varepsilon = (\varepsilon_1, \varepsilon_2, ..., \varepsilon_n) \in R^n_+$ be given. Then x_ε is an ε-Pareto solution of the problem (P) if and only if:*

1. $x_\varepsilon \in X$,

2. $\sum_{i=1}^{n} f_i(x_\varepsilon) \leqq \sum_{i=1}^{n} f_i(x) + \sum_{i=1}^{n} \varepsilon_i$
for all $x \in X(x_\varepsilon, \varepsilon) = \{x \in X \mid f(x) \leqq f(x_\varepsilon) - \varepsilon\}$.

PROOF. As x_ε is not in $X(x_\varepsilon, \varepsilon)$, x_ε it is called an *almost* $\left(\sum_{i=1}^{n} \varepsilon_i\right)$*-solution* of the following optimization problem [see Loridan (1980)]:

$$(Q) : \inf \left\{ \sum_{i=1}^{n} f_i(x) \mid x \in X(x_\varepsilon, \varepsilon) \right\}.$$

If we suppose that x_ε satisfies (1) and (2), then from (2) we have:

$$\sum_{i=1}^{n} f_i(x_\varepsilon) = \sum_{i=1}^{n} f_i(x) + \sum_{i=1}^{n} \varepsilon_i,$$

and we can verify that

$$f_i(x_\varepsilon) = f_i(x) + \varepsilon_i,\ 1 \leq i \leq n.$$

Hence from previous remark, we have that x_ε is an ε-Pareto solution of the problem (P).

If x_ε is an ε-Pareto solution of the problem (P), then for every $x \in X$ such that

$$f_i(x) \leq f_i(x_\varepsilon) - \varepsilon_i,\ 1 \leq i \leq n,$$

we obtain $f_i(x) + \varepsilon_i = f_i(x_\varepsilon)$ and x_ε is an almost $\left(\sum_{i=1}^{n} \varepsilon_i\right)$-solution of the problem (Q). The proof is complete. ■

We suppose again that $\varepsilon \in R^n_+ \setminus \{0\}$.

THEOREM 13.2. *If there exists $\alpha \in intR^n_+$ such that $\sum_{i=1}^{n} \alpha_i f_i(x)$ is bounded from below on X, then we have*

1. *$\underline{\varepsilon\text{-}Eff}(P)$ is non-empty*

2. *every $\left(\sum_{i=1}^{n} \varepsilon_i\right)$-solution of the minimization problem*

$$\inf\left\{\sum_{i=1}^{n} \alpha_i f_i(x) \mid x \in X\right\}$$

is an ε-Pareto solution of the problem (P).

PROOF. If we denote by $A = \{f(x) \mid x \in X\}$, the by definition we have

$$\varepsilon\text{-}E_m[A, R^n_+] = \underline{\varepsilon\text{-}Eff}(P)$$

and applying Proposition 13.1 we have that $\underline{\varepsilon\text{-}Eff}(P)$ is non-empty.

Let x_ε be a $\left(\sum_{i=1}^{n} \varepsilon_i\right)$-solution of the minimization problem

$$\inf\left\{\sum_{i=1}^{n} \alpha_i f_i(x) \mid x \in X\right\}.$$

If we suppose that x_ε is such that

$$f(x_\varepsilon) \notin \underline{\varepsilon\text{-}Eff}(P),$$

then there exists $x \in X$ such that

$$f_i(x) \leqq fi(x_\varepsilon) - \varepsilon_i,\ 1 \leqq i \leqq p$$

with at least one strict inequality and

$$\sum_{i=1}^{n} \alpha_i f_i(x) < \sum_{i=1}^{n} \alpha_i f_i(x_\varepsilon) - \sum_{i=1}^{n} \varepsilon_i,$$

which contradicts the fact that x_ε is a $\left(\sum_{i=1}^{n} \varepsilon_i\right)$-solution and thus completes the proof. ■

COROLLARY 13.1. *If for each $i = 1, 2, ..., n$, the mapping f_i is bounded from below on X, then $\underline{\varepsilon\text{-}Eff}(P)$ is non-empty.*

PROOF. Obviously, for every $\alpha \in \text{int} R_+^n$ we have that the mapping $\left(\sum_{i=1}^{n} \alpha_i f_i(x)\right)$ is bounded from below on X and we apply Theorem 13.2. ■

THEOREM 13.3. *If for each $i = 1, 2, ..., n$, the mapping f_i is bounded from below on X, then for every $x_0 \in X$ which is not a Pareto solution of the problem (P), there exists an ε-Pareto solution x_ε dominated by x_0, i.e. $f(x_\varepsilon) \leqq f(x_0)$.*

PROOF. We consider the set

$$X(x_0) = \{x \in X \mid f_i(x) \leqq f_i(x_0), 1 \leqq i \leqq n\} =$$
$$= \{x \in X \mid f(x) \leqq f(x_0)\}.$$

We observe that there exists $x_\varepsilon \in X(x_0)$ such that

$$\sum_{i=1}^{n} f_i(x_\varepsilon) \leqq \inf_{x \in X(x_0)} \sum_{i=1}^{n} f_i(x) + \sum_{i=1}^{n} \varepsilon_i.$$

If we suppose that x_ε is not an ε-Pareto solution, then there exists $u \in X$ such that $f(u) \leq f(x_\varepsilon) - \varepsilon$, that is, $\sum_{i=1}^{n} f_i(u) \leqq f_i(x_\varepsilon) - \varepsilon_i, 1 \leqq i \leqq n$, with at least one strict inequality. Hence,

$$\sum_{i=1}^{n} f_i(u) < \sum_{i=1}^{n} f_i(x_\varepsilon) - \sum_{i=1}^{n} \varepsilon_i \leqq \inf_{x \in X(x_0)} \sum_{i=1}^{n} f_i(x).$$

As $x_\varepsilon \in X(x_0)$ we also have $u \in X(x_0)$ and hence the inequality

$$\sum_{i=1}^{n} f_i(u) < \inf_{x \in X(x_0)} \sum_{i=1}^{n} f_i(x)$$

is a contradiction and the proof is complete. ■

The next result is similar to Theorem 13.3.

THEOREM 13.4. *If for each $i = 1, 2, ..., n$, the mapping f_i is bounded from below on X, then for every $x_0 \in X$ which is not an ε-Pareto solution of the problem (P), there exists an ε-Pareto solution x_ε which is dominated by x_0, up to ε, that is $f(x_\varepsilon) \leqq f(x_0) - \varepsilon$.*

PROOF. The proof is similar to the proof of Theorem 13.3 by making use of the set $X(x_0, \varepsilon) = \{x \in X \mid f(x) \leqq f(x_0) - \varepsilon\}$. ∎

The results presented in this section are due to P. Loridan. An interesting generalization of the notion of ε-efficiency will be presented in the next section.

13.2 H-EFFICENCY

The notion of *H-efficiency* is due to A.B. Németh and takes place between *Pareto ε-efficiency* and *Pareto efficiency.* The notion of H-efficiency extends formally that of ε-efficiency, when the element ε in the positive cone is replaced by a set H in it.

Let E be a real vector space and $K \subset E$ a pointed convex cone. When E is a topological vector space, the cone K will be supposed to be closed. Putting $x \leqq y$ whenever $y - x \in K$, we obtain an order relation in E and we say that (E, K) is an ordered vector space.

Let (E, K) be an ordered vector space and $A \subset E$ a non-empty subset.

We recall that a point $x_* \in A$ is said to be an efficient point (or *Pareto minimum point*) if and only if $(x_* - K) \cap A = \{x_*\}$, or equivalently, if and only if there is no $x \in A$ such that $x \leqq x_*$ and $x \neq x_*$ (i.e. there is no $x \in A$ such that $x \leq x_*$). The set of all *efficient* (Pareto minimum) points of A will be denoted by $E_m[A, K]$. We can show that $E_m[A, K] = E_m[A + K, K]$.

Let ε be an element in K. Then $x_\varepsilon \in A$ is said to be an ε-efficient point (or Pareto ε-efficient point, or ε near to minimum point) of A if there is no $x \in A$ such that $x \leq x_\varepsilon - \varepsilon$. Obviously, the point x_ε is an ε-efficient point of A if and only if $x_\varepsilon \in A$ and $(x_\varepsilon - \varepsilon - K) \cap A = \emptyset$. The set of ε-efficient points of A will be denoted by ε-$E_m[A, K]$. We can show that ε-$E_m[A, K] \subset \varepsilon$-$E_m[A + K, K]$, but generally the converse is not true.

We observe also that for the existence of ε-efficient points for some A, it is necessary that $\varepsilon \neq 0$. Hence, in this section we suppose implicitely that $\varepsilon \neq 0$. The following notion, which is due to A.B. Németh, is obtained replacing ε by a subset H of K.

Let (E, K) be an ordered vector space, H a non-empty subset of $K \setminus \{0\}$ and A a non-empty subset of E.

DEFINITION 13.4. We say that a point $x_* \in A$ is an *H-efficient point* of A if there is no $x \in A$ such that $x \leqq x_* - h$ for some $h \in H$.

Obviously, x_* is an H-efficient point of A if and only if

$$x_* \in A \text{ and } (x_* - H - K) \cap A = \emptyset. \tag{13.1}$$

The set of H-efficient points of A will be denoted by $H\text{-}E_m[A, K]$. When we speak about H-efficiency we will suppose that H is a non-empty subset of $K \setminus \{0\}$.

Considering the characterization (13.1) we can show the following relations:

$$H\text{-}E_m[A,K] = (H+K)\text{-}E_m[A,K] \subset H\text{-}E_m[A+K,K] =$$

$$= (H+K)\text{-}E_m[A+K,K]. \tag{13.2}$$

Certainly, when $H = \{\varepsilon\}$, where $\varepsilon \in K \setminus \{0\}$, then H-efficiency becomes ε-efficiency.

It is also elementary to show the following relations:

$$H\text{-}E_m[A,K] = \bigcap_{h \in H} (h\text{-}E_m[A,K]) \tag{13.3}$$

and

$$(H_1 \cup H_2)\text{-}E_m[A,K] = (H_1\text{-}E_m[A,K]) \cap (H_2\text{-}E_m[A,K]). \tag{13.4}$$

From (13.3) we deduce that

$$H_1 \subset H_2 \implies H_2\text{-}E_m[A,K] \subset H_1\text{-}E_m[A,K], \tag{13.5}$$

which implies

$$(H_1 \cap H_2)\text{-}E_m[A,K] \supset (H_1\text{-}E_m[A,K]) \cup (H_2\text{-}E_m[A,K]). \tag{13.6}$$

Remark. It is possible to have $H_1\text{-}E_m[A,K] = H_2\text{-}E_m[A,K] = \emptyset$ while $(H_1 \cap H_2)\text{-}E_m[A,K] \neq \emptyset$.

In this sense we cite the following example given in (Németh, 1989).

Consider the vector space R^2 ordered by the cone K – the convex cone generated by $\{u, v\}$, where u and v are two linearly independent vectors in R^2. Consider the set $A = R^2 \setminus \text{int}(-K)$. In this case we have:

$$u\text{-}E_m[A,K] = \{t_1 v + t_2 u \mid t_1 < 0, t_2 \in [0,1[\},$$

$$v\text{-}E_m[A,K] = \{t_1 u + t_2 v \mid t_1 < 0, t_2 \in [0,1[\}.$$

If $H_1 = \{u, 2v, u+v\}$ and $H_2 = \{2u, v, u+v\}$, then $H_1\text{-}E_m[A,K] = H_2\text{-}E_m[A,K] = \emptyset$, while $(H_1 \cap H_2)\text{-}E_m[A,K] = (u+v)\text{-}E_m[A,K] \supset \partial A \neq \emptyset$, where ∂A is the boundary of A. We consider on R^2 the Euclidean topology.

It is interesting to present some results related to the comparison of H-efficiency. The relation (13.5) shows that the correspondence $H \to H\text{-}\ E_m[A,K]$

is antitonic, that is, the larger H implies the smaller H-$E_m[A,K]$. Hence, H-efficiency for a larger H is more restrictive.

Let H_1 and H_2 be non-empty subsets of $K\setminus\{0\}$. We say that H_1-efficiency is *coarser* than H_2-efficiency if for every set $A \in E$ it holds H_1-$E_m[A,K] \supset H_2$-$E_m[A,K]$. In this case we will write H_1-$eff \leqq H_2$-eff.

Relation (13.5) shows that $H_1 \subset H_2$ implies H_1-$eff \leqq H_2$-eff.

Remark. The relation H_1-$eff = H_2$-eff does not imply any inclusion relation for the sets H_1 and H_2.

THEOREM 13.5. *If H_1 and H_2 are non-empty subsets in $K \setminus \{0\}$, then the following assertions are equivalent:*

1. *H_1-$eff \leqq H_2$-eff,*
2. *$H_1 \subset H_2 + K$,*
3. *for every $x \in H_1$ we have $(x - K) \cap H_2 \neq \emptyset$.*

PROOF.

(1) $\Longrightarrow$ (2). We suppose that (2) is not true. If this is the case, let $x \in H_1 \setminus (H_2 + K)$ be an arbitrary element. We consider the set $A = \{0, x\}$. Then obviously, $0 \in (x - H_1 - K) \cap A$, that is $x \notin H_1$-$E_m[A,K]$.

On the other hand we have that $x \in H_2$-$E_m[A,K]$. Indeed, if this is not true, it holds either $0 \in x - H_2 - K$ or $x \in x - H_2 - K$. In the first case one has $x \in H_2 + K$ contradicting the fact that $x \in H_1 \setminus (H_2 + K)$.

In the second case it would follow $0 \in H_2 + K$, which contradicts the condition $H_2 \subset K \setminus \{0\}$.

Therefore we have a set A such that H_2-$E_m[A,K] \not\subset H_1$-$E_m[A,K]$ which shows that (1) cannot hold.

(2) $\Longrightarrow$ (1). If (2) is true it is sufficient to observe that (13.2) and (13.5) imply for an arbitrary set $A \subset E$ that,

$$H_2\text{-}E_m[A,K] = (H_2 + K)\text{-}E_m[A,K] \subset H_1\text{-}E_m[A,K]$$

and consequently (1) is true. The equivalence (2) $\Leftrightarrow$ (3) is elementary. ∎

COROLLARY 13.2. *If H_1, H_2 are two non-empty subsets in $K \setminus \{0\}$ then we have H_1-$eff = H_2$-eff if and only if $H_1 + K = H_2 + K$.*

PROOF. The second relation in the corollary implies that $H_1 \subset H_2 + K$ and $H_2 \subset H_1 + K$. ∎

COROLLARY 13.3. *If H_1, H_2 are two non-empty subsets in $K \setminus \{0\}$ and H_1-$eff = H_2$-eff, then $E_m[H_1,K] = E_m[H_2,K]$.*

PROOF. We know that $E_m[H_i,K] = E_m[H_i + K, K], i = 1,2$. The corollary is now a consequence of Corollary 13.2. ∎

We will give now some existence theorems for H-efficiency.

Let (E, K) be again an ordered vector space. The following proposition is a technical result necessary for the proofs of other results.

PROPOSITION 13.3. *Let H be a non-empty subset in $K \setminus \{0\}$. Then the following assertions are equivalent:*

1. *Every non-empty lower bounded subset in E possesses H-efficient points,*
2. *Any set of the form $\left\{\sum_{i=1}^{n} h_i \mid n \in N\right\}$, with h_i (not necessarily distinct) elements in H is not upper bounded.*

PROOF. (1) $\Longrightarrow$ (2). Suppose that (2) does not hold. Then there exists $\{h_i\} \subset H$ with $\left\{\sum_{i=1}^{n} h_i \mid n \in N\right\}$ upper bounded.

Then $A = \left\{-\sum_{i=1}^{n} h_i \mid n \in N\right\}$ has a lower bound. Let $x \in A$ be an arbitrary element. Then $x = -\sum_{i=1}^{n} h_i$ for some $m \in N$.

We have

$$x - h_{m+1} = -\sum_{i=1}^{m+1} h_i \in A$$

and

$$x - h_{m+1} \in x - H \subset x - H - K.$$

Hence for arbitrary $x \in A$ we have that

$$(x - H - K) \cap A \neq \emptyset$$

which means that (1) is not true.

(2) $\Longrightarrow$ (1). Assume that a subset A of E is lower bounded and is without H-efficient points, i.e.

$$(x - H - K) \cap A \neq \emptyset \text{ for every } x \in A.$$

Let $x_1 \in A$ be an arbitrary element and construct the sequence $\{x_n\}_{n \in N}$ by taking

$$x_{i+1} \in (x_i - H - K) \cap A, i \in N. \tag{13.7}$$

From (13.7) we have that $x_{i+1} = x_i - h_i - k_i$ with some $k_i \in K$. Then we have

$$h_i \leqq x_i - x_{i+1},\ i \in N$$

which implies

$$\sum_{i=1}^{n} h_i \leqq x_1 - x_{n+1} \leqq x_1 - y,\ n \in N,$$

with y some lower bound of A.

Therefore we have a contradiction to condition (2) and the proof is complete. ■

COROLLARY 13.4. *If each non-empty lower bounded subset of E has H_i-efficient points for H_i, a non-empty subset in $K \setminus \{0\}$, and $i = 1, 2, ..., n$, then each non-empty lower bound subset of E has H-efficient points where*

$$H = \left\{ \lambda_1 h_1 + \lambda_2 h_2 + ... + \lambda_n h_n \mid h_i \in H; \lambda_i \geq 0; i = 1, 2, ..., n; \sum_{i=1}^{n} \lambda_i = 1 \right\}.$$

PROOF. Assume the contrary. In this case by Proposition 13.3, there exists a sequence $\{y_i\}_{i \in N} \subset H$ such that the set $\left\{ \sum_{i=1}^{m} y_i \mid m \in N \right\}$ is upper bounded.

From the definition of the set H we have that

$$y_i = \lambda_1^{(i)} h_1^{(i)} + ... + \lambda_n^{(i)} h_n^{(i)}, \text{ with } h_j^{(i)} \in H_j,$$

$\lambda_j^{(i)} \geq 0$ and $\sum_{j=1}^{n} \lambda_j^{(i)} = 1$, for every i.

Since for every i there is some $j \in \{1, 2, ..., n\}$ such that $\lambda_j^{(i)} \geqq \frac{1}{n}$, it follows that there exists some subscript, say j_0 and a subsequence $\left\{ h_{j_0}^{(i_k)} \right\}_{k \in N}$ of $\{h_{j_0}^{(i)}\}_{i \in N}$ such that $\lambda_{j_0}^{(i_k)} \geqq \frac{1}{n}$ for each k. For this subsequence we have

$$\frac{1}{n} \sum_{k=1}^{m} h_{j_0}^{(i_k)} \leqq \sum_{i=1} i_m y_i, \text{ for each } m.$$

From our hypothesis on the sequence $\{y_i\}$ we deduce that the set $\left\{ \sum_{k=1}^{m} h_{j_0}^{(i)} \mid m \in N \right\}$ is order bounded. This conclusion contradicts, by Proposition 13.3, the assumption on H_{j_0}, and the proof is complete. ■

COROLLARY 13.5. *If every non-empty lower bounded set in E has H_i-efficient points for $i = 1, 2, ..., n$, then every such set has $H_1 \cup H_2 \cup ... \cup H_n$ -efficient points too.*

PROOF. Indeed, because $H_1 \cup H_2 \cup ... \cup H_n \subset H$, where H is the set defined in Corollary 13.4 we have that the corollary is a consequence of Corollary 13.4 and of relation (13.5). ■

Remark. An interesting consequence of Corollary 13.4 is also the fact that, if H is a finite set and if every lower bounded set in E has H-efficient points, then the same is true for $(conv(H))$-efficient points. (We denoted by $conv(H)$ the convex hull of H).

COROLLARY 13.6. *If H is a non-empty finite dimensional set in $K \setminus \{0\}$ and if every non-empty lower bounded set in E has H-efficient points, then every such set has also $(conv(H))$-efficient points.*

PROOF. If $\dim H = n$, then by Caratheodory's Theorem, for every $y \in conv(H)$ there exist some $h_1, h_2, ..., h_{n+1}$ in H such that y is their convex combination.
Therefore we can apply Corollary 13.4 with $n+1$ instead of n and $H_1 = ... = H_{n+1} = H$. ∎

Now we will show an interesting relation between the *semi-Archimedianity* and the H-efficiency of non-empty lower bounded sets in E. In this sense we have the following result due also to A.B. Németh.

Let (E, K) be an ordered vector space. We say that this space is *semi-Archimedian* if, from $0 \leqq ty \leq z$ for some $z \in K$ and all $t > 0$ it follows that $y = 0$. We can show that the condition in the definition of the semi-Archimedian property is equivalent to the following property: *from $0 \leqq ny \leqq z$ for some $z \in K$ and all $n \in N$ it follows that $y = 0$.*

We have the following result.

THEOREM 13.6. *Let (E, K) be an ordered vector space. The following assertions are equivalent:*

1. *(E, K) is semi-Archimedian,*
2. *Every non-empty lower bounded set in E has H-efficient points for each H with the property that there exists a finite set $H_0 \subset K \setminus \{0\}$ such that $H \subset conv(H_0) + K$,*
3. *Every non-empty lower bounded set in E has ε-efficient points for every $\varepsilon \in K \setminus \{0\}$.*

PROOF.

(1) $\Longrightarrow$ (2). Applying the relations (13.2), (13.5) and Corollary 13.4 we can reduce the proof to the case of a finite set $H_0 = \{h_1, h_2, ..., h_m\}$. Considering Proposition 13.10, we have to show that if (E, K) is semi-Archimedian, then no set of the form $\left\{ \sum_{k=1}^{n} h_{i_k} \mid i_k \in \{1, 2, ..., m\}, n \in N \right\}$ can be upper bounded.
If this is not true, then in this case, there would exist an index $j_0 \in \{1, 2, ..., m\}$ such that the set $\{nh_{j_0} \mid n \in N\}$ would be upper bounded. The semi-Archimedianity implies that $h_{j_0} = 0$, which is impossible since $H_0 \subset K \setminus \{0\}$.

(2) $\Longrightarrow$ (3). This implication is obvious.

(3) $\Longrightarrow$ (1). Suppose that (E, K) is not semi-Archimedian, i.e., there exist some x and y in K, $x \neq 0$ such that

$$0 \leqq tx \leqq y, \text{ for any } t > 0. \tag{13.8}$$

Consider the set $A = \{\alpha y - \beta x \mid \alpha, \beta > 0\}$. Then by (13.8), $A \subset K$ and hence it is lower bounded. Let u be an arbitrary element of A. Then $u = \alpha y - \beta x$, for some positive real numbers α and β. Obviously we have $u - x \in u - x - K$ and $u - x = \alpha y - (\beta + 1)x \in A$. Therefore for each $u \in A$ we have $(u - x - K) \cap A \neq \emptyset$, which shows that condition (3) does not hold for $\varepsilon = x$, and the proof is complete. ■

Let $(E(\tau), K)$ be an ordered locally convex space. We recall that K is *regular* if every decreasing lower bounded net in E is convergent. Inspired by this notion A.B. Németh introduced the following notion.

We say that K is *quasi-regular* if every decreasing lower bounded net in E is a Cauchy net.

The following proposition due also to A.B. Németh is useful for the proofs of other results.

PROPOSITION 13.4. *Let $(E(\tau), K)$ be an ordered locally convex space. The following assertions are equivalent:*

1. *K is quasi-regular*
2. *Every decreasing sequence in K is Cauchy-sequence,*
3. *For any sequence $\{x_n\}_{n \in N}$ in K, the condition $x_n \notin U$ for every n and for some neighbourhood U of 0, implies that the set $\left\{\sum_{i=1}^{n} x_i \mid n \in N\right\}$ cannot be bounded.*

PROOF.
(1) $\Longrightarrow$ (2). This implication is obvious.

(2) $\Longrightarrow$ (1). Suppose that (1) does not hold. In this case there exists a decreasing net $\{x_i\}_{i \in I}$ in K which is not Cauchy-net. This means that there exists some neighborhood U of 0 such that for any index i, there exist $j \leqq i$ and $k \geqq i$ such that $x_j - x_k \notin U$. Fix an arbitrary index and denote the corresponding element by x_1.

Then there exist some elements x_2 and x of the net such that $x_2 \leqq x_1, x_3 \leqq x_2$ and $x_2 - x_3 \notin U$.

Start now with x_3 in place of x_1 to get two elements x_4 and x_5 of the net with $x_5 \leqq x_4 \leqq x_3$ and $x_4 - x_5 \notin U$. Continuing this process we find a decreasing sequence $\{x_n\}_{n \in N}$ in K which is not Cauchy-sequence. This contradicts (2) and the implication is proved.

(2) $\Longrightarrow$ (3). Suppose that (3) is not true. In this case there would exist a sequence $\{y_n\}_{n \in N}$ in K and a neighbourhood U of 0 such that $y_n \notin U$ for

any $n \in N$, for which $\left\{\sum_{i=1}^{n} y_i\right\}_{n\in N}$ would be an increasing upper bounded sequence.

Because a such sequence is not Cauchy-sequence we have that (2) is not true.

(3) $\Longrightarrow$ (2). If K is not quasi-regular, then we can contradict (3) by the following reason. The negation of (2) implies the existence of a decreasing sequence $\{u_n\}_{n\in N}$ in K which is not Cauchy-sequence.

Hence, there exists some neighborhood V of 0 such that for each index i_0 there exist $i \geqq i_0$ and $j \geqq i_0$ such that $u_i - u_j \notin V$. We take a balanced neighbourhood U of 0 such that $U + U \subseteq V$. If u_{i_1} is an arbitrary element of the above sequence then there is $i > i_1$ for which $u_i - u_{i_1} \notin U$, for if this would not be true, $u_i - u_{i_1} \in U$ and $u_j - u_{i_1} \in U$ for any i and $j > i_1$, and this would contradict the above assumption since it would imply that $u_j - u_i \in U + U \subseteq V$ for any such i and j.

Therefore we can construct a subsequence of $\{u_n\}_{n\in N}$, denoted by $\{v_n\}_{n\in N}$ such that $v_n - v_{n+1} \notin U$, $v_{n+1} \leqq v_n$ for every $n \in N$. Taking $y_n = v_n - v_{n+1}, n \in N$, we have that the sequence $\{y_n\}_{n\in N}$ is in K and it contradicts (3) and the proof is complete. ■

THEOREM 13.7. *If (E,K) is an ordered locally convex space, then the following assertions are equivalent:*

1. *The cone K is quasi-regular;*
2. *Every non-empty lower bounded set A in E possesses H-efficient points for every non-empty subset H in K such that $E \setminus H$ is a neighborhood of 0;*
3. *Every non-empty denumerable lower bounded set A in E possesses H-efficient points for every non-empty denumerable set H in K suvh that $E \setminus H$ is a neighborhood of the origin.*

PROOF.

(1) $\Longrightarrow$ (2). We assume that some non-empty lower bounded subset A in E does not have H-efficient points for some non-empty subset H in $K \setminus U$, where U is a neighborhood of 0. Then by Proposition 13.3 there exists some sequence $\{h_i\}_{i\in N}$ of (not necessarily distinct) elements in H such that the set $\left\{\sum_{i=1}^{n} h_i \mid n \in N\right\}$ is upper bounded. Considering Proposition 13.4 we obtain that (1) does not hold.

(2) $\Longrightarrow$ (3). The proof is trivial.

(3) $\Longrightarrow$ (1). We assume that K is not quasi-regular. Applying Proposition 13.4 again we conclude that there exists a neighborhood U of 0 and a sequence $\{x_i\}_{i\in N}$ with $x_i \in K \setminus U$, for each $i \in N$, such that the sums $\sum_{i=1}^{n} x_i$, $n \in N$

have an upper bound x_*. That is, the set $A = \left\{ -\sum_{i=1}^{n} x_i \mid n \in N \right\}$ is bounded from below by $-x_*$. Let x be an arbitrary element in A. Then $x = -\sum_{i=1}^{m} x_i$, for some m. We have further

$$x - x_{m+1} = -\sum_{i=1}^{m+1} x_i \in A$$

and

$$x - x_{m+1} \in x - H \subset x - H - K,$$

with $H = \{x_i | i \in N\}$. That is, we have that $(x - H - K) \cap A \neq \emptyset$, for every $x \in A$. Therefore for the denumerable lower bounded set A in E and the denumerable set $H \subset K \setminus U$ the condition (3) does not hold. ∎

13.3 INFINITESIMAL EFFICIENCY

Let $(F, ||\cdot||)$ and $(E, ||\cdot||)$ be Banach spaces, E being ordered by a closed pointed convex cone $K \subset E$. Given a non-empty subset X in F and a mapping $f : X \longrightarrow E$, we consider the vector optimization problem:

$$\min_{x \in X} f(x) \tag{13.9}$$

For problem (13.9) we are interested in finding all solutions that are *efficient* (minimal) points, i.e., the elements of the set

$$E_m[f(X), K] = \{f(x_0) \mid (f(x_0) - K) \cap f(X) = \{f(x_0)\}\}.$$

We know that in this case $x_0 \in X$ is called a Pareto (minimal) point. In many existence theorems for efficiency it is supposed that $f(X)$ is a closed set. In this section we will suppose that $f(X)$ is not necessarily a closed set.

To study problem (13.9) under this assumption we need to introduce the following notion.

DEFINITION 13.5. Given $\varepsilon > 0$, we say that $y_\varepsilon \in f(X)$ is an *ε-approximative efficient point with respect to K*, if there exists an element $b \in E$ such that $||b|| < \varepsilon$ and $f(X) \cap [y_\varepsilon - b - K \setminus \{0\}] = \emptyset$.

The following theorem is due to G. Isac.

THEOREM 13.8. *Let $(F, ||\cdot||)$ and $(E, ||\cdot||)$ be Banach spaces and $K \subset E$ a closed pointed convex cone. Let $\Psi : E \longrightarrow R$ be a continuous linear mapping such that*

$$-K \subseteq \{x \in E \mid \Psi(x) + ||x|| \leq 0\},$$

X a non-empty subset in F and $f : X \longrightarrow E$ a mapping. If there exists $x_\alpha \in D = \overline{f(X)}$ such that Ψ is bounded from below on the set $(x_\alpha - K) \cap D$,

then for every $\varepsilon > 0$ there exists an ε-approximative efficient point $y_\varepsilon \in f(X)$ with respect to K.

PROOF. Because Ψ is bounded from below on the set $(x_\alpha - K) \cap D$, we have that $D \neq E$, In the first step we will show that there exists a $(-K)$-efficient point $x_* \in D$ in the sense that $(x_* - K) \cap D = \{x_*\}$.

Indeed, we consider on D the generalized dynamical system $\Gamma(x) = (x - K) \cap D$ for all $x \in D$ and we define a sequence $\{x_n\}_{n=0}^{\infty} \subset D$ inductively as follows.

We take $x_0 = x_\alpha$ and we suppose that $x_n \in D$ is defined and $\Gamma(x_{k+1}) \subseteq \Gamma(x_k)$ for all $k = 0, 1, 2, ..., n-1$.

We have two possibilities:

(i). either $\Gamma(x_n) = (x_n - K) \cap D = \{x_n\}$ or,

(ii). there exists $x \neq x_n$ with $x \in \Gamma(x_n)$.

If (i) is the case we have $x_n = x_*$. Suppose that (2) is true. Since

$$-K \subseteq \{x \in E \mid \Psi(x) + ||x|| \leq 0\},$$

we deduce from (ii) and considering the linearity of Ψ that

$$\Psi(x) - \Psi(x_n) + ||x - x_n|| = \Psi(x - x_n) + ||x - x_n|| \leq 0$$

and

$$\Psi(x) \leq \Psi(x_n) - ||x - x_n|| < \Psi(x_n).$$

Thus,

$$\Psi(x_n) - \inf_{v \in \Gamma(x_n)} \Psi(v) \geq \Psi(x_n) - \Psi(x) > 0.$$

We pick a point $x_{n+1} \in \Gamma(x_n)$ such that

$$\Psi(x_{n+1}) \leq \inf_{v \in \Gamma(x_n)} \Psi(v) + \frac{1}{2}\left[\Psi(x_n) - \inf_{v \in \Gamma(x_n)} \Psi(v)\right], \tag{13.10}$$

and we have $\Gamma(x_{n+1}) = (x_{n+1} - K) \cap D$ and so on. Since $-K$ is a convex cone, we can show that $\Gamma(x_{n+1}) \subseteq \Gamma(x_n)$, for all $n = 0, 1, 2, ...$ or the process is stopped after a finite number of steps. The sequence $\{\Gamma(x_n)\}_{n=0}^{\infty}$ (if it is not stopped after a finite number of steps) is decreasing. From (13.10) we obtain

$$\Psi(x_{n+1}) - \inf_{v \in \Gamma(x_{n+1})} \Psi(v) \leq \Psi(x_{n+1}) - \inf_{v \in \Gamma(x_n)} \Psi(v) \leq$$

$$\frac{1}{2}\left[\Psi(x_n) - \inf_{v \in \Gamma(x_n)} \Psi(v)\right] \tag{13.11}$$

If $z \in \Gamma(x_n)$,then applying (13.11) n times we obtain

$$\Psi(x_n) - \Psi(z) \leq \Psi(x_n) - \inf_{v \in \Gamma(x_n)} \Psi(v) \leq$$

$$\frac{1}{2} \leq \left[\Psi(x_{n-1}) - \inf_{v \in \Gamma(x_{n-1})} \Psi(v)\right] \leq ... \leq$$

$$\leq \frac{1}{2^n}\left[\Psi(x_\alpha) - \inf_{v\in\Gamma(x_\alpha)} \Psi(v)\right].$$

Since $-K \subseteq \{x \in E \mid \Psi(x) + ||x|| \leq 0\}$ and $\Gamma(x_n) = (x_n - K) \cap D$, we have for any $z \in \Gamma(x_n)$,

$$||z - x_n|| \leq -\Psi(z - x_n) = \Psi(x_n) - \Psi(z)$$

and finally,

$$||z - x_n|| \leq \Psi(x_n) - \Psi(z) \leq \frac{1}{2^n}\left[\Psi(x_\alpha) - \inf_{v\in\Gamma(x_\alpha)} \Psi(v)\right].$$

Therefore $\delta(\Gamma(x_n)) \leq \frac{1}{2^{n-1}}\left[\Psi(x_\alpha) - \inf_{v\in\Gamma(x_\alpha)} \Psi(v)\right]$, which implies that $\delta(\Gamma(x_n)) \to 0$ as $n \to \infty$.

Applying Theorem 12.56 [Chapter 12], we obtain an element $x_* \in D = \overline{f(X)}$ such that $(x_* - K) \cap D = \{x_*\}$.

Now, for a given $\varepsilon > 0$ we consider the open set $B = \{b \in E \mid ||b|| < \varepsilon\}$. Because $x_* \in \overline{f(X)}$ we have that

$$(x_* + B) \cap \overline{f(X)} \neq \emptyset \tag{13.12}$$

From (13.12) we deduce the existence of an element $b \in B$ and of an element $y_\varepsilon \in f(X)$ such that $x_* + b = y_\varepsilon$. Therefore, we have

$$\overline{f(X)} \cap [y_\varepsilon - b - K] = \{y_\varepsilon - b\}$$

or equivalently

$$\overline{f(X)} \cap [y_\varepsilon - b - K \setminus \{0\}] = \emptyset$$

which implies

$$f(X) \cap [y_\varepsilon - b - K \setminus \{0\}] = \emptyset$$

and the proof is complete. ■

13.4 ε-EFFICIENCY WITH RESPECT TO AN ARBITRARY SET

In this section we will study the existence and some properties of ε-approximate elements defined not necessarily by an convex cone but by an arbitrary set.

The main results are based on a nonconvex separation theorem with a nonlinear continuous monotone functional as separating functional. The origin of this separation theorem is the following nonlinear functional introduced in 1983 in Gerstewitz (Tammer) (1983).

Let $E(\tau)$ be a topological vector space, $D \subset E$ an open convex set and $K \subset E$ a closed convex cone with the property $clD - K \setminus \{0\} \subset D$. Let

$k_0 \in K \setminus \{0\}$ be an arbitrary element. we define $s_\lambda := clD + \lambda k_0$ and the mapping

$$\phi(y) := \inf\{\lambda | y \in S_\lambda, \lambda \in R\}.$$

We denoted by clD the closure of D. Properties of the functional ϕ and some preliminary nonconvex separation theorems were proved in Gerstewitz (Tammer) and Ivanov (1985) and Gerstewitz (Tammer) (1986).

Some results from these papers were used in the following separation theorem due to Chr. Tammer and P. Weidner. Before giving this separation theorem we need to introduce the following notions.

Let $B \subset E$ be a non-empty subset.

DEFINITION 13.6. A mapping $\phi : E \to R$ is called:
(i). *B-monotone* if $x_1 \in x_2 + B$ implies $\phi(x_2) \leq \phi(x_1)$,

(ii). *strictly B-monotone* if $x_1 \in x_2 + (B \setminus \{0\})$ implies $\phi(x_2) < \phi(x_1)$

(iii). *subadditive* if $\phi(x_1 + x_2) \leq \phi(x_1) + \phi(x_2)$, for all $x_1, x_2 \in E$.

The following separation theorem is useful in this section.

THEOREM 13.9. *Let $E(\tau)$ be a topological vector space, K a cone in E with non-empty interior and C a proper subset of E with non-empty interior such that $clC - \text{int}K \subset \text{int}C$. If $A \subset E$ is a non-empty subset then we have:*

1. *$A \cap \text{int}C = \emptyset$ if and only if there exists a continuous mapping $\phi : E \to R$, which is strictly $(\text{int}K)$-monotone with range $]-\infty, +\infty[$ and $\phi(A) \geq 0, \phi(\text{int}A) > 0, \phi(C) \leq 0, \phi(\partial C) = 0, \phi(\text{int}C) < 0$ (where ∂C is the boundary of C).*

2. *If clC is a convex set, then the mapping ϕ in (1) can be chosen such that it is also convex.*

3. *One can construct the mapping ϕ in (1) such that ϕ is B-monotone for each set $B \subset E$ with $\partial C - B \subset clC$ and strictly B-monotone if $\partial C - (B \setminus \{0\}) \subset \text{int}C$.*

4. *If $\partial C + \partial C \subset clC$ then ϕ in (1) can be chosen such that it is subadditive in E.*

PROOF. For a proof of this theorem the reader is referred to Gerth (Tammer) and Weidner (1990). ∎

Let X and B be proper subsets of E, $k_0 \in E \setminus \{0\}$ and $\varepsilon \in R_+$. We assume $B + \alpha k_0 \subset B$ for all $\alpha \in R_+ \setminus \{0\}$, and we define $B_{\varepsilon k_0} := \varepsilon k_0 + B \setminus \{0\}$. The following notion is due to Tammer [1] (1993).

DEFINITION 13.7. We say that an element x_ε of X is an *εk_0-efficient element of X with respect to B* if there is no element x of X with $x_\varepsilon \in x + B_{\varepsilon k_0}$.

Remark. If $\varepsilon = 0$ we have the definition of efficient elements of a set X with respect to B.

If $\varepsilon = 0$ and $B = R^n_+$ (in this case $E = R^n$), we have the classical notion of Pareto efficient element. We will denote the set of εk_0-efficient elements of X with respect to B by $E_m[X, B_{\varepsilon k_0}]$, and the set of efficient elements of X with respect to B by $E_m[X, B]$.

PROPOSITION 13.5. *Let X and B be proper subsets of a topological vector space $E(\tau)$, $\varepsilon \in R_+ \setminus \{0\}$ and $k_0 \in E \setminus \{0\}$. Suppose that*

$$B_{\varepsilon k_0} := \varepsilon k_0 + (B \setminus \{0\}) \subset \varepsilon k_0 + (B \cup \{0\}) \subset B.$$

If for a continuous, strictly B-monotone mapping $\phi : E \to R$ with the range $]-\infty, +\infty[$ we have $\phi(0) = 0$ and

$$\inf_{x \in X} \phi(x) > -\infty$$

then $E_m[X, B_{\varepsilon k_0}]$ is non-empty.

PROOF. We associate to $k_0 \in E \setminus \{0\}, \varepsilon \in R_+ \setminus \{0\}$ and B the set

$$X(B, k_0, \varepsilon) := \left\{ x_* \in X \mid \phi(x_*) \leq \inf_{x \in X} \phi(x) - \phi(-\varepsilon k_0) \right\}.$$

Since $\phi(0) = 0, \varepsilon k_0 \in B \setminus \{0\}$ and ϕ is strictly B-monotone we have

$$\phi(-\varepsilon k_0) < \phi(0) = 0.$$

Now, $\inf_{x \in X} \phi(x) > -\infty$ and $0 < -\phi(-\varepsilon k_0)$ imply that $X(B, k_0, \varepsilon) \subset E_m[X, B_{\varepsilon k_0}]$.

Indeed, if we assume that there exists an element $x_* \in X(B, k_0, \varepsilon)$ such that $x_* \notin E_m[X, B_{\varepsilon k_0}]$, then there must be an element $x' \in X$ with $x' \in x_* - \varepsilon k_0 - (B \setminus \{0\})$.

Considering the strict B-monotonicity, the subadditivity of ϕ and the definition of $X(B, k_0, \varepsilon)$ we have

$$\phi(x') < \phi(x_* - \varepsilon k_0) \leq \phi(x_*) + \phi(-\varepsilon k_0) \leq \inf_{x \in X} \phi(x),$$

which is impossible since $x' \in X$, and the proof is complete. ∎

Remark.

1. If we consider the special case $E = R^n, B = R^n_+, \varepsilon k_0 \in R^n_+$ and $\phi(x) = \sum_{i=1}^{n} a_i x_i$ with $a = (a_1, ..., a_n) \in \text{int} R^n_+$, we obtain proposition 13.1 due to P. Loridan (1984).

2. In Proposition 13.5 the compactness is not used, as in many existence theorems for Pareto efficiency.

PROPOSITION 13.6. *Let X and B be proper subsets of a topological vector space $E(\tau)$. Let $K \subset E$ be a convex cone with $intK$ non-empty. Suppose that B is an open set with $0 \in clB \setminus intB$ and such that $clB + intK \subset B, \partial B + (K \setminus \{0\}) \subset B, k_0 \in intK$ and $\varepsilon \in R_+ \setminus \{0\}$. If $\phi : E \to R$ is a continuous strictly B-monotone and subadditive mapping with $\phi(0) = 0$, with range $]-\infty, +\infty[$ and $-\infty < \inf_{x \in X} \phi(x)$, then for each $x_0 \in X$ with $x_0 \notin E_m[X, K_{\varepsilon k_0}]$ there exists an element $x_\varepsilon \in E_m[X, K_{\varepsilon k_0}]$ with $x_0 \in x_\varepsilon + (K \setminus \{0\})$.*

PROOF. We define the set

$$X^0 = \{x \in X \mid x_0 \in x + K\}.$$

From $x_0 \notin E_m[X, K_{\varepsilon k_0}]$ it follows that X^0 is non-empty. Since $\phi(0) = 0$ and $-\infty < \inf_{x \in X} \phi(x) \leq \inf_{x \in X^0} \phi(x)$, we have that Proposition 13.5 implies the existence of an element $x_\varepsilon \in E_m[x^0, B_{\varepsilon k_0}]$ with $x_\varepsilon \in x^0, x_\varepsilon \neq x_0$, i.e. $x_\varepsilon \in x_0 - (K \setminus \{0\})$. Applying the Separation Theorem (Theorem 13.9) on $A := x^0$ and $C := x_\varepsilon - \varepsilon k_0 - B$, we obtain the existence of a continuous strictly K-monotone mapping $\psi : E \to R$ with

$$\psi(x_\varepsilon - \varepsilon k_0) \leq \psi(x), \text{ for all } x \in X^0 \tag{13.13}$$

If we assume that $x_\varepsilon \notin E_m[X, K_{\varepsilon k_0}]$, then there is an $x_* \in X$ with $x_\varepsilon \in x_* + \varepsilon k_0 + (K \setminus \{0\})$. From the strict K-monotonicity of the mapping ψ we have

$$\psi(x_*) < \psi(x_\varepsilon - \varepsilon k_0). \tag{13.14}$$

From (13.13) and (13.14) it follows

$$\psi(x_*) < \psi(x_\varepsilon - \varepsilon k_0) \leq \psi(x), \text{ for all } x \in X^0,$$

and finally,

$$\psi(x_*) < \inf_{x \in X^0} \psi(x). \tag{13.15}$$

Moreover, we conclude from

$$x_\varepsilon \in x_0 - (K \setminus \{0\} \text{ and}$$

$$x_* \in x_\varepsilon - \varepsilon k_0 - (K \setminus \{0\}) \subset x_\varepsilon - (K \setminus \{0\}),$$

that $x_* \in x_0 - K$ and considering the definition of X^0 we have $x_* \in X^0$ which is a contradiction to (13.15) and the proof is complete. ∎

A special concept of *subdifferential* can be used to give another necessary condition for εk_0-efficiency.

Let $E(\tau)$ be a topological vector space, E^* the dual of E and F a real Banach space. Let $K \subset E$ be a closed convex cone. Denote by K^* the topological dual of K and by $\hat{K}^*$ the strong dual of K (which is also named the *quasi-interior* of K^*).

It is known that

$$\hat{K}^* = \{y^* \in E^* \mid \langle y^*, x\rangle > 0, \text{for all } x \in K \setminus \{0\}\}.$$

We suppose that $\hat{K}^*$ is non-empty.

Let $L(F, E)$ be the space of linear continuous operators from F into E. We assume that $f : F \to E$ is a convex vector-valued function. The following definition is due to Thierfelder (1991).

DEFINITION 13.8. Let $k_0 \in K \setminus \{0\}$ and $\varepsilon \in R_+$. An element $T \in L(E, F)$ belongs to the (εk_0)-*subdifferential* $\partial^{\leq}_{\varepsilon k_0} f(x_0)$ of f at the point $x_0 \in F$ if

$$f(x_0 - T(x_0) \in f(x) - T(x) + \varepsilon k_0 - K, \text{ for all } x \in F.$$

We introduce also the following notations.

$$\partial^{\not>}_{\varepsilon k_0} f(x_0) := \{T \in L(E, F) \mid f(x_0 - T(x_0) \notin f(x) - T(x) + \varepsilon k_0 + K, \text{ for all } x \in F\},$$

and

$$\partial^{z^*}_{\varepsilon k_0} f(x_0) := \{T \in L(E, F) \mid \text{ there exists } z^* \in \hat{K}^* \text{ with}$$
$$z^*(f(x_0) - T(x_0)) \leq z^*(f(x) - T(x)) + \varepsilon z^*(k_0), \text{ for all } x \in F\}.$$

We have the following result.

PROPOSITION 13.7. *The following relations are true:*

1. $f(x_\varepsilon) \in f(x) + \varepsilon k_0 - K$, *for all* $x \in F$, *if and only if* $0 \in \partial^{\leq}_{\varepsilon k_0} f(x_\varepsilon)$,

2. $f(x_\varepsilon) \in E_m[f(F), \varepsilon k_0 + K]$, *if and only if* $0 \in \partial^{\not>}_{\varepsilon k_0} f(x_\varepsilon)$,

3. $z^*(f(x_\varepsilon)) \leq \inf_{x \in F} z^*(f(x)) + \varepsilon z^*(k_0)$, *if and only if* $0 \in \partial^{z^*}_{\varepsilon k_0} f(x)$.

PROOF. The relations are consequences of definitions. ∎

13.5 EKELAND VARIATIONAL PRINCIPLE TYPES FOR VECTOR VALUED MAPPINGS, EFFICIENCY AND APPROXIMATIVE EFFICIENCY

Since its appearance in 1972 the Ekeland's variational principle has been the subject of many papers and it has found many applications in analysis and in applied mathematics: *cf.* Ekeland (1972,1974,1979, 1983, 1990), Brezis and Browder (1976), Dancs e.a. (1983), Hyers, Isac, and Rassias (1997), Isac (1983,1996), Isac [3] (to appear) among others. The Ekeland's principle is equivalent to the Caristi-Kirk's Fixed Point Theorem, to the Drop Theorem and to the Petral Theorem and by these equivalences it has interesting applications in nonlinear analysis and to the study of the geometry of Banach spaces [see for example Hyers, Isac and Rassias (1997) and the references cited in Isac (1996)].

Some extensions of this principle were also proved by several authors (see for example Isac (1996), and its references). The Ekeland's principle has also many applications to the study of Pareto optimization. As we presented in Chapter 12 of this book, we remarked in 1983 that an extension of Caristi-Kirk's Fixed Point Theorem can be used to study the existence of critical points for generalised dynamical systems and hence in particular to study the existence of Pareto efficient points (Isac, 1983). After 1983 other variants of Ekeland's principle for Pareto optimization were presented in Khan (1986), Tammer (1992), Tammer [1],[2],[5] (1993), Tammer [3] (to appear), Isac (1996), Chen and Huang (1998), Chen e.a. (2000).

Our interest now, in this section, is to present several kinds of applications of some generalizations to vector-valued mappings of Ekeland's principle, to the study of efficiency and to the approximative efficiency.

First, we recall the *Ekeland's Principle (the Weak Form)* and the *Caristi-Kirk's Fixed Point Theorem.*

Let (X, d) be a complete metric space. We say that a mapping $f : X \to R \cup \{\infty\}$ is *not improper* if it is not identically equal to $+\infty$.

THEOREM 13.10[EKELAND'S PRINCIPLE-WEAK FORM]. *If $f : X \to R \cup \{\infty\}$ is a lower semicontinuous, bounded from below and not improper mapping, then for every $\varepsilon > 0$ there exists $x_\varepsilon \in E$ such that:*

1. $f(x_\varepsilon) \leq \inf_{x\in E} f(x) + \varepsilon$,
2. $f(x_\varepsilon) < f(x) + \varepsilon d(x, x_\varepsilon)$ *for all* $x \in E \setminus \{x_\varepsilon\}$.

■

It is well known that *Ekeland's Principle-Weak Form* is equivalent to the following fixed point theorem.

THEOREM 13.11[CARISTI-KIRK'S FIXED POINT THEOREM]. *Let (X, d) be a complete metric space and let $f : X \to R \cup \{\infty\}$ be a lower semicontinuous, bounded from below and not improper mapping. If $T : X \to X$ is a set-valued maping such that*

$$f(y) \leq f(x) - d(x, y), \text{ for all } x \in E, y \in T(x)$$

then there exists $x_* \in E$ such that $x_* \in T(x_*)$. ■

For a proof of this equivalence the reader is referred to Hyers, Isac, and Rassias (1997).

Let (X, d) be a complete metric space and $E(\tau)$ a topological vector space. We suppose the following assumptions.

(I). $K \subset E$ is a closed pointed convex cone with non-empty interior, $k_0 \in \text{int}K$ and $B \subset E$ is another cone with non-empty interior such that $clB+(K\setminus\{0\}) \subset \text{int}B$.

(II). $f : X \to E$ is a lower semicontinuous mapping with respect to k_0 and B in the sense that

$$X_r = \{x \in X \mid f(x) \in k_0 - clB\}$$

is a closed set for each $r \in R$. The mapping f is supposed also to be bounded from below, i.e. $f(X) \subset y + B$ for a certain $y \in E$.

Under the assumptions (I) and (II) we consider the *vector optimization problem:*

$$(VOP): \begin{cases} \text{find } x_* \in X & \text{with } f(x_*) \in E_m[f(X), K] \text{ where} \\ E_m[f(X), K] = & \{f(x_*) | x_* \in X, f(X) \cap (f(x_*) - (K \setminus \{0\})) = \emptyset\}. \end{cases}$$

We will give an existence result for the problem (VOP). The following two results are necessary for the proof of this existence result.

THEOREM 13.12. *Let (X, d) be a complete metric space and $\psi : X \to R$ a mapping, lower semicontinuous and bounded from below. If for each $u \in X$ with $\inf_{x \in X} \psi(x) < \psi(u)$, there is an element $v \in X$ with $v \neq u$ and*

$$\psi(v) + d(u, v) \leq \psi(u),$$

then there exists an element $x_* \in X$ such that $\psi(x_*) = \inf_{x \in X} \psi(x)$.

PROOF. This result is due to W. Takahashi and it was proved in Takahashi (1991). ∎

The next existence result for efficiency is based on the following result.

THEOREM 13.13. *Let $E(\tau)$ be a topological vector space. Suppose given two cones K and B in E with non-empty interior and such that $clB + (K \setminus \{0\} \subset intB$ and $k_0 \in intK$. Then the mapping $\phi : E \to R$ defined by*

$$\phi(x) = \inf\{\lambda \in R \mid \ x \in -clB + \lambda k_0\}$$

is continuous, subadditive and strictly K-monotone.

PROOF. This result is a consequence of Theorem 2.1 and Corollary 2.1 proved in Gerth (Tammer) and Weidner (1990). ∎

The following two results are due to Chr. Tammer and for this, we need to introduce the following assumption:

(III). For each $u \in X$ with $f(u) \notin E_m[f(X), K]$ there is a $v \in X$ with $v \neq u$ and $f(v) + k_0 d(u, v) \in f(u) - K$.

THEOREM 13.14. *If the assumptions (I), (II) and (III) are satisfied, then there exists an element $x_* \in X$ such that $f(x_*) \in E_m[f(X), K]$.*

PROOF. Let $w \in X$ be an arbitrary element such that $f(w) \notin E_m[f(X), K]$. We consider the mappings $f_w : X \to E$ defined by

$$f_w(x) = f(x) - f(w),$$

and $\phi : E \to R$ defined by

$$\phi(x) = \inf\{\lambda \in R \mid x \in cl(f_w(w) - B) + \lambda k_0\}, \text{ for all } x \in E \tag{13.16}$$

Assumption (III) implies the existence of an element $v \in X$ with $v \neq w$ and

$$f_w(v) \in -k_0 d(w, v) + f_w(w) - K. \tag{13.17}$$

Because $f_w(w) = 0$, Theorem 13.26 implies that the mapping ϕ in (13.16) is subadditive and strictly K-monotone. The properties of ϕ together with (13.17) imply

$$\phi(f_w(v)) \leq \phi(-k_0 d(w, v) + f_w(w)) \leq \phi(-k_0 d(w, v)) + \phi(f_w(w)).$$

Applying Lemma 4.4 proved in Tammer (1992) we have for $\alpha > 0$ that $\phi(-\alpha k_0) = -\alpha$, and hence

$$\phi(-k_0 d(w, v)) = -d(w, v).$$

Therefore we have

$$\phi(f_w(v)) + d(w, v) \leq \phi(f_w(w)).$$

Because of assumption (III) we obtain for ach $x \in X$ with $f(x) \notin E_m[f(X), K]$ (by using also the subadditivity and the strict K- monotonicity of ϕ), that there exists an element $v \in X$ with $v \neq x$ and

$$\phi(f_w(v)) + d(x, v) \leq \phi(f_w(x)).$$

Further the lower semicontinuity of f with respect to k_0 and B (and hence also of f_w) implies the lower semicontinuity of $\phi \circ f_w$ [see also Lemma 4.1 proved in Tammer (1992)].

Because the assumptions of Theorem 13.12 are satisfied, there exists an element $x_* \in X$ such that

$$\phi(f_w(x_*)) = \min_{x \in X} \phi(f_w(x)).$$

The strict K-monotonicity of ϕ implies

$$f_w(x_*) \in E_m[f_w(X), K],$$

which means,

$$f_w(X) \cap (f_w(x_* - (K \setminus \{0\})) = \emptyset.$$

Hence

$$f(X) \cap (f_w(x_*) - (K \setminus \{0\})) = \emptyset,$$

which means that $f(x_*) \in E_m[f(X), K]$ and the proof is complete. ■

Now we will show that Theorem 13.14 implies the following fixed point theorem which is a generalization to vector-valued mappings of Caristi-Kirk's Fixed Point Theorem.

THEOREM 13.15. *If the assumptions (I) and (II) are satisfied, then any mapping $h : X \longrightarrow X$ satisfying*

$$k_0 d(x, h(x)) \in f(x) - f(h(x)) - K, \text{ for all } x \in X \tag{13.18}$$

has a fixed point $x_* \in X$, i.e. $h(x_*) = x_*$.

PROOF. If we supose that $h(x) \neq x$ for all $x \in X$, then considering (13.8) we have that for all $x \in X$ there exists an element $v \in X$ (for example $v = h(x)$) such that $x \neq v$ and

$$f(v) + k_0 d(x, v) \in f(x) - K.$$

Applying Theorem 13.14 we conclude that there exists an element $x_* \in X$ such that $f(x_*) \in E_m[f(X), K]$.

The definition of the set $E_m[f(X), K]$ implies that for all $x \in X$ we have

$$f(x) \notin f(x_*) - (K \setminus \{0\}) \tag{13.19}$$

and for $x = h(x_*)$ it follows (considering (13.19)) that

$$f(h(x_*)) \notin f(x_*) - (K \setminus \{0\}). \tag{13.20}$$

On the other hand, for $k_0 \in \text{int} K$ (especially $k_0 \neq 0$ and $x_* \neq h(x_*)$) it follows

$$k_0 d(x_*, h(x_*)) \in K \setminus \{0\}. \tag{13.21}$$

Relation (13.18) implies

$$k_0 d(x_*, h(x_*)) \in f(x_*) - f(h(x_*)) - K$$

and hence

$$f(x_*) - f(h(x_*)) \in k_0 d(x_*, h(x_*)) + K \subset K \setminus \{0\},$$

since K is a convex cone with $K \cap (-K) = \{0\}$. Therefore we have $f(h(x_*)) \in f(x_*) - K \setminus \{0\}$, which is impossible because of relation (13.20). ■

Remark.

1. Other Caristi-Kirk type Fixed Point Theorems for vector-valued mappings were also proved in Németh (1989), Khan (1986) and Isac (1996).
2. If $E = R, K = R_+$ and $f : X \to R$ then Theorem 13.15 is exactly the Caristi-Kirk fixed point theorem.

Now we will give an existence theorem for approximately efficient points, applying Theorem 13.15. This approximative efficiency is obtained perturbing the mapping f.

THEOREM 13.16. *If assumptions* (I) *and* (II) *are satisfied, then for every* $\varepsilon > 0$ *there exists an element* $x_\varepsilon \in X$ *such that*

$$f_{\varepsilon k_0}(x_\varepsilon) \in E_m[f_{\varepsilon k_0}(X), K], f_{\varepsilon k_0}(x_\varepsilon) \neq f_{\varepsilon k_0}(x), \text{ for all } x \neq x_\varepsilon \tag{13.22}$$

where $f_{\varepsilon k_0} := f(x) + k_0 d(x, x_\varepsilon)\sqrt{\varepsilon}$.

PROOF. Suppose that condition (13.22) does not hold. In this case, for a certain $\varepsilon > 0$ there exists no element x_ε with

$$f_{\varepsilon k_0}(x_\varepsilon) \in E_m[f_{\varepsilon k_0}(X), K]$$

and

$$f_{\varepsilon k_0}(x_\varepsilon) \neq f_{\varepsilon k_0}(x), \text{ for all } x \neq x_\varepsilon.$$

Hence for all $x \in X$ there exists an element $y \neq x$ with

$$f(y) + k_0 d_1(x, y) \in f(x) - K, \tag{13.23}$$

where $d_1 = \sqrt{\varepsilon} d$.

Now we consider the mapping $h : X \to X$ defined by $h(x) = y$, where y is given according to (13.23). Then obviously

$$k_0 d_1(x, h(x)) \in f(x) - f(h(x)) - K$$

and consequently assumption (13.18) in Theorem 13.15 is satisfied.

Therefore, by Theorem 13.15 we have that the mapping h has a fixed point which contradicts our assumption that $h(x) = y \neq x$ for all $x \in X$ and the proof is complete. ■

We note that Theorem 13.16 is also due to Chr. Tammer.

For the next results we suppose that $(X, ||\cdot||)$ is a Banach space, $E(\tau)$ is a topological vector space and $f : X \to E$ a mapping. We suppose assumptions (I) and (II) satisfied where in assumption (I) the metric space (X, d) is replaced by the Banach space $(X, ||\cdot||)$.

Given the mapping f, we consider the following vector optimization problem

$$(VOP)(f, B) : \begin{cases} \text{evaluate the set } E_m[f(X), K] \text{ where} \\ E_m[f(X), K] = \{f(x_*) | x_* \in X \text{ and} \\ f(X) \cap (f(x_*) - (K \setminus \{0\})) = \emptyset\}. \end{cases}$$

To derive in this general case some existence results we introduce the following definition due to Chr. Tammer.

DEFINITION 13.9. We say that an element $f(x_\varepsilon) \in f(X)$ is an *approximately efficient point* of $f(X)$ with respect to $B, k_0 \in \text{int}K$ and $\varepsilon > 0$ if $f(X) \cap (f(x_\varepsilon) - \varepsilon k_0 - (B \setminus \{0\})) = \emptyset$.

The approximately efficient point set of $f(X)$ with respect to B, k_0 and ε will be denoted by $E_m[f(X), B_{\varepsilon k_0}]$, where $B_{\varepsilon k_0} = \varepsilon k_0 + B$.

We need also to recall the following separation theorem proved in Gerth (Tammer) and Weidner (1990).

THEOREM 13.17. *If the following assumptions are satisfied:*

1. *$E(\tau)$ is a topological vector space,*
2. *$C \subset E$ is a convex cone with non-empty interior,*
3. *$A \subset E$ is a non-empty subset,*

then, $A \cap (-intC) = \emptyset$ if and only if there exists a continuous sublinear functional $\phi : E \to R$ which is strictly $(intC)$-monotone with the range $]-\infty, \infty[$ and $\phi(A) \geq 0, \phi(intA) > 0, \phi(-C) \leq 0, \phi(-\partial C) = 0, \phi(-intC) < 0, \phi(C) \geq 0$ and $\phi(intC) > 0$. ■

Remark. The separating functional $\phi : E \to R$ constructed in the proof of Theorem 13.17 by

$$\phi(x) := \inf\{\lambda \in R \mid x \in \lambda k_0 - clC\} \tag{13.24}$$

has the following properties:

(p_1). $\phi(x) < r \Leftrightarrow x \in -\text{int}C + rk_0$

(p_2). $\phi(x) \leq r \Leftrightarrow x \in -clC + rk_0$

(p_3). $\phi(x) = r \Leftrightarrow x \in \partial C + rk_0$

(p_4). $\phi(x) \geq r \Leftrightarrow x \notin -\text{int}C + rk_0$

(p_5). $\phi(x) > r \Leftrightarrow x \notin -clC + rk_0$

PROPOSITION 13.8. *If assumption (I) is satisfied with the metric space (X,d) replaced by a Banach space $(X, \|\cdot\|)$ and $f : X \to E$ is lower semicontinuous with respect to k_0 and B and $\phi : E \to R$ is the mapping defined in (13.24) with $K = B$, then $(\phi \circ f)(x) = \phi(f(x))$ is lower semicontinuous at each point $x \in X$.*

PROOF. For a proof of this result the reader is referred to Tammer (1992).

■

PROPOSITION 13.9. *Assume that condition (I) is satisfied with the metric space (X,d) replaced by a Banach space* $(X, ||\cdot||)$ *and suppose given a real number* $\alpha > 0$*. Then the mapping* $\phi : E \to R$ *defined in (13.24) with* $K = B$ *has the following properties:*

1. $\phi(0) = 0,$
2. $\phi(k_0) = 1,$
3. $\phi(\alpha k_0) = \alpha$ *and*
4. $\phi(-\alpha k_0) = -\phi(\alpha k_0) = -\alpha.$

PROOF. A proof of this result is in Tammer (1992). ■

We need also to recall the following variant of Ekeland's principle.

THEOREM 13.18[EKELAND'S PRINCIPLE IN STRONG FORM]. *Let* $(X, ||\cdot||)$ *be a Banach space and* $f : K \to \bar{R}$ *a proper lower semicontinuous real-valued mapping, bounded from below. If* $\varepsilon > 0$ *and* $u \in X$ *are such that* $f(u) \leq \inf_{x \in X} f(x) + \varepsilon$*, then for any* $\lambda > 0$ *there exists an element* $x_* \in dom(f)$ *such that*

1. $f(x_*) \leq f(u) - \lambda ||x_* - u||,$
2. $||x_* - u|| \leq \frac{\varepsilon}{\lambda},$
3. $f(x_*) < f(x) + \lambda ||x - x_*||,$ *for all* $x \neq x_*.$

PROOF. For a proof of this classical result the reader is referred to Ekeland (1974) or Hyers, Isac and Rassias (1997). ■

The following existence result establishes an interesting relation between the approximative efficiency obtained by the perturbation of the cone B and the approximative efficiency obtained by the perturbation of the mapping f.

THEOREM 13.19. *Suppose assumptions (I) and (II) to be satisfied, where in (I) the metric space (X,d) is replaced by a Banach space* $(X, ||\cdot||)$*. Then, for any* $\varepsilon > 0, \lambda > 0$ *and any* $f(x_*) \in E_m[f(X), B_{\varepsilon k_0}]$*, there exists an element* $x_\varepsilon \in X$ *such that:*

1. $f(x_\varepsilon) \in f(x_*) - \lambda ||x_* - x_\varepsilon|| k_0 - clB,$ *hence* $f(x_\varepsilon) \in E_m[f(X), D_{\varepsilon k_0}]$*, where* D *is an open subset of* E *with* $K \backslash \{0\} \subset D, 0 \in \partial D$ *and* $clD + (K \backslash \{0\}) \subset D$
2. $||x_* - x_\varepsilon|| \leq \dfrac{\varepsilon}{\lambda},$
3. $f_{\lambda k_0}(x_\varepsilon) \in E_m[f_{\lambda k_0}(X), K],$ *where* $f_{\lambda k_0}(x) = f(x) + \lambda ||x_* - x_\varepsilon|| k_0.$

PROOF. Let $x_* \in X$ be an element such that

$$f(X) \cap (f(x_*) - \varepsilon k_0 - (B \setminus \{0\})) = \emptyset.$$

If we put $h(x) := f(x) - f(x_*)$ we obtain

$$h(X) \cap (h(x_*) - \varepsilon k_0 - B \setminus \{0\})) = \emptyset$$

and

$$(h(X) + \varepsilon k_0) \cap (h(x_*) - (B \setminus \{0\})) = \emptyset.$$

We observe that the assumptions of Theorem 13.17 are satisfied for

$$A = h(X) + \varepsilon k_0 \text{ and } C = h(x_*)(B \setminus \{0\}).$$

Therefore, applying Theorem 13.17 we obtain the existence of a continuous sublinear K-monotone mapping $\phi : E \to R$ defined as in (13.24) with $C = B$. [We used the fact remarked in Gerth (Tammer) and Weidner (1990) that if $K \neq E$ is a convex cone with nonempty interior then the strict (int$K$)-monotonicity implies the K-monotonicity].

Moreover, we have

$$\phi(h(x_*)) \leq \phi(h(x) + \varepsilon k_0) \leq \phi(h(x)) + \phi(\varepsilon k_0),$$

for all $x \in X$.

From Proposition 13.9 we obtain $\phi(\varepsilon k_0) = \varepsilon$ and so

$$\phi(h(x_*)) \leq \inf_{x \in X} \phi(h(x)) + \varepsilon, \varepsilon > 0.$$

From the given assumptions $\phi \circ h$ is a lower semicontinuous functional (see Proposition 13.8) which is bounded from below.

Applying Theorem 13.18 we get for any $\varepsilon > 0$ and $\lambda > 0$ that there exists an element $x_\varepsilon \in X$ such that

(i). $\phi(h(x_\varepsilon)) \leq \phi(h(x_*)) - \lambda ||x_* - x_\varepsilon||$,

(ii). $||x_* - x_\varepsilon|| \leq \dfrac{\varepsilon}{\lambda}$

(iii). $\phi(h(x_\varepsilon)) < \phi(h(x)) + \lambda ||x_* - x_\varepsilon||$, for all $x \neq x_\varepsilon, x \in X$.

From (i) we deduce

$$\phi(h(x_\varepsilon)) + \phi(-\varepsilon k_0) \geq \phi(h(x_\varepsilon) - \varepsilon k_0) = \psi(0) \qquad (13.25)$$

Because of Theorem 13.17 we have for $h(x_*) = 0$ that $\phi(h(x_*)) = 0$ and so (13.25) implies

$$\phi(h(x_\varepsilon)) \leq -\lambda ||x_* - x_\varepsilon||.$$

The last inequality and the property (p_2) imply

$$h(x_\varepsilon) \in -clB - \lambda ||x_* - x_\varepsilon|| k_0$$

and

$$f(x_\varepsilon) \in f(x_*) - \lambda||x_* - x_\varepsilon||k_0 - clB.$$

On the other hand, from (i) we get

$$\phi(h(x_\varepsilon)) \leq \phi(h(x_*)) - \lambda||x_* - x_\varepsilon|| \leq \phi(h(x_*)) \leq$$

$$\leq \inf_{x \in X} \phi(h(x)) + \varepsilon \leq \phi(h(x)) + \phi(\varepsilon k_0), \text{ for all } x \in X \tag{13.26}$$

Now we define a set

$$D := \{y \in E \mid \phi(-y + h(x_\varepsilon) - \varepsilon k_0) < \phi(h(x_\varepsilon) - \varepsilon k_0)\} \tag{13.27}$$

and a mapping $\psi : E \to R$ by

$$\psi(y) := \phi(y + h(x_\varepsilon) - \varepsilon k_0) \tag{13.28}$$

Considering (13.28), (13.26), Proposition 13.9 and the sublinearity of ϕ we get

$$\psi(h(X) + \varepsilon k_0 - h(x_\varepsilon)) = \phi(h(X)) \geq$$

$$\geq \phi(h(x_\varepsilon)) + \phi(-\varepsilon k_0) \geq \phi(h(x_\varepsilon) - \varepsilon k_0) = \psi(0). \tag{13.29}$$

Because (13.27) and (13.28) we can conclude that:

$$\psi(-D) = \phi(-D + h(x_\varepsilon) - \varepsilon k_0) < \phi(h(x_\varepsilon) - \varepsilon k_0) = \psi(0).$$

The last inequality and (13.29) imply

$$(h(X) + \varepsilon k_0 - h(x_\varepsilon) \cap (-D) = \emptyset,$$

and

$$f(X) \cap (f(x_\varepsilon) - \varepsilon k_0 - D) = \emptyset.$$

The strict K-monotonicity of the mapping ϕ implies $K \setminus \{0\} \subset D$. Since ϕ is continuous the functional ψ is also continuous and so D is open. In order to show that $clD + K \setminus \{0\} \subset D$, we choose $x_1 \in clD$ and $x_2 \in x_1 + (K \setminus \{0\})$. Then, because of the strict K-monotonicity of ψ we have

$$\psi(-x_2) < \psi(-x_1) \leq \psi(0)$$

and

$$x_1 \in clD \subset \{x \mid \psi(-x) \leq \psi(0)\}.$$

This means that $x_2 \in clD + (K \setminus \{0\})$ implies $x_2 \in D$. Since $k_0 \in \text{int}K$ and $clD+(K\setminus\{0\}) \subset D$ it follows that $\varepsilon k_0+D \subset D$. Also, $0 \in cl(K\setminus\{0\}), K\setminus\{0\} \subset D$ and $0 \notin D$ imply that $0 \in clD \setminus D$. Therefore we obtain assertion (1) of Theorem 13.19.

We observe that (ii) coincides with assertion (2). In order to prove assertion (3) of our theorem, we suppose that there exists an element $x \in X$ such that

$$f(x) + \lambda||x - x_\varepsilon||k_0 \in f(x_\varepsilon) - (K \setminus \{0\}).$$

Setting $h(x) = f(x) - f(x_*)$ we obtain $h(x) + \lambda||x - x_\varepsilon||k_0 \in h(x_\varepsilon) - (K \setminus \{0\})$. Then we have $f(x) \in h(x_\varepsilon) - \lambda||x - x_\varepsilon||k_0 - (K \setminus \{0\})$.

Considering the strict K-monotonicity of the sublinear mapping ϕ defined in (13.24) with $C = B$ we deduce

$$\phi(h(x)) < \phi(h(x_0) - \lambda||x - x_\varepsilon||k_0 \leq \phi(h(x_\varepsilon)) + \phi(-\lambda||x - x_\varepsilon||k_0).$$

Applying Proposition 13.9 we get with

$$\phi(-\lambda||x - x_\varepsilon||k_0) = -\lambda||x - x_\varepsilon||$$

such that

$$\phi(h(x)) + \lambda||x - x_\varepsilon|| \leq \phi(h(x_\varepsilon)),$$

which is a contradiction to (iii) and the proof is complete. ∎

An interesting variant of Theorem 13.19 where the space X is replaced by a closed non-empty subset U is Theorem 4.1 proved in Tammer (1992). Now we will give an extension to vector valued mappings of Ekeland's principle. This extension is due to G. Isac and it was recently considered by several authors as Chen and Huang (1998), Chen e.a. (1997), Chen e.a. (2000) among others. We note that our extension has interesting applications to the study of efficiency and ε-efficiency (Isac, 1996).

The following two results are useful for our extension of Ekeland's principle.

PROPOSITION 13.10. *Let $E(\tau)$ be a locally convex space and $K \subset E$ a convex cone. If K is normal then every monotone increasing and order bounded sequence in E is weakly Cauchy-sequence. The converse is true if E is a Banach space. In general, the converse implies that every order interval is topologically bounded.*

PROOF. We suppose that K is normal and $\{x_n\}_{n\in N} \subset E$ is a sequence such that

$$x_1 \leq x_2 \leq \cdots \leq u$$

where u is an element in E.

The normality of K implies that $\{x_n\}_{n\in N}$ is topologically bounded (since $\{x_n\}_{n\in N} \subset [x_1, u]_0$) where $[x_1, u]_0 = \{x \in E \mid x_1 \leq x \leq u\}$.

For every $f \in K^*$ we have that $\{f(x_n)\}_{n\in N}$ is convergent and because $E^* = K^* - K^*$, we deduce that, for every $h \in E^*$, the sequence $\{h(x_n)\}_{n\in N}$ is convergent, that is $\{x_n\}_{n\in N}$ is weakly Cauchy-sequence.

To prove the converse, we suppose that not any order interval is topologically bounded. Then there exist $u, w \in E$ such that $u < w$ and $[u, w]_0$ is not topologically bounded.

Using the translation we can suppose that there exists $v \in K$ such that the order interval $[0, v]_0$ is not topologically bounded. Hence, there exists a continuous seminorm p and a sequence $\{x_n\}_{n\in N} \subset [0, v]_0$ such that $7^n \leq p(x_n)$, for every $n \in N$.
Now we consider the sequence

$$y_n = \frac{x_1}{2} + \frac{x_2}{2^2} + \cdots + \frac{x_n}{2^n}, \text{for every } n \in N.$$

We have

$$y_1 \leq y_2 \leq \cdots \leq y_n \leq \ldots \leq v$$

and hence $\{y_n\}_{n\in N}$ is order bounded.
Therefore by assumption $\{y_n\}_{n\in N}$ is a weakly Cauchy-sequence and consequently it is topologically bounded.
On the other hand we have,

$$p(y_n) \geq \frac{1}{2^n}p(x_n) - \frac{1}{2^{n-1}}p(x_{n-1}) - \cdots - \frac{1}{2}p(x_1) \geq$$

$$\geq \left(\frac{7}{2}\right)^n - \left(\frac{7}{2}\right)^{n-1} - \ldots - \frac{7}{2} = \left(\frac{7}{2}\right)^n - \frac{\left(\frac{7}{2}\right)^n - \frac{7}{2}}{\frac{7}{2} - 1} =$$

$$= \frac{3}{5}\left(\frac{7}{2}\right)^n + \frac{7}{5}, \text{ for every } n \in N,$$

which implies that $p(y_n) \to +\infty$, that is we have a contradiction.
When E is a Banach space we have that our condition implies that K is normal, since for a Banach space the cone K is normal if and only if any order interval is topologically bounded, fact which is not true for any locally convex space. The proof is complete. ■

Let (X, d) be a complete metric space. We recall that a set-valued mapping $\Gamma : X \to X$ (i.e. $\Gamma : X \to 2^X$) is a dynamical system if $\Gamma(x)$ is non-empty for any $x \in X$, and we say that an element $x_* \in X$ is a critical point of Γ if $\Gamma(x_*) = \{x_*\}$.

Every critical point of Γ is a fixed point, but the converse is not true. To prove our extension of Ekeland's principle we will use the following result, which is a reformulation, for dynamical systems, of Dancs-Hegedus-Medvegyev's theorem. [See (Dancs e.a. (1983)].

THEOREM 13.20. *Let (X, d) be a complete metric space and $\Gamma : X \to 2^X$ a dynamical system. If the following assumptions are satisfied:*

(i). $\Gamma(x)$ is closed for every $x \in X$,

(ii). $x \in \Gamma(x)$ for every $x \in X$.,

(iii).$x_2 \in \Gamma(x_1)$ implies $\Gamma(x_2) \subseteq \Gamma(x_1)$ for all $x_1, x_2 \in X$,

(iv). for every sequence $\{x_n\}_{n\in N} \subset X$ with the property $x_{n+1} \in \Gamma(x_n)$, for every $n \in N$ we have $\lim_{n\to\infty} d(x_n, x_{n+1}) = 0$,

then Γ has a critical point $x_* \in X$. Moreover, for every $x_0 \in X$, there is a critical point of Γ in $\Gamma(x_0)$. ■

Let $(E(\tau), K$ be an ordered locally convex space, where $K \subset E$ is a closed convex cone.

DEFINITION 13.10. We say that a mapping $\Phi : X \times X \to E$ is a *halfdistance* if the following properties are satisfied:

1. $\Phi(x,x) = 0$ for all $x \in X$,
2. $\Phi(x,y) \leq \Phi(x,z) + \Phi(z,y)$, for all $x, y, z \in X$.

Remark. The class of halfdistances is not empty, since for every mapping $f : X \to E$, the mapping $\Phi(x,y) = f(y) - f(x)$, for all $x, y \in X$ is a halfdistance.

Also, if F is an arbitrary vector space and $T : F \to E$ is a subadditive mapping such that $T(0) = 0$, then for every mapping $f : X \to F$, we can verify that

$$\Phi(x,y) = T(f(y) - f(x)), \text{ for all } x, y \in X$$

is a halfdistance.

THEOREM 13.21. *Let (X, d) be a complete metric space $(E(\tau), K)$ an ordered locally convex space, with K-normal and Φ a halfdistance from X into E. If for an element $k^0 \in K \setminus \{0\}$ the following assumptions are satisfied:*

1. *for every $x \in X$, the set*
$$\{y \in X \mid \Phi(x,y) + k^0 d(x,y) \in -K\}$$
is closed,
2. *there exist $v_0 \in X$ and $w_0 \in E$ such that $\Phi(v_0, x) \geq w_0$, for every $x \in X$,*

then there exists $x_* \in X$ such that

$$\Phi(x_*, x) + k^0 d(x_*, x) \notin -K, \text{ for all } x \in X \setminus \{x_*\}.$$

PROOF. First, we remark that if we consider the dynamical system

$$\Gamma_\Phi(x) = \{y \in X \mid \Phi(x,y) + k^0 d(x,y) \in -K\},$$

for every $x \in X$, then the theorem will be proved if we show that Γ_Φ has a critical point in X. To show this, it is sufficient to verify the assumptions (i)-(iv) of Theorem 13.20.

Indeed, from assumption (1) we have that $\Gamma_\Phi(x)$ is closed for every $x \in X$, that is (i) is satisfied.

Using the properties of d and Φ we have immediately that $x \in \Gamma_\Phi(x)$, for every $x \in X$ which means that (ii) is also satisfied. To verify (iii) we consider two elements $x_1, x_2 \in X$ such that $x_2 \in \Gamma_\Phi(x_1)$. We must show that $\Gamma_\Phi(x_2) \subseteq \Gamma_\Phi(x_1)$. Indeed, since $x_2 \in \Gamma_\Phi(x_1)$ we have

$$\Phi(x_1, x_2) + k^0 d(x_1, x_2) \in -K \tag{13.30}$$

If $z \in \Gamma_\Phi(x_2)$ is an arbitrary element, we have

$$\Phi(x_2, z) + k^0 d(x_2, z) \in -K \tag{13.31}$$

We will have $z \in \Gamma_\Phi(x_1)$ if we show that

$$\Phi(x_1, z) + k^0 d(x_1, z) \in -K.$$

From (13.30) there is an element $k_1 \in K$ such that $\Phi(x_1, x_2) + k^0 d(x_1, x_2) = -k_1$ and from (13.31) there is an element $k_2 \in K$ such that $\Phi(x_1, x_2) + k^0 d(x_1, x_2) = -k_2$.

Since Φ is a halfdistance there is an element $k_3 \in K$ such that

$$\Phi(x_1, z) = \Phi(x_1, x_2) + \Phi(x_2, z) - k_3$$

and since $k^0 d(x_1, z) \leq k^0 d(x_1, x_2) + k^0 d(x_2, z)$, there is an element $k_4 \in K$ such that

$$k^0 d(x_1, z) = k^0 d(x_1, x_2) + k^0 d(x_2, z) - k_4.$$

Finally we obtain that

$$\Phi(x_1, z) + k^0 d(x_1, z) = \Phi(x_1, x_2) + \Phi(x_2, z) - k_3 + k^0 d(x_1, z)$$

$$= \Phi(x_1, x_2) + \Phi(x_2, z) - k_3 + k^0 d(x_1, x_2) + k^0 d(x_2, z) - k_4$$

$$= -k_1 - k_2 - k_3 - k_4 \in -K, \text{ which implies that } \Gamma_\Phi(x_2) \subseteq \Gamma_\Phi(x_1),$$

that is, (iii) is satisfied. To verify (iv) we consider a sequence $\{x_n\}_{n \in N} \subset X$ such that $x_{n+1} \in \Gamma_\Phi(x_n)$ for every $n \in N$, with x_1 arbitrary element in X. We have

$$\Phi(x_n, x_{n+1}) + k^0 d(x_n, x_{n+1}) \in -K,$$

which implies

$$k^0 d(x_n, x_{n+1}) = -\Phi(x_n, x_{n+1}) - k_5, \text{ where } k_5 \in K \tag{13.32}$$

Since Φ is a halfdistance we have,

$$\Phi(v_0, x_{n+1}) \leq \Phi(v_0, x_n) + \Phi(x_n, x_{n+1}),$$

which implies

$$-\Phi(x_n, x_{n+1}) \leq \Phi(v_0, x_n) - \Phi(v_0, x_{n+1})$$

and hence, there is an element $k_6 \in K$ such that

$$-\Phi(x_n, x_{n+1}) = \Phi(v_0, x_n) - \Phi(v_0, x_{n+1}) - k_6. \tag{13.33}$$

From (13.32) and (13.33) we obtain

$$k^0 d(x_n, x_{n+1}) = \Phi(v_0, x_n) - \Phi(v_0, x_{n+1}) - k_5 - k_6$$

and if we denote $k = k_5 + k_6$, we have

$$k^0 d(x_n, x_{n+1}) = \Phi(v_0, x_n) - \Phi(v_0, x_{n+1}) - k$$

that is,

$$k^0 d(x_n, x_{n+1}) \leq \Phi(v_0, x_n) - \Phi(v_0, x_{n+1}) \tag{13.34}$$

If we consider the series $\sum_{n=1}^{\infty} k^0 d(x_n, x_{n+1})$ and we denote by $S_m = \sum_{n=1}^{m} k^0 d(x_n, x_{n+1})$, we remark that $S_m \leq \Phi(v_0, x_1) - w_0$, for every $m \in N$, that is the sequence $\{S_m\}_{m \in N}$ is order bounded in E.

Since the sequence $\{S_m\}_{m \in N}$ is monotone increasing and K is normal, we have that it is a weakly Cauchy-sequence. (We used Proposition 13.35). Since $\{S_m\}_{m \in N}$ is a weakly Cauchy-sequence on the line $L(k^0)$ generated in E by the element k^0, it is convergent in $L(k^0)$ and since $L(k^0)$ is homeomorphic to the real field by the homeomorphism $\phi(tk^0) = t$, we obtain that $\sum_{n=1}^{\infty} k^0 d(x_n, x_{n+1})$ is a convergent series.

Hence, the sequence $\{d(x_n, x_{n+1})\}_{n \in N}$ is convergent to zero and assumption (iv) is satisfied.

Applying finally Theorem 13.36 we have that Γ_Φ has a critical point and the proof is complete. ∎

If we apply also the second part of the conclusion of Theorem 13.20 we obtain the following corollary.

COROLLARY 13.7. *If all the assumptions of Theorem* 13.21 *are satisfied, then there exists an element* $x_* \in \Gamma_\Phi(v_0)$ *such that* $\Phi(x_*, x) + k^0 d(x_*, x) \notin -K$, *for all* $x \in X \setminus \{x_*\}$. *(The element* v_0 *is the same as in Theorem* 13.21*).* ∎

The next result is the analogue for vector-valued functions of the *strong form of Ekeland's principle.*

THEOREM 13.22. *Let (X,d) be a complete metric space,* $(E(\tau), K)$ *an ordered locally convex space with* K *a normal closed convex cone and* $f : X \to E$ *a mapping.If* $\varepsilon > 0$ *is an arbitrary real number and there exist* $a \in X$ *and* $k^0 \in K \setminus \{0\}$ *such that:*

1. *$f(a) \leq f(x) + \varepsilon k^0$, for all $x \in X$,*

2. *for every $x \in X$ and every real number $\alpha > 0$ the set*

$$\{y \in X | f(y) - f(x) + \alpha d(x,y)k^0 \in -K\} \text{ is closed },$$

then for every $\lambda > 0$ there exists $x_\lambda \in X$ such that:

(i). $f(x_\lambda) \leq f(a)$,

(ii). $d(x_\lambda, a) \leq \lambda$,

(iii). $f(x) - f(x_\lambda) + \frac{\varepsilon}{\lambda} d(x_\lambda, x)k^0 \notin -K$, for all $x \in X \setminus \{x_\lambda\}$.

PROOF. We consider the halfdistance $\Phi : X \times X \to E$ defined by $\Phi(x,y) = f(y) - f(x)$ and we remark that all the assumptions of Theorem 13.21 are satisfied for the distance $\frac{\varepsilon}{\lambda} d$, if we consider $v_0 = a$ and $w_0 = -\varepsilon k^0$.

Then by Corollary 13.7 there exists

$$x_\lambda \in \Gamma_\Phi(a) = \left\{ y \in X \mid f(y) - f(a) + \frac{\varepsilon}{\lambda} d(a,y)k^0 \in -K \right\}$$

such that

$$f(x) - f(x_\lambda) + \frac{\varepsilon}{\lambda} d(x_\lambda, x)k^0 \notin -K, \text{ for all } x \in X \setminus \{x_\lambda\},$$

which is exactly conclusion (iii).

Since $x_\lambda \in \Gamma_\Phi(a)$, we have that there is $k \in K$ such that

$$\frac{\varepsilon}{\lambda} d(a, x_\lambda)k^0 = -k + f(a) - f(x_\lambda),$$

that is

$$\frac{\varepsilon}{\lambda} d(a, x_\lambda)k^0 + k = f(a) - f(x_\lambda)$$

and hence

$$\frac{\varepsilon}{\lambda} d(a, x_\lambda)k^0 \leq f(a) - f(x_\lambda) \leq \varepsilon k^0,$$

which implies $d(a, x_\lambda) \leq \lambda$, that is the conclusion (ii) is satisfied.

Finally, since $f(a) - f(x_\lambda) - k = \frac{\varepsilon}{\lambda} d(a, x_\lambda)k^0$ we deduce $f(a) = f(x_\lambda) + k + \frac{\varepsilon}{\lambda} d(a, x_\lambda)k^0$ which implies $f(x_\lambda) \leq f(a)$, that is (i) is also satisfied and the proof is complete. ∎

THEOREM 13.23. *Let (X,d) be a complete metric space, $(E(\tau), K)$ an ordered locally convex space with K normal and let Φ be a halfdistance from X into E. Given $k^0 \in K \setminus \{0\}$, we suppose the set $\{y \in X \mid \Phi(x,y) + k^0 d(x,y) \in -K\}$ to be closed for every $x \in X$. For every $v_0 \in X$ such that there exists $w_0 \in E$ with the property that $\Phi(v_0, x) \geq w_0$, for every $x \in X$, we denote by*

$$\mathcal{D}_0 = \{x \in X \mid \Phi(v_0, y) + k^0 d(v_0, y) \in -K\}.$$

Then under the hypotheses indicated before, the following assertions are equivalent:

(A). [EKELAND'S PRINCIPLE]
There exists $x_* \in \mathcal{D}$ such that

$$\Phi(x_*, x) + k^0 d(x_*, x) \notin -K, \text{ for all } x \in X \setminus \{x_*\}.$$

(B). [CARISTI-KIRK TYPE THEOREM]
If $T : X \to X$ is a multivalued mapping such that, for every $\bar{x} \in \mathcal{D}_0$ there exists $x \in T(\bar{x})$ such that

$$\Phi(\bar{x}, x) + k^0 d(\bar{x}, x) \in -K,$$

then there exists $x_* \in \mathcal{D}_0$ such that $x_* \in T(x_*)$.

(C). If $M \subset X$ has the property that for every $\bar{x} \in \mathcal{D}_0 \backslash M$ there exists $x \in T(\bar{x})$ such that $x \neq \bar{x}$ and

$$\Phi(\bar{x}, x) + k^0 d(\bar{x}, x) \in -K$$

then there exists $x_* \in \mathcal{D}_0 \cap M$.

PROOF. For a proof of this theorem the reader is referred to Isac (1996). ■

Now, we will give an application to the study of ε-efficiency.

THEOREM 13.24. *Let (X,d) be a complete metric space, $(E(\tau), K)$ a locally convex space ordered by the closed convex normal cone K and $f : X \to E$ a mapping. If the following assumptions are satisfied:*

1. *for an element $k^0 \in K \setminus \{0\}$ the set*

$$\{y \in X \mid f(y) - f(x) + k^0 d(x, y) \in -K\}$$

is closed for every $x \in X$,

2. *there exist $v_0 \in X$ and $w_0 \in E$ such that*

$$f(x) - f(v_0) \geq w_0, \text{ for all } x \in X,$$

then for every $\varepsilon > 0$ there exists $x_\varepsilon \in X$ such that for $f_{\varepsilon k^0}(x) := f(x) + \sqrt{\varepsilon} d(x, x_\varepsilon) k^0$ we have

(i). $f_{\varepsilon k^0}(x_\varepsilon) \in E_m[f_{\varepsilon k^0}(X), K]$ and

(ii) $f_{\varepsilon k^0}(x_\varepsilon) \neq f_{\varepsilon k^0}(x)$ for all $x \in X \setminus \{x_\varepsilon\}$.

PROOF. We suppose that assertions (i), (ii) do not hold, that is, for a certain $\varepsilon > 0$ there exists no element x_ε with

$$f_{\varepsilon k^0}(x_\varepsilon) \in E_m[f_{\varepsilon k^0}(X), K] \text{ and}$$

$$f_{\varepsilon k^0}(x_\varepsilon) \neq f_{\varepsilon k^0}(x) \text{ for all } x \in X \setminus \{x_\varepsilon\}.$$

This implies that for each $x \in X$ there exists an element $y \neq x$ with the property

$$f(y) + k^0 d_*(x,y) \in f(x) - K, \text{ where } d_* := \sqrt{\varepsilon} d. \tag{13.35}$$

Now we define the mapping $T : X \to X$ by $T(x) = y$, where y is given according to (13.35). Then obviously we have $k^0 d_*(x, T(x)) + f(T(x)) - f(x) \in -K$ and we remark that all the assumptions of Theorem 13.23 (B) [Caristi-Kirk] are satisfied considering

$$\Phi(x,y) = f(y) - f(x) \text{ and}$$

$$\mathcal{D}_0 = \{x \in X | f(x) - f(v_0) + k^0 d_*(v_0, x) \in -K\}.$$

Hence we obtain that T has a fixed point x_* which is impossible, since $T(x) = y \neq x$ for all $x \in X$ and the theorem is proved. ∎

Remark. Our Theorem 13.23 was recently generalized in the papers by Chen and Huang (1998), Chen e.a. (2000), Chen e.a. (1997).

It is interesting to apply these generalizations to the study of ε-efficiency.

13.6 APPROXIMATIVE EFFICIENCY BY A PERTURBATION OF CONE

In this section, we will show that if we pass from the cone K to another cone $K_\lambda \subseteq K$ (where $\lambda \in R_+ \setminus \{0\}$), we obtain the existence of an efficient point with respect to K_λ. The cone K_λ is defined such that, when λ tends to zero, the cone K_λ tends to K. Moreover, the efficient point with respect to K_λ is obtained by a special construction, i.e., we have a constructive existence result.

The idea to use this special cone K_λ and the main result of this section are due to Göpfert and Tammer (1995).

Let $(E, ||\cdot||)$ be a Banach space and $K \subset E$ a closed convex cone. We recall that given a closed subset A of E, we say that a point $x_0 \in A$ is a K-support point (or conical support point) if $A \cap (K + x_0) = \{x_0\}$.

Obviously, a K-support point is a Pareto (maximal) efficient point.

The existence of K-support points has been studied by several authors. In particular the Bishop-Phelps Theorem is an important existence theorem for K-support points (Bishop and Phelps, 1962; Phelps, 1974).

The Bishop-Phelps Theorem gives the existence of support points of a closed set A in the convex hull of A. We will give now an interesting existence theorem for conical support points when the set A is only closed. Moreover this result is a constructive existence theorem and it is due to Göpfert and Tammer (1995).

Given $\lambda \in]0,1[$ and $k^0 \in intK$, we consider the convex cone

$$K_\lambda = \{x \in E|\ x - \lambda||x||k^0 \in K\}.$$

We have that $K_\lambda \subseteq K$. Indeed, if $x \in K_\lambda$, then there exists $k \in K$ such that $x - \lambda||x||k^0 = k$, which implies $x = k + \lambda||x||k^0 \in K$.

We can also show that if $0 < \lambda_1 < \lambda_2 < 1$, then $K_{\lambda_2} \subseteq K_{\lambda_1}$.

Since $k^0 \in intK$, by a classical separation theorem we have that there exists $y^* \in K^*$ such that $y^*(k^0) = 1$.

If $x \in K_\lambda$ then we have

$$y^*(x - \lambda||x||k^0) = y^*(x) - \lambda||x||y^*(k^0) = y^*(x) - \lambda||x|| \leq 0,$$

which implies that

$$x \in K(y^*, \lambda) = \{x \in E|\ \lambda||x|| \leq y^*(x)\},$$

that is $K_\lambda \subseteq K(y^*, \lambda)$. We recall that $K(y^*, \lambda)$ is a Bishop-Phelps cone.

Now, we give the main result.

THEOREM 13.25. *Let $(E, ||\cdot||)$ be a Banach space and $K \subset E$ a pointed closed convex cone with non-empty interior and $A \subset E$ a colsed subset. Suppose that A is bounded by K, that is, $A \subset y - K$ for an arbitrary $y \in E$. Then, given $0 < \lambda < 1$, we have that for any $u \in A$ there exists an element $x_* \in A$ with $x_* \in K_\lambda + u$ and $A \cap (K_\lambda + x_*) = \{x_*\}$.*

PROOF. We define the following sequence of sets:

$$A_n = A \cap (K_\lambda + x_n). \tag{13.36}$$

From our assumptions we deduce that the sets A_n are closed. The sequence $\{x_n\}_{n \in N}$ is defined by induction.

Let $x_1 = u$ and, if we have obtained $x_1, x_2, ..., x_n$, we choose $x_{n+1} \in A$ such that:

$$\text{does not exist } x \in A_n \text{ with } x \in x_{n+1} + \frac{1}{n+1}k^0 + (K \setminus \{0\}). \tag{13.37}$$

The element x_{n+1}exists because A is bounded from above by K. Since $x_{n+1} \in A_n$, we have that $x_{n+1} \in K_\lambda + x_n$, which implies

$$K_\lambda + x_{n+1} \subseteq K_\lambda + K_\lambda + x_n \subseteq K_\lambda + x_n,$$

where the last inclusion follows from the fact that K_λ is a convex cone.

From (13.36) we have

$$A_{n+1} = A \cap (K_\lambda + x_{n+1}) \subset A \cap (K_\lambda + x_n) = A_n$$

and hence $A_{n+1} \subseteq A_n$.

Further, for an arbitrary $x \in A_{n+1}$ it holds $x \in x_{n+1} + K_\lambda$. This implies $x - x_{n+1} \in K_\lambda$. From the definition of the cone $K_\lambda$it follows

$$(x - x_{n+1}) - \lambda||x - x_{n+1}||k^0 \in K,$$

which implies

$$x \in x_{n+1} + \lambda||x - x_{n+1}||k^0 + K. \tag{13.38}$$

On the other hand $x \in A_{n+1} \subseteq A_n$ implies $x \in A_n$. Moreover, because of (13.37) we have

$$x \notin x_{n+1} + \frac{1}{n+1}k^0 + K \setminus \{0\}.$$

Hence, considering (13.38) we have

$$\lambda||x - x_{n+1}|| < \frac{1}{n+1},$$

and so, $diam(A_{n+1}) \leq \dfrac{1}{\lambda(n+1)}$.

Since A is closed in E it is complete. Hence, according to Cantor's Theorem we have that $\bigcap_{n \in N} A_n$ contains exactly one element.

Let x_* be this element. Obviously, $x_* \in A_1 = A \cap (K_\lambda + u)$, thus $x_* \in K_\lambda + u$. Moreover,

$$x_* \in A_n = A \cap (K_\lambda + x_n), \text{ for all } n \in N,$$

which implies that $x_* \in K_\lambda + x_n$, for all $n \in N$, and hence

$$x_* + K_\lambda \subseteq K_\lambda + x_n + K_\lambda \subset K_\lambda + x_n, \text{ for all } n \in N.$$

Therefore we have

$$(K_\lambda + x_*) \cap A \subset \bigcap_{n \in N} (K_\lambda + x_n) \cap A = \bigcap_{n \in N} A_n = \{x_*\}$$

and the proof is complete. ■

13.7 EFFICIENCY IN PRODUCT SPACES AND ε-EFFICIENCY

In this section we will give some maximal point theorems in a product space $X \times Y$, where X and Y are Banach spaces, in order to derive some variational principles for vector optimization problems, in general spaces without any scalarization.

The maximal point theorems are generalizations of Phelps' maximal point theorem and are due to A. Göpfert, Chr. Tammer and C. Zalinescu.

We note also that the variational principles for vector optimization problems obtained by maximal point theorems in product spaces, are different from the variational principles given by Loridan, Nemeth, Khan, Tammer and Isac.

Let $(X, \|\cdot\|)$ be a Banach space. Given $\varepsilon > 0$ we consider the convex cone

$$K(\varepsilon, X, R) = \left\{(x, r) \in X \times R \mid r + \sqrt{\varepsilon}\|x\| \leq 0\right\}.$$

The cone $K(\varepsilon, X, R)$ is closed. Using the cone $K(\varepsilon, X, R)$, R.R. Phelps showed that Ekeland's variational principle (the strong form) (Theorem 13.18) is a direct consequence of the following maximal point theorem.

THEOREM 13.26. *Let $(X, \|\cdot\|)$ be a Banach space and suppose that $A \subset X \times R$ is a non-empty closed set such that the set*

$$\{r \in R \mid (x, r) \in A, \text{ for some } x \in X\}$$

is bounded from below.
Then for any $(\alpha, \beta) \in A$ there exists a point $(x_*, r_*) \in A$ such that

(i). $(x_*, r_*) \in A \cap (K(\varepsilon, X, R) + (\alpha, \beta))$

(ii). $\{(x_*, r_*)\} = A \cap (K(\varepsilon, X, R) + (x_*, r_*))$.

PROOF. A proof of this theorem is given in Phelps (1993). ■

We will give now a *new maximal point theorem* due to Göpfert and Tammer (1995). In this new maximal point theorem we have that conclusions (i) and (ii) of Theorem 13.43 hold even in $X \times Y$, where Y is a Banach space ordered by a given cone K. To obtain this, we must replace the cone $K(\varepsilon, X, R)$ by an appropriate cone in $X \times Y$. From this new maximal point theorem we will obtain a variant of Ekeland's principle applicable to multicriteria optimization.

Let $(X, \|\cdot\|)$ be an arbitrary Banach space and $(Y, \|\cdot\|)$ a Banach space, ordered by a closed pointed convex cone $K \subset Y$, such that $\text{int}K$ is non-empty.

We choose $k^0 \in \text{int}K$ such that $\|k^0\| = 1$. For any $\varepsilon > 0$ we define the pointed convex cone

$$K_\varepsilon = \left\{(x, y) \in X \times Y \mid y + \sqrt{\varepsilon} k^0 \|x\| \in -K\right\}. \tag{13.39}$$

We have that $(0, -k^0) \in K_\varepsilon$. Also considering K as embedded in $X \times Y$ we have that

$$K = \{(0, -y) \mid (x, y) \in K_\varepsilon, \text{ for some } x\}.$$

Indeed, $(x, y) \in K_\varepsilon$ implies $y \in -\sqrt{\varepsilon} k^0 \|x\| - K \subset -K$ and $(0, y) \in \{0\} \times (-K)$ implies $y + \sqrt{\varepsilon} k^0 \|0\| \in -K$ that means $(0, y) \in K_\varepsilon$ for all $y \in -K$.

If we take $Y = R, K = R_+$ and $k^0 = 1$, we have that K_ε is exactly the cone $K(\varepsilon, X, R)$. The new maximal point theorem is the following result.

THEOREM 13.27. *Let $(X, \|\cdot\|)$ and $(Y, \|\cdot\|)$ be Banach spaces and $A \subset X \times Y$ a closed subset. Let $K \subset Y$ be a pointed, closed, convex cone, with nonempty interior. If the following assumptions are satisfied:*

1. $k^0 \in \text{int}K, y_0 \in Y$ and $||k^0|| = 1$,
2. $\{y \in Y | (x,y) \in A \text{ for some } x \in X\} \subset y_0 + K$,
3. $B \subset Y$ is a pointed convex cone with $clB + (K \setminus \{0\}) \subset \text{int}B$,
4. $\varepsilon > 0$

then, for any $(x,y) \in A$ there exits a point $(x_*,y_*) \in A$ such that

$$(x_*,y_*) \in A \cap (K_\varepsilon + (x,y)) \text{ and } \{(x_*,y_*)\} = A \cap (K_\varepsilon + (x_*,y_*)).$$

PROOF. We follow the ideas of the proof proposed by A. Göpfert and Chr. Tammer. Consider a sequence of sets $\{A_n\}_{n\in N}$ of the form

$$A_n = A \cap (K_\varepsilon + (x_n,y_n)) \tag{13.40}$$

From our assumption we deduce that A_n is closed for any $n \in N$. The sequence $\{(x_n,y_n)\}$ is defined inductively as follows:

$$(x_1,y_1) = (x,y)$$

We supposed that we obtained $(x_1,y_1),(x_2,y_2),...,(x_n,y_n)$. If this is the case, we choose $(x_{n+1},y_{n+1}) \in A_n$ such that

$$\text{does not exist } (x,y) \in A_n \text{ with } y \in y_{n+1} - \frac{1}{n+1}k^0 - (B \setminus \{0\}). \tag{13.41}$$

The element (x_{n+1},y_{n+1}) must exist because of the inclusion

$$\{y \in Y \mid (x,y) \in A \text{ for some } x \in X\} \subset y_0 + K$$

The inclusion $(x_{n+1},y_{n+1}) \in A_n$ implies the inclusion $(x_{n+1},y_{n+1}) \in K_\varepsilon + (x_n,y_n)$ and hence we can conclude that

$$K_\varepsilon + (x_{n+1},y_{n+1}) \subset K_\varepsilon + K_\varepsilon + (x_n,y_n) \subset K_\varepsilon + (x_n,y_n).$$

Therefore, we have from the definition of the set A_n in (13.40) that

$$A_{n+1} = A \cap (K_\varepsilon + (x_{n+1},y_{n+1})) \subset A \cap (K_\varepsilon + (x_n,y_n)) = A_n,$$

and hence

$$A_{n+1} \subseteq A_n \tag{13.42}$$

Since for an arbitrary element $(x,y) \in A_n$ we have $(x,y) \in K_\varepsilon + (x_n,y_n)$ we obtain the inclusions

$$(x,y) - (x_n,y_n) \in K_\varepsilon \text{ and } y - y_n + \sqrt{\varepsilon}||x - x_n||k^0 \in -K$$

which imply

$$y \in y_n - \sqrt{\varepsilon}||x - x_n||k^0 - K. \tag{13.43}$$

Because of (13.41) and the inclusion $(x,y) \in A_n \subset A_{n-1}$ we get the inequality

$$\sqrt{\varepsilon}||x - x_n|| < \frac{1}{n}$$

and so

$$||x - x_n|| < \frac{1}{n\sqrt{\varepsilon}}. \tag{13.44}$$

Now, we introduce a functional $\phi : Y \to R$ defined by

$$\phi(y) = \inf\{t \in R \mid y \in -clK + tk^0\}, \tag{13.45}$$

which is a continuous, sublinear, strictly $intK$-monotone with range $(-\infty, +\infty)$ and $\phi(\alpha k^0) = \alpha$, for all $\alpha \in R$. [About the functional ϕ see Gerth (Tammer) and Weidner (1990)].

Then for $(x,y) \in A_n$ from the sublinearity of the functional ϕ in (13.41) and (13.42) it follows that

$$\phi(y) + \phi\left(\frac{1}{n}k^0\right) \geq \inf_{y \text{ with } (x,y)\in A_n} \phi\left(y + \frac{1}{n}k^0\right) \geq$$

$$\geq \inf_{y \text{ with } (x,y)\in A_{n-1}} \phi\left(y + \frac{1}{n}k^0\right) \geq \phi(y_n),$$

and hence

$$\phi(y) + \phi\left(\frac{1}{n}k^0\right) \geq \phi(y_n)$$

which implies

$$\phi(y_n) - \phi(y) \leq \frac{1}{n}. \tag{13.46}$$

We have also $0 \leq \phi(y_n) - \phi(y)$. Indeed, for an arbitrary $(x,y) \in A_n$ we have (because of (13.43)),

$$y_n - y \in \sqrt{\varepsilon}||x - x_n||k^0 + K \text{ and so}$$

$$y_n - \sqrt{\varepsilon}||x - x_n||k^0 \in y + K.$$

Further, the K-monotonicity of the functional ϕ in (13.45) [See also (Gerth (Tammer) and Weidner (1990)], implies the inequality

$$\phi\left(y_n - \sqrt{\varepsilon}||x - x_n||k^0\right) \geq \phi(y).$$

Considering the sublinearity of the functional ϕ we obtain

$$\phi(y_n) + \phi\left(-\sqrt{\varepsilon}||x - x_n||k^0\right) \geq \phi(y).$$

Therefore we can conclude $\phi(y_n) - \phi(y) \geq \sqrt{\varepsilon}||x - x_n|| \geq 0$ and regarding (13.46), we have $0 \leq \phi(y_n) - \phi(y) \leq \frac{1}{n}$. For $n \to \infty$ we obtain $\phi(y_n) - \phi(y) \to 0$.

Because of (13.41) it follows from the definitions of the functional ϕ in (13.45) that $y_n - y \to 0$ as $n \to \infty$ and hence $||y_n - y|| < \delta_n$ for all $n, \delta_n \to 0$ as $n \to \infty$.

Using (13.44) we get

$$||x - x_n|| + ||y_n - y|| \leq \frac{1}{n\sqrt{\varepsilon}} + \delta_n$$

and so $diam(A) \to 0$. The completeness of Y and the closedness of the sets A_n yield that $\bigcap_{n\in N} A_n$ contains only one element. Let (x_*, y_*) be this element.

Obviously, we have

$$(x_*, y_*) \in A_1 = A \cap (K_\varepsilon + (x, y)),$$

which implies

$$(x_*, y_*) \in K_\varepsilon + (x, y).$$

Further,

$$(x_*, y_*) \in A_n = A \cap (K_\varepsilon + (x_n, y_n)), \ (n \in N)$$

implies

$$(x_*, y_*) \in K_\varepsilon + (x_n, y_n), \ (n \in N)$$

and hence

$$(x_*, y_*) + K_\varepsilon \subset K_\varepsilon + (x_n, y_n) + K_\varepsilon \subset K_\varepsilon + (x_n, y_n), \ (n \in N).$$

Finally we have

$$A \cap (K_\varepsilon + (x_*, y_*)) \subset A \cap (K_\varepsilon + (x_n, y_n)) = A_n, \text{for} \ \ (n \in N),$$

which implies $A \cap (K_\varepsilon + (x_*, y_*)) = \{(x_*, y_*)\}$ and the proof is complete. ■

Now, we will show that a variational principle for an optimization problem with an objective function which takes its values in a partially ordered space is a consequence of Theorem 13.27.

Let $(X, ||\cdot||)$ and $(Y, ||\cdot||)$ be Banach spaces and $K \subset Y$ a pointed, closed, convex cone with non-empty interior. Given a mapping $f : X \to Y$, we recall that f is *bounded from below* on X if there exists an element $y \in Y$ with $f(X) \subset y + K$.

We recall also that f is *lower semicontinuous* with respect to $y \in Y$ and K (on short: f is *K-lower semicontinuous*), if the set

$$D(X; r) = \{x \in X \mid f(x) \in y - K\}$$

is closed for each $y \in Y$.

We know that an element $f(x_\varepsilon) \in f(X)$ is called an *approximately efficient point* of $f(X)$ with respect to $K, k^0 \in \text{int}K$ and $\varepsilon \geq 0$, if

$$f(X) \cap (f(x_\varepsilon) - \varepsilon k^0 - (K \setminus \{0\})) = \emptyset.$$

The set of approximately efficient points of $f(X)$ with respect to K, k^0 and ε will be denoted by $E_m[f(X), K_{\varepsilon k^0}]$, where $K_{\varepsilon k^0} = \varepsilon k^0 + K$.

If $\varepsilon = 0$, the set $E_m[f(X), K_{\varepsilon k^0}]$ coincides with the set of efficient points of $f(X)$ with respect to K.

THEOREM 13.28. *Let $(X, ||\cdot||)$ be an arbitrary Banach space, $(Y, ||\cdot||)$ a Banach space ordered by a closed, pointed, convex cone $K \subset Y$ with non-empty interior and $f: X \to Y$ a K-lower semicontinuous mapping, bounded from below. Suppose given $k^0 \in intK$ with $||k^0|| = 1$ and a convex pointed cone $B \subset Y$ with $clB + (K \setminus \{0\}) \in intB$.*
Then, for any $\varepsilon > 0$ and any $f(x_0) \in E_m[f(X), K_{\varepsilon k^0}]$ there exists an element $x_\varepsilon \in X$ with the following properties:

1. $f(x_\varepsilon) \in f(x_0) - \sqrt{\varepsilon}||x_\varepsilon - x_0||k^0 - K$,
2. $||x_\varepsilon - x_0|| \le \sqrt{\varepsilon}$,
3. $f_{\varepsilon k^0}(x_\varepsilon) \in E_m[f(X), K_{\varepsilon k^0}]$, *where* $f_{\varepsilon k^0}(x) := f(x) + \sqrt{\varepsilon}||x - x_\varepsilon||k^0$.

PROOF. The assumption that f is K-lower semicontinuous and $\text{int}K \neq \emptyset$ imply that the set

$$epif = \{(x,y) \in X \times Y \mid y \in f(x) + K\}$$

is closed. Now, we take $A = epif$ and consider the set

$$K_\varepsilon + (x_0, f(x_0)) = \{(x,y) \in X \times Y | \sqrt{\varepsilon}||x - x_0||k^0 \in -y + f(x_0) - K\}.$$

We observe that

$$(x_0, f(x_0)) \in epif \cap (K_\varepsilon + (x_0, f(x_0))) \neq \emptyset.$$

Since the assumptions of Theorem 13.27 are satisfied, there exists a pair $(x_\varepsilon, y_\varepsilon)$ with

$$(x_\varepsilon, y_\varepsilon) \in epif \cap (K_\varepsilon + (x_0, f(x_0))) \tag{13.47}$$

and

$$\{(x_\varepsilon, y_\varepsilon)\} = epif \cap ((K_\varepsilon + (x_\varepsilon, y_\varepsilon)). \tag{13.48}$$

From (13.48) we get because of the inclusion $(x_\varepsilon, y_\varepsilon) \in epif$ that we must have

$$epif \cap (K_\varepsilon + (x_\varepsilon, f(x_\varepsilon))) = \{(x_\varepsilon, f(x_\varepsilon))\}.$$

This means that there is no $x \in X$, $x \neq x_\varepsilon$ with

$(x, f(x)) \in K_\varepsilon + (x_\varepsilon, f(x_\varepsilon)) = \left\{(x,y) \in X \times Y | y - f(x_\varepsilon) + \sqrt{\varepsilon}||x - x_\varepsilon||k^0 \in -K\right\}$.
Hence, we can conclude that

$$f(x) - f(x_\varepsilon) + \sqrt{\varepsilon}||x - x_\varepsilon||k^0 \notin -K$$

for all $x \in X, x \neq x_\varepsilon$, which implies that for all $x \in X$ it holds

$$f_{\varepsilon k^0}(X) \cap (f_{\varepsilon k^0}(x_\varepsilon) - (K \setminus \{0\})) = \emptyset,$$

where

$$f_{\varepsilon k^0}(x) := f(x) + \sqrt{\varepsilon}||x - x_\varepsilon||k^0.$$

Therefore we get $f_{\varepsilon k^0}(x_\varepsilon) \in E_m[f_{\varepsilon k^0}(X), K]$ and so we have shown that statement (3) holds. Moreover, (13.47) implies

$$\sqrt{\varepsilon}||x_\varepsilon - x_0||k^0 \in f(x_0) - y_\varepsilon - K, \text{ with } y_\varepsilon \in f(x_\varepsilon) + K$$

and hence we have

$$\sqrt{\varepsilon}||x_\varepsilon - x_0||k^0 \in f(x_0) - f(x_\varepsilon) - K - K$$

and

$$f(x_\varepsilon) \in f(x_0) - \sqrt{\varepsilon}||x_\varepsilon - x_0||k^0 - K,$$

which means that statement (1) is true.

Finally, since $f(x_0) \in E_m[f(X), K_{\varepsilon k^0}]$ it holds that there exists no $x \in X$ such that

$$f(x) \in f(x_0) - \varepsilon k^0 - (K \setminus \{0\}).$$

Considering statement (1) we have that $\sqrt{\varepsilon}||x_\varepsilon - x_0||k^0 \leq \varepsilon$ and the proof is complete. ■

The reader can find another variant of Theorem 13.28 in Göpfert, Tammer, and Zalinescu (1999).

An interesting research subject is to find new maximal point theorems in product spaces, since following this way we can find new Ekeland's type variational principle applicable to the study of ε-efficiency in Pareto optimization.

REFERENCES

Bishop E and Phelps RR. 'The support functionals of a convex set'.- In: *Proc. Symp. on Math.*, Amer. Math. Soc., Providence, RI. 1962; 7: 27-35.

Brezis H and Browder FE. A general principle on ordered set in nonlinear functional analysis. Adv. Math. 1976; 21: 777-787.

Caristi J. Fixed point theorems for mappings satisfying inwardness condition. Trans. Amer. Math. Soc. 1976; 215: 241-251.

Chen GY and Huang XX.

1. A unified approach to the existing three types of variational principles for vector valued functions. Math. Meth. Oper. Res. 1998; 48: 349-357.

Chen GY, Huang XX and Hou SH. General Ekeland's variational principle for set-valued mappings. J. Optim. Theory Appl. 2000; 106: 151-164.

Chen GY, Huang XX and Lee GM. "Equivalents of approximate variational principle for vector valued functions and applications". Preprint, Academia Sinica, Institute of Systems Science, Beijing, 1997.

Dancs S, Hegedus M. and Medvegyev P. A general ordering and fixed point principle in complete metric space. Acta Sci. Math. (Szeged). 1983; 46: 381-388.

Ekeland I.

1. Sur les problèmes variationnels. C.R. Acad. Sci. Paris, 1972; 275: A1057-A1059.

2. On some variational principle. J. Math. Anal. Appl. 1974; 47: 324-354.

3. Nonconvex minimization problems. Bull. Amer. Math. Soc. 1979; 1(3): 443-474.

4. Some lemmas about dynamical systems. Math. Scan. 1983; 52: 262-268.

5. 'The ϵ-variational principle revised', (notes by S. Terracini). -In: *Methods on Nonconvex Analysis*, A. Cellina, ed., Lecture Notes in Mathematics, Berlin-heidelberg: Springer-Verlag Nr.1446, 1990. - P. 1-15

Gerth (Tammer) Chr. "Näherungslösungen in der Vektoroptimierung", Seminarbereichte der Sektion Mathematik der Humblodt-Universität zu Berlin, 1978; 90: 67-76.

Gerth (Tammer) Chr. and Weidner P. Nonconvex separation theorems and some applications in vector optimization. J. Optim. Theory Appl. 1990; 67: 297-320.

Gerstewitz (Tammer) Chr.

1. Nichtkonvexe Dualität in der Vektoroptimierung. Wiss. Zeitschrift TH., Leuna-Merseburg. 1983; 25(3): 357-364.

2. Nichtkonvexe Trennungssätze und deren Anwendung in der Theorie der Vektoroptimierung. Seminarbereicht Nr. 80 der Sektion Mathematik der Humboldt Universität zu Berlin, 1986, 19-31.

Gerstewitz (Tammer) Chr. and Ivanov E. Dualität für nichtkonvexe Vektoroptimierungsprobleme. Wiss. Zeitschrift der TH Ilmenau. 1985; 31: 61-81.

Göpfert A and Tammer Chr.

1. A new maximal point theorem. J. Functional Anal. Appl. 1995; 14: 379-390.

2. "Maximal point theorems in product spaces and applications for multicriteria approximation problems". Preprint Nr.26, Institute of Optimization and Stochastics, Martin-Luther Universität, Halle-Wittenberg, Germany, 1998.

3. ϵ-Approximate solutions and conical support points. A new maximal point theorem. Z. Angew. Math. Mech. 1995; 75: 595-596.

Göpfert A, Tammer Chr. and Zalinescu C. "Maximal point theorems in product spaces." Preprint, Institute of Optimization and Stochastics, Martin-Luther Universität, Halle-Wittenberg, Germany, 1999.

Helbig S, Georgiev P, Pateva AD and Todorov M. 'ε-Efficiency elements.' Lecture on the Conference *Mehrkriterielle Entscheidung*, Fehrenbach, Germany, 1992.

Henkel EC and Tammer Chr.

1. ε-Variational inequalities in partially ordered spaces. Optimization. 1996; 36: 105-118.

2. ε-Variational inequalities for vector approximation problems. Optimization. 1996; 38: 11-21.

Hyers DH, Isac G and Rassias TM. *Topics in Nonlinear Analysis and Applications.* Singapore, New Jersey, London, Hong-Kong: World Scientific, 1997.

Isac G.

1. Sur l'éxistence de l'optimum de Pareto. Rèv. Mat. Univ. Parma. 1983; (4)9: 303-325.

2. 'The Ekeland's principle and Pareto ϵ-efficiency'. - In: *Multi-Objective Programming and Goal Programming*, M. Tamiz, ed. Lecture Notes in Econom. Math. Systems Nr.432, Springer-Verlag, 1996, 148-162.

3. 'On Pareto efficiency. A general constructive existence principle'. (To appear in the volume: *Combinatorial and Global Optimization*, P.M. Pardalos, A. Migdalas and R. Burkard, eds.)

Khan PQ. "On Caristi-Kirk's theorem and Ekeland's principle for Pareto extrema". Preprint Nr. 357, Institute of Mathematics, Polish Academy of Science, 1986.

Kutateladze SS. Convex ε-programming. Soviet Mathematics Doklady. 1979; 20: 1048-1050 (*in Russian*).

Loridan P.

1. ε-Solutions in vector minimization problems. J. Optim. Theory Appl. 1984; 42: 265-276.

2. Necessary conditions for ε-optimality. Math. Programming Study. 1982; 19.

Németh AB.

1. Between Pareto efficiency and Pareto ϵ-efficiency. Optimization. 1989; 20: 615-637.

2. Near to minimality in ordered vector spaces. Mathematica (Cluj). 1981; 23(46): 239-243.

3. "Summation criteria for regular cones with applicatons." Babes-Bolyai Univ., Faculty of Math., Research Seminar, Preprint Nr.4, 1981, 99-124.

4. "Simultaneous transformation of the order and the topology by nonlinear operators." Babes-Bolyai Univ., Faculty of Math., Research Seminar, Preprint Nr.1, 1984, 135-158.

5. A nonconvex vector minimization problem. Nonlinear Anal. Theory, Math. Appl. 1986; 10: 669-678.

Phelps RR.

1. Support cones in Banach spaces and their applications. Advances in Math. 1974; 13: 1-19.

2. *Convex Functions, Monotone Operators and Differentiability*, 2nd ed., Lecture Notes in Math. Nr. 1364, Springer-Verlag, 1993.

Staib T. On two generalization of Pareto minimality. J. Opt. Theory Appl. 1988; 59: 289-306.

Strodiot JJ, Nguyen VH and Heukenes N. "ε-Optimal solutions in nondifferentiable convex programming and some related questions". Faculté Universitaires de Namur, Belgium, Report Nr. 80/12, 1980.

Takahashi W. 'Existence theorems, generalized fixed point theorems for multivalued mappings'.- In: *Fixed Point Theory and Applications*, M.A. Thera and J.B. Baillon, eds. Logman, Notes in Mathematics Series 252, 1991.

Tammer Chr.

1. 'Existence results and necessary conditions for ϵ-efficient elements'. - In: *Multicriteria Decision*, Brosowski e.a., eds. Proceedings of the 14-th Meeting of the German Working Group "Mehrkriterielle Entscheidigung", Frankfurt/Main- Bern: Lang-Verlag, 1993. - P.97-110.

2. 'A variational principle and fixed point theorem'. Proceedings of the 16-th IFIP Conference on System Modelling and Optimization, Compiegne, France, 1993.

3. A variational principle and applications for vectorial control approximation problems. (To appear: Math. Journ. Univ. Bacau, Romania.)

4. A genaralization of Ekeland's variational principle. Optimization. 1992; 25: 129-141.

5. 'Necessary conditions for approxiamtely efficient solutions of vector approximation'.- In: *Proceedings of the 2nd Conference on Approximation and Optimization*, Havana, Cuba, 1993.

Tanino T and Sawaragi Y. Conjugate maps and duality in multiobjective optimization. J. Optim. Theory Appl. 1980; 31: 473-499.

Thierfelder J. Subdifferentiale and Vektoroptimierung. Wiss. Zeitschrift TH Ilmenau. 1991; 37(3): 89-100.

Valyi I. Approximate saddle-point theorems in vector optimization. J. Optim. Theory Appl. 1985; 55: 435-448.

Index

Nonconvex Optimization and Its Applications

1. D.-Z. Du and J. Sun (eds.): *Advances in Optimization and Approximation.* 1994 ISBN 0-7923-2785-3
2. R. Horst and P.M. Pardalos (eds.): *Handbook of Global Optimization.* 1995 ISBN 0-7923-3120-6
3. R. Horst, P.M. Pardalos and N.V. Thoai: *Introduction to Global Optimization* 1995 ISBN 0-7923-3556-2; Pb 0-7923-3557-0
4. D.-Z. Du and P.M. Pardalos (eds.): *Minimax and Applications.* 1995 ISBN 0-7923-3615-1
5. P.M. Pardalos, Y. Siskos and C. Zopounidis (eds.): *Advances in Multicriteria Analysis.* 1995 ISBN 0-7923-3671-2
6. J.D. Pintér: *Global Optimization in Action.* Continuous and Lipschitz Optimization: Algorithms, Implementations and Applications. 1996 ISBN 0-7923-3757-3
7. C.A. Floudas and P.M. Pardalos (eds.): *State of the Art in Global Optimization.* Computational Methods and Applications. 1996 ISBN 0-7923-3838-3
8. J.L. Higle and S. Sen: *Stochastic Decomposition.* A Statistical Method for Large Scale Stochastic Linear Programming. 1996 ISBN 0-7923-3840-5
9. I.E. Grossmann (ed.): *Global Optimization in Engineering Design.* 1996 ISBN 0-7923-3881-2
10. V.F. Dem'yanov, G.E. Stavroulakis, L.N. Polyakova and P.D. Panagiotopoulos: *Quasidifferentiability and Nonsmooth Modelling in Mechanics, Engineering and Economics.* 1996 ISBN 0-7923-4093-0
11. B. Mirkin: *Mathematical Classification and Clustering.* 1996 ISBN 0-7923-4159-7
12. B. Roy: *Multicriteria Methodology for Decision Aiding.* 1996 ISBN 0-7923-4166-X
13. R.B. Kearfott: *Rigorous Global Search: Continuous Problems.* 1996 ISBN 0-7923-4238-0
14. P. Kouvelis and G. Yu: *Robust Discrete Optimization and Its Applications.* 1997 ISBN 0-7923-4291-7
15. H. Konno, P.T. Thach and H. Tuy: *Optimization on Low Rank Nonconvex Structures.* 1997 ISBN 0-7923-4308-5
16. M. Hajdu: *Network Scheduling Techniques for Construction Project Management.* 1997 ISBN 0-7923-4309-3
17. J. Mockus, W. Eddy, A. Mockus, L. Mockus and G. Reklaitis: *Bayesian Heuristic Approach to Discrete and Global Optimization.* Algorithms, Visualization, Software, and Applications. 1997 ISBN 0-7923-4327-1
18. I.M. Bomze, T. Csendes, R. Horst and P.M. Pardalos (eds.): *Developments in Global Optimization.* 1997 ISBN 0-7923-4351-4
19. T. Rapcsák: Smooth Nonlinear Optimization in R^n. 1997 ISBN 0-7923-4680-7
20. A. Migdalas, P.M. Pardalos and P. Värbrand (eds.): *Multilevel Optimization: Algorithms and Applications.* 1998 ISBN 0-7923-4693-9
21. E.S. Mistakidis and G.E. Stavroulakis: *Nonconvex Optimization in Mechanics.* Algorithms, Heuristics and Engineering Applications by the F.E.M. 1998 ISBN 0-7923-4812-5

Nonconvex Optimization and Its Applications

22. H. Tuy: *Convex Analysis and Global Optimization*. 1998 ISBN 0-7923-4818-4
23. D. Cieslik: *Steiner Minimal Trees*. 1998 ISBN 0-7923-4983-0
24. N.Z. Shor: *Nondifferentiable Optimization and Polynomial Problems*. 1998 ISBN 0-7923-4997-0
25. R. Reemtsen and J.-J. Rückmann (eds.): *Semi-Infinite Programming*. 1998 ISBN 0-7923-5054-5
26. B. Ricceri and S. Simons (eds.): *Minimax Theory and Applications*. 1998 ISBN 0-7923-5064-2
27. J.-P. Crouzeix, J.-E. Martinez-Legaz and M. Volle (eds.): *Generalized Convexitiy, Generalized Monotonicity: Recent Results*. 1998 ISBN 0-7923-5088-X
28. J. Outrata, M. Kočvara and J. Zowe: *Nonsmooth Approach to Optimization Problems with Equilibrium Constraints*. 1998 ISBN 0-7923-5170-3
29. D. Motreanu and P.D. Panagiotopoulos: *Minimax Theorems and Qualitative Properties of the Solutions of Hemivariational Inequalities*. 1999 ISBN 0-7923-5456-7
30. J.F. Bard: *Practical Bilevel Optimization*. Algorithms and Applications. 1999 ISBN 0-7923-5458-3
31. H.D. Sherali and W.P. Adams: *A Reformulation-Linearization Technique for Solving Discrete and Continuous Nonconvex Problems*. 1999 ISBN 0-7923-5487-7
32. F. Forgó, J. Szép and F. Szidarovszky: *Introduction to the Theory of Games*. Concepts, Methods, Applications. 1999 ISBN 0-7923-5775-2
33. C.A. Floudas and P.M. Pardalos (eds.): *Handbook of Test Problems in Local and Global Optimization*. 1999 ISBN 0-7923-5801-5
34. T. Stoilov and K. Stoilova: *Noniterative Coordination in Multilevel Systems*. 1999 ISBN 0-7923-5879-1
35. J. Haslinger, M. Miettinen and P.D. Panagiotopoulos: *Finite Element Method for Hemivariational Inequalities*. Theory, Methods and Applications. 1999 ISBN 0-7923-5951-8
36. V. Korotkich: *A Mathematical Structure of Emergent Computation*. 1999 ISBN 0-7923-6010-9
37. C.A. Floudas: *Deterministic Global Optimization: Theory, Methods and Applications*. 2000 ISBN 0-7923-6014-1
38. F. Giannessi (ed.): *Vector Variational Inequalities and Vector Equilibria*. Mathematical Theories. 1999 ISBN 0-7923-6026-5
39. D.Y. Gao: *Duality Principles in Nonconvex Systems*. Theory, Methods and Applications. 2000 ISBN 0-7923-6145-3
40. C.A. Floudas and P.M. Pardalos (eds.): *Optimization in Computational Chemistry and Molecular Biology*. Local and Global Approaches. 2000 ISBN 0-7923-6155-5
41. G. Isac: *Topological Methods in Complementarity Theory*. 2000 ISBN 0-7923-6274-8
42. P.M. Pardalos (ed.): *Approximation and Complexity in Numerical Optimization: Concrete and Discrete Problems*. 2000 ISBN 0-7923-6275-6
43. V. Demyanov and A. Rubinov (eds.): *Quasidifferentiability and Related Topics*. 2000 ISBN 0-7923-6284-5

Nonconvex Optimization and Its Applications

44. A. Rubinov: *Abstract Convexity and Global Optimization.* 2000 ISBN 0-7923-6323-X
45. R.G. Strongin and Y.D. Sergeyev: *Global Optimization with Non-Convex Constraints.* 2000 ISBN 0-7923-6490-2
46. X.-S. Zhang: *Neural Networks in Optimization.* 2000 ISBN 0-7923-6515-1
47. H. Jongen, P. Jonker and F. Twilt: *Nonlinear Optimization in Finite Dimensions.* Morse Theory, Chebyshev Approximation, Transversability, Flows, Parametric Aspects. 2000 ISBN 0-7923-6561-5
48. R. Horst, P.M. Pardalos and N.V. Thoai: *Introduction to Global Optimization.* 2nd Edition. 2000 ISBN 0-7923-6574-7
49. S.P. Uryasev (ed.): *Probabilistic Constrained Optimization.* Methodology and Applications. 2000 ISBN 0-7923-6644-1
50. D.Y. Gao, R.W. Ogden and G.E. Stavroulakis (eds.): *Nonsmooth/Nonconvex Mechanics.* Modeling, Analysis and Numerical Methods. 2001 ISBN 0-7923-6786-3
51. A. Atkinson, B. Bogacka and A. Zhigljavsky (eds.): *Optimum Design 2000.* 2001 ISBN 0-7923-6798-7
52. M. do Rosário Grossinho and S.A. Tersian: *An Introduction to Minimax Theorems and Their Applications to Differential Equations.* 2001 ISBN 0-7923-6832-0
53. A. Migdalas, P.M. Pardalos and P. Värbrand (eds.): *From Local to Global Optimization.* 2001 ISBN 0-7923-6883-5
54. N. Hadjisavvas and P.M. Pardalos (eds.): *Advances in Convex Analysis and Global Optimization.* Honoring the Memory of C. Caratheodory (1873-1950). 2001 ISBN 0-7923-6942-4
55. R.P. Gilbert, P.D. Panagiotopoulos† and P.M. Pardalos (eds.): *From Convexity to Nonconvexity.* 2001 ISBN 0-7923-7144-5
56. D.-Z. Du, P.M. Pardalos and W. Wu: *Mathematical Theory of Optimization.* 2001 ISBN 1-4020-0015-4
57. M.A. Goberna and M.A. López (eds.): *Semi-Infinite Programming. Recent Advances.* 2001 ISBN 1-4020-0032-4
58. F. Giannessi, A. Maugeri and P.M. Pardalos (eds.): *Equilibrium Problems: Nonsmooth Optimization and Variational Inequality Models.* 2001 ISBN 1-4020-0161-4
59. G. Dzemyda, V. Šaltenis and A. Žilinskas (eds.): *Stochastic and Global Optimization.* 2002 ISBN 1-4020-0484-2
60. D. Klatte and B. Kummer: *Nonsmooth Equations in Optimization.* Regularity, Calculus, Methods and Applications. 2002 ISBN 1-4020-0550-4
61. S. Dempe: *Foundations of Bilevel Programming.* 2002 ISBN 1-4020-0631-4
62. P.M. Pardalos and H.E. Romeijn (eds.): *Handbook of Global Optimization, Volume 2.* 2002 ISBN 1-4020-0632-2
63. G. Isac, V.A. Bulavsky and V.V. Kalashnikov: *Complementarity, Equilibrium, Efficiency and Economics.* 2002 ISBN 1-4020-0688-8

KLUWER ACADEMIC PUBLISHERS – DORDRECHT / BOSTON / LONDON

MIX
Papier aus verantwortungsvollen Quellen
Paper from responsible sources
FSC® C105338

If you have any concerns about our products,
you can contact us on
ProductSafety@springernature.com

In case Publisher is established outside the EU,
the EU authorized representative is:
Springer Nature Customer Service Center GmbH
Europaplatz 3, 69115 Heidelberg, Germany

Printed by Libri Plureos GmbH
in Hamburg, Germany